*Jon M-Query*

# SCIENCE, CONFLICT, and SOCIETY

*Jon M-Query*

*Readings from*
**SCIENTIFIC
AMERICAN**

# SCIENCE, CONFLICT and SOCIETY

*With Introductions by*
**GARRETT HARDIN**
*University of California, Santa Barbara*

**W. H. FREEMAN AND COMPANY**
*San Francisco*

# Preface

He who finds himself in the middle of a maelstrom has only a poor idea of where he is going, or where he will end up. History at times seems like a maelstrom, and this is one of those times. Since the first Battle of Berkeley in the fall of 1964 it has been quite apparent that something is going on. Attempts to exorcise the devils of disorder by identifying them with devils we have conquered in the past have failed. We are in the middle of some sort of a revolution, but what sort it is we don't know. Some of us suspect that the activists themselves are as little aware as their adversaries of the nature of the social changes of which they are agents. More passive commentators on the social scene may be no wiser; certainly the many answers they come up with cannot all be correct.

The years since 1964 have brought thinking men to a renewed respect for the factors of change. We all have a natural and understandable tendency to regard a static condition as normal, or (at a more sophisticated level of analysis) to regard a steady-state as the normal situation. Change, conflict, disorganization, mutation, revolution and reorganization tend to be regarded as abnormal. In the study of biological functions we distinguish between physiology and pathology. The changes of the sort just mentioned we are inclined to identify as part of pathology.

It is certainly defensible to consider physiology as the normal, and pathology as the abnormal state of affairs. This does not need belaboring. But, in a deep sense, pathology is also normal. The phenomena of pathology are capable of being described; under similar conditions, they are repetitive. We may in fact view pathology as the consequence of normal processes taking place in abnormal settings. In any event we should not ignore pathology; our survival may literally depend on our taking up the study of pathology courageously and seeing what we can make of it.

In the minds of many people in the past, and in not a few even today, science has seemed to be a somewhat pathological force in society. Science and its handmaiden, technology, have persistently disrupted the established order of society. In contrast to an earlier day this disruption is now viewed by most people as being, on the whole, good. But no one denies that science and technology are disruptive influences.

Until about the beginning of the Second World War science was an outside force that impinged upon what we now call "the Establishment." The mammoth financing of the Manhattan Project that produced the nuclear bomb set a new pattern of large-scale expenditure of money for scientific purposes—a pattern that has been followed ever since. The total "R & D" (research and development) budget of the federal government of the United States now runs into billions of dollars each year. On such a scale, science and technology are no longer forces operating outside the Establishment—they are themselves an establishment that constitutes a significant part of the Establishment. After being so long the initiators of disruptive actions, science and technology now find themselves on the receiving end of disruptive actions. Willy-nilly, scientists have become political agents; and like all political agents, they are subjected to a steady barrage of doubts, suspicions, insinuations, and attacks from many directions. Unused to such treatment, they sometimes complain. Their complaints evoke more amusement than sympathy from politicians of the traditional sort, who say, in effect, "If you don't like the heat, get out of the kitchen." Any establishment that receives billions of dollars a year in tax money must expect to be forever justifying its existence and its activities.

Science is now both an agent of change and a recipient of change. Concepts of what it can do, and what it should do, are being steadily revised, but the revision is not a part of any grand over-all program of action. No one person, no agency, and no school is laying out the guidelines to the future. I do not think it is possible at this time to produce a book that successfully takes a global view of the complex interactions of science and society. A more fruitful approach, I think, is to examine some of the numerous apparently unrelated (or loosely related) controversies that are still live issues in our day. The collection of articles and essays here assembled supplies us with no philosopher's stone, but to thoughtful minds it may be an effective goad to fruitful study.

*Santa Barbara*
*July 1969*

GARRETT HARDIN

# Contents

## IV. WHAT PRICE PROGRESS?

## V. WAR: THE ANGUISH OF RENUNCIATION

NOTE ON CROSS-REFERENCES

Cross-references within the articles are of three kinds. A reference to an article included in this book is noted by the title of the article and the page on which it begins; a reference to an article that is available as an offprint but is not included here is noted by the article's title and offprint number; a reference to a SCIENTIFIC AMERICAN article that is not available as an offprint is noted by the title of the article and the month and year of its publication.

# I

## SCIENTISTS AND SOCIETY

# I

## Scientists and Society

### Introduction

This collection of writings from *Scientific American*, unlike preceding volumes, makes extensive use of book reviews. The departure from tradition is needed because of the exceedingly broad nature of the subject. Many of the areas of conflict between science and society have not been adequately surveyed in brief articles. Often, however, important books have been written about them, and many of these have been dealt with in the pages of *Scientific American* in long reviews that are in themselves first-rate critical essays. It is to these that I have turned in setting the stage in this section for the articles found in the later sections.

From the renaissance of the magazine in 1948 until his death in 1966, James R. Newman was the book review editor. By training and temperament he was exceptionally well fitted to discuss the impingement of science and society on each other. Newman was a lawyer with an unusually deep interest in mathematics and philosophy. In 1940 he collaborated with the mathematician Edward Kasner in writing the ground-breaking book *Mathematics and the Imagination*. It was a spendid example of popularization in the best sense, and every attempted popularization of mathematics written since that time shows its influence. In the opinion of many, it is still the best book of its sort available.

At the conclusion of the Second World War, Newman became involved in what was his most important historical role. After the work of the Manhattan Project was revealed in the bombing of Hiroshima and Nagasaki a monumental struggle for control of atomic energy developed in the halls of Congress. The military wanted to continue to control it. Most scientists, both those who had worked in the Manhattan Project as well as those outside, were adamantly opposed to military control. Droves of scientists descended on Washington, science lobbies were created, and many articles and books were written.

In introducing Congressmen to the new world that science had discovered, many scientists played an important part, notably Robert Oppenheimer and Leo Szilard. James R. Newman was probably no less important. He had a distinct advantage over the scientists: he spoke the language of legislators. The majority of the members of Congress are and always have been lawyers, and a lawyer's view of truth and reality is somewhat different from that of a scientist. (Whether this is necessarily so will not be argued here.) Newman, trained as a lawyer, but with a lifelong interest in mathematics and science, could speak to both groups. In the most hectic days of the power struggle centered on the atom, Newman gave unstintingly of himself. In the end, the civilians won. To those of us who think that this was a good thing—and this includes almost all scientists—Newman's work in Washington was of great national importance.

As book review editor of *Scientific American*, Newman chose wisely the few books out of the vast flood that might have been reviewed. His frequent lead-reviews set a high standard for others to aspire to. In his review of Kingsley Amis' *New Maps of Hell* Newman discusses the position of science fiction in the world of scientists in the first part of the twentieth century. There is considerable evidence that during this time a large proportion of the men who became scientists and engineers did so because of a boyish enthusiasm for science fiction. Implicit in this literature is the idea that anything that can be dreamed of can be invented. Most of us find it hard to believe that this is true, but belief in this myth may serve a useful function in aiding the development of the ego-strength of a scientist or engineer, particularly the latter. For good or ill, much of the scientific world we live in has been created by believers in the premises of science fiction.

An important question that now faces us is this: "Are we coming to the end of this era?" Though no studies have been made of the matter, there is a general impression that the young people who are growing up now are less interested in science fiction that emphasizes new technological marvels than they are in works that deal with the social implications of scientific discoveries and the social responsibilities of science. If this impression is well founded we can reasonably expect that the world that will be created by a new generation of scientists in the latter half of the twentieth century will be qualitatively different from that created in the first half. The nineteenth-century poet Alfred de Vigny once asked, "What is a great life?"; to which he answered, "It is a thought of youth wrought out in ripening years." What are the thoughts that youth is thinking today? If we knew, we might be able to predict the developments that will take place twenty years from now.

A pervasive dream of science fiction has been the conquest of space. Space is regarded as the next frontier, and it has been assumed (by analogy) that this frontier will be conquered just as surely as earthly frontiers have been. The second review of Newman's, the review of a collection of articles assembled by A.G.W. Cameron, should suffice to explode this dream. The possibility that man might ever travel from the earth to a planet of another star seems remote, to put the matter conservatively. Even communication with other worlds seems a dubious business. How you choose your assumptions is a matter of taste, but it is notable that what one might call a sympathetic study carried out by the RAND Corporation came to the conclusion that worlds inhabited by intelligent beings (if more than one such world exists) must be on the average about forty light years apart. How do you carry on an intelligent conversation with another being when the time taken for a round-trip question and reply takes eighty years?

Communication under these circumstances would require a communal vision and a persistence that does not notably characterize a society that believes in Science and Progress. In fifteenth-century Florence, Lorenzo Ghiberti did not hesitate to devote more than forty of the best years of his life to the creation of the bronze doors of the baptistery. Who in our science-dominated society will devote so much of his time to a single project? Who would devote his best efforts to creating a system of communication with another world, when the fruits of his labors could be reaped only by his children or his grandchildren? This is the irony of the so-called "Space Age": though made possible by the restless and novelty-seeking minds that create scientific and technological progress, it could be brought to a full fruition only by people of an utterly different temperament, by men and women who possessed a persistence and patience now so rare as to be almost freakish. To

3

achieve the dream of the space age that we have conceived, it would be necessary—in the anthropological sense—to destroy ourselves psychologically and culturally. That is the price tag on the "Space Age." Are the worshippers of science and technology willing to pay it?

With the review of Arthur Koester's *The Sleepwalkers* we are taken back to the origins of modern science, when the groundwork for man's belief in the possibility of a space age was being laid. The reviewer, I. Bernard Cohen, is a professor of the history of science at Harvard University. Koestler is best known as a novelist and political commentator, but, like Newman, has long had a keen interest in science and its implications. His analysis of the significance of Kepler's life and work stirred up a good deal of controversy in scientific quarters. Cohen shows the grounds on which *The Sleepwalkers* may be justly criticized; yet he presents Koestler's analysis sympathetically.

Many scientists resent it when an outsider interprets science for the layman, even though they are aware that large areas of potential interest remain to be brought to the attention of the general public. Until scientists themselves exploit these riches more thoroughly, it seems a bit ungracious to protest the actions of interested and intelligent outsiders who take on this important task. (Of course, any important errors should be duly noted.) Whatever we may think of his conclusions, Koestler's insistence on the importance of the nonrational, mysterious, obscure, and nonrigorous origins of scientific discovery is a healthy antidote to the all-too-common view of the scientist as a rigorously rational, nonemotional, thinking machine.

The story of Kepler is a story of cloistered science. The public ordeal of his contemporary, Galileo, was an almost unique event, and perhaps more connected with the civil war between clerical orders than with the heretical aspects of science. For several centuries scientists were, for the most part, believed to be rather harmless. The arrival of the atomic bomb completely changed public opinion on this matter. Seeking a scapegoat, some humanists quite naturally pounced upon the atomic scientists who had created the horror. This is the tack that Robert Jungk took in his *Brighter Than a Thousand Suns*. The physicist Robert R. Wilson, who was a participant in the creation of the atomic bomb, points out numerous errors in Jungk's history—errors that are particularly unfortunate because it is the most widely read history of the bomb. It has no doubt played an important part in creating the average man's view of the moral character of scientists, a view that must be recognized (we need not say *accepted*) by those who seek to explain science to others.

In sharp contrast to the somewhat dubious history presented by historian Jungk, Sir Charles Snow's *Science and Government* gives a really professional view of how history is made. Snow is the man who created the catch phrase "The Two Cultures" to stand for the two dangerously separated subcultures of our time, science and the humanities. Snow himself, remarkably enough, stands in both camps: he is both a scientist and a novelist, both a successful administrator and (as this book shows) a shrewd and knowing historian. As the physicist P.M.S. Blackett points out in his review of Snow's book, the story that Snow recounts shows vividly the practical difference between operating under the spell of a blind faith in a particular technological gadget and being committed to the logical analysis of alternatives that we now call "operations research," a discipline of which Blackett is himself one of the founding fathers. The contrast between blind faith and operations research is a contrast between superstition and rationality. That superstition *may* lead to desirable results is not an acceptable defense. It certainly cannot be depended on to do so, as the post mortem analysis of the strategic-bombing plan clearly shows.

Established beliefs are seductive. In the early 1960's, the introduction of the methods of operations research into the American Pentagon produced an upsurge of emotional—one might even say moral—revulsion among the belief-prone militarists, a revulsion that is still affecting national policy. Nonetheless, in military circles as elsewhere, each fresh invasion by rationality is viewed as subversion. It is characteristic of the romantic spirit, whether operating in the Pentagon or in Greenwich Village, that it is appalled by coldblooded rationality. In the Pentagon, romanticism takes the form of unquestioning belief in the ability of military hardware to solve political problems.

No attempt to understand the relation of scientists to society can omit a discussion of The Bomb. What we did in Hiroshima and Nagasaki is beyond recall and—some say—beyond redemption. After nearly a quarter of century we are still too close to these events to know their significance. Those who are willing to wrestle with the puzzle will find Robert Lifton's *Death in Life: Survivors of Hiroshima* a help. The Bomb taugh us a truth that apparently has to be learned anew in every generation: namely, that civilization is only the thinnest of veneers covering an ever-threatening anarchy. Jacob Bronowski, scientist, literary artist, and philosopher, who reviews Lifton's book, says: "We have to get it into our heads that the atomic weapons do not create casualties but chaos—a lasting chaos of the human values." When the crunch comes, each of us takes care of himself and of himself alone: that simple and biologically necessary fact engenders chaos.

To those who survive and reaffiliate with society, a sense of guilt comes later. And persists. Surviving an overwhelming catastrophe of this sort creates many problems that we could not dream of beforehand. Not the least of these is the problem of variety in humankind. We often praise variety—at least some of us do. But there is no question that we all, at times, find it difficult to tolerate. This basic fact, central to some of the worst unsolved problems of our time, appears in ironic form in the bombed cities of Japan. Humanity resents the variety of human forms and distorted temperaments that man's science has created here. Where charity is called for, resentment seems both cruel and irrelevant. But there it is.

This first section closes with another review by James R. Newman, a review that tackles the large issue of human survival, as discussed in books by Bertrand Russell and Erich Fromm. Countless presuppositions and thought-patterns contribute to our difficulty in laying out a path to survival. Even if we insist that science is ethically neutral—and not all men agree to this—it is nevertheless clear that the unexamined traditions of science, developed in an earlier day to attack earlier problems, can impede our progress toward a solution of the problem of survival of the human race.

No one doubts that one of the great glories of science is its objectivity. Yet such an abstraction as science can, like concrete human personalities, have the "defects of its qualities." In discussing wars, the reasons for wars, and the methods by which they may be prevented, seemingly objective language may be dangerous. Newman cites Lord Wootton, who rightly castigates a radio announcer for saying "if war comes. . . ." Wars, as Wootton points out, do not come. They are produced by men. Natural scientists, studying the world of nature outside of man, have developed a tradition of conveying their findings in language that is impersonal, passive, unemotional, "objective." Though this approach produces a rhetoric that is not wholly felicitous, it can be defended when it is confined to the world of natural science. But in the world of the behavioral sciences, the world in which facts are influenced by the behavior of men (including their rhetorical statements) such an approach is most dangerous. To say, "war comes," is to leave out of the assertion the true subject, the true doer of the action, namely, willful men. War is the result of human actions; it is not the subject of the sentence, "War comes."

This is no trivial matter. In the behavioral sciences what one says about the truth influences the truth. (This is in sharp contrast to the natural sciences where one's assertions, right or wrong, do not influence fact: gravity exists, whether we say it exists or not.) Behavioral scientists who habitually resort to the impersonal, passive rhetoric of such statements as "wars come" promote in their auditors very real and dangerous tendencies toward passivity, negativity, pessimism, and fatalism. Rhetorical modes that can claim the name of objectivity in the natural sciences are irresponsible in the behavioral. That *men*, in all their complexity and mystery, are the subject of whatever occurs in the behavioral sciences must never be forgotten.

# 1

# The History and Present State
# of Science Fiction

JAMES R. NEWMAN

*July 1960*

A review of *New Maps of Hell*, by Kingsley Amis. New York: Harcourt, Brace and World, 1960.

One either likes science fiction or one doesn't; there's no ought about it either way. Not to like it is no proof in itself that one is a cultivated chap, above that sort of thing; nor is being a fan proof that one has a low mind or that one is a great man seeking relaxation and escape in this corner rather than in Westerns or detective stories. I myself am not an addict, and I have never quite understood, I must confess, how one gets hooked. It happens, I have now learned, in adolescence or not at all, like addiction to jazz. It is not unusual, according to my source, for the same person to have contracted both addictions. This is not surprising if one considers how much the two modes have in common. They came to flowering at about the same time—the 1920's—and underwent mutations around 1940; in their present form they are characteristically American products "with a large audience and a growing band of practitioners in Western Europe." Both have a "radical tinge." That is to say, science fiction shows the tinge in its social caricatures; and jazz, though its cacophonies and offbeats may not represent the brand of subversion that literal-minded Congressional investigators can get their teeth into, shows its radicalism in the attitudes of those connected with it. Neither science fiction nor jazz has produced absolutely first-rate artists (I shall doubtless have to take cover on this), but they have turned up gifted and interesting practitioners; both "have arrived at a state of anxious and largely naïve self-consciousness; both, having decisively and for something like half a century separated themselves from the main streams of serious music and serious literature, show signs of bending back towards those streams." Finally, both science fiction and jazz have established themselves so firmly as to attract the attention of the "cultural diagnostician" or "trend hound." Liking science fiction for itself is one thing; coming to it like a microbe hunter, to see where it breeds, who is most susceptible to it and so on, is quite another thing. Not always a bad thing, to be sure, but not necessarily a good thing either, and apt to be pretentious.

The source to which I am indebted for these insights is Kingsley Amis's elegant and entertaining little book, based on a series of lectures he gave in 1959 as one of the Christian Gauss seminars at Princeton University. Amis has written some very funny novels, and he is a wit in the best sense: perceptive, sympathetic and impatient of intellectual heavies. Collegiate seminars on criticism are not often favored with his kind of felicitous impudence; the audience must have been delighted. What he gave them was a thoughtful, wholly unpretentious, amusing and self-amused survey of a branch of writing that deserves just such treatment. The lecturer neither goes overboard nor goes slumming; science fiction, as he makes clear, is not "tomfool sensationalism, but neither is it a massive body of serious art destined any moment to engulf the whole of Anglo-American writing." It is a story form that provides entertainment to many, possesses, at its best, elements of genuine literary quality, and is an interesting social and psychological phenomenon.

How is the form defined? "Science fiction is that class of prose narrative treating of a situation that could not arise in the world we know, but which is hypothesised on the basis of some innovation in science or technology, or pseudo-science or pseudo-technology, whether human or extra-terrestrial in origin." Science fiction, as Amis emphasizes, is to be distinguished from fantasy. To get to the stars in a hurry you apply a "space-warp" (which Bernhard Riemann and Albert Einstein would have found quite natural), or you put your craft into "hyper-drive" (which sounds more reasonable than what is promised for a new Buick). To spend an evening with Madame DuBarry, you turn on your time machine, and zip—there you are. Mr. Midas, who owns a Greek restaurant in the Bronx, saves the life of an extra-terrestrial who is living in New York anonymously as an observer for the Galactic Federation; in gratitude, the extraterrestrial, who is "a master of sciences far beyond ours, makes a machine which alters the molecular vibrations of Mr. Midas's body so his touch will have a transmuting effect upon other objects." Thus the Midas myth in modern garb. All this is science fiction. Fantasy, on the other hand, deals in elves, broomsticks, pixies, cromlechs, occult powers, incantations. Every pretense at a rational explanation is abandoned. (The difference is important, and one might be tempted to say portentous things about it. Amis isn't.) To be sure, there are borderline cases. The vampire, for example, is a staple of 19th-century fantasy; but in his novel *I Am Legend* Richard Matheson makes use of the myth for science-fiction purposes, whereby everything is rationally explained. In our day, in any case, fact is stranger than science fiction. The universe is changing, as is our notion of it; even our tiny environment has queernesses and monstrous possibilities. A nuclear war would produce some strange shapes; maybe not giant ants or "armour-plated, radioactive, supersonic pterodactyls," but can one be absolutely sure? Put it, then, that the borderline between science fiction and fantasy is hard to draw. Still, there is a difference.

Amis sketches the ancestry of the form. The learned dispute the origins. Some point to the Atlantis items in Plato's *Timaeus* and *Critias*, or to Aristophanes' *The Birds*; others prefer the interplanetary voyage recounted by a second-century Greek prose romancer, Lucian of Samosata. On his trip to the moon there is an encounter with some women who are grapevines from the waist down; the men in the moon are of fantastic appearance and habits, but they are not menacing. (The notion of "nasty aliens"—aliens are non-earthlings—is a comparatively recent one.) Johannes Kepler describes a moon voyage, as does Bishop Godwin in 1638, the latter's craft being drawn by wild swans; Cyrano de Bergerac's traveler gets to the moon in a chariot powered by rockets. The first visit to the earth by an alien is in Voltaire's *Micromégas*. *Gulliver's Travels* clearly has a place in the line, for two reasons. First, Swift was at pains to make the details of his stories rational, that is, non-magical; like Jules Verne he "counterfeited verisimilitude." Second, he invented satirical Utopias, blending ingenious invention and powerful social criticism, which is characteristic of much of contemporary science-fiction. (On this point Thomas More's *Utopia* and Francis Bacon's *New Atlantis* also deserve notice. Bacon's island, particularly, boasts all kinds of technological marvels.)

The paleontologist of science fiction may sniff at many of these earlier examples, but he will concede that Mary Shelley's *Frankenstein* was a major progenitor. His descendants, and those of his monster, are all over the place: mad and malevolent scientists as well as mild eccentrics who mean well but lose control over their creatures. Two big themes are descended from the original book: robots which turn on their master (Karel Capek's *R.U.R.*, published in 1920, was the classic modern treatment of this notion) and "functionalized sermons on the dangers of overgrown technology." Amis also mentions as a *Frankenstein* descendant the stock figure of the "morally irresponsible researcher indifferent to the damage he may cause." I am not sure that the figure is confined to fiction.

Verne, with his technological Utopias and his pessimism about invention overreaching itself, was of course a pioneer of modern science-fiction. The literary quality of his stuff is not high, but some of his yarns are fascinating, and he never ran out of ideas. His foresight was remarkable. In *Five Weeks in a Balloon* (1862) there is the following dialogue:

"'Besides,' said Kennedy, 'the time when industry gets a grip on everything and uses it to its own advantage may not be particularly amusing. If men go on inventing machinery, they'll end by being swallowed up by their own inventions. I've often thought that the last day will be brought about by some colossal boiler heated to three thousand atmospheres blowing up the world.'

"'And I bet the Yankees will have a hand in it,' said Joe."

The other major creator of the modern form was H. G. Wells. Wells was much less interested in pseudo-science than Verne was, and he was also an incomparably better writer. For him the story was the thing; scientific verisimilitude, a minor matter. Time machines and Martians (*The War of the Worlds*), monsters (*The Food of the Gods*), invisible men, carnivorous plants—these were used to entrance the reader, to create suspense and excitement rather than to satirize or weave an allegory. Their influence on science fiction, especially of the pre-1940 vintage, was immense. But there were other Wells stories, little mentioned and apparently not influential in shaping modern developments, that forecast the modern satirical Utopia with "fantastic exactness." "A Story of the Days to Come" has such features as advertising bawled out of loud-speakers, phonographs replacing books, vast technological unemployment, with the idle supported by a kind of "international poorhouse called the Labor Company," huge trusts that own everything, children brought up in state crèches, hypnosis as a cure for antisocial traits, dreams that can be obtained to order; and, as an especially edifying touch, men use depilatories instead of shaving.

In 1926 Hugo Gernsback founded *Amazing Stories*, the first magazine entirely dedicated to science fiction. The event was part of a struggle for recognition and existence, because science fiction was still beleaguered by fantasy stories and by space operas, both of which attracted more readers. Space operas, as Amis explains, are like horse operas, with Mars taking the place of Arizona, ray guns replacing six-guns, bad aliens with green skins substituting for outlaws, and bug-eyed monsters, known as Bems, for Indians. (Bems, by the way, which have tentacles and V-shaped mouths, were a Wellsian invention.) Gradually, however, science fiction edged ahead in the competition with Bems, galactic hoodlums and scantily clad virgins.

Around 1940 the present era began. Where there had been five science-fiction magazines in 1938, there were 22 in 1941. There was, of course, no sudden transformation of imbecility into literary excellence. Plenty of mad scientists were still running around; also on the scene were king-sized ants, king-sized sirens in "abbreviated shorts and light cotton sweaters," handsome supermen, and monsters with yellow fangs and massive paws that could rip away a 15-year-old girl's clothing. But the adult was at least "co-present with the stupidly or nastily adolescent."

Amis admits that a science-fiction magazine, even in the new era, does not at first glance look too promising. A title such as *Fantastic Universe* or *Astounding Science Fiction* is not calculated to tempt the fully developed brain. The magazine cover is likely to be either crudely sensational or crudely whimsical; it may picture a multi-armed alien Santa Claus. The inner pages advertise trusses ("Wouldn't trade mine for a farm," says the testimonial of an agricultural user), Rosicrucianism, royal jelly (prolongs life), hand grenades ($1 for an exact replica), "stuffed girls' heads." The contents, however, are at least in part on a somewhat higher level.

Amis samples the October, 1958, issue of *Astounding Science Fiction*. One story, "The Yellow Pill," introduces a psychiatrist in New York who is trying to cure a patient of the delusion that they are both on a space-ship in flight: the psychiatrist takes the yellow pill (an "anti-delusion compound"), presumably to encourage the patient to follow suit, and finds himself aboard a space ship; the patient, taking no pill, imagines himself cured and walks out of the door—leading into empty space instead of the outer office. "Big Sword" is about a small boy who sets up telepathic communication with the tiny aliens of a distant planet and then persuades his earthling elders, who are about to set off on an expedition to destroy the little creatures, that they are friendly and should not be murdered but helped. Another story is about some "amiable visiting aliens" who run out of food; another "shows alien and man agreeing to overlook the differences of appearance and habit that repel or frighten each of them and so coming to an understanding."

Most of this may strike you as moderately dull, but it isn't vicious or incendiary or corrupting. There are no mad scientists, no Bems, giant newts or huge hairy spiders. The alien races are all friendly. There is a moral, however faint, in most of the stories: communication breeds understanding, which breeds tolerance. There is not much science, pseudo-science, technology or sex. This is fairly representative of the content of the other magazines, says Amis, and of the new science fiction as a whole. (The books—the current annual crop is between 150 and 200—are either anthologies of short stories reprinted from magazines, or novels that can be traced back to the magazines, either in serial or in rudimentary form.)

It may be that the trend toward uplift has held down the number of readers; at any rate, science fiction has no mass audience. *Galaxy* sells about 125,000 copies per an issue in the U. S. (there are also foreign editions; the Swedes are special fans); *Astounding* sells 100,000 copies in this country, 35,000 copies in England, and has subscribers in the U.S.S.R. and China; *Amazing* sells 50,000. The total science-fiction readership, then, comes to around half a million. But it is a faithful audience, given to banding together in fan clubs. These are found all over the U. S., the highest concentration being on the West Coast. They often meet weekly and even print their own fan magazines. They have such names as "The Elves', Gnomes' and Little Men's Science Fiction, Chowder and Marching Society," and "The Little Monsters of America." Every year there are regional conferences and a three-day world convention. The clubs are inclined to be politically liberal, especially on racial questions; once a Communist group from Brooklyn formed an Association for the Political Advancement of Science Fiction.

Many more men than women read science fiction, and the audience has a technological or scientific bias. Amis's studies show that engineers and scientists account for perhaps 40 per cent of the readers. That there is a healthy spread in occupational distribution, however, is evidenced by L. Sprague de Camp's anecdote (he has written an authoritative handbook on science fiction) which tells how a science-fiction writer, "happening to visit a New Orleans bordello, found his works so popular with the staff that he was asked to consider himself their guest for the evening."

The anecdote should not be taken as proof of widespread dissoluteness among science-fiction writers. Licentiousness is apt to be expensive, and the financial rewards of science fiction are not high. It is therefore often a secondary occupation of men who teach science or do research. The University of Cambridge astronomer Fred Hoyle is a conspicuous example. Most of the writers and editors, ac-

cording to Amis, treat their calling with "great, sometimes excessive, seriousness." Their claims include such points as that science fiction is "the last refuge of iconoclasm in American literature," that it exists "to afford objectivity to the reader, for better consideration of himself and his species," that "it helps mankind to be humble." These are respectable views, and even the hyperboles—Robert A. Heinlein, for example, says that science fiction is superior to most modern fiction, is "the most difficult of all prose forms" and is "the only form of fiction which stands even a chance of interpreting the spirit of our times"—are self-respecting. "Few things," says H. L. Gold, the editor of *Galaxy*, "reveal so sharply as science fiction the wishes, hopes, fears, inner stresses and tensions of an era, or define its limitations with such exactness." Perhaps this too is hyperbole, but every kind of literature—even the comic strip—is like a mythology in the sense of embodying man's fears and aspirations.

Amis attempts to distinguish between what the literature reveals of these hopes and fears, and the ideas and morals that it consciously and deliberately sets forth. Sex, as I have already indicated, is no longer an important theme in science fiction, although it strongly persists in the fantasy stories; for that matter, they also continue to celebrate mad scientists, mutilation and such winsome items as jellyfish which, when you step on them, call you by name. The sense of insecurity exhibits itself in many different forms: the race of men is doomed, with or without a nuclear war; men are puppets, at the mercy of the quirks of time, the vagaries of matter and galactic flaws; individuality and free will are an illusion. Robots are still a menace, but less than they used to be. Philip K. Dick has a delightful story, "The Defenders," about robots sent out to fight a war on the surface by a human race now living underground; the robots keep sending down reports about great battles and massive destruction, but when, after several decades, a party of men comes up to see how things are going, they find that the robots made peace almost immediately and, "arguing that mankind could never be trusted not to start things again, have been spending their time faking their reports as interestingly as possible." Nostalgia for the rural way of life is very strong; this is another aspect of the anti-science strain in the new science fiction. Arcadian pleasures, simple things, escape from the world of too much electronics—from what, in short, is called the abundant life—is a heartfelt, recurring

emotion. But it must not be supposed that a science-fiction hero is a fellow who simply wants to escape; if he doesn't like things as they are, he means to do something about it, not merely run away. In George Orwell's *1984* O'Brien says to Winston: "If you want a picture of the future, imagine a boot stamping on a human face—forever." No orthodox science-fiction writer, says Amis, would ever concede such abject defeat. In general, then, while no clear-cut, unified picture emerges, the main attitudes are "reassuring": reasonable fears, sensible wishes, admirable hopes. We are not splendid, but neither are we hopelessly degraded; *something* will help us if we help ourselves.

Amis examines the Utopias and the new maps of hell, the sphere of social diagnosis and warning. In William Tenn's story "Null-P" (the P stands for Plato and his notion that political leaders must be men of merit) the U. S., after an atomic war that has produced a zoo of mutations, elects as president George Abnego, the apotheosis of the average man, "median in his flesh and the condition and quantity of the very teeth in his mouth." Mediocrity is enthroned; the non-elite or "unbest" are in charge. Eventually a race of intelligent Newfoundland retrievers domesticate "*Homo abnegus*" and "prize him for his stick-throwing abilities and initiate selective breeding to this end." When a machine is developed that throws a stick better than he can, man disappears "except in the most backward canine communities." The satire is sharp, but general; no recognizable personalities appear, and this seems to be characteristic of science fiction.

Another conformist hell is described in Ray Bradbury's *Fahrenheit 451*. (Reverting to his jazz comparison, Amis calls Bradbury the Louis Armstrong of science fiction, because he is "the one practitioner well known by name to those who know nothing whatever about his field.") The suppression of books—451 degrees Fahrenheit is supposedly the temperature at which book paper ignites—is a major activity of this particular society. A book is considered a "loaded gun in the house next door." Books interfere with sport, pleasure, group spirit, fun; they make men feel unequal because some read and some do not. Books hinder the pursuit of happiness by making us worry; they make us indecisive when we conclude that there are two sides to a question. The cure is obvious. The hero of the story, Montag, is a fireman; when an alarm is turned in, he and his mates charge out of the station on their wagon to burn somebody's house down—one with books in it—under the regulations of the "Firemen of America," "established, 1790, to burn English-influenced books in the Colonies. First Fireman: Benjamin Franklin." While Montag is busy on his rounds ("Monday Millay, Wednesday Whitman, Friday Faulkner"), his wife lies on the bed in her room with the "little Seashells"—thimble radios—plugged into her ears, rapturously oblivious to everything but an "electronic ocean of sound, of music and talk and music and talk coming in, coming in on the shore of her unsleeping mind." I need hardly add that these days we see grown men and women walking down the street carrying little radios that are heard through a tiny earpiece. This presumably is what we are fighting for.

Utopias have other desirable features. In Robert Sheckley's "A Ticket to Tranai" wives are put into "stasis" or suspended animation, and are taken out by their husbands only when they are needed. The only rule is that you must take your wife out a few hours a week. The wives' compensation for this deep-freeze celibacy is that they age much more slowly than men, so they have many husbands and gain the financial benefit of multiple widowhood. Disimprovement engineering is another Utopian plus which furthers the commercial aim of built-in obsolescence; so is the new abundance installment economics that permits the citizen with the helicopter and subterranean swimming pool who is $200,000 in debt to have even more playthings by mortgaging the first 30 years of his son's earnings, and by doing the same with his grandchildren.

I should like to give one more choice example, from the writings of the able science-fiction practitioner Frederik

Pohl, who is concerned with the sublimities of production and consumption. A new advertising procedure backed by the Establishment consists of driving a van into a residential area playing "at top volume a tape recording of fire engines answering an alarm. Then . . . a harsh, sneering voice, louder than the archangel's trumpet, howled: 'Have you got a freezer? *It stinks!* If it isn't a Feckle Freezer, it stinks! Only this year's Feckle Freezer is any good at all! You know who owns an Ajax Freezer? Fairies own Ajax Freezers! You know who owns a Triplecold Freezer? Commies own Triplecold Freezers! Every freezer but a brand-new Feckle Freezer *stinks!*'

"The voice screamed inarticulately with rage: 'I'm warning you! Get out and buy a Feckle Freezer right away! Hurry up! Hurry for Feckle! Hurry, hurry, hurry, Feckle, Feckle, Feckle, Feckle, Feckle, Feckle!'"

The hysteria, the minatory tone, the indecent assaults on privacy are not only fairly close to what advertisers do to us via radio and television every feckling moment of the day, but even closer, as Amis says, to what they would like to do to us if they dared.

Where is science fiction going? Is it marked for higher destinies? There are two things to consider: science fiction simply as readable literature, and science fiction in its concern with political or economic man. From the standpoint of its merit as literature, science fiction has had its ups and downs. It now takes itself very seriously, but this does not mean that its best stories are up to the best of H. G. Wells, say, who himself certainly did not take the form too seriously, at least not in today's sense. No one can pretend that science fiction goes very deep into human personality and motivations. The human heart is a fea-

ture only when it is a cavity resonator enclosed by plastics and titanium, and activated by microwaves. The ingenious gadgets are themselves beginning to run out. There are, as I recall from group theory, something like 275 basic patterns of wallpaper; why should science fiction have a larger range? Of course science fiction—the highbrow stuff, at any rate—is getting less scientific. It is still pretty special, so that one has to judge it by its own rules (however difficult these may be to define), just as one judges chess problems not as actual game situations but as esoteric intellectual contrivances bound by peculiar conventions. But the specialness is getting less special, which is to say that a number of science-fiction writers are turning their product in the direction of the mainstream of serious literature, and that certain writers of stature—William Golding, for instance—are invading the science-fiction domain. Each group, as Amis observes, can learn from the other, though admittedly the mainstream group has vastly more to teach its science-fiction counterparts than to learn from them. Their common ground, I should think, is dissent, revulsion, indignation, even bewilderment, over the less lovely aspects of society. Science fiction may run out of scientific ideas, but never out of instances of imbecility and wickedness that can easily be transformed into comic infernos.

It is important not to come to any grand conclusions about what science fiction is or may yet become. It does not exist in isolation, however queer it may seem; it is very much a piece of ourselves; its future is uncertain, but whose is not? It has, I am now convinced from Amis's book, something of value. I hope it prospers.

# 2

# Communicating with Intelligent Life Elsewhere in the Universe

JAMES R. NEWMAN

*February 1964*

A review of *Interstellar Communication*, by A. G. W. Cameron. New York: W. A. Benjamin, 1963.

"We are in great haste to construct a magnetic telegraph from Maine to Texas; but Maine and Texas, it may be, have nothing important to communicate." Now, a little more than a century since Thoreau entertained this irony, we are once more confronted with the possibility of enlarging the universe of human jabber—not, thank heaven, between Maine and Texas, but between our small planet and the possible inhabitants of other planets. A contemporary cynic might inquire: "What have we got to tell that is worth sending billions of miles into space?"; but before we can indulge ourselves in the luxury of such skepticism there is much to be found out. Is there life on other planets in our solar system? Are there other stars in our galaxy that have planets? Is it reasonable to assume that, if such planets exist, some are life-bearing? If some are, have their organisms evolved to the stage of man? If so, do these organisms have advanced societies that are capable of communicating with us or of receiving signals we might be capable of sending them? Or are they so far beyond us they would no more waste time in talking to us than we would seek to confer with a midge? To be practical, if someone is out there, how do we engage his attention?

These are the main questions dealt with in this engrossing anthology assembled and edited by a Canadian astrophysicist now working at the Institute for Space Studies in New York. Not too long ago such an anthology would have been pure science fiction. The developments of the past 20 or 30 years, however, have drastically altered the outlook

of even hardheaded theoreticians, experimentalists and engineers. The reprints and original articles constituting this book come from astronomers, physicists, mathematicians, biochemists, electrical engineers and practitioners of other specialties, all of them concerned with the possibility of interstellar communication. Some are more imaginative than others, more philosophically disposed, more poetic, more awed, more scared of what might pop out at them. Not all scientists, after all, are speculative or reflective; there are lots of clever fellows who merely want to make the key with which to open the lock. But a high proportion of the contributors to Cameron's collection are aware of the implications for man of this branch of study, and all seem to think there is a pretty good chance we are not alone in the galaxy and that in the foreseeable future evidence may be found to complete this final phase of the Copernican intellectual revolution.

A long chain of reasoning and observation has brought us to the portal of the new adventure. For centuries, perhaps millenniums, men could freely speculate, as did Teng Mu, a scholar of the Sung dynasty, about "other skies and other earths." But knowledge, like the Lord, giveth and taketh away, and so, at the same time as our universe expanded and our windows on it became more ample and numerous, certain theories arose concerning scientific limits and impossibilities. With respect to the possibility of life, not to say intelligent life, in other worlds, a limit was imposed by the respected cosmogonic hypothesis of Sir James Jeans. In his view the solar system was the product of a chance encounter between two stars. In this catastrophe the earth and the other planets were spun off as debris. But

such encounters must be rare, because stars are tiny compared with the distances between them. Therefore planetary systems such as ours must be extremely rare.

Jeans's hypothesis, however, came under attack in the 1930's. Several nearly unanswerable points were raised against it. Astrophysicists calculated that fragments drawn out of the sun by a grazing star would fall back into the sun. It was suggested by Henry Norris Russell (and later mathematically demonstrated by the Soviet astronomer N. N. Pariysk) that a collision or near collision could not explain the circumstance that 98 per cent of the angular momentum of the system is contained in the planets and only 2 per cent resides in the sun. The hypothesis was gradually supplanted by the notion that sun and planets condensed simultaneously out of one cloud of gas and dust. It is now believed that this is the usual process of star formation, and the inference has been drawn that many stars are accompanied by planets. This is the first link of the chain.

How many systems physically resembling ours are there? Our galaxy is estimated to have a population of the order of 100 billion stars. Many of these, it is thought, either never had planets or swallowed them as they were being born. The very massive stars formed from the larger gas clouds, for example, rotate much more rapidly than the sun and exert a formidable gravitational attraction. They seem to contain the entire angular momentum of the parent cloud, which is to say they share none of it with planets. This still leaves a vast number of stars as likely candidates. It is a plausible deduction that these stars have, in Philip Morrison's words, "bequeathed their spin to planets." If this reasoning is correct, it is probable that there are at least several billion planets in our galaxy alone.

We have seen none of them. Why? Because the existence of planetary systems comparable to ours cannot, as the Soviet astrophysicist I. S. Shklovsky points out, be detected by astronomical observation, even in the nearest stars. He relates the late Otto Struve's example of an imaginary observer perched in the orbital plane of Jupiter at a distance from the sun of 10 parsecs (a little more than 30 light-years). With the best astronomical instruments now available he would not be able to detect even that giant planet as an object adjacent to the sun. Moreover, when Jupiter passed across the solar disk, as it does once every 11.8 years, it would dim the sun's light no more than a hundredth of 1 per cent, an amount that would be impractical to detect by the best contemporary photometers.

Not every planet would be hospitable to life; an even smaller fraction could support highly organized and intelligent life—as we like to think of ours. Su-Shu Huang has demonstrated that the habitable zone of a planet increases with the luminosity of its sun. There must be enough radiation, otherwise the temperature would be too low to sustain the chemistry of life, and not too much radiation, otherwise the temperature would be too high for such chemistry. In short, the eligible planets must lie in the "Mars-earth region."

We are whittling down the possible number of earthlike planets but we must go further. Our interest, when we envision communication, is in planets that may have intelligent life *now*, not in planets that may once have sheltered it but whose "psychozoic era" has passed, nor in those that may attain this level in the future. Intelligent life on earth has had a span of perhaps a million years. We may be nearing the end, having nibbled too fast on the apple of knowledge and too little on the fruit of the owl. This may be the fate of all advanced societies; dinosaurs lasted longer. Pessimism aside, several time-limiting factors must be taken into account. First, it is highly probable that even under favorable conditions it takes several billion years for the complex beings we are looking for to evolve from the primordial consommé. Second, in the long history of our galaxy it is likely that many stars have played out their role as hosts to life. Third, there is some reason to believe that societies as a whole have a finite span, just as their individual members do.

These considerations point to the conclusion that today only a small fraction of the 100 billion stars of our galaxy have planets that can give rise to life and intelligent beings. The proportion is small, but the congregation of which it is a part is huge; thus a number of the order of 100 million eligible planets may not be wildly wrong.

In our solar neighborhood about 5 per cent of the stars, according to Huang's estimate, may have life associated with them; 41 of these lie within five parsecs (about 16.7 light-years) of the solar system and two, Epsilon Eridani (10.8 light-years) and Tau Ceti (11.8 light-years), have "a good chance" of possessing planets bearing life.

At this point a most important question enters. We have been proceeding on a grand assumption: that given certain simple ingredients and certain sources of energy, life will begin and will evolve in any appropriate cradle. Is this assumption sound? That it probably is, is demonstrated in an admirable article, the longest and best in this collection, by Melvin Calvin.

The essential ingredients are carbon, nitrogen, oxygen and hydrogen. These, together with a few others, are the principal atoms of living material. The primeval atmosphere of the earth, it is conjectured, consisted mainly of four substances: molecular hydrogen, carbon combined with hydrogen as methane, oxygen combined with hydrogen in the form of water and nitrogen combined with hydrogen in the form of ammonia. Out of these everything has to be cooked: organic molecules, trilobites, water lilies, fleas, man.

The beginning of the life process involves tearing atoms from one another, recombining them, tearing them apart and recombining them again and again to form more complex substances. Large quantities of free energy are required for tearing things apart. Among the forms that were available on earth and must be available on other life-bearing planets are ultraviolet radiation (in our case from the sun), cosmic rays, radiation from radioactive elements, and lightning. Calvin and his colleagues performed experiments in 1950 in which a cyclotron was a source of ionizing radiation, to see if they could effect certain basic transformations of simple molecules. The experiments were successful. Starting with carbon dioxide (instead of methane) and hydrogen, the experimenters were able to make reduced, or hydrogenated, carbon compounds such as formic acid and formaldehyde. Since that time the theory of a reducing pri-

meval atmosphere "has become increasingly accepted," and by a wide variety of experiments using ionizing radiation, ultraviolet radiation and electric discharges (lightning) simple molecules have been converted into more complex ones such as acetic acid, succinic acid and the amino acid glycine.

But the achievement itself immediately posed an even more formidable question: Are there more efficient ways to transform these molecules in a manner that will be selective? If these early processes were mere chance affairs, there would be no reason to believe that the complex substances would have endured, let alone have achieved a higher architecture. In the absence of a selective operator the free energy would break up the complexes as quickly as they were being formed, and this random shuttling between construction, destruction and reconstruction would proceed indefinitely and futilely.

It was soon discovered that in the presence of certain substances acting as catalysts the transformations become more selective. The process might run something like this. If substance $A$ has a choice of going to substances $B$, $C$ or $D$, and if $D$ is the kind of substance that for chemical reasons accelerates the rate of the transformation from $A$ to $D$, then more of $A$ will become $D$ than will become $C$ or $B$. This is called autocatalysis. $D$ makes itself and selects the transformations leading to its own creation.

The next stage requires the passage from complex but still small molecules to the giants: the polymers. The amino acids "constitute one of the principal classes of compounds in which giant molecules can be made by hooking small ones together, end to end. There is an amino (basic) end and an acid end, and these can combine indefinitely to make a long chain which is called polypeptide and which constitutes one of the principal structural materials (protein) of all living things." The biochemist Sidney W. Fox has shown that under what he calls prebiological conditions some of the amino acids could be converted into protein-like substances. By heating up a mixture of some 20 amino acids in molten glutamic acid a little above the boiling point of water he got the amino acids to hook together and make polypeptides.

Calvin describes further researches that have thrown light on the primordial formation of nucleic acids. The twin backbones of these giant molecules are joined by hydrogen bonds between their constituent bases; on the outside of the bases rows of sugar molecules linked by phosphate groups act "as though two ribbons were tying together the edges of a row of playing cards." If these ribbons are twisted, the playing cards will instead of lying flat turn sideways and stack up. Flat molecules do stack up under certain conditions: they crystallize. Although the stacking up is spontaneous, the order that the four kinds of base (thymine, adenine, cytosine and guanine) can follow "apparently contains the necessary information . . . which tells the present-day organism what to become. This arrangement of bases contains the genetic information of a modern organism."

The path from chaos to order is exquisitely intricate, but we have vital clues to it. The staircase to the giant molecule is understood, but as yet the organization of these giant molecules into larger structures can only be conjectured. Again, Calvin suggests, crystallization is the means of ascent. Even so, even though certain macroscopic structures can under the proper conditions be precipitated out of solutions of giant molecules, the structures are far different from living cells. They lack, among other things, the membrane that separates the cell from its environment. There are, however, physicochemical mechanisms for producing enclosing membranes. For example, if a film of oil is lying on a layer of protein that in turn is resting on water, the film rolls up into droplets, some enclosing air, some water. F. Millich and Calvin have shown that, when one takes a fairly dilute solution of a synthetic polymer and adds a trace of iron to it, the material separates into oily droplets that contain the iron. This process, called coacervation, is regarded as a "primary phenomenon" in the development of cellular structures. Giant molecules in solution have the ability to congregate or to form into droplets. They also tend to pack themselves in ordered arrays, "provided that they themselves have ordered structures."

Before leaving Calvin's account I must say a few words about another of his investigations, one having to do with the formation of the bases that constitute the nucleic acids. One can trace a sequence of chemical and physical events that might have led to living substance. Proteins and nucleic acids are the central actors both in the development of the living substance and in conferring on it the properties regarded as defining life. Self-reproduction is of course one of these properties, and it stems from the serial array of the bases

in a nucleic acid, which array contains the information that directs the manufacture of proteins. It is important to know how this close and essential relation arose in the evolutionary symphony, and whether the nucleic acid or the protein came first.

An experimental approach to these questions was an attempt in the laboratory to synthesize nucleic acid bases under primitive terrestrial conditions. Electron radiation resembling that from the abundant radioactive isotope potassium 40 was used to bombard a mixture of gases believed to resemble the earth's primeval atmosphere. One product of the bombardment was hydrocyanic acid. This was not unexpected for two reasons: hydrocyanic acid is a stable compound and it is not uncommon in the universe. What was more significant was that out of the mixture containing hydrocyanic acid Calvin and his collaborator C. Palm could get the bases needed for making nucleic acids. It was already known that when hydrocyanic acid is put into a water solution with ammonia (believed to have been present in the primeval atmosphere), adenine, one of the chief bases, is formed; indeed, adenine is simply a molecule made up of five hydrocyanic acid molecules put together in a certain way. Now it was discovered that electron bombardment will produce "respectable amounts" of hydrocyanic acid, adenine and other bases. (Lightning, it has since been learned, will also do the trick.) Once again experiments had confirmed the hypothesis that certain simple substances—methane, ammonia, water and hydrogen—torn apart by ionizing radiations would recombine; that the new combinations would not be random but instead would be specific substances, some of which are the bricks of life.

The setting, then, is something like this. A large number of well-tempered planets, neither too hot nor too cold, sedately travel their stable orbits. The planets vary in size, their radii lying between 1,000 and 20,000 kilometers. Their atmosphere is benign; they are old enough to have lost much of the hydrogen that originally enveloped them, and their mass and mean density are such as to enable them to hold on to an adequate amount of the relatively heavy oxygen molecules. The stars with which they are associated are of an average sort, ranging in age between a billion and 10 billion years. These planets, it is thought, are habitable, and some of them are quite possibly inhabited by intelligent beings. It is proposed that we search for them.

There are two possible methods: signaling and space travel. The second can be disposed of without straining. In the past few years we have been exposed to both high- and low-level blather about space travel. Politicians, Government officials, military philosophers and rocketeers have been preaching the holy mission of traveling around the universe in space suits. National honor and national security are said to depend on our going somewhere and getting there first. Besides, it would be fun and the preparations are good business. Leaving aside the usefulness and feasibility of local exploration—a modest trip to the moon, say—the notion of vast space voyages belongs, the physicist Edward Purcell says, "back where it came from, on the cereal box." Both he and Sebastian von Hoerner, a German astrophysicist, expose the hoax.

Assume that a journey is planned to a nearby star—12 light-years away. We do not want to take many generations to do it so let us arbitrarily say we will make the round trip in 28 years of earth time. This means reaching a speed 99 per cent of the velocity of light in the middle of the journey, slowing down at our destination and then returning. The special theory of relativity shows that this 28-year journey will cost the voyagers only 10 years of aging time. Forget about the twin paradox; you have troubles enough (and anyway, if its implications are *not* accepted, the conclusions to be drawn from this example become, as Purcell says, even stronger).

The problem is to design a rocket to perform this mission. The elementary laws of mechanics "inexorably impose" a certain relation between the initial and the final mass of the rocket in the "ideal" case. It is plain that the propellant selected must have a very high exhaust velocity. The relation is given by this simple equation:

$$\frac{\text{initial mass}}{\text{final mass}} = \left[\frac{c + V_{max}}{c - V_{max}}\right]^{c/2V_{ex}}$$

Here $c$ is the speed of light; $V_{max}$, the rocket's maximum velocity; $V_{ex}$, the exhaust velocity. Obviously if $V_{max}$ is near the speed of light, the denominator of the equation becomes very small and the exponent very large. So far, so good. Practical questions aside, we can use a nuclear-fusion propellant that burns hydrogen to helium with 100 per cent efficiency. This will give an exhaust velocity about an eighth of the speed of light. Purcell calculates that to attain a speed 99 per cent of the velocity of light from this kind of push one needs an initial mass (mostly the fuel) that is a little over a billion times the final mass of the rocket. To put up a ton, therefore, one must start off with a million tons.

This will never do. So one may as well go for broke, take the ultimate step and use the perfect propellant, namely matter-antimatter. This helps. To attain 99 per cent of the velocity of light a ratio of only 14 is needed between the initial and the final mass. But what with having "to slow down to a stop, turn around, get up to speed again, come home [if you care] and stop," the tame 14 becomes $14^4$, which is some 40,000. Thus by Purcell's reckoning, to take a 10-ton payload on the 28-light-year trip requires a 400,000-ton rocket, half matter and half antimatter —the latter admittedly not easy to put in tanks.

A small point about the antimatter rocket. To achieve the required acceleration the rocket will have to radiate about $10^{18}$ watts near the beginning of its journey. This turns out to be "only a little more than the total power the earth receives from the sun." Purcell reminds us, however, that the rocket radiation is not sunshine but gamma rays. Therefore there are two protection problems: to shield the occupants of the vehicle and to shield the earth itself from the rocket's radiation.

Von Hoerner offers a similar item. The power-mass ratio ($P$) needed to get close to the velocity of light, using an antimatter photon-thrust device, comes to something like this (in units of watts or horsepower):

$$P = 3 \times 10^6 \text{ watts/gram}$$

$$= 4 \times 10^3 \text{ h.p./gram}$$

Our most efficient fission reactors, used for ship propulsion, have an output of 15,000 kilowatts and weigh 800 tons; for them $P$ is equal to .02 watt per gram. To achieve the power-mass ratio of von Hoerner's equation the reactor can weigh no more than five grams (about the weight of ten paper clips). To express it another way, to fulfill the equation the engine of a car producing 200 horsepower cannot weigh more than a tenth of a paper clip. If 150,000 kilowatts is all one can get out of each power plant, 40 million annihilation power plants are required, plus six billion transmitting stations of 100 kilowatts each, all together having no more mass than 10 tons, to approach the velocity of light within 2 per cent with-

in 2.3 years of the crew's time. And if, for example, one fails to fulfill this requirement by a factor of a million (so that one has "only" 40 power plants and 6,000 transmitters, weighing in all 10 tons), it would take 2.3 million years for the crew to approach the velocity of light within 2 per cent.

Communication by signals is what is left to us. Here it is at least conceivable that we might succeed without the combined help of Aladdin and the Wizard of Oz.

Should we broadcast or should we simply listen? Other worlds may be broadcasting. The science and art of electromagnetic devices are advanced and we could, for example, using a reasonably large antenna, send a 10-word telegram a distance of 12 light-years with a dollar's worth of electrical energy (Purcell). Would our message be understood? "AUNT IDA VERY ILL COME AT ONCE LOVE WERNHER" might not be. Moreover, the reply, assuming it came, would be at least 24 years delayed. A hurry-up people like us might get impatient, or forget they had ever sent a message. What about the level of technology among our planetary listeners? Wireless telegraphy is only about 50 years old and highly discriminating receivers are very recent. If, says Purcell, "we look for people who are able to receive our signals but have not surpassed us technologically, i.e., people who are not more than 20 years behind us but still not ahead, we are exploring a very thin slice of history." By all odds it would be better to listen— although it would be well to notice that if all societies follow this practice, everyone has a long wait.

In an imaginative paper published in *Nature* in 1959, Giuseppe Cocconi and Morrison suggested a listening program. According to their assumptions there are other civilized societies. They are probably well ahead of us scientifically and therefore know at least as much about the possibilities of life on our planet as we know about life on other planets; they have for a long time been sending out signals and have been "patiently" awaiting answering signals that would make known to them "that a new society has entered the community of intelligence." It is likely that the channel they have been using is one "that places a minimum burden of frequency and angular discrimination on the detector." Appropriate to our beginner status as interplanetary readers, they can be expected to give us the equivalent of big letters, easy words, wide spacing and broad margins. With

this in mind and also various technical considerations relating to background noise, the absorption of certain frequencies in planetary atmospheres and the like, the question is: At what frequency should we look for signals?

The task at first glance seems enormously difficult, but on closer examination it turns out that we are the beneficiaries of circumstance. "In the most favored radio region there lies a unique, objective standard of frequency, which must be known to every observer in the universe: the outstanding radio emission line at 1,420 Mc/sec ($\lambda = 21$ cm) of neutral hydrogen. It is reasonable to expect that sensitive receivers for this frequency will be made at an early stage of the development of radioastronomy. That would be the expectation of the operators of the assumed source, and the present state of terrestrial instruments indeed justifies the expectation. Therefore we think it most promising to search in the neighborhood of 1,420 Mc/sec."

Cocconi and Morrison proposed the examination of certain stars within 15 light-years, lying in a direction with low background noise. About a year later the proposal was acted on. A program called Project Ozma (queen of the land of Oz) was carried on for three months with an 85-foot radio telescope at the National Radio Astronomy Observatory. The two target stars were Tau Ceti and Epsilon Eridani. No signals of extraterrestrial origin were picked up, but there was an almost unbearably exciting false alarm when, within a minute or so of turning the antenna toward Tau Ceti, unmistakably strong signals began to pour in. Soon afterward they faded and could not be rediscovered. The observer, Frank D. Drake, concluded they had been terrestrial signals, but one may indulge one's fancy if one likes. Thirty years ago Carl Størmer and Balthus Van der Pol caught similar signals that were never explained.

The search will go on. It may take decades, centuries. We may have to go out 1,000 light-years. The quest may very well fail, particularly in its present form. Optical masers have been suggested as a possible means of communication; this would entail spectral resolution to distinguish between starlight and a possible separate optical signal. As far as I know, no one has proposed enlisting extrasensory perception. Thought may travel faster than light. (It is, however, appropriate to recall J. B. Priestley's observation that those who say they are getting messages from the void are probably getting them from the void within.

We have not the remotest notion what the signals we are seeking would say. It seems to be agreed that they would have to have non-Gaussian characteristics; that is, they would have to be orderly, patterned, nonrandom. But "orderly" by the standards of very advanced beings may not be discernibly orderly by our standards. If one's life expectancy is 10,000 years, say, one may design communication patterns to unfold slowly in order to guard against dying of boredom by repetition. Morrison speculates that number sequences might be sent. A string of primes, the number pi, logarithms, an algebraic formula. This does not strike me as very imaginative. B. M.

Oliver describes a fanciful way of sending bits of information that could then be decoded and rearranged into pictures. Always it is assumed that we will be treated like half-wits or small children. Morrison says he cannot conceive that our earth is the concern of the great enterprises of knowledge among "far societies" but that it falls, rather, under the aegis of a "Department of Anthropology." The prevailing motives among us for space travel tend to justify this view, and a few of the contributors to this book are silly enough to impute to the societies of other planets the same murderous impulses, military ambitions, obsessions and follies that shape our own civilization.

One thing is certain: All men will have to co-operate in this unknown journey. It is an adventure for mankind, requiring unlimited patience, the highest skill and shared good will. If signals come to us, we shall want to reply. And then in turn we must wait for signals showing that we have been heard and understood. It is the perfect, deliberate philosophical discourse. We will bequeath to our children the answers to the questions we ask. If the venture is conceived in that spirit, we may have something important to communicate.

# 3

# The History of Man's Picture of the Universe

I. BERNARD COHEN

*June 1959*

A review of *The Sleepwalkers: A History of Man's Changing View of the Universe*, by Arthur Koestler. New York: Macmillan, 1959.

"In the index to the 600-odd pages of Arnold Toynbee's *A Study of History*, abridged version, the names of Copernicus, Galileo, Descartes and Newton do not occur." These words, with which the preface of *The Sleepwalkers* begins, exemplify both the tone and the purpose of Arthur Koestler's new book. He is concerned with "the gulf that still separates the Humanities from the Philosophy of Nature." Koestler prefers the latter expression to "Science," because it carries with it the "rich and universal associations" of the 17th century, the "days when Kepler wrote his *Harmony of the World* and Galileo his *Message from the Stars*." In fact, the greater part of Koestler's book deals with the 17th century, the age known now for the "Scientific Revolution" but then for the "New Philosophy," which had as its aim not a revolution in technology but the "understanding of nature." Koestler regards the 17th century nostalgically as an age that still had a place in it for the universal man, and that had not seen either the present fragmentation of knowledge or the erection of "academic and social barriers" between the "Sciences" and the "Humanities."

One result of this "cold war," according to Koestler, is that there exist histories of the sciences but "no comprehensive survey of man's changing vision of the universe which encloses him," a deficiency that Koestler believes his own efforts to have remedied. He has not written "a history of astronomy, though astronomy comes in where it is needed to bring the vision into sharper focus; and, though aimed at the general reader, it is not a book of 'popular science' but a personal and speculative account of a controversial subject."

Koestler asks the reader to keep in mind that despite the time limit he has imposed—to begin with the Babylonians and to end with Newton—the subject was still so vast that to keep it "within manageable limits" he "could attempt only an outline." The result is thus "sketchy in parts, detailed in others, because selection and emphasis of the material was guided by my interest in certain specific questions, which are the *leitmotifs* of the book." These are two in number. First there are "the twin threads of Science and Religion, starting with the undistinguishable unity of the mystic and the savant in the Pythagorean Brotherhood." These threads run through history together, "falling apart and reuniting again, now tied up in knots, now running on parallel courses, and ending in the polite and deadly 'divided house of faith and reason' of our day." Koestler believes that a "study of the evolution of cosmic awareness in the past may help to find out whether a new departure is at least conceivable, and on what lines." Such a statement strongly suggests, as the text confirms, that what Koestler objects to is not merely the fragmentation of science but fragmented science itself, which has failed to achieve for man the goals that the 17th-century natural philosophers held to be the fruits of their inquiries.

The second *leitmotif* is the "psychological process of discovery," which Koestler conceives "as the most concise manifestation of man's creative faculty." He is equally concerned with "that converse process" which blinds man "towards truths which, once perceived by a seer, become so heartbreakingly obvious." Koestler calls this negative aspect of discovery a "blackout shutter" and observes its presence in scientific men of the highest genius: Aristotle, Ptolemy, Galileo, Kepler. It is suggested that a study of the greatest scientific men of the past shows that "while part of their spirit was asking for more light, another part had been crying out for more darkness." Koestler believes that to understand science one must be concerned with the character and personality of scientists, since "all cosmological systems . . . reflect the unconscious prejudices, the philosophical or even the political bias of their authors." Whether the subject be physics or physiology, "no branch of Science, ancient or modern, can boast freedom from metaphysical bias of one kind or another." This is Koestler's explanation for the fact that the evolution of science follows a bewildering series of zigzags rather than a simple and logical linear progression. And so we are led to Koestler's main theme: "The history of cosmic theories, in particular, may without exaggeration be called a history of collective obsessions and controlled schizophrenias; and the manner in which some of the most important individual discoveries were arrived at reminds one more of a sleepwalker's performance than an electronic brain's."

That it should be Arthur Koestler who has chosen this particular assignment may be an occasion for some surprise. His reputation, at least in this country, was made by the novel *Darkness at Noon,* in which he explored the psychological reasons for the confessions of old-line Bolsheviks in the Russian trials of the 1930s. Five years ago, however, Koestler decided to give up political matters for science. This decision was not quite as odd as it might seem; in pre-Hitler Germany Koestler had been science editor of the Ullstein chain of newspapers, and he had started his career as a science student in Vienna.

That his present writing on science should concentrate so much on the psychological aspects of scientific discovery is in keeping with Koestler's psychological approach to political problems. His *Insight and Outlook* dealt with questions similar in meaning but not in kind to those considered in *The Sleepwalkers*. His portrayal of conflicting attitudes in *The Yogi and the Commissar* may help to prepare the reader for his present reflections on the mysticism of the early cosmologists, as opposed to the hard-boiled positivism of modern science.

It will be apparent to anyone who picks up this book that Koestler has done a solid job of research. He provides a wealth of quotations from original source-material and buttresses them with references to some of the major results of historical scholarship. The very bulk of the book may deter some readers; it should be said at once that the narrative is fascinating and the style engaging and urbane. The author's personal prejudices and predilections will variously delight, enrage, stimulate, repel, attract, astound and titivate the scientist, the historian and the general reader. The book is stuffed with all kinds of recondite information about the lives of the early scientists and their contributions to the science of the universe. At the same time Koestler propagates old errors and adds new ones of his own; in general he presents a highly biased and/or inaccurate view of the nature and growth of science, and a distorted account of some of the major episodes in the history of astronomy.

Koestler is most successful in his discussion of Johannes Kepler, to whom about half *The Sleepwalkers* is devoted. One almost has the feeling that what began as a biography of Kepler ended up as a history of cosmology only by virtue of having Copernicus, Tycho Brahe and Galileo appear so that Kepler might be made a more sympathetic figure; and that a wholly inadequate account of ancient and medieval cosmology was added only to point up the originality of Kepler's contributions. Koestler claims that his own scholarly contribution lies mainly in the sections dealing with Kepler, "whose works, diaries and correspondence have so far not been accessible to the English reader; nor does a serious English biography exist." This is true, and it may be added that the only two major studies of Kepler's astronomy are all but inaccessible even to scientists and historians. Indeed, they were not consulted by Koestler.

Kepler makes an ideal subject for Koestler. Because one of Kepler's main concerns was the influence of the stars on the lives of men, he kept meticulous records of both the great and the minute events of his own life. How many biographers could begin, as Koestler does: "Johannes Kepler, Keppler, Khepler, Kheppler or Keplerus was conceived on 16 May, A.D. 1571, at 4:37 a.m., and was born on 27 December at 2:30 p.m., after a pregnancy lasting 224 days, 9 hours and 53 minutes. The five different ways of spelling his name are all his own, and so are the figures relating to conception, pregnancy and birth, recorded in a horoscope which he cast for himself." Koestler remarks: "The contrast between his carelessness about his name and his extreme precision about dates reflects, from the very outset, a mind to whom all ultimate reality, the essence of religion, of truth and beauty, was contained in the language of numbers."

Guided by Kepler's self-portrait, Koestler sketches for us the early years and education of a man who described himself as "born destined to spend much time on difficult tasks from which others shrunk." Quite early he became a Copernican, at a time when very few scientists had read that author's difficult exposition of a sun-centered system of the universe. But Kepler began his career more as an astrologer than a defender of Copernicus. His first job as teacher was marked by a prediction: From the stars he foresaw a very cold winter and an invasion by the Turks! Luckily for him he proved to be right about both. Although he knew that much of astrology was quackery, he remained convinced to his death that the stars and planets determine the fate of man.

Soon Kepler produced his first major work, the full title of which is: *A Forerunner to Cosmological Treatises, containing the Cosmic Mystery of the admirable proportions between the Heavenly Orbits and the true and proper reasons for their Numbers, Magnitudes, and Periodic Motions.* Kepler believed he had discovered why there are only six planets, and why they are placed as they are in the Copernican system. This discovery he considered to be a divine revelation, he being only the instrument directed by God's will and plan. Kepler's discovery relates numbers and geometry: six planets and five regular solids. He conceived that within the sphere containing Saturn's orbit there is inscribed a cube, within which is inscribed another sphere containing the orbit of Jupiter. Within the sphere of Jupiter is a tetrahedron containing the sphere of Mars; within the sphere of Mars is a dodecahedron containing the sphere of the earth; within the sphere of the earth is an icosahedron containing the sphere of Venus; within the sphere of Venus is an octahedron containing the sphere of Mercury. Given a certain thickness for the surface of each sphere, the distances in fact come out right. For the young Copernican, steeped in the Pythagorean ideal and convinced that everywhere in nature there are harmonies of number and shape, it was a significant stroke. To the end of his life Kepler considered it a more important discovery than the celebrated laws that bear his name.

It was Kepler's good fortune (he held it to be an act of divine Providence) that his book brought his skills in computation to the attention of Tycho Brahe. This Danish astronomer (for some reason Koestler insists on Frenchifying his name to "Tycho de Brahe") had reformed astronomy by making continuous rather than occasional observations of the planets, and by creating new instruments that made it possible to determine the planetary positions within less than eight minutes of arc. To appreciate what this figure means, one must recall that Copernicus had said that if he could ever predict the planetary positions within 10 minutes of arc, he would be as happy as Pythagoras was when he discovered his theorem. To Tycho, whose reforms made an exact science of astronomy, whose system of the world briefly became a rival of Ptolemy's and Copernicus's, and whose observations provided Kepler with the materials for his own deep discoveries, Koestler devotes an unilluminating pair of short chapters.

Using Tycho's observations of Mars, Kepler found his first two laws; these he published in his *New Astronomy*, which he described as "a physics of the sky, derived from investigations of the motions of the star Mars." Kepler's first law states that the planets move around the sun in elliptical orbits, one focus of the ellipse being occupied by the sun. His second law states that along this orbit each planet moves in such a manner that a line drawn from the sun to the planet sweeps out equal areas in equal times. Our modern assignment of the words "first" and "second" here suggests a logical progression: Is it not obvious that to determine the speed with which a planet moves along its orbit, and to compute the area swept out by a line drawn from sun to planet, there must be a prior determination of the shape of the orbit?

But it is just this point that provides Koestler with a chief illustration of his central thesis. For the fact of the matter is that in this instance (though it is not unique in the development of science) the logical order of the discovery is just the reverse of the chronological order.

Kepler's *New Astronomy* gives us almost every stage in the development of his ideas—even calculations based on suppositions he later knew to be false. To the exhausted reader he says (in Chapter 16): "If thou art bored with this wearisome method of calculation, take pity on me who had to go through at least seventy repetitions of it, at a very great loss of time." By Chapter 19 he is at last willing to give up all the variations he had introduced into older hypotheses, and to approach his "goal according to my own ideas"; at last the reader could proceed to Kepler's own hypotheses. But 13 chapters later Kepler is still showing the consequences of physical assumptions, chiefly that some kind of *anima motrix* emanates from the sun, diminishes according to the inverse proportion of the distance and is active only in the plane of the planet's motion. In this way Kepler found the "second" law, knowing it was based on errors which he said "like a miracle . . . cancel out in the most precise manner." On this score Koestler remarks: "The correct result is even more miraculous than Kepler realized, for his explanation of the reasons *why* his errors cancel out was once again mistaken, and he got, in fact, so hopelessly confused that the argument is practically impossible to follow—as he himself admitted. And yet, by three incorrect steps and their even more incorrect defense, Kepler stumbled on the correct law. It is perhaps the most amazing sleepwalking performance in the history of science—except for the manner in which he found his First Law." In the course of this new adventure Kepler turned to an egg-shaped planetary orbit and repudiated the previously announced second law; he said that he regretted that the orbit was not "a perfect ellipse [for then] all the answers could be found in Archimedes's and Apollonius's work." Then by chance he suddenly found a clue to the fact that the orbit was indeed elliptical. He tells us: "I felt as if I had been awakened from a sleep." But search as he might ("until I went nearly mad"), he could never find "a reason why the planet preferred an elliptical orbit."

From antiquity up to the time of Kepler's book everyone was agreed that planetary motions were either circles or some combination of circles. No one was able to understand why a planet should move along an ellipse until Newton showed that this orbit was a consequence of a central force of gravitation dependent on the inverse square of the distance. It is not surprising that Kepler's contemporaries (including Galileo) never accepted elliptical orbits, if indeed they had ever heard of them. Koestler cannot forgive Galileo his treatment of Kepler. "In his works," says Koestler, Galileo "rarely mentions Kepler's name, and mostly with intent to refute him. Kepler's three Laws, his discoveries in optics, and the Keplerian telescope, are ignored by Galileo, who firmly defended to the end of his life circles and epicycles as the only conceivable form of heavenly motion." This statement is repeated a number of times with variations. Yet according to Koestler's own presentation even to find the first two laws within Kepler's *New Astronomy* is a heroic exercise. Moreover, the reader who has found them will quickly discover that they are based on false and untenable assumptions. Koestler writes: "The ellipse had nothing to recommend it in the eyes of God and man; Kepler betrayed his bad conscience when he compared it to a cartload of dung which he had to bring into the system as a price for ridding it of a vaster amount of dung. The Second Law he regarded as a mere calculating device, and constantly repudiated it in favour of a faulty approximation; the Third as a necessary link in the system of harmonies, and nothing more." If this is so, it is hardly surprising that Galileo did not accept Kepler's laws, even if he had succeeded in finding them within Kepler's prolix and confused presentation.

Koestler emphasizes the fact that Kepler was the first man to base astronomy on physics, even that he was the first man to make astronomy a causal physical science. But Koestler denies the reader equal insight into the relationship between Galileo's Copernican astronomy and Galileo's new science of dynamics. So the link between the work of Kepler and Galileo on the one hand and of Newton on the other is only half-presented. Koestler altogether gives Galileo a hard time. Kepler is "disarming"; according to Koestler, no one who met him "in the flesh or by correspondence could seriously dislike him." Galileo, in contrast, is characterized by "arrogance minus humility"; he is a man held in "aversion"; he is "ambiguous" and "subtle"; he confounds "freedom of thought" with "sophistry, evasion, and plain dishonesty." Galileo is condemned by Koestler on the ground that at the age of 70 he was frightened in the prison of the Inquisition, from which he was released (according to a report of the time) "more dead than alive." Koestler ascribes Galileo's "panic" to "psychological causes." Says Koestler: "It was the unavoidable reaction of one who thought himself capable of outwitting all and making a fool of the Pope himself, on suddenly discovering that he has been 'found out.'" But there is no reason whatever to believe that Galileo intended to make a fool of the Pope. I would submit that Galileo's panic (if he felt it at all) in a prison with that particular history needs no "psychological" explanation. That Koestler, who so sensitively described the thoughts of the prisoner Rubashov in *Darkness at Noon*, should find it necessary to invoke such an explanation is rather odd.

Why is it that Koestler can find no sympathy for Galileo, even in the trial room of the Inquisition? Koestler writes that although he has from early childhood retained "vivid impressions" of the "wholesale roasting alive of heretics by the Spanish Inquisition" (which is neither relevant nor exactly true), he finds "the personality of Galileo equally unattractive, mainly on the grounds of his behaviour towards Kepler." This absurd bias makes a travesty of the concluding portion of Koestler's book by forcing him to ignore or to belittle Galileo's profound contribution to physics and astronomy. All this because Galileo did not immediately answer Kepler's letters or accept his laws! Furthermore, Koestler makes it plain that his bias is not based "on affection for either party in the conflict" between Galileo and the Roman Church, but on his own personal "resentment that the conflict did occur at all." For Koestler the conflict was only in the nature of "a clash of individual temperaments," and not "a fatal collision between opposite philosophies of existence." Because of his belief in the "unitary source of the mystical and scientific modes of experience" and "the disastrous results of their separation," Koestler treats Galileo as the villain of his book.

The presentation of Copernicus is equally unsympathetic; the section devoted to his work is entitled "The Timid Canon." Koestler calls him "a stuffy pedant," lacking "the sleepwalking intuition of the original genius." Having "got hold of a good idea," Copernicus "expanded it into a bad system." I find myself wondering how different Koestler's book might have been if he had brought to bear on Copernicus the insight with which he portrayed the old Bolshevik Rubashov. Surely the reader would have been fascinated and rewarded by a thoughtful consideration of the mind of a man who stood at the threshold of the world of modern science and yet was not free of the shackles of his medieval heritage, who presented a new system of the world and yet knew that there was no physics to account for it.

Since Koestler presents Copernicus without understanding, it is hardly surprising that his account of the pre-Copernican period is a collection of platitudes. Indeed, he almost completely ignores astronomical and cosmological thought from Ptolemy to Copernicus, which has been considerably illuminated by recent scholarship. Because he finds Ptolemy "profoundly distasteful," Ptolemy's conceptions are reduced to "the work of a pedant with much patience and little originality, doggedly piling 'orb in orb.' " A handful of pages contemptuously dismiss the greatest achievement of ancient astronomy because of what is held to be an unforgivable failure to adopt the heliocentric system proposed by Aristarchus, a position in which Koestler finds it necessary to ignore the fact that Aristarchus's heliocentric astronomy was never worked out by its author but was only a suggestion whose implications were never—and could not possibly have been—realized. To go back farther in time, Koestler's presentation of the world of the Babylonians and the Egyptians reads as if it had been written decades ago, before serious investigation had made any dent in our ignorance of their respective astronomical systems. One must conclude that any statement in the book that does not refer to Kepler's life or writing must be taken with a grain of salt.

Koestler's epilogue contains the moral of his tale. It requires the whole salt-cellar. He declares that the divorce of science from religion—the freeing of science from its "mystical ballast"—enabled it to "sail ahead at breathtaking speed to its conquest of new lands beyond every dream." The price paid for

this rapid advance was that "it carried the species to the brink of physical self-destruction, and into an equally unprecedented spiritual impasse. Sailing without ballast, reality gradually dissolved between the physicist's hands; matter itself evaporated from the materialist's universe." This evil is imputed in the first instance to the villain of the book, Galileo, because it was he who "banished the qualities which are the very essence of the sensual world—colour and sound, heat, odour, and taste—from the realm of physics to that of subjective illusion." In the end the very "hard atoms of matter" have gone "up in fire-works; the concepts of substance, force, of effects determined by causes, and ultimately the very framework of space and time turned out to be as illusory as the 'tastes, odours and colours' which Galileo had treated so contemptuously." Says Koestler: "Compared to the modern physicist's picture of the world, the Ptolemaic universe of epicycles and crystal spheres was a model of sanity."

Koestler tries to make out that 20th-century physics is just like the pre-Keplerian cosmology. The fact that quantum mechanics works is for him irrelevant; it reminds him only of "Urban VIII's famous argument which Galileo treated with scorn: that a hypothesis which works must not necessarily have anything to do with reality for there may be alternative explanations of how the Lord Almighty produces the phenomena in question."

Koestler thus comes to the depressing conclusion of his reflections. Modern science—and here Koestler departs from cosmology to deal entirely with physics—is a "modern version of scholasticism." Koestler condemns the men of science for their reaction to what he calls "the phenomena of 'extra-sensory perception' " even as he bemoans the "rigorous banishment of the word 'purpose' " from the vocabulary of science. Science is divorced from faith, with the result that "neither faith nor science is able to satis-

fy man's intellectual cravings." Science, since the days of the villainous Galileo, has allegedly "claimed to be a substitute for, or the legitimate successor of, religion; thus its failure to provide the basic answers produced not only intellectual frustration but spiritual starvation." If only science could regain "its mystical inspiration," if only a new Kepler would arise to confute Heisenberg and Schrö-dinger, then the world might be better in some way. In the absence of a Kepler, Koestler concludes with the hope that the "muddle of inspiration and delusion, of visionary insight and dogmatic blind-ness, of millennial obsessions and disci-plined double-think, which this narrative has tried to retrace, may serve as a cau-tionary tale against the *hubris* of science —or rather of the philosophical outlook based on it."

For me the most melancholy aspect of Koestler's conclusion is that it bears so little relation to the book which it serves as an epilogue. Five years of contempla-tion of the peculiar qualities of Kepler's mind, of the work of his predecessors and successors in the development of cosmology, have resulted only in an empty lament over 20th-century physics. So little has Koestler learned from his re-searches that the epilogue is actually based to a considerable degree on his earlier writings, borrowed, as he says, "without quotation marks."

Fortunately for the reader, Koestler is frank about his prejudices. And he has done his bit to correct the old-fashioned notion that science progresses only by orderly, logical steps. Science is an ac-tivity of human beings, and human be-ings do not always behave exactly the way history expects them to. There is no doubt that the scientific activity of man has wellsprings that we do not fully un-derstand. When all is said and done, Koestler's eloquence in describing just how disorderly and illogical the progress of science can be may even outweigh his deficiencies as a historian.

# 4

# The Scientists Who Made
# the Atom Bomb

ROBERT R. WILSON

*December 1958*

A review of *Brighter Than A Thousand Suns: A Personal History of the Atomic Scientists*, by Robert Jungk. New York: Harcourt, Brace and World, 1958.

The title of Robert Jungk's *Brighter than a Thousand Suns* is adapted from the epic poem of India, the Bhagavad-Gita. Jungk's book is also a kind of epic, complete with tragically flawed heroes and a broad scene extending from the Old World villages of Göttingen and Cambridge to the pink and purple desert of New Mexico. Indeed, the birth of nuclear energy and the atomic bomb—its effect upon the men who developed it, upon the political environment which determined it and then was determined by it, upon a world which does not quite know what to do with it—is surely a subject fit for an epic.

The subtitle of the book is *A Personal History of the Atomic Scientists*. There is no doubt that the account is personal, but though Jungk is trained as a historian it is not history. It is marred by inaccuracies on nearly every page. Jungk has gone deep enough into his subject to catch something of the spirit of physics and of physicists (he has dug up a wealth of amusing anecdotes), but he has not gone sufficiently deep to acquire the understanding necessary to qualify him for the kind of analysis he attempts of awesome events and of complex personalities.

Although Jungk has interviewed quite a number of people and read quite a few documents, there is little new material in the volume except for the anecdotes. What is new is the synthesis of many stories, the tying up of many strands—a job so artfully done that the account reads like a novel and is as exciting as a murder mystery. What is also new is

the interpretation of events and the analysis of the actors in the atomic drama. It is here that Jungk's subject is too big for him; his conclusions are frequently irresponsible, sensational and in questionable taste.

It is all to the good that Jungk comes to grips with the heart of the matter: the moral dilemma faced by those scientists attending the birth of the bomb. I cannot censure the author for concluding that his heroes are less than perfect; it may be the decision of history that they were evil personified. It is still too early to view the matter in its true perspective. Inasmuch as I, along with so many others, played my own role in this real-life epic, due allowance must be made for my defensive bias. Reading this book was an emotional experience for me; I found many old doubts, questions and uncertainties awakened. Sometimes Jungk touched old sores (he also succeeded in opening a few new ones), but usually I was moved not by remorse but by indignation.

Jungk takes us on a journey through the years to the quiet times when physics was a pleasant, intellectual subject, not unlike the study of Medieval French in its popular interest. At that time physicists were nonentities for the man on the street. In presenting rather carefully selected but sharply drawn vignettes of the period, Jungk is at his best. He has picked the right people and the right places at the right time to portray the beginnings of modern physics and modern physicists. He also shows a remarkable talent for discussing in adult but simple language some of the actual physics that was involved. Having sketched these idyllic times and then, I might add parenthetically, neglecting to give any account of the exceedingly important American work which bridged

the period between early atomic physics and the later flowering of nuclear physics, Jungk comes directly to the atomic bomb. One of his main themes is the contrast between the wartime work of nuclear physicists in this country and of their counterparts in Germany. His extraordinary thesis seems to be this: Physicists in Germany managed, by their great diligence, *not* to build a bomb physicists in the U. S., on the other hand, failed—they did *not* manage *not* to build a bomb. It is at this point that my indignation is aroused. Jungk's argument involves relationships of fact and moral value that are beyond the wisest of us to perceive at this stage of history. Furthermore, there is considerable artificiality about a discussion of the moral aspects of a scientific and technological development such as the atomic bomb. In the first place, there is no single-minded group which may be called "American physicists." Physicists do not look alike or think alike, nor are they of the same political persuasion. In the second place, important though physicists may have been in the making of the atomic bomb, there were a host of others—chemists, engineers, technicians, industrialists, military men—all of whom played important roles, and without whose cooperation and interaction the bomb would never have been built. Consequently it is presumptuous for me, a physicist, to discuss the ethics of the physicists who built the bomb. Obviously, however, each individual who participates in such a project does have a personal moral problem. One might also assign a measure of responsibility to each group of men in the project. If the physicists, for example, had decided not to build the atomic bomb, the bomb would probably not have been built. We physicists cast our lot with building it. We therefore bear the responsibility for our choice. Were we moral? Jungk thinks not, and by contrast praises the virtue of physicists working in Germany.

Let us consider the case of the German physicists. Because I take the optimistic view of human nature I should like to accept Jungk's views, but I cannot. The willing suspension of disbelief can go just so far. Jungk is the first person to make a serious case for the German physicists, and he makes it badly. A very different view of the matter has been presented by the Dutch-born U. S. physicist Samuel A. Goudsmit in his well-documented book *Alsos,* the result of his wartime study of German progress in developing nuclear energy. He has commented there, and elsewhere, on the conversations of the German physicists, temporarily in custody after the war, when they learned of the explosion of the atomic bomb. They were surprised. They had not believed that it was technically feasible. Goudsmit describes the German failure in terms of ineptitude, political interference and lack of support. Even if we choose not to believe this and accept Jungk's tale of a moral sit-down strike, the comparison is grossly unfair. Although it is now 15 years after the event, I still feel there is no comparison between physicists working for Nazi tyranny, which was hostile to intellectualism, and our activities in a desperate defense of a decent world. We are asked to believe that German nuclear physicists sabotaged the war effort while German rocket physicists worked enthusiastically, and successfully delivered to their armed forces what might have been a crucial weapon. Do blunted ethical perceptions go hand in hand with an interest in classical physics, and does morality march only with nuclear physics? It seems more logical that political and economic support was given to one field rather than another, and that this was decisive.

The other side of the argument, namely our failure not to make the weapon, is more complex. One should recall the atmosphere when work on the bomb began. The Germans were successful on all fronts; they were ruthless; there was a good chance that they would win the war. We knew that they were working on nuclear energy. I held pacifist views at the time, but in the light of the Nazi danger I felt that I had little moral choice. I suspect that the lines of the children of light and of the children of darkness were drawn as clearly as they ever will be. Neutrality would have been a selfish luxury.

Even if it was right to work on the Manhattan Project when the Germans were winning, was it right to continue the work after the Germans surrendered? In retrospect this question has bothered me greatly. Not at the time, however! Perhaps I should now respect myself more if I had left the work on V-E Day. The possibility simply did not occur to me. Events tumbled upon each other too quickly; the German surrender coincided with our preparations for the critical first test of the bomb in the New Mexico desert. We were the heroes of our epic, and there was no turning back. We were working on a problem to which we were completely committed; there was little time to re-examine our moral position from day to day.

In this connection I do remember calling in 1944 a meeting of my cyclotron group at Los Alamos. The subject of the meeting was rather pretentiously announced as "The Impact of the Gadget [the bomb] on Civilization." The meeting was advertised to other groups at Los Alamos, and between 50 and 100 scientists attended. We discussed what the world might be like as a result of our endeavors. At the time our work was not going very well; it was not at all evident that the bomb would be finished before the war was over. Secrecy was a necessary evil with which we lived uneasily. We imagined a world of conventional weapons and conventional world politics where the military kept the atomic bomb a secret. The best thing that we could think of—and we were all in agreement —was that it was absolutely essential to finish the bomb before the meeting in San Francisco at which the United Nations was formed. Then, it was our hope, the new world organization would be able to deal with the atomic bomb from the beginning. We failed that rendezvous. There was no consideration given at San Francisco to nuclear energy.

Those who criticize us for making the bomb might reflect on the kind of world which would have resulted had we not finished the bomb before the war was over. It seems almost certain that if we on the Manhattan Project had not built the bomb, the job would have been done by others within five or 10 years, either in this country or elsewhere. Nuclear energy was a time bomb set for the human race, and eventually the human race would have had to reckon with it. Was it not for the best that the public demonstration of the bomb, which ended World War II, came at a time when new relationships among nations were being established?

Jungk goes to considerable lengths to demonstrate the international origins of nuclear physics. Yet by implication he would have U. S. physicists bear the full responsibility for the bomb. During the war Los Alamos was as international a community of scientists as has ever existed. Our agreement in recognizing the hazards of nuclear energy was as complete as our agreement in purely scientific matters.

If Jungk is critical of the physicists who made the bomb, he is positively censorious of the decision to drop the bomb on Hiroshima and Nagasaki. There are many who share this view; I count myself among them. It is only with his conclusion that I am in agreement, however, for here, as usual, Jungk oversimplifies and distorts history. I recognize that my own feeling about the military use of the bomb was and remains largely sentimental rather than rational. I felt betrayed when the bomb was exploded over Japan without discussion or some peaceful demonstration of its power to the Japanese. More common, and possibly as rational, is the opinion held by many of my colleagues: that the war was ended by the atomic bomb and that more lives, both American and Japanese, would have been lost had the war been permitted to continue. Henry L. Stimson, who had more to do with the decision to drop the bomb than anyone else, perhaps expressed their sentiments when he wrote: "The face of war is the face of death; death is an inevitable part of every order that a wartime leader gives. The decision to use the atomic bomb was a decision that brought death to over 100,000 Japanese. No explanation can change that fact and I do not wish to gloss it over. But this deliberate, premeditated destruction was our least abhorrent choice. The destruction of Hiroshima and Nagasaki put an end to the Japanese war. It stopped the fire raids and the strangling blockade; it ended the ghastly specter of a clash of great land armies. The bombs dropped on Hiroshima and Nagasaki ended a war. They also made it wholly clear that we must never have another war. This is the lesson men and leaders everywhere must learn, and I believe that when they learn it they will find a way to lasting peace. There is no other choice."

*Brighter than a Thousand Suns,* like the Bhagavad-Gita, has its Krishna— Robert Oppenheimer. Although the author has never interviewed his subject, he has evidently discussed him at length with some physicists and has attempted what amounts to a skeletal biography. Oppenheimer is of course a fascinating figure, but one so complex and with such depths that I, who worked intimately with him during the war, would quail before the task that Jungk approaches so capriciously. It is not a flattering portrait, nor is it a true one. Once more the author stoops to what must be deliberate distortion by selecting what conforms to his thesis and ignoring the rest. Unfortunately he has been aided here by the more professional efforts of those who revoked Oppenheimer's clearance in 1954 and published the proceedings. Here were indecently exposed all manner of irrelevant private details about the life of a fine scientist who had devoted his best years to serving his country. The trial of Oppenheimer was one of the last great *autos-da-fé* of the McCarthy era. Perhaps a European journalist like Jungk cannot be blamed too much for his erroneous judgment so long as this ungrateful nation's verdict, made during hysterical times, stands unrescinded.

In view of Jungk's profound errors of judgment, it is perhaps pedantic to add that he has mangled the names of many minor characters of his epic, and that he has a positively abandoned manner with dates. He has not been able to copy with any degree of accuracy a simple list from *Atomic Energy for Military Purposes*, the classic history of the Manhattan Project. He indicates that liquor was banned at Los Alamos, when in fact the only fluid in short supply there was water. (When Klaus Fuchs was arrested, many of us not only asked ourselves, "Why did he do it?," but also, "How did he dare to drink so much at Los Alamos?") Oppenheimer was not educated at the Los Alamos Ranch School for Boys. It is simply not true that "barely a dozen physicists on the bomb projects had an over-all view." Robert Serber did not collect the pool money for the most accurate prediction of the energy released by the first atomic explosion, nor was he a "visitor" to Los Alamos. I. I. Rabi won the pool, and Serber was a regular member of the Los Alamos staff. The Wilsons lived across the hall from the Serbers, and often said: "The Serbers are our nerbers." The Serbers now live across the hall from the Rabis in New York. The world of nuclear physics is rather chummy, and anyone who confuses Robert Serber and I. I. Rabi has not studied that world very carefully.

All this is not, of course, particularly important. It suggests, however, that even Jungk's anecdotal history is weak. What is more to the point, the factual errors do indicate a lighthearted and uncritical approach toward his material, which becomes serious when he attacks men and interprets events. The book lacks the documentation one expects of a historian; one never knows whether Jungk obtained some particular fact from someone's lips or from an unpublished paper or from his own vivid imagination.

Although I have criticized Jungk's book, I respect his motives in attempting to understand the forces that shape science and scientists. Assuming that civilization survives the invention of this ultimate weapon, the human spirit will be challenged often and perhaps even more seriously by future developments in science and technology. For me the question is: Are scientists to attain the heightened sense of moral values which will enable them to determine the direction of these developments with humanistic and humanitarian ends in view, or will humanists and humanitarians attain an understanding of scientific values that will allow them to determine the wise direction of science? I suspect that civilization will best be served by a true fusion—or at least a close mutual understanding—of science, the humanities and politics. There are many who are presently preoccupied with exactly this problem. Jungk has struck out boldly, if not carefully, in this direction by his provocative analysis of this case history of science, scientists and civilization.

# 5

# The Psychological Wreckage
# of Hiroshima and Nagasaki

J. BRONOWSKI

*June 1968*

A review of *Death in Life: Survivors of Hiroshima*, by Robert Jay Lifton. New York: Random House, 1968.

The second world war began in the summer of 1939 with a pact between Hitler and Stalin and ended six years later with the dropping of two atomic bombs on Japan. To almost everyone at the time the second of these events was as unexpected and shocking as the first had been. Evidently the huge enterprise of inventing, building and mounting the atomic bomb had been the best-kept secret of the war. And for most people, scientists and nonscientists alike, it also turned out to be the most grisly secret. After the first days of triumphant wonder a kind of shudder went through the world, a swell of fear and revulsion together, which 20-odd years have now smoothed out of our memories. Everyone at the time had a sense of guilt about the atomic bomb, and although most people naturally shifted the blame to science, that desperate disclaimer was also a sign of penitence.

Twenty years is too long for sorrow, which time does not so much heal as blunt. The bombs that wiped out Hiroshima and Nagasaki have been allowed to become modest weapons in the tactical armory, and people read with resigned indifference that hydrogen bombs more than a hundred times as powerful are carried overhead in clusters 24 hours of the day. In that ebb tide of conscience, in which we let governments argue *about* disarming, the moral impulse of 1945 has been eroded. We might have supposed that the sense of guilt had been washed away without a trace, had not Professor Lifton discovered it still haunting (of all people) the survivors of Hiroshima. The discovery gives his quiet and penetrating book a kind of cosmic irony that, more than any burst of righteousness, ought to shake us all out of our somnambulism.

Professor Lifton had spent four years in Japan, off and on since the war, before he paid a visit to Hiroshima. For the last two of those years he had worked on a psychological and historical study of Japanese youth. There he found (among other things) what we all know and resolutely forget:

"The great majority had either no memory of the war at all or only the most meager recollections of it. But what became clear when I explored with them their sense of themselves and their world was the enormous significance for them, however indirectly expressed, of the fact that Japan alone had been exposed to atomic bombs."

So he decided early in 1962 to stay on for another six months and to spend them in Hiroshima, with a few days in Nagasaki added.

His method of study was to interview 75 people, usually for two periods of two hours each. The interviews were carried out in Japanese through a research assistant, although Professor Lifton does speak some Japanese. Forty-two of those interviewed were chosen for their known and articulate prominence in atomic bomb matters, and the other 33 were taken at random from the official list. All were *hibakusha*, which means that either they were within the city limits when the bomb fell, or entered the inner city in the next 14 days, or came into close contact with bomb victims, or that their mother was in one of these groups and was pregnant with them at the time.

Several different estimates have been made of the number of people killed by the bombs in Hiroshima and Nagasaki. I shall stick to the figures that my colleagues and I in the British Mission to Japan worked out in November, 1945, there. We computed that in Hiroshima when the bomb fell on August 6 there were 320,000 people, of whom 80,000 were killed, and that in Nagasaki on August 9 there were 260,000 people, of whom 40,000 were killed. Large numbers of people have died since then from the aftereffects of radiation from the atomic bombs. Allowing for these and for normal deaths over 20 years, the figures are in fair agreement with the official number of *hibakusha* now, namely 160,000 from Hiroshima and 130,000 from Nagasaki.

The scenes after the bombs fell have been described in eerie detail by John Hersey from the testimony of survivors in Hiroshima and by Dr. Takashi Nagai at first hand in Nagasaki. Every victim believed the bomb had exploded directly above him. Those in the open were badly burned by the flash, so that often their own families could not recognize them. (This happened to two of Professor Lifton's survivors.) Those indoors were buried and when they struggled out, almost naked, they found themselves surrounded by fires. Everyone who could move filed in numb silence to the rivers; the injured were abandoned and could be heard in the still heat crying for water. Professor Lifton quotes one of his survivors who was a schoolgirl at the time:

"I felt my body to be so hot that I thought I would jump into the river.... The teacher from another class, a man whose shirt was burning, jumped in. And when I was about to jump, our own class teacher came down and she suddenly jumped into the river.... Since we had always looked up to our teachers, we wanted to ask them for help. But the teachers themselves had been wounded and were suffering the same pain we were."

Another of Professor Lifton's survivors, a professor of history, saw the destruction from a hill overlooking the city:

"That experience, looking down and finding nothing left of Hiroshima—was so shocking that I simply can't express what I felt.... Hiroshima didn't exist— that was mainly what I saw—Hiroshima just didn't exist."

But those inside the city saw a greater destruction in the breakdown of human consciousness, so that the refugees seemed to them to be "a people who walked in the realm of dreams":

"Those who were able walked silently toward the suburbs in the distant hills, their spirits broken, their initiative gone. When asked whence they had come, they pointed to the city and said, 'That way,' and when asked where they were going, pointed away from the city and said, 'This way.' They were so broken and confused that they moved and behaved like automatons."

So much for what we knew and have been diligently burying in the backs of our minds. What have 20 years done to those 290,000 people who were there and who cannot so easily tidy away the memory of what they saw and how they behaved? Is there indeed a single theme in their lives? Are they still dominated by the atomic bombs?

Certainly Hiroshima has become the atomic bomb city. It has a Peace Park, of course, a memorial hall, a children's monument and a Peace Museum. There is a cenotaph with the inscription: "Rest in peace. The mistake shall not be repeated." The word "mistake" is ambiguous enough to make many citizens feel that they are being blamed for the bomb. (This is a common complaint among hibakusha, which they express in the ironic sentence "I apologize for having been exposed to the atomic bomb.") A reinforced concrete building almost under the bomb that stood up to the blast pretty well has been preserved as a permanent Dome of Peace, even though some survivors are still distressed by the sight of it.

But these sober memento mori are put in the shade by the other showplace of the peace industry, which is the Hiroshima entertainment district. Nowhere else in Japan is there such a splendor of bars, cafés, restaurants, geisha houses, dance halls and what Professor Lifton politely calls "transient quarters for various kinds of illicit sex." Those who saw the film Hiroshima, Mon Amour will remember the contrast between the two aspects of the new Hiroshima, and the strange implication that nevertheless ran behind it that they are inseparable. And so they are: the bomb is entwined in the

lives of those who survived it, whether they lead a children's march or show their keloid scars in a Hiroshima brothel. The hibakusha leaders may rage at those who "sell the bomb," but their anger is also a form of self-accusation; they cannot help themselves, they must exploit the bomb like any crippled postcard-seller.

The hibakusha have not been able to escape the ambivalence that always dogs the victim of misfortune. They would like other people to treat them as though they were normal, and at the same time they are hurt if they are not given special sympathy. The effect of these contradictory demands is to frighten those who were not exposed to the bomb, so that survivors find it hard to get jobs, to marry and even to mix with others. Twenty years ago their neighbors were afraid because the hibakusha were still psychologically numb, were often ill and (who knows?) might beget monsters. But now the alienation is different in temper, and simply puts the hibakusha aside as people who *have something else on their minds.*

The clear August day in 1945 possesses the mind of the survivor as an experience unlike any other, which totally overthrew his inner ordering of the world. He had woken that morning with the confidence, built so carefully through the years of growing up, that things go like this and not otherwise—that teachers help you, that the city is a solid home and that people act together by choice. By nightfall that framework of unwritten laws that had seemed to be laws of nature had fallen apart into meaningless pieces—the teacher had jumped into the river, Hiroshima didn't exist and its people behaved like automatons. Just as sometimes a great man has the order in the world revealed to him in a vision (René Descartes on November 10, 1619, for example, and Blaise Pascal on November 23, 1654), so ordinary men and women in Hiroshima that day had a direct vision of a counterrevelation: the failure of the human order in the world. Professor Lifton calls this "the replacement of the natural order of living and dying with an unnatural order of death-dominated life." Thus his stress is on "the indelible imprint of death immersion," when I would stress the fatal immersion in the collapse of human values. But in principle Professor Lifton and I are at one, and what he finds is the same withdrawal of confidence, a cutting of the roots of conduct, that I felt there three months after the bomb was dropped.

Somehow the roots have to heal, of

course; the delinquents disappear from Hiroshima's station, the widows move into the entertainment district and the men whose assurance has been sapped ask their firms without fuss to let them work at lesser jobs. But a psychological wound that is healed is still a wound, a kind of bomb scar or keloid of the personality, which expresses in another form the ambiguity of the hibakusha. They have been the victims of a disaster that was manufactured by others, yet their sense of failure is more anguished than we expect, and we see that there is another wound under the scar. The victim has not only been a victim; he has also witnessed and (by his inaction) has condoned the action of turning one's back on other victims. Every survivor has memories of those he did not help:

"I heard many voices calling for help, voices calling their fathers, voices of women and children.... I felt it was a wrong thing not to help them, but we were so much occupied by running away ourselves that we left them."

"His head was covered with blood, and when he saw us he called to us.... 'Yano [he said to my daughter], please take my child with you. Please take him to the hospital over there.'... I heard later that he survived...but that the child died.... And when I think of not helping him despite his begging me to help, I can only say that it is a very pitiful thing."

Because family ties are strong in Japan the memories are particularly painful in those who feel they neglected their parents. For example, a girl who was 14 at the time is full of remorse for her father's death, although she only complained that his wounds smelled.

The keloid on the personality of the survivor at Hiroshima and Nagasaki is the sense of guilt. "I apologize for having been exposed to the atomic bomb" is not altogether a joke after all, if we read it to mean "I apologize for having *survived* the atomic bomb." Even when the victim was helpless at the time and could have done nothing to help, he is disrupted by doubt and a feeling of inadequacy. All around him people died; why did he deserve to live? Will that act of *hubris* be visited on his children? "Those who died are dead," a survivor from Nagasaki said to Professor Lifton, "but the living must live with this dark feeling." There is no shaking off the divided feeling between one's own fate and the fate of others, between suffering and fear, between pity and revulsion. It is symbolized by the memory that comes to one hibakusha at a Japanese feast:

"The color of my brother's keloid—the color of his burns—mixes together with my feeling…what I saw directly—that is, the manner in which he died—that's what I remember.... The color was similar to that of a dried squid when broiled —so that I think of it whenever I see dried squid.... I have…a very lonely feeling."

Professor Lifton is masterly in his analysis of the ambiguities that make the *hibakusha* so disturbed in himself and disturbing to us. He writes without jargon and without hectoring, with a personal air of courtesy that invites the reader's assent but does not presume on it. It is notoriously hard to persuade scientists that there is any point in psychological analyses that do not give unique answers. (It is Karl Popper's complaint about Freud and Adler.) Yet I think Professor Lifton will persuade even a skeptical scientist that it is the nature of tragedy to divide the human personality against itself, and that the survivors of Hiroshima act out this division symbolically in their sense of guilt.

The *hibakusha* resents his own feeling of guilt, yet he cannot resolve it; he is equally unhappy as victim and as survivor. The inner struggle between the roles of victim and survivor is displayed particularly plainly by the Japanese, for two reasons. First, they are brought up in a rigid code of family and social propriety, so that the breakdown of conduct at Hiroshima and Nagasaki was very stark for them. And second, the Japanese pay much attention to symbolic meanings, so that the division in the *hibakusha*'s feelings is constantly reinforced by the ambivalence in the symbols that express them. (For example, the citizens of Hiroshima are wild supporters of their baseball team, but it usually comes in last.)

It was therefore natural for Professor Lifton to think of applying his theme also to some other body of survivors who are not Japanese. In his last chapter, "The Survivor," he makes some comparisons with the men who went through the Nazi concentration camps and lived. What he says here is worth reading, and is evidently consistent with his findings in Japan. But since he has had to take his evidence at second hand from other people's writings we now miss the force of direct speech that makes the rest of the book so convincing. We can see that those who survived the camps are troubled by guilty memories as the *hibakusha* are, but the analysis has become more formal, and there is just a hint of a classroom thesis in the argument.

For the rest, I am convinced by Professor Lifton's powerful book, and I am at variance with his analysis at only one point. As his choice of title shows, he is preoccupied with death as the visible cause and symbol of the survivor's guilt, and in fact he commonly uses the single phrase "death guilt" for it. He says early on, "I shall use the term 'death guilt' throughout the book to encompass all forms of self-condemnation associated with literal or symbolic exposure to death and dying."

Now, it is true that the victims of the man-made disasters at which we have stood by were much exposed to death and dying. I have quoted the casualty figures at Hiroshima and Nagasaki—and as for the concentration camps, "a person who fully adhered to all the ethical and moral standards of conduct of civilian life on entering the camp in the morning would have been dead by nightfall." Nevertheless, I think that the distress of the survivors was caused, more profoundly than by the encounter with death, by the dissolution of "the ethical and moral standards of conduct of civilian life" that robbed them of their bearings.

Certainly the fear of death (and of the unpredictable way death would strike in Hiroshima and Auschwitz) is the infecting virus that begins the social dissolution. Yet the girl in Hiroshima filled with remorse for her father's death had not killed him; what she blamed herself for was disrespect. Or think of one of the men I wrote about in *The Face of Violence*, say Joseph Wiener, once professor of international law at the University of Vienna, whom the Nazis drove mad and put in charge of the camp pigs. He informed against prisoners who stole food from the pig swill; yet it was not the presence of death that deranged him but (as he knew in his sane moments) the debasement in his person of human dignity. Such examples seem to me to make plain that what unhinges the victim's self and leaves it rudderless is not an inner split between life and death but a split between himself and the social order. From childhood he has been taught to submit his own demands to those of society (this is my theme in *The Face of Violence*) and suddenly not life but society dies; he sees walk into the open the anarchy he once dreamed of in secret.

I have left to last the question that must be uppermost for many readers: What did the Japanese think of the Americans, who, after all, had dropped the atomic bombs? I found this the most puzzling issue to get to grips with 20-odd years ago, and it seems to be so still. The Japanese seldom spoke out in plain resentment; perhaps they were either too stunned or too polite. They seemed mostly to be aware that we knew as little about the effects of the bomb in advance as they did; one or two professors spoke to me ironically about the "experiment," and the deepest anger Professor Lifton finds is at having been treated like guinea pigs.

Many Japanese feel they were singled out as no white enemy would have been, as something less than human; some of them may have seen the pictures of Japanese looking like vermin (like the Nazi pictures of the Jews) that were current in America during the war. There was a rumor in Nagasaki when I was there that the atomic bomb burned only dark-skinned people; it was not true, but it put the right pinch of scientific fact in the racial stew.

But the dominant feeling was and is that Americans are simply insensitive. Professor Lifton gives many examples to justify that: the early censorship, the display of power and wealth, the clinical examinations and the policy of studying the victims without treating them. One wonders what the Japanese would now think if the first bomb had in fact been dropped on Kyoto, as General Groves's target group had proposed. I invented my own rumor at the time, which is not true but which I am glad to spread in *Scientific American,* that a member of the target group was told this would be like bombing Florence and that he asked "Florence who?"

There is nothing to say about the prospect that has not been said already. As Professor Lifton's book demonstrates, we have to get it into our heads that the atomic weapons do not create casualties but chaos—a lasting chaos of the human values. The *hibakusha* are haunted and lamed by a shame that is not their own, and that is 20 years old—and so are we. Like them we have an ambivalence between self and society, nation and humanity, that prevents us from forming any policy of right conduct. And the guilt that comes from facing two ways bites deeper, because it prevents us from crystallizing any *principle* of right conduct. Professor Lifton's last message is that we should learn what he calls survivor wisdom, and no one can now doubt that survivor wisdom says that only men of principle survive whole.

# 6

# The Role of
# Two Scientists in Government

P. M. S. BLACKETT

*April 1961*

A review of *Science and Government*, by Sir Charles Snow. Cambridge, Mass.: Harvard University Press, 1961.

To the people making it, history is a long series of decisions: whether to do this or to do that; whether to take one road or another; whether to be bold or whether to be cautious; whether to make war or to make peace; or, in this nuclear age, whether to live or to die. H-bombs have made us all decision-conscious. All decisions must start with an analysis of the past, for without understanding the past it is not possible to predict the future, and without some prediction of the future no rational decisions can be made. The decision to take one course of action rather than another implies the prediction that the one chosen will produce more favorable results than the one rejected. Unless one can make some such prediction, one has to rely on a guess; when this is successful, history calls it inspired.

How have the important decisions of the tumultuous past three decades actually been made? What sort of people made them? How much part did calm and detached thought play, and how much instinctive feelings or plain emotion? To what extent did personal loyalties and personal hates dictate the pattern of world events? To what extent did the decision makers think out the complex consequences of their actions and plan accordingly? Or did they stake their country's—or indeed mankind's—future on a gamble? Or were the decision makers perhaps more like players of chess than of poker?

All these questions came into my mind as I read C. P. Snow's *Science and Government*, originally presented at Harvard University as the Godkin Lectures. Snow is primarily concerned with understanding how some of the important decisions of our time were in fact made. His training as an experimental scientist, his years as a civil servant in close touch with the British scientific effort during the war and after, his experience as a director of a major engineering firm and finally his authorship of many successful novels—have given him a background that no contemporary, either in Britain or the U. S., has had. Moreover, his main interests as a novelist have been concerned less with the relationship between men and women than with the relationship between men and men, as they live their professional lives in government departments, in scientific laboratories, in the board rooms of industry or the common rooms of universities. The interplay of personalities and policies, of abilities and ambitions, the actual functioning of that remarkable abstraction, the so-called British Establishment, were among the interests that have made Snow the novelist of committees and court politics in this scientific age.

Snow analyzes in detail two major decisions of British war policy: the decision made between 1935 and 1937 to give the development of radar the highest possible priority, and the decision in 1942 to make the bombing of German cities a major part of the British war effort. In the conflicts that preceded these two fateful decisions, two outstanding and very different scientists, Henry Tizard and Frederick Lindemann, played a major role. Much of Snow's book is concerned with the clash between these two strong personalities. By various accidents I was personally involved in both conflicts, and I can vouch for the fundamental truth of Snow's account of what went on. Moreover, I think that his description of the conflicts and his penetrating insight into the characters of the two men is brilliantly carried out: this is a first-rate piece of writing. One quotation must suffice here:

"Judged by the simple criterion of getting what he wanted, Lindemann was the most successful court politician of the age. One has to go back a long way, at least as far as Père Joseph, to find a gray eminence half as effective. Incidentally, there exists a romantic stereotype of the courtier—as someone supple, devoid of principle, thinking of nothing except keeping his place at court. Now Lindemann was, in functional terms, a supreme courtier; and yet no one could be more unlike that stereotype. Life is not as simple as that, nor as corrupt in quite that way. Throughout his partnership with Churchill, Lindemann remained his own man. A remarkable number of the ideas came from him. It was a two-sided friendship. There was admiration on Lindemann's side, of course, but so there was on Churchill's. It was a friendship of singular quality—certainly the most selfless and admirable thing in Lindemann's life, and in Churchill's, much richer in personal relations, it nevertheless ranked high. It is ironical that such a friendship, which had much nobility and in private showed both men at their human best, should in public have led them into bad judgments."

There is no doubt that Tizard must be given a major part of the credit, and Lindemann none, for the radar chain. When war came, Britain had an operational early-warning radar system all around its east and south coasts; moreover, it had fighter squadrons trained to intercept the German bombers by using radar plots. Our edge over the enemy was more in massive deployment and operational training than in the basic knowledge of electronics. Tizard, above all others, was responsible for the high priority that led to the rapid development and installation of the radar system. Without it the Battle of Britain in 1940—a near thing at best—might have been lost, with incalculable historic consequences. As Snow points out, this particular decision was not technically a difficult one, being in effect a choice of doing something that might work as against doing nothing. The conflict over the decision, which was very real and in slightly different circumstances might have gone the wrong way, appears in retrospect to have been at the bottom purely personal. At that time Lindemann opposed anything suggested by his former friend Tizard.

Tizard's second vital contribution related directly to the U. S. In the summer of 1940, soon after the fall of Paris, he persuaded a reluctant British Government to send to the U. S. a mission headed by himself with the famous black box containing samples, blueprints and reports on nearly all important new British war devices, including the magnetron. The contents of this box were later called by an American writer "the most valuable cargo ever brought to our shores." This imaginative act of trust, which Tizard initiated and forced through (Tizard described it as "bringing American scientists into the War before their Government"), had immensely beneficial effects on the scientific aspects of the Allied war effort.

By 1941 Tizard was widely recognized as the ablest British scientist to apply himself to the problems of war. He was popular both with scientists and with the armed services; he had two major achievements to his credit—the radar chain and the American mission—and in addition he had done much to make the services scientifically minded. Two years later he was effectively out of the war effort. How this came about is a major theme of Snow's book. The cause of this disastrous turn of fortune—in my view disastrous for the whole British war effort—was another conflict of judgment on priorities, this time about the bombing offensive. As I was deeply involved in this, I can add something to Snow's vivid account. I will also say something of the historic background and of the aftermath of the decision to concentrate a major part of the British war effort on the destruction of German housing. So far as I know, it was the first time that a modern nation had deliberately planned a major military campaign against the enemy's civilian population rather than against his armed forces. During my youth in the Navy in World War I such an operation would have

been inconceivable. Incidentally, the German air attacks on London from September, 1940, to May, 1941, were undertaken with little serious planning, and they were called off when the Germans attacked the U.S.S.R.

I remember fire-watching on the roof of a block of flats in Westminster in September, 1940, on the evening of the day the "blitz" began. We were watching the glow from the burning East London docks, and bombs were falling on central London. A young bomber pilot by me said: "I can hardly bear to wait till we can do it back to them." Such understandable sentiments do not necessarily make good strategy, nor does the commonly used argument: What else could we have done?

The origin of the Allied bombing offensive goes much further back. It was a product of the rise of the air forces of the world and of their determination to evolve a strategic role for air power that would make them independent of the two older services, the army and the navy. Since this requirement excluded co-operation with either of these two services as its major role, the Air Force sought the strategic role of attacking the sources of economic and military power in the enemy country. When this policy was first put into effect in the early summer of 1940, it was gradually realized that the accuracy of navigation was far too poor to allow our night bombers to hit anything smaller than a fair-sized town—and generally not even that. So the attempt to hit military installations, factories and transport centers was abandoned for a general attack on the centers of civilian population. Until 1943 the effort was on such a small scale and was so ineffective as to have negligible military effect. The decision to make the dehousing of the German working-class population, with the object of lowering its morale and will to fight, a major part of the British war effort was made in the spring of 1942, as Snow relates.

In March, Prime Minister Churchill asked the Air Staff how they thought they could best help our Russian allies, then engaged in desperate land battles, by stepping up the bombing offensive against Germany. The Air Staff report to the Prime Minister was sent to the Admiralty, and they sent it to me for comment. I was then Director of Operational Research at the Admiralty. Since the Air Staff estimates in this paper of what the bombing campaign had hitherto accomplished seemed to me to be based on very flimsy evidence, I set about checking the claims by the simple

method of noting what the German blitz on England had accomplished and calculating from this what our much weaker attack should have done to Germany. The result was startling. It seemed probable that the German casualties from our bombs could not have been much more than the loss of trained air crews. Further, I estimated that the reduction of industrial production was less than 1 per cent. So on these two counts the direct effect of the British bombing offensive of 1940 to 1942 seemed nearly a dead loss. I took these calculations personally to Lindemann, then Churchill's personal adviser. In effect he accepted my calculations and realized that neither by the casualties inflicted nor by the interruption of production was our **bombing offensive likely to bring much relief to the Russians. So he switched the objective to the destruction of German housing.** Together with the Air Staff a new paper was prepared claiming that, if all possible priority were given to bomber production and training, it should be possible in the next critical 18 months to destroy some 50 per cent of all houses in all towns with a population of more than 50,000. In April a Cabinet paper setting out these estimates was prepared and contained the recommendation that this strategy was the best available to help the Russians.

This was the paper which Snow mentions as having come to both Tizard and me; we independently agreed that the claims of destruction of houses was probably about five times too high. Our forebodings went unheeded and the bombing policy was adopted. After the war we learned that the estimates were 10 times too high.

From my talks with Lindemann at this time I became aware of that trait of character which Snow so well emphasizes: this was his almost fanatical belief in some particular operation or gadget to the almost total exclusion of wider considerations. Bombing to him then seemed the one and only useful operation of the war. He said to me (unfortunately I have no record of this conversation, but he probably said the same to others) that he considered any diversion of aircraft production and supply to the antisubmarine campaign, to army co-operation or even to fighter defense—in fact, to anything but bombing—as being a disastrous mistake. Lindemann even suggested that the building up of strong land forces for the projected invasion of France was wrong. Never have I encountered such fanatical belief in the efficacy of bombing.

The high priority given thereafter to

everything pertaining to the bombing offensive made it very difficult to get adequate air support for the vital Battle of the Atlantic. If this had got worse there would have been no more bombing offensive for lack of fuel and bombs, and no invasion of France in 1944. I remember that during the winter of 1942 and 1943 the Admiralty had to enlist President Roosevelt's personal influence to ensure that a squadron of that admirable antisubmarine aircraft, the B-24, was allocated to Coastal Command (where they were brilliantly successful) and not, as the Air Staff wanted, sent to bomb Berlin, for which they were not very suitable. However, at the Casablanca Conference in January, 1943, a combined American and British bombing offensive was formally adopted as a major part of the British war strategy.

No part of the war effort has been so well documented as this campaign, which had as its official objective "the destruction and dislocation of the German military, industrial and economic system and the undermining of the morale of the German people to the point where their capacity for armed resistance is fatally weakened." Immediately after the war the U. S. Strategic Bombing Survey was sent to Germany to find out what had been achieved. A very strong team (which included two men who are now advisers to President Kennedy, J. K. Galbraith and Paul Nitze) produced a brilliant report, which was published in September, 1945. Without any doubt the area-bombing offensive was an expensive failure. About 500,000 German men, women and children were killed, but in the whole bombing offensive 160,000 U. S. and British airmen, the best young men of both countries, were lost. German war production went on rising steadily until it reached its peak in August, 1944. At this time the Allies were already in Paris and the Russian armies were well into Poland. German civilian morale did not crack.

Perhaps it is not surprising that the report of the Strategic Bombing Survey seems to have had a rather small circulation; it is to be found in few libraries and does not appear to have been directly available, even to some historians of the war.

If the Allied air effort had been used more intelligently, if more aircraft had been supplied for the Battle of the Atlantic and to support the land fighting in Africa and later in France, if the bombing of Germany had been carried out with the attrition of the enemy defenses in mind rather than the razing of cities to the ground, I believe the war could have been won half a year or even a year earlier. The only major campaign in modern history in which the traditional military doctrine of waging war against the enemy's armed forces was abandoned for a planned attack on its civilian life was a disastrous flop. I confess to a haunting sense of personal failure, and I am sure that Tizard felt the same way. If we had only been more persuasive and had forced people to believe our simple arithmetic, if we had fought officialdom more cleverly and lobbied ministers more vigorously, might we not have changed this decision?

Snow devotes the last part of his book to extracting from these two cautionary tales, as he calls his accounts of the radar and the bombing conflicts, some lessons for the future. He wisely warns us of the danger of what he calls the euphoria of gadgets, meaning by this the tendency on the part of some scientists—and not only scientists—to believe that a new device, or a new tactic, is a solution of all our defense problems. This was fundamentally the error behind the overconcentration during the war on the area bombing of enemy cities. It is worth remembering that Germany never did this. Her remarkable military successes of the first years of the war were achieved by brilliant co-ordination of armor, artillery, infantry and close air support. The same was true of Russia. When she finally drove the German armies back from Stalingrad into Germany, this was achieved by the co-ordi-nated use of land and air power. In fact, Germany was eventually defeated primarily by the methods that had brought her such startling successes earlier. Of the three million German war dead and missing up to November, 1944, 75 per cent were on the Russian front. This is an indication of the extent to which World War II was primarily a land war. The air operations of the bombing offensive carried on independently of military operations did begin to have an important effect during the summer of 1944. However, by this time the German armies had been decisively defeated both in the East and in the West.

This is not the place to attempt to apply in detail some of these lessons to postwar defense problems. There is, however, one comment that must be made. Never have Snow's twin warnings, of the danger of thinking that one weapon will solve our problems, and of the illusion that one can rely on maintaining technical superiority, been more vividly illustrated by the early years of nuclear weapons. Here the euphoria both of gadgets and of secrecy reached their highest and most disastrous intensity. Through a blind obeisance to a single weapon the West let down the strength of its conventional forces and failed even to develop prototypes of modern weapons for land warfare. In spite of the vast technological strength of the Western world, its ground armies in Europe are not only much smaller but also much inferior in equipment to those of the Soviet army. This has led the West to a reliance on nuclear weapons that is certainly dangerous and could be suicidal. A calm contemplation of the last 15 years makes one remember the cynical comment that the only lesson ever learned from history is that no one ever learns from history. But unless we do, there will be no more history. Snow's little book, with its wisdom and penetration, should do much to stimulate serious thought on these vital problems of decision making.

# 7

# The Probability of Human Survival

JAMES R. NEWMAN

*February 1962*

A review of *Has Man a Future?*, by Bertrand Russell. New York: Simon and Schuster, 1962; and *May Man Prevail?*, by Erich Fromm. New York: Doubleday, 1961.

When the patriarch Abraham pleaded with Jehovah to spare the wicked cities of Sodom and Gomorrah, he gained from him a promise that if 10 righteous men could be found in Sodom it would not be destroyed. Ten righteous men could not be found. "Then the Lord rained upon Sodom and upon Gomorrah brimstone and fire from the Lord out of heaven; and He overthrew those cities, and all the plain, and all the inhabitants of the cities, and that which grew upon the ground."

An Abraham interceding for man today would not have a much easier task. Virtuous and kindly men are rare; even rarer are those not afraid to speak out and denounce evil. Here, then, is an occasion for celebration: two books by two upright men, men of wisdom and courage, who know what to do and do it. One is by the world's most distinguished philosopher and jailbird, who for 70 years has benefited mankind by refusing to hold his tongue; the other is by an eminent psychoanalyst and social critic whose writings have been a lamp to sanity and decency.

Bertrand Russell's little book was written only a few months ago. The proofs of it that I saw bore corrections that he, as I understand it, made in prison while serving a term for the grave offense of endangering the safety of Her Majesty's realm by sitting down in Parliament Square. Some 43 years earlier he was incarcerated for engaging in pacifist propaganda. On that occasion the prison sentence was long enough for him to write his famous *Introduction to Mathematical Philosophy*, a work which the warden found perplexing but no clear threat to the security of the state. *Has Man a Future?* is a very different sort of thing. It is a brilliant tract against war, a devastating exposé of folly and as pungent a piece of subversion as has brightened the drab scene of political literature in a long while.

Man has come up from the apes. "Up" means that he differs a little from apes in appearance and has developed many skills. An ape can learn to get a banana with a stick; a man can learn this too. Man, however, has greater capacities. For example, he can write music, grind a lens, aviate, prevent smallpox. He has also learned how to organize unfriendliness on a vast scale and has perfected marvelous techniques of mass homicide. These are much respected and cultivated—so much so, in fact, that man has arrived at a point where he can exterminate the whole of his kind. Will he do it? If the name he gives himself, viz., Homo sapiens, means that he possesses even a modicum of wisdom, he will not. The question remains open whether human history up to the atomic age will be considered prologue or epilogue.

Russell describes the setting in which the question is to be decided. He reviews the story of nuclear weapons; how they came to be made, how they work, what scientists felt about their use against Japan, what generals and others have said of our chances of surviving a nuclear war. On this last point there is obvious disagreement. The exchange of a large number of atomic bombs between the U.S. and the U.S.S.R. may not kill everyone, even in the Northern Hemisphere. Millions may survive and procreate generations of idiots and monsters. Still, this merciful prospect is based on the use only of existing weapons. Since scientific ingenuity is endless, we may confidently expect improvements. Herman Kahn, the publicist of nuclear war, has suggested a "doomsday machine" that could in a moment destroy the whole population of the world; he regards this, however, as undesirable. But not everyone is so forbearing, and Hitler in the last days in his bunker would doubtless have preferred the end of man to the ignominy of surrender. Let the machine be made, let fears, hatreds and international anarchy continue, and the chances are that a group of dedicated lunatics will press the doomsday button. But death has other doors for men to take their exits. Chemical and biological warfare are still in their infancy. Artificial satellites carrying atomic bombs could inflict enormous slaughter. Cobalt bombs—a mere $6 billion worth—would, according to the chemist Linus Pauling, guide every living creature to the grave.

What men think of each other and of themselves, what they can be driven to do by fear, their moral sense, their respect for life, their true values and their clichés—these are of course the factors which will decide the future. Weapons lead no lives of their own. Wars do not simply come. Recently, in a letter to *The Times* of London, Lord Wootton of Abinger answered another baron (Lord Coleraine), who in castigating Russell referred to him as an "aging adolescent." Among other things Wootton referred to a B.B.C. interviewer who used the phrase: "If war comes. . . ." "Wars, Sir," Wootton wrote the editor, "do not come. They are not acts of God; they are not even acts of the devil. They are made by the conscious and deliberate decisions of men; and they cannot be made by one side alone. The time has come for the statesmen of the world to speak realistically; for them to speak not of the risk of war coming but of the possibility that they will themselves choose the wholesale massacre of innocent people and the potential extermination of the human race as lesser evils than any alternative."

Russell is much concerned with certain slogans which are used to manipulate and inflame opinion. Patrick Henry's famous exhortation about liberty or death is a perfect example. For Henry and for his listeners this had meaning. It was an individual decision based on moral conviction. It said: "I will give my life, if need to be, to promote a just cause, to assure liberty to others." It was based on devotion to his fellow men, and it represents one of the noblest ideals which man can achieve. But what can such a slogan mean when used to justify a nuclear war? The national leader who says that death is preferable to slavery is not merely asserting that he would personally rather perish than submit to tyranny. He is asking others to do so—men, women, children, by the tens of millions. He is saying that whatever the cost, even if it is the end of the human species, freedom must be preserved. Freedom of what? Mount Monadnock? Horseshoe crabs? The gold in Fort Knox? As for the plain man who utters the slogan, what is he saying? "I would rather have my little Sidney and Dolores dead than be ordered about by foreigners"? Or: "If necessary, the people of Luxembourg and Siam must be thrown into the fire to keep the torch of liberty lit"? This is not an individual decision. It is not born of ethical convictions. It is not intended to promote a good. It is a piece of grandiloquent blather. Nothing exposes so clearly the absurdity of the slogan when advanced on behalf of a nuclear war than the crisis over Berlin. For even if we assume that a nuclear war will not eradicate mankind, that it will merely destroy the civilization we know, and lead, after "recovery," to governmental despotisms more severe than any that have yet existed, no one can doubt that the people of Berlin will be vaporized within a few hours after the missiles begin to fly. What assurances, then, are we giving to the inhabitants of this freedom-loving city? Perhaps, for them at least, the slogan should be modified. We rise to say to the leaders of the U.S.S.R.: "Give the people of Berlin liberty or we will give them death." I find it hard to believe the Berliners would embrace us for this ultimatum.

Among sane men as well as lunatics there can be disagreement about anything. Beauty, fish soup, the nature of reality, the writings of Franz Kafka, politics, fallout, afterlife, crime and punishment, bullfighting, virtue, even mathematical truth. But among sane men there can be no disagreement as to

nuclear war. It is the worst disaster that could possibly happen. If it is to be averted in the long run, if we are to survive, certain conditions, says Russell, must be achieved. All the major weapons of war and all the other means of mass destruction must be yielded up by individual nations and transferred to a world authority. National armed forces must be reduced to a level no greater than that needed for internal police action. A world government must take over not only those activities which make wars possible between individual states but also positive functions to promote international order, political freedom, economic equality, social cohesion.

Obviously an immense task of re-education would confront the world authority. Consider the prevailing opinions about the importance of national freedom. It is a vague concept, but on that very account it has a powerful emotional appeal, expressed in everything from anthems and odes to dances and monuments. Yet, as Russell points out, those who argue in favor of unrestricted national freedom do not realize that the same reasons would justify unrestricted individual freedom. No one supposes that the laws which make murder illegal are an unreasonable infringement on personal liberty. The state requires such laws to maintain safety and order, to keep its citizens whole, to permit them to go about their business and sleep tranquilly. But there is "immense reluctance" to admit that such a code must apply equally to the relations of national states to each other and to the world at large. Political philosophers have uttered a good deal of nonsense about sovereignty. Even the admirable attempts to create a body of international law and to establish an effective world court have been thwarted by the accepted principle that it is optional with each national state to respect the law or submit to jurisdiction. The prevalent cliché in a democracy such as ours is that when the peoples' representatives vote to have a war, it is right and just and heaven will support it. Clearly no international authority can prevent us from killing foreigners, or foreigners from killing us, so long as we and they possess vast armaments. Equally clearly it will be difficult to persuade states to give up their armaments until people come to recognize that the present "anarchic national freedoms are likely to result in freedom only for corpses"; and to surrender a certain amount of sovereignty, as well as armaments, to an international institution which could prevent war would produce much more freedom in the world than now exists—"just as there

is more freedom owing to the prevention of individual murder." Modern weapons have made certain political forms obsolete; and modern weapons are likely to make the human race obsolete if no substitutes for these forms are found.

Much else is illuminated in these pages. The problems of disarmament negotiations, territorial questions which have to be decided if peace is to be achieved, the urgency of taking certain first steps to reduce tensions, the danger inherent in the present policy of instant retaliation, the abandonment of nuclear tests, the prevention of the spread of nuclear weapons to powers which do not yet possess them. Russell's wit, eloquence, passion and human sympathy are as fresh as ever; there is not a dull passage, not a canned thought. I found only one train of argument unconvincing. He defends scientists against the charge that they are morally to blame for the peril to which nuclear weapons expose the world. As evidence he offers, among others, the ill-fated Franck report, the Mainau statement, the Pugwash movement, the exertions on behalf of peace by such men as Linus Pauling. It is plain enough, however, that while many scientists have played an honorable, front-running role in this sphere, many more have devoted their full energies to researches directly or indirectly connected with the improvement of weapons. Not infrequently they have disclaimed social responsibility for the ultimate product of their labors by asserting disingenuously that decisions as to use lay outside their control.

It has been said that wars are made in the minds of men. This has a fine, impressive ring, but it is not a very helpful notion, since everything else that men will to do, and do, is also made in their heads. In so far, however, as the statement implies that man is not fate's cockleshell, that the wars he makes are his responsibility, it is to be encouraged. But there is another side to the matter which deserves more attention than it usually gets. The drive to war is a social disease. It is a psychopathological disorder which afflicts individuals, feeds on itself and spreads to entire groups. Only a lunatic would claim that human happiness can be achieved by inflicting human suffering. Yet, unfortunately, there are strains of lunacy in all of us which can be awakened. It is this dangerous fact which underscores the importance of Erich Fromm's book. For as a man who combines two skills, profound understanding of the development of political thought over the last century and sensitive insight into human motivations, despairs, fears, longings, he is singularly qualified to discuss how we got into this mess and how we can get out of it.

His argument runs along these lines: Societies cling tenaciously to their ways. Men in each society believe that the way in which they exist is natural and inevitable. Change leads to chaos and destruction. Yet in time societies do change. Growth of population, economic needs, political conquests, technological advances force change; so do man's growing self-awareness, his striving for freedom and independence, his ambitions and hopes for himself and his children. Most social changes have occurred in "violent and catastrophic ways." The cake of custom is hard-baked and hard to break. To adapt is to live; to resist adaptation is to perish; and history is the graveyard of dinosaurs and cultures. There have, however, been nonviolent changes, in which societies have anticipated the need for reform to meet fundamentally new conditions. Today we again face "one of the crucial choices in which the difference between violent versus anticipatory solution may spell the difference between destruction and fertile growth of our civilization." The Western bloc and the Russian bloc confront each other with apparently implacable hatred and suspicion. What Cromwellians thought of papists and Jacobeans of Girondists, Americans think of Communists: they are of the Devil. Ours, we are convinced, is the right way, the only way; theirs means slavery, brutalization, the extinction of freedom and all decencies. Both sides are fully armed and prepared to go to glory. Conciliation is regarded as the coward's part; negotiation is halfway to appeasement, appeasement halfway to surrender. So the armaments increase, bitterness becomes bitterer, and the threat of nuclear war becomes more ominous.

Is there a way out? Admittedly nothing can be settled by a nuclear war. We can no longer afford a Thirty Years' War to achieve a peace of exhaustion. We can no longer afford what is so prettily called a war with "conventional weapons." We cannot afford even a 30-minute exchange of missiles carrying multimegaton loads of atomic explosive.

If, then, no salubrious bloodletting can be permitted, nothing remains but a course of rational anticipatory behavior. To be sure, the chances that this will be adopted are for the moment bleak. Our common enemy, the weapons, stand in the way. So much treasure has been poured into them, so much reliance is placed on their awesomeness that they have become idols. But a much more important obstacle is the thought barrier built on "clichés, ritualistic ideologies, and even a good deal of common craziness that prevents people—leaders and led—from seeing sanely and realistically what the facts are, from separating the facts from the fictions and, as a consequence, from recognizing alternative solutions to violence." The removal of this obstacle requires a critical examination of our assumptions about such matters as the nature of communism, the future of the underdeveloped countries, the value of the deterrent for preventing war. It also requires a serious analysis of our own biases and of "certain semipathological forms of thinking which govern our behavior."

Fromm examines the evolution of the Soviet system from its beginnings. Its revolutionary leaders were, in various degrees, imbued with the ideas and ideals of Marxism. But these ideas, as so often happens in history, soon degenerated into a ritual, an ideology which could serve to bind people together, to rationalize and justify irrational and immoral acts. Indeed, this ideology, supported by terror, has been used by the new bureaucracy to manipulate opinion, to further the very antithesis of the ideas from which the ideology sprang and on which the state continues to lavish its rites and ceremonies. Whatever it may once have been, the U.S.S.R. is now, Fromm argues, a conservative, state-controlled, industrial managerialism and not, as the official ideology asserts, a democratic, classless society. The only large group of people, says Fromm, who take the communist ideology seriously are we in the U.S.; the Russian leaders "have the greatest trouble in shoring it up with nationalism, moral teaching, and increased material satisfactions." The ideals of socialism have been essentially abandoned, as have the hopes for a revolution in the West. The goal was set by Stalin. He liquidated the socialists' revolution in the name of socialism. Cynical, ruthless, insatiably ambitious, he permitted nothing to stand in the way of transforming the U.S.S.R. from a weak, backward, disorganized, have-not nation into one of the two strongest military powers in the world and the strongest industrial power in Europe. It is true that, like Hitler, his methods "corroded the sense of humanity in the rest of the world"; it is also true that, in spite of the setback of a ruinous war, he bequeathed to his heirs a strong economic and political system.

If we are not to become victims of our own clichés and those of the Russians, we must see the adversary plain: a fat cat, concerned mainly with preserving itself and resisting the onslaughts of lean cats. With tremendous territory, with immense resources of raw material and with no need for markets, the U.S.S.R., Fromm argues, has no desire for world domination. What Khrushchev seeks is a preservation of the status quo, which includes the recognition of the western borders of the Soviet sphere of influence (including East Germany). He seeks to prevent West German rearmament. He seeks an understanding with the U.S., an end to the cold war and world disarmament. (It would probably not displease him either if, by some grand mutation, all the Chinese were overnight turned into shrubs.) Fromm is unimpressed with Khrushchev's felicitous boast: "We will bury you." This is an ideological pronouncement, a piece of bogeyman gabble. That the Soviet Union is imperialist Fromm does not doubt; that it is bent on world domination he dismisses emphatically. The conclusion fits neither the facts nor the rational inferences which one can draw from them. It suits Soviet policy to have us stub our toes in Asia and Africa; and the Kremlin is not averse to supplying arms and technical aid under circumstances which will multiply both our headaches and our commitments. But, on the large scale, beyond ringing their country with buffer satellites—as we attempt to ring it with airfields and missile pads—today's Soviet leaders do not aim at world conquest. So capable and disinterested a student as Barbara Ward recently pointed out in *The New York Times* a "most remarkable instance of Soviet restraint," namely Finland. Would world conquerors have left this country outside their orbit? she asks.

We see the Soviet Union, Fromm asserts, through distorting spectacles. We see a "blend between a revolutionary Lenin and an imperialist Czar" and therefore mistake Khrushchev's "rather conventional and limited movements for signs of the 'Communist-imperialist drive for world domination.'" If we could remove these spectacles we would see that the way is open for an accommodation between the two blocs. It is not an accommodation that entails surrender; nor does it imply that we must come to embrace their ways or they ours. It requires adaptation and change; it requires giving up the picture of a world divided between God and the Devil, between the Goods and the Bads. It is hard to give up cherished convictions. But if life is better than death and our convictions shape our fate, it would be well to make sure that what we believe makes sense. A man who offers to walk across the gorge of Niagara Falls on a tightrope to prove something should at least make sure that what he seeks to prove is provable.

There is another side to Fromm's book which deserves comment. It has to do with paranoid thinking and is closely connected with the possibility of re-examining calmly our convictions about the Soviet menace. Just as there is a *folie à deux*, there is an insanity of millions. The Children's Crusade, the extreme psychological reactions to the Black Death, witch-hunting during the Counter Reformation, the hatred of the Huns during World War I are typical examples.

Paranoid thinking is a bad business. The man who says that everyone is out to get him, that his friends, colleagues, even his wife are conspiring to murder him is very likely insane. Of course it is *possible* that his fears are well founded. It is still unlikely; as unlikely as finding a Chesterfield cigarette in a newly opened package of Camels. The victim, however, is hard to persuade by reasoned arguments. For him possibility rather than probability is the guide of life. This is the essence of his illness. His contact with reality is thin and brittle. He is able to demonstrate with virtuosity that what *might* happen *will* happen. "Reality, for him, is mainly what exists within himself, his own emotions, fears, and desires."

Such an unfortunate is easily recognized. But when masses of people adopt paranoid thinking, it is another matter. Most Americans, Fromm says, think about Russia in a paranoid fashion. They ask what is *possible* rather than what is *probable*. Surely it is possible that Khrushchev wants to conquer us, to bury us, just as it is possible that when I have indigestion it is because my wife and children have been taking turns putting arsenic in my stew. But to ask what they hope to gain by this roundelay, as to ask what Khrushchev hopes to gain by flattening our cities and incinerating our people, is a disturbing question that no paranoid wants to hear.

Another pathological mechanism is projection. One accuses the person one hates and fears of having toward oneself the very designs one has on him. There is no better outlet for one's own hostilities than to pretend that they belong to the person they are directed against. Again, this is readily perceived in individual cases, but not when the same projective mechanism is shared by millions and supported by their leaders. Thus "the enemy"—whatever vague, faceless mass this may be—becomes the embodiment of all the horrors we ourselves wish upon it. The Soviet Union and Communist China are the enemy to whom our projective thinking is directed; yet somehow we find their terror more terrible, their inhumanity more inhumane, their cruelty crueller than that of a Trujillo, a Batista or even a Hitler. This is not to say that Stalin was a kindly man, only that lovers of freedom are not always consistent.

Fromm points to still another type of pathology that plays a great role in political thought—that of fanaticism. The fanatic's pathology resembles that of a depressed person who suffers from numbness rather than sadness. He is incapable of feeling anything. The fanatic builds for himself an idol, "an absolute, to which he surrenders completely but of which he also makes himself a part." This gives him an illusion of feeling. History, both ancient and modern, offers many examples of the passionate, grandiose, fanatical leader who was able to hypnotize masses of people, to persuade them to follow any course, however mad or cruel or self-destructive. The fanatic is particularly effective if the content of his idol is "love," "brotherliness," "God," "salvation," "race," "honor" and so on, rather than frank destructiveness or hostility. He is so seductive, and hence so dangerous politically, "because he *seems* to feel so intensely and to be so convinced. Since we all long for certainty and passionate experience, is it surprising that the fanatic succeeds in attracting so many with his counterfeit faith and feelings?"

Pathological thinking in all its forms is unquestionably the single greatest threat to peace. When, with the aid of modern methods of mass communication, it is deliberately and systematically provoked, entire nations can be made to behave insanely, to embrace folly as wisdom, to exalt death over life. Technically this is the atomic age; emotionally it is in many respects the Stone Age. We look down on the Aztecs who on a feast day sacrificed 20,000 men to their gods so as to keep the universe in its proper course. Yet we sacrifice millions of men in wars and entertain the possibility of extinguishing the human race for causes *we* think are noble. The facts are the same, "only the rationalizations are different."

Has man a future? Perhaps. It is up to him. Does he deserve to prevail? Would it not be better if the hatred and fear and suffering and cruelty which constitute such a large part of the record of history were to come to a final end, leaving the planet "peaceful at last, sleeping quietly"? It is to an old and a wise man that one turns for the answer of youth and hope and courage.

"When," Lord Russell writes, "as in the Egyptian Book of the Dead, the possibly last man comes before the Judge of the Underworld, and pleads that the extinction of his species is a matter for regret, what arguments will he be able to offer?

"If I were the pleader to Osiris for the continuation of the human race, I should say: 'O just and inexorable judge, the indictment of my species is all too well deserved, and never more so than in the present day. But we are not all guilty, and few of us are without better potentialities than those that our circumstances have developed. Do not forget that we have but lately emerged from a morass of ancient ignorance and agelong struggle for existence. Most of what we know we have discovered during the last twelve generations. Intoxicated by our new power over nature, many of us have been misled into the pursuit of power over other human beings. This is an *ignis fatuus*, enticing us to return to the morass from which we have been partially escaping. But this wayward folly has not absorbed all our energies. What we have come to know about the world in which we live, about nebulae and atoms, the great and the small, is more than would have seemed possible before our own day. You may retort that knowledge is not good except in the hands of those who have enough wisdom to use it well. But this wisdom also exists, though as yet sporadically and without the power to control events. Sages and prophets have preached the folly of strife, and if we listen to them we shall emerge into new happiness.

"'It is not only what to avoid that great men have shown us. They have shown us also that it is within human power to create a world of shining beauty and transcendent glory. Consider the poets, the composers, the painters, the men whose inward vision has been shown to the world in edifices of majestic splendor. All this country of the imagination might be ours. And human relations, also, could have the beauty of lyric poetry. At moments, in the love of man and woman, something of this possibility is experienced by many. But there is no reason why it should be confined within narrow boundaries; it could, as in the Choral Symphony, embrace the whole world. These are things which lie within human power, and which, given time, future ages may achieve. For such reasons, Lord Osiris, we beseech Thee to grant us a respite, and a chance to emerge from ancient folly into a world of light and love and loveliness.'"

# II

## THE ROOTS OF
## SOCIAL BEHAVIOR

# II

## The Roots of
## Social Behavior

### *Introduction*

Reviewing the progress of scientific knowledge we see two contrasting ways of regarding science.

> As the discovery of things not known at all.
> As a rediscovery and critical validation of things long known, but not universally conceded as true because of the approximately equal plausibility of contrary beliefs.

Such recondite matters as the quantum theory or the "central dogma" of molecular biology fall under the first heading. The most conspicuous discoveries that fall under the second heading are those having to do with the social nature of man. This is not surprising. Social behavior is the life blood of drama, poetry, some branches of philosohy, and homely anecdote and gossip. In the last analysis it may turn out that everything that scientists find out about the social nature of man was recorded long before by nonscientists. Because expensive scientific research so often retraces paths laid out at no public expense by nonscientists, legislators frequently protest the allocation of public funds for the rediscovery of the obvious. To this objection there is a relevant reply: the nonscientific literature includes contradictory assertions. To evaluate them and to sort out the true from the false is the proper business of critical scientific investigation; it is costly, but worth every penny of the cost.

Because the practical decisions of everyday life cannot wait upon a time-consuming scientific study, most of the decisions we make in social situations have necessarily to be based on folk wisdom unsupported by scientific studies. The scientific study of social phenomena is a comparatively late development in the house of Science. Investigation of social phenomena is, in many respects, more difficult than the investigation of the behavior of electrons or nucleotides.

Another factor has delayed the scientific discovery of social principles, a factor internal to the scientist himself. Michael Polanyi—at various stages in his life a physicist, a social scientist, and a philosopher—once remarked that the situation of the scientist can be "pictured by using Milton's simile, which likens truth to a shattered statue, with fragments lying widely scattered and hidden in many places. Each scientist on his own initiative pursues independently the task of finding one fragment of the statue and fitting it to those collected by others." If he is to succeed in his enterprise, the scientist must, at some stage or other, commit himself to a vision of the statue that he is trying to reconstruct (for its form is never self-evident). And, says Polanyi, "to the extent to which a discoverer has committed himself to a new vision of reality, he has separated himself from others who still think on the old lines."

The ability to separate one's self from others is not common; it may even, in some deep sense, be properly considered to be pathological. Whether natural or not, a measure of alienation has characterized all the great natural scientists who have created our conceptual world. There is ample biographical evidence that many of the best of the natural scientists have been antagonistic, or at least antipathetic, to the phenomena studied by the social sciences. Fortunately, this antagonism is now slowly diminishing. But the phenomena of basic social reactions and their human significance are only slowly becoming part of the concern of natural scientists.

In the light of this history we can understand why Harry F. Harlow's study of "Love in Infant Monkeys" created such a stir when it first appeared. Previous studies in animal behavior had been characterized by a marked determination to exclude such "soft" and "fuzzy" intellectual concepts as "love." Without wrestling with the difficult problem of defining this old idea, Harlow showed that the word stands for something that we must take into account if we are going to understand behavior and personality in all of the higher animals, including ourselves. Scholars who object to the sloppiness of the concept should refine it, not jettison it.

In "Early Experience and Emotional Development" Victor H. Denenberg describes some methods of studying the comparative importance and the interactions of heredity and experience in determining emotional characteristics. The studies were made with rats. We cannot, of course, crawl inside the rat's skull and know what emotions he is experiencing, so we must infer them from such behavior as defecation and cowering in a corner. Inferring emotional unease from extra defecation is defensible because it corresponds with reports of human behavior in stressful situations, e.g., on the battlefield.

Most of what Denenberg reports is what "common sense" could have predicted; but not all. Particularly significant in the present day is his report that mixing emotionally disturbed rats with emotionally stable rats does not improve the emotional health of the disturbed individuals, but on the contrary makes it worse. It is obvious that such an experiment—although not definitive—is relevant to the growing practice of "busing" students between school districts in order to create a greater mixture of individuals from different socio-economic groups, who may differ statistically in their emotional development. The proper response of those who feel the implications of these experiments to be unwelcome is not a rhetorical attack, but the carrying out of more varied and more sophisticated experiments.

We will probably continue for a long time to discover surprising facts in even the simplest situations, not so much because making observations is inherently difficult as because of a human tendency to think as our neighbor does, no matter how much variance there is between the facts and what Kenneth Galbraith has called "the conventional wisdom." Even though the successful practice of science requires that the practitioner separate himself from others (as Polanyi put it) it is a separation that never comes easily, and can never be total. At a time when Charles Darwin was nourishing in secret his vision of evolution through natural selection, he wrote the following revealing lines: "Though I, of course, believe in the truth of my own doctrine, I suspect that no belief is vivid until shared by others. When I think of the many cases of men who have studied one subject for years, and have persuaded themselves of the truth of the foolishest doctrines, I feel sometimes a little frightened, whether I may not be one of these monomaniacs."

It is perhaps necessarily true that "no belief is vivid until shared by others." Darwin's cousin, Francis Galton, in 1881 remarked that: "In early life it seems to be a hard lesson to an imaginative child to distinguish between the real and visionary world. If the fantasies are habitually laughed at and otherwise discouraged, the child soon acquires the power of distinguishing them; any incongruity or nonconformity is quickly noted, the vision is found out and discredited, and is not further attended to. In this way the natural tendency to see them is blunted by repression."

We may say also that the natural tendency of an observant child to commit himself to a vision of reality that is not shared by others is likewise repressed. Solomon E. Asch, in his "Opinions and Social Pressure," describes an interesting experimental study of what happens to a personal vision of reality when it is exposed as being unsupported by one's peers. It is significant that the support of even a single other observer immensely strengthens the will of the individual to be true to his own vision of reality. In childhood we necessarily learn to use the reactions of others to our reports as one of the touchstones of truth. Science, by its nature, is engaged in promulgating heterodoxy; but even heterodoxy is hardly bearable unless mixed with a little bit of "togetherness."

The German philosopher Georg Hegel (1770-1831) said that "freedom is the recognition of necessity." This is one of those truths of social science that has constantly to be rediscovered and reaffirmed, because its implications (or at any rate what are thought to be its implications) are viewed as unacceptable by so many "true believers." In any case Hegel's dictum, like any aphorism, does not encompass the whole of truth. Leon Festinger's "Cognitive Dissonance" explores many of the ramifications of Hegel's aphorism. It is gratifying to see that the experimental method can be used to expand what are otherwise mere philosophic insights. The field of investigation is strikingly open-ended. A biologist in particular cannot help wondering how the undoubted differences between people affect their reactions to cognitive dissonance. What will be the results of competition between different temperamental types? What will be the long-term effects of the results of such competition? If temperamental differences are, in part, determined by heredity, what will be the consequences of natural selection? Or, to put it another way: in what way is the design of a society correlated with its selective effects on people whose tolerance for cognitive dissonance differs?

In his *Essay on Man* Alexander Pope (1688-1744) wrote: "Whatever is, is right." Festinger's study gives us one modern interpretation of this aphorism. Edward T. Hall, Jr., shows us another in "The Anthropology of Manners." Pope's line does not express something that is often said openly and consciously, but something that is deeply felt within us, quite unconsciously—and that's the trouble. Unaware of assumptions unconsciously made, assumptions we might deny or at any rate modify significantly if they were overtly spoken, we blunder into conflict in our dealings with other peoples whose "whatever" is different but nonetheless "right." Of the many differences that create conflict between peoples, differences in manners are, in principle, one of the least troublesome. The trick is to uncover them. This is not easy but it can be done; and once it is done, the path to the removal of the abrasive is relatively easily seen and comparatively painlessly taken. In the end, we are more amused than alarmed by this source of conflict.

Muzafer Sherif's "Experiments in Group Conflict" shows how the use of children as subjects facilitates meaningful investigations of the dynamics of human interactions. For various social reasons it is difficult to carry out such experiments with grown men and women. For one thing, it costs more to use adults as experimental subjects. They are also more sophisticated and thus more likely to see through contrived crises, and when they find out what has happened they may raise quite an uproar about having been used as guinea pigs. (Does this justify using minors as guinea pigs? A pretty ethical problem!) One of the more suggestive results of Sherif's investigation was the discovery of the role that group processes play in creating leaders. The cognitive dissonance that is created by the perception of shortcomings in a leader was minimized by false perceptions: where the leader fell short, his followers gave him credit for more than he deserved. Symmetrically, at the other end of the dominance scale, underdogs were perceived to have performed worse than they actually did.

Involved here is a phenomenon which was spoken of many years ago by the zoologist Paul Scott as the "magnification of difference by a threshold." In a society like ours, which is officially committed to a philosophy of equalitarianism, we tend to attribute any perception and open identification of social differences to malevolence. But it is quite clear that there are completely unconscious, wholly natural social mechanisms that tend to augment whatever differences may be naturally present. Is this bad? Should we combat such natural tendencies? The answers to these questions depend on how we view social organization and social stability generally. The more we value social stability the more we tend to value leadership, even though it conflicts with some of our ideas of justice.

It is the nature of advancing scientific knowledge that it threatens not only the institutions and folk-beliefs of the common man but also those of the most sophisticated seekers after knowledge. The world of science and scholarship is deeply committed to the belief that all censorship is evil. The origins of this commitment trace back to two noble passages in *Areopagitica* of the English poet John Milton (1608-1674):

> What wisdom can there be to choose, what continuence to forbear, without the knowledge of evil? He that can apprehend and consider vice with all her baits and seeming pleasures, and yet abstain, and yet distinguish, and yet prefer that which is truly better, he is the true warfaring Christian. I cannot praise a fugitive and cloistered virtue unexercised and unbreathed, that never sallies out and seeks her adversary, but slinks out of the race, where that immortal garland is to be run for, not without dust and heat.
>
> And though all the winds of doctrine were let loose to play upon the earth, so Truth be in the field, we do injuriously by licencing and prohibiting to misdoubt her strength. Let her and Falsehood grapple; who ever knew Truth put to the worse in a free and open encounter?

With respect to ideas expressed in print—which is what Milton had in mind— it is fair to say that no present-day investigators of the unknown disagree with Milton. The prohibition of censorship is a necessary cornerstone of the developing edifice of knowledge. That which today we cannot say, tomorrow we cannot think. This is the tragedy of effective censorship; this is why we reject it.

But all this refers to ideas in print. What if the ideas—and it is not certain that the word "ideas" is the proper one—are expressed not in cold print but in the warm imagery of movies accompanied by the emotion-arousing medium of sound? Is the optimism of Milton then justified? Is it certain that Right will win out, that Virtue will triumph, that Tenderness will displace Aggression? Not self-evidently so. This is the message of Leonard Berkowitz's "The Effects of Observing Violence." Although not definitive, his experiments clearly indicate that we can no longer lightly assume that a catharsis in ourselves will always result from observing violent behavior in others. Observed violence may increase overt violence in the observer. This is a disturbing thought because it forces us to grapple with the problem of censorship in a new context.

The evidence is far from being sufficient to justify sounding a clarion call of alarm. But it is sufficiently suggestive to make us realize that "communication" does not merely communicate—it also *acts* on the observer, causing him (with a certain probability) to do things that he might not otherwise do. When the medium of communication has a high emotional valence we must hold the communicator *responsible* in some degree for the actions that follow. This, in principle, seems clear: but what are we to do about it?

# 8

# Love in Infant Monkeys

HARRY F. HARLOW

*June 1959*

The first love of the human infant is for his mother. The tender intimacy of this attachment is such that it is sometimes regarded as a sacred or mystical force, an instinct incapable of analysis. No doubt such compunctions, along with the obvious obstacles in the way of objective study, have hampered experimental observation of the bonds between child and mother.

Though the data are thin, the theoretical literature on the subject is rich. Psychologists, sociologists and anthropologists commonly hold that the infant's love is learned through the association of the mother's face, body and other physical characteristics with the alleviation of internal biological tensions, particularly hunger and thirst. Traditional psychoanalysts have tended to emphasize the role of attaining and sucking at the breast as the basis for affectional development. Recently a number of child psychiatrists have questioned such simple explanations. Some argue that affectionate handling in the act of nursing is a variable of importance, whereas a few workers suggest that the composite activities of nursing, contact, clinging and even seeing and hearing work together to elicit the infant's love for his mother.

Now it is difficult, if not impossible, to use human infants as subjects for the studies necessary to break through the present speculative impasse. At birth the infant is so immature that he has little or no control over any motor system other than that involved in sucking. Furthermore, his physical maturation is so slow that by the time he can achieve precise, coordinated, measurable responses of his head, hands, feet and body, the nature and sequence of development have been hopelessly confounded and obscured. Clearly research into the infant-mother relationship has need of a more suitable laboratory animal. We believe we have found it in the infant monkey. For the past several years our group at the Primate Laboratory of the University of Wisconsin has been employing baby rhesus monkeys in a study that we believe has begun to yield significant insights into the origin of the infant's love for his mother.

Baby monkeys are far better coordinated at birth than human infants. Their responses can be observed and evaluated with confidence at an age of 10 days or even earlier. Though they mature much more rapidly than their human contemporaries, infants of both species follow much the same general pattern of development.

Our interest in infant-monkey love grew out of a research program that involved the separation of monkeys from their mothers a few hours after birth. Employing techniques developed by Gertrude van Wagenen of Yale University, we had been rearing infant monkeys on the bottle with a mortality far less than that among monkeys nursed by their mothers. We were particularly careful to provide the infant monkeys with a folded gauze diaper on the floor of their cages, in accord with Dr. van Wagenen's observation that they would tend to maintain intimate contact with such soft, pliant surfaces, especially during nursing. We were impressed by the deep personal attachments that the monkeys formed for these diaper pads, and by the distress that they exhibited when the pads were briefly removed once a day for purposes of sanitation. The behavior of the infant monkeys was reminiscent of the human infant's attachment to its blankets, pillows, rag dolls or cuddly teddy bears.

These observations suggested the series of experiments in which we have sought to compare the importance of nursing and all associated activities with that of simple bodily contact in engendering the infant monkey's attachment to its mother. For this purpose we contrived two surrogate mother monkeys. One is a bare welded-wire cylindrical form surmounted by a wooden head with a crude face. In the other the welded wire is cushioned by a sheathing of terry cloth. We placed eight newborn monkeys in individual cages, each with equal access to a cloth and a wire mother [*see the illustration on page 41*]. Four of the infants received their milk from one mother and four from the other, the milk being furnished in each case by a nursing bottle, with its nipple protruding from the mother's "breast."

The two mothers quickly proved to be physiologically equivalent. The monkeys in the two groups drank the same amount of milk and gained weight at the same rate. But the two mothers proved to be by no means psychologically equivalent. Records made automatically showed that both groups of infants spent far more time climbing and clinging on their cloth-covered mothers than they did on their wire mothers. During the infants' first 14 days of life the floors of the cages were warmed by an electric heating pad, but most of the infants left the pad as soon as they could climb on the unheated cloth mother. Moreover, as the monkeys grew older, they tended to spend an increasing amount of time clinging and cuddling on her pliant terry-cloth surface. Those that secured their nourishment from the wire mother showed no tendency to spend more time on her than feeding required, contradicting the idea that affection is a response that is learned or derived in asso-

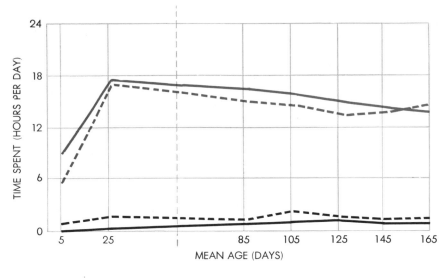

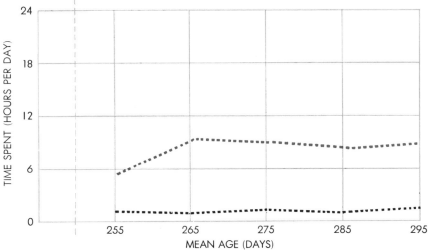

**STRONG PREFERENCE FOR CLOTH MOTHER was shown by all infant monkeys.** Infants reared with access to both mothers from birth (*top chart*) spent far more time on the cloth mother (*colored curves*) than on the wire mother (*black curves*). This was true regardless of whether they had been fed on the cloth (*solid lines*) or on the wire mother (*broken lines*). Infants that had known no mother during their first eight months (*bottom chart*) soon came to prefer cloth mother, but spent less time on her than the other infants.

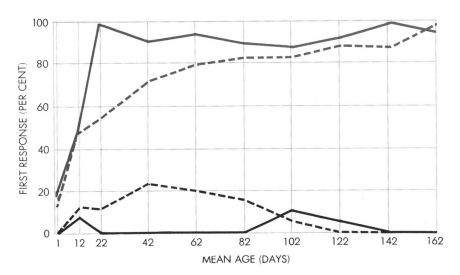

**RESULTS OF "FEAR TEST"** (*see the photographs on page 44*) showed that infants confronted by a strange object quickly learned to seek reassurance from the cloth mother (*colored curves*) rather than from the wire mother (*black curves*). Again infants fed on the wire mother (*broken lines*) behaved much like those fed on cloth mother (*solid lines*)

ciation with the reduction of hunger or thirst.

These results attest the importance—possibly the overwhelming importance—of bodily contact and the immediate comfort it supplies in forming the infant's attachment for its mother. All our experience, in fact, indicates that our cloth-covered mother surrogate is an eminently satisfactory mother. She is available 24 hours a day to satisfy her infant's overwhelming compulsion to seek bodily contact; she possesses infinite patience, never scolding her baby or biting it in anger. In these respects we regard her as superior to a living monkey mother, though monkey fathers would probably not endorse this opinion.

Of course this does not mean that nursing has no psychological importance. No act so effectively guarantees intimate bodily contact between mother and child. Furthermore, the mother who finds nursing a pleasant experience will probably be temperamentally inclined to give her infant plenty of handling and fondling. The real-life attachment of the infant to its mother is doubtless influenced by subtle multiple variables, contributed in part by the mother and in part by the child. We make no claim to having unraveled these in only two years of investigation. But no matter what evidence the future may disclose, our first experiments have shown that contact comfort is a decisive variable in this relationship.

Such generalization is powerfully supported by the results of the next phase of our investigation. The time that the infant monkeys spent cuddling on their surrogate mothers was a strong but perhaps not conclusive index of emotional attachment. Would they also seek the inanimate mother for comfort and security when they were subjected to emotional stress? With this question in mind we exposed our monkey infants to the stress of fear by presenting them with strange objects, for example a mechanical teddy bear which moved forward, beating a drum. Whether the infants had nursed from the wire or the cloth mother, they overwhelmingly sought succor from the cloth one; this differential in behavior was enhanced with the passage of time and the accrual of experience. Early in this series of experiments the terrified infant might rush blindly to the wire mother, but even if it did so it would soon abandon her for the cloth mother. The infant would cling to its cloth mother, rubbing its body against hers. Then, with its fears assuaged through intimate contact with the moth-

**CLOTH AND WIRE MOTHER-SURROGATES** were used to test the preferences of infant monkeys. The infants spent most of their time clinging to the soft cloth "mother," (*foreground*) even when nursing bottles were attached to the wire mother (*background*).

 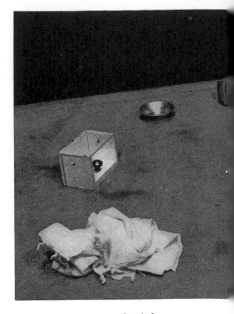

"OPEN FIELD TEST" involved placing a monkey in a room far larger than its accustomed cage; unfamiliar objects added an addi- tional disturbing element. If no mother was present, the infant would typically huddle in a corner (*left*). The wire mother did

er, it would turn to look at the previously terrifying bear without the slightest sign of alarm. Indeed, the infant would some- times even leave the protection of the mother and approach the object that a few minutes before had reduced it to abject terror.

The analogy with the behavior of hu- man infants requires no elaboration. We found that the analogy extends even to less obviously stressful situations. When a child is taken to a strange place, he usually remains composed and happy so long as his mother is nearby. If the moth- er gets out of sight, however, the child is often seized with fear and distress. We developed the same response in our infant monkeys when we exposed them to a room that was far larger than the cages to which they were accustomed. In the room we had placed a number of unfamiliar objects such as a small arti- ficial tree, a crumpled piece of paper, a folded gauze diaper, a wooden block and a doorknob [*a similar experiment is depicted in the illustrations on these two pages*]. If the cloth mother was in the room, the infant would rush wildly to her, climb upon her, rub against her and cling to her tightly. As in the previous experiment, its fear then sharply di- minished or vanished. The infant would begin to climb over the mother's body and to explore and manipulate her face. Soon it would leave the mother to inves- tigate the new world, and the unfamiliar objects would become playthings. In a typical behavior sequence, the infant might manipulate the tree, return to the mother, crumple the wad of paper, bring it to the mother, explore the block, ex-

plore the doorknob, play with the paper and return to the mother. So long as the mother provided a psychological "base of operations" the infants were unafraid and their behavior remained positive, exploratory and playful.

If the cloth mother was absent, how- ever, the infants would rush across the test room and throw themselves face- down on the floor, clutching their heads and bodies and screaming their distress. Records kept by two independent ob- servers—scoring for such "fear indices" as crying, crouching, rocking and thumb- and toe-sucking—showed that the emo- tionality scores of the infants nearly tripled. But no quantitative measure- ment can convey the contrast between the positive, outgoing activities in the presence of the cloth mother and the stereotyped withdrawn and disturbed behavior in the motherless situation.

The bare wire mother provided no more reassurance in this "open field" test than no mother at all. Control tests on monkeys that from birth had known only the wire mother revealed that even these infants showed no affection for her and obtained no comfort from her presence. Indeed, this group of animals exhibited the highest emotionality scores of all. Typically they would run to some wall or corner of the room, clasp their heads and bodies and rock convulsively back and forth. Such activities closely re- semble the autistic behavior seen fre- quently among neglected children in and out of institutions.

In a final comparison of the cloth and wire mothers, we adapted an experiment originally devised by Robert A. Butler

at the Primate Laboratory. Butler had found that monkeys enclosed in a dimly lighted box would press a lever to open and reopen a window for hours on end for no reward other than the chance to look out. The rate of lever-pressing de- pended on what the monkeys saw through the opened window; the sight of another monkey elicited far more activi- ty than that of a bowl of fruit or an emp- ty room [see "Curiosity in Monkeys," by Robert A. Butler; SCIENTIFIC AMERICAN Offprint 426]. We now know that this "curiosity response" is innate. Three- day-old monkeys, barely able to walk, will crawl across the floor of the box to reach a lever which briefly opens the window; some press the lever hundreds of times within a few hours.

When we tested our monkey infants in the "Butler box," we found that those reared with both cloth and wire mothers showed as high a response to the cloth mother as to another monkey, but dis- played no more interest in the wire mother than in an empty room. In this test, as in all the others, the monkeys fed on the wire mother behaved the same as those fed on the cloth mother. A con- trol group raised with no mothers at all found the cloth mother no more inter- esting than the wire mother and neither as interesting as another monkey.

Thus all the objective tests we have been able to devise agree in showing that the infant monkey's relationship to its surrogate mother is a full one. Com- parison with the behavior of infant mon- keys raised by their real mothers con- firms this view. Like our experimental monkeys, these infants spend many

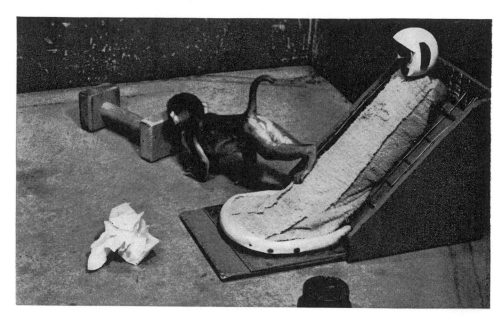

not alter this pattern of fearful behavior, but the cloth mother provided quick reassurance. The infant would first cling to her (*center*) and then set out to explore the room and play with the objects (*right*), returning from time to time for more reassurance.

hours a day clinging to their mothers, and run to them for comfort or reassurance when they are frightened. The deep and abiding bond between mother and child appears to be essentially the same, whether the mother is real or a cloth surrogate.

While bodily contact clearly plays the prime role in developing infantile affection, other types of stimulation presumably supplement its effects. We have therefore embarked on a search for these other factors. The activity of a live monkey mother, for example, provides her infant with frequent motion stimulation. In many human cultures mothers bind their babies to them when they go about their daily chores; in our own culture parents know very well that rocking a baby or walking with him somehow promotes his psychological and physiological well-being. Accordingly we compared the responsiveness of infant monkeys to two cloth mothers, one stationary and one rocking. All of them preferred the rocking mother, though the degree of preference varied considerably from day to day and from monkey to monkey. An experiment with a rocking crib and a stationary one gave similar results. Motion does appear to enhance affection, albeit far less significantly than simple contact.

The act of clinging, in itself, also seems to have a role in promoting psychological and physiological well-being. Even before we began our studies of affection, we noticed that a newborn monkey raised in a bare wire cage survived with difficulty unless we provided it with a cone to which it could cling. Re-

cently we have raised two groups of monkeys, one with a padded crib instead of a mother and the other with a cloth mother as well as a crib. Infants in the latter group actually spend more time on the crib than on the mother, probably because the steep incline of the mother's cloth surface makes her a less satisfactory sleeping platform. In the open-field test, the infants raised with a crib but no mother clearly derived some emotional support from the presence of the crib. But those raised with both showed an unequivocal preference for the mother they could cling to, and they evidenced the benefit of the superior emotional succor they gained from her.

Still other elements in the relationship remain to be investigated systematically. Common sense would suggest that the warmth of the mother's body plays its part in strengthening the infant's ties to her. Our own observations have not yet confirmed this hypothesis. Heating a cloth mother does not seem to increase her attractiveness to the infant monkey, and infants readily abandon a heating pad for an unheated mother surrogate. However, our laboratory is kept comfortably warm at all times; experiments in a chilly environment might well yield quite different results.

Visual stimulation may forge an additional link. When they are about three months old, the monkeys begin to observe and manipulate the head, face and eyes of their mother surrogates; human infants show the same sort of delayed responsiveness to visual stimuli. Such stimuli are known to have marked ef-

fects on the behavior of many young animals. The Austrian zoologist Konrad Lorenz has demonstrated a process called "imprinting"; he has shown that the young of some species of birds become attached to the first moving object they perceive, normally their mothers [see "'Imprinting' in Animals," by Eckhard H. Hess; SCIENTIFIC AMERICAN Offprint 416]. It is also possible that particular sounds and even odors may play some role in the normal development of responses or attention.

The depth and persistence of attachment to the mother depend not only on the kind of stimuli that the young animal receives but also on when it receives them. Experiments with ducks show that imprinting is most effective during a critical period soon after hatching; beyond a certain age it cannot take place at all. Clinical experience with human beings indicates that people who have been deprived of affection in infancy may have difficulty forming affectional ties in later life. From preliminary experiments with our monkeys we have found that their affectional responses develop, or fail to develop, according to a similar pattern.

Early in our investigation we had segregated four infant monkeys as a general control group, denying them physical contact either with a mother surrogate or with other monkeys. After about eight months we placed them in cages with access to both cloth and wire mothers. At first they were afraid of both surrogates, but within a few days they began to respond in much the same way as the other infants. Soon they were

**FRIGHTENING OBJECTS** such as a mechanical teddy bear caused almost all infant monkeys to flee blindly to the cloth mother, as in the top photograph. Once reassured by pressing and rubbing against her, they would then look at the strange object (*bottom*).

spending less than an hour a day with the wire mother and eight to 10 hours with the cloth mother. Significantly, however, they spent little more than half as much time with the cloth mother as did infants raised with her from birth.

In the open-field test these "orphan" monkeys derived far less reassurance from the cloth mothers than did the other infants. The deprivation of physical contact during their first eight months had plainly affected the capacity of these infants to develop the full and normal pattern of affection. We found a further indication of the psychological damage wrought by early lack of mothering when we tested the degree to which infant monkeys retained their attachments to their mothers. Infants raised with a cloth mother from birth and separated from her at about five and a half months showed little or no loss of responsiveness even after 18 months of separation. In some cases it seemed that absence had made the heart grow fonder. The monkeys that had known a mother surrogate only after the age of eight months, however, rapidly lost whatever responsiveness they had acquired. The long period of maternal deprivation had evidently left them incapable of forming a lasting affectional tie.

The effects of maternal separation and deprivation in the human infant have scarcely been investigated, in spite of their implications concerning child-rearing practices. The long period of infant-maternal dependency in the monkey provides a real opportunity for investigating persisting disturbances produced by inconsistent or punishing mother surrogates.

Above and beyond demonstration of the surprising importance of contact comfort as a prime requisite in the formation of an infant's love for its mother—and the discovery of the unimportant or nonexistent role of the breast and act of nursing—our investigations have established a secure experimental approach to this realm of dramatic and subtle emotional relationships. The further exploitation of the broad field of research that now opens up depends merely upon the availability of infant monkeys. We expect to extend our researches by undertaking the study of the mother's (and even the father's!) love for the infant, using real monkey infants or infant surrogates. Finally, with such techniques established, there appears to be no reason why we cannot at some future time investigate the fundamental neurophysiological and biochemical variables underlying affection and love.

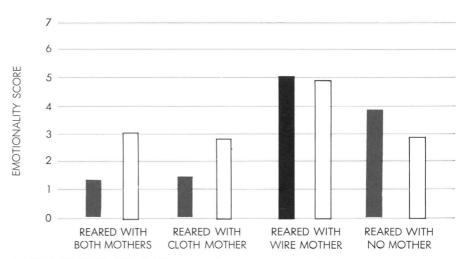

SCORES IN OPEN FIELD TEST show that all infant monkeys familiar with the cloth mother were much less disturbed when she was present (*color*) than when no mother was present (*white*); scores under 2 indicate unfrightened behavior. Infants that had known only the wire mother were greatly disturbed whether she was present (*black*) or not (*white*).

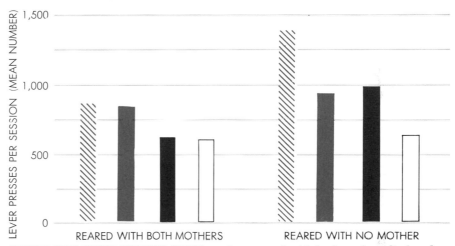

"CURIOSITY TEST" SHOWED THAT monkeys reared with both mothers displayed as much interest in the cloth mother (*solid color*) as in another monkey (*hatched color*); the wire mother (*black*) was no more interesting than an empty chamber (*white*). Monkeys reared with no mother found cloth and wire mother less interesting than another monkey.

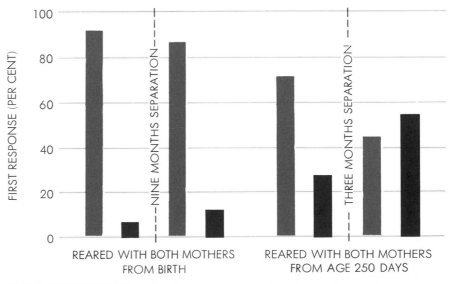

EARLY "MOTHERING" produced a strong and unchanging preference for the cloth mother (*color*) over the wire mother (*black*). Monkeys deprived of early mothering showed less marked preferences before separation and no significant preference subsequently.

# 9

# Early Experience and Emotional Development

VICTOR H. DENENBERG

*June 1963*

A substantial body of clinical evidence supports the general observation that the experiences of early infancy play a paramount role in the growth and development of the individual personality. In particular it has been observed that the emotionality or anxiety of the mother is strongly reflected in the anxiety or lack of it exhibited by the child. With the aim of bringing the process of emotional development under experimental study, investigators in recent years have been observing the relation between early experience and adult behavior in a variety of laboratory animals. They have been gratified to discover that the effects of various modes of treating and handling the infant animal show up consistently and reliably in the adult. To the extent that the analogy with human behavior can be credited, these experiments can be taken as confirmation of what has been observed in the clinic. This work has greater significance, however, because it has shown that the difficult realm of emotional development can be studied in the laboratory on the same terms as, for example, learning or motor development.

During the past few years my colleagues and I at Purdue University have been engaged in work along this line using as our laboratory subjects infant rats of the Purdue Wistar stock. In general we have found that social experiences prior to and immediately after weaning bring about more or less permanent changes in emotional behavior at maturity. As might be expected, our experiments show that the mother-infant relation has decisive effects on the later emotional development of the infant. We have also found—as investigators working with other animals have observed—that early infant-infant relations may have comparably decisive effects.

Our primary method of measuring differences in the emotional behavior of individual rats has been the "open field" test [*see illustration on opposite page*]. The animal to be tested is placed in a box about four feet square for three minutes. The floor of the box is marked off into nine-inch squares to provide a convenient index for measuring the animal's activity. A rat that is capable of flexible and adaptive behavior responds to this unfamiliar situation by running about and exploring its new surroundings. Its activity can be scored by counting the number of squares it enters during the period of observation. An emotionally disturbed rat—that is, one with a low or underdeveloped capacity for adapting to unfamiliar situations—responds by cowering in a corner of the box or by creeping about timidly. The disturbed rat also tends to defecate and urinate frequently, a characteristically maladaptive stress response mediated by the autonomic, or involuntary, nervous system. The open-field test thus provides two objective and quantitative indexes for measuring the emotional state of the test animal: the number of squares entered and the number of boluses defecated during the three-minute test period.

In our experiments involving the mother-infant relation we set out first to measure the degree to which the emotional state of a mother rat affects the emotional state of her offspring. With performance in the open-field test as our standard we were able to classify nonpregnant female rats as scoring "high" or "low" on the scale of emotionality or anxiety. All the animals were then mated and put into individual laboratory cages where, about 21 days later, they gave birth to litters of eight or more pups. In order to investigate whether or not the emotional behavior of the young was influenced by the behavior of the mother, we set up an experiment designed to simulate the effects of inconsistent mothering. Seizing the opportunity presented when two females with the same emotional classification happened to give birth on the same day, we exchanged these mothers every 24 hours until the pups were weaned at 21 days. The differences in the maternal behavior of the mothers in this experiment, we thought, would have effects comparable to the inconsistent maternal care of a single emotionally disturbed mother. In the case of mothers not "rotated" in this fashion, we took care to remove each one from its cage for a short period each day in order to eliminate the possible effects of the handling process itself.

After weaning (at 21 days) the infants were in all cases raised in laboratory cages with littermates of the same sex and were subjected to the open-field test at the age of 50 days. Their performance furnished conclusive evidence that there is a direct relation between the emotional state of a mother and that of her offspring. The offspring of emotionally disturbed mothers showed much lower activity scores and much higher defecation scores than the offspring of normal mothers. The offspring of both classes of mothers that had been subjected to rotated mothering showed a corresponding decrease in activity scores and increase in defecation scores.

Because a direct relation was found between the emotional state of a mother and the emotional state of her offspring, the experiment could be interpreted as demonstrating the effect of hereditary rather than social influences on the emotionality of the offspring. In a control experiment we accordingly exposed the offspring of each class of mother to the maternal care of mothers of both classes on the same 24-hour rotational schedule. If hereditary factors prevailed, one would expect that the offspring of normal mothers would show higher activity scores than the offspring of disturbed mothers. When infants subjected to this treatment had been allowed to mature and were tested in the open field, however, we discovered that this was not the case. The scores of these offspring were not related in any significant way to the emotional classification of their biological mothers [*see top illustration on next page*].

The results of this control experiment were so one-sided that we undertook still another, designed to find out whether or not the infant's early social experience is the sole determinant of its later emotional behavior. This time when a normal and a disturbed female gave birth on the same day, we switched mothers so that the mothers of each class were made to foster the offspring of the other. The pups were not touched by the experimenter during this exchange. After this neither the mothers nor the young were handled again until the young were weaned at 21 days. The scores of these offspring, when tested in the open field at the age of 50 days, show that genetic constitution is not without influence on later emotional behavior. Independent of such influence, however, the mother in her interaction with the young between birth and weaning brings about a relatively permanent change in the emotional behavior of the offspring. The ways in which this postnatal maternal influence is communicated to the infants remain to be explored in more detail.

At this point the design and the results of our experiments suggested that we ought to look into another question: How does the behavior of offspring influence the behavior of mothers? In human families it is well known that a cranky infant can arouse anxiety and emotional tension in the mother, whereas a peaceful baby that eats, sleeps and smiles has the capacity to charm and soothe all concerned. In order to investigate this reciprocal relation it was necessary to produce infant rats with emotionality scores as well established as those of the mother rats used in our experiments. Although the open-field test provided a reliable means of classifying normal and disturbed mothers, it was not practicable to differentiate the emotional types of very young pups by this means. Therefore we undertook another experiment, which established that the emotionality of rats can be fixed at will

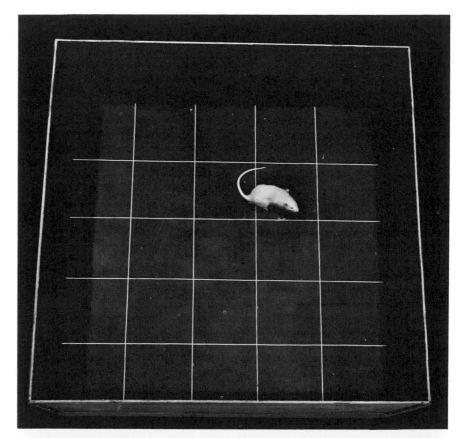

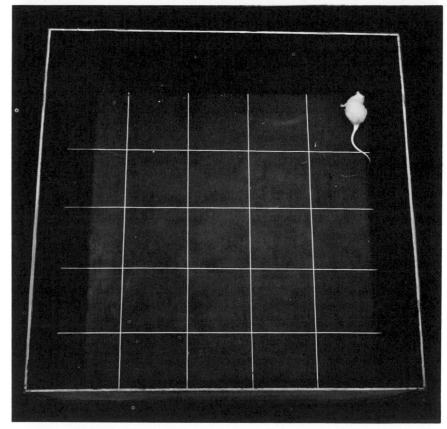

**EMOTIONAL BEHAVIOR** of a rat can be measured on the "open field": an area four feet square marked off into smaller squares. When a normal rat (*top*) is placed in this unfamiliar environment, it generally responds by running about and exploring its new surroundings. An emotionally disturbed rat (*bottom*) responds to the same situation by cowering in a corner and defecating. The principal indices of "emotionality" are the number of squares entered and the number of boluses defecated during three-minute test period.

| BIOLOGICAL MOTHER | ROTATED MOTHERS | ACTIVITY SCORE ON OPEN-FIELD TEST (50 DAYS) |
|---|---|---|
| EMOTIONALLY DISTURBED | BOTH EMOTIONALLY DISTURBED | 48 |
| EMOTIONALLY DISTURBED | EMOTIONALLY DISTURBED AND NORMAL | 74 |
| NORMAL | EMOTIONALLY DISTURBED AND NORMAL | 74 |
| NORMAL | BOTH NORMAL | 95 |

| BIOLOGICAL MOTHER | FOSTER MOTHER | ACTIVITY SCORE ON OPEN-FIELD TEST (50 DAYS) |
|---|---|---|
| EMOTIONALLY DISTURBED | EMOTIONALLY DISTURBED | 61 |
| EMOTIONALLY DISTURBED | NORMAL | 88 |
| NORMAL | EMOTIONALLY DISTURBED | 96 |
| NORMAL | NORMAL | 106 |

**MOTHERS WERE EXCHANGED** between litters according to two different schedules. Some were rotated between their own and another's litter every 24 hours from a common birth date until weaning 21 days later (*graph at left*). Other pairs were exchanged at birth and were not handled again until weaning (*graph at right*). When the offspring of four different maternal combinations in the first group were tested in the open field at 50 days, their perform-ances reflected the strong residual influence of the mother-infant social relation but did not show any evidence of a comparable hereditary influence. When offspring in the second group were tested, their emotional behavior revealed the influence of their genetic backgrounds as well as of their early social experience. Scores are generally lower in the graph at left because the daily rotation of mothers in itself has a disturbing effect on the infant rats.

by suitable treatment at an early age. We found that rats subjected to electric shock during the first few days of life later showed extremely adaptive and accordingly high activity scores in the open-field test. Seymour Levine, who is now at Stanford University, similarly found that almost any routine of daily handling in early life is reflected in a higher degree of emotional stability at maturity. Conversely, rats that spend their early life in the comparative isolation of laboratory cages tend to be emotionally disturbed and adapt poorly in the open-field test [see "Stimulation in Infancy," by Seymour Levine; SCIENTIFIC AMERICAN Offprint 436].

Using our electric-shock techniques, we were able to supply infants of both emotional classifications to mothers of both kinds and thereby set up four dif-

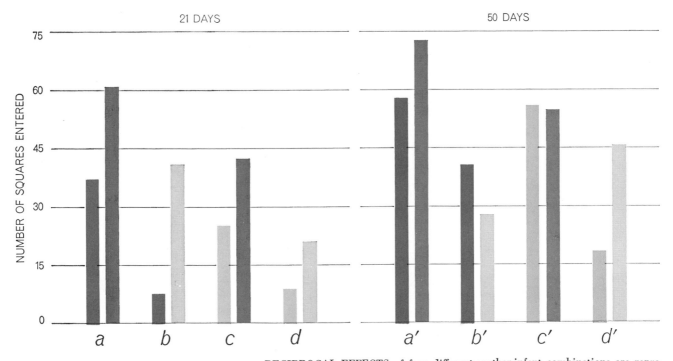

21 DAYS       50 DAYS

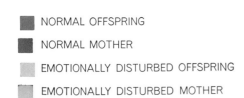

NORMAL OFFSPRING

NORMAL MOTHER

EMOTIONALLY DISTURBED OFFSPRING

EMOTIONALLY DISTURBED MOTHER

**RECIPROCAL EFFECTS** of four different mother-infant combinations are represented by the paired bars on this graph. Open-field activity scores were recorded for both mothers and infants when the infants were weaned at 21 days (*left*) and again at 50 days (*right*). Normal and emotionally disturbed mothers were initially sorted out according to their prepregnancy open-field test scores. Normal infants were obtained by subjecting them to mild electric shocks in early infancy. Infants that had not received this treatment were classified as emotionally disturbed. The results indicate that the emotional behavior of both the mothers and the infants can be altered significantly as a result of the social interaction inherent in the rearing process.

ferent mother-infant combinations. At weaning, and again at 50 days, both the mothers and the young of each combination were tested in the open field. In general the results of these tests showed that significant changes in the emotional behavior of both mothers and young occurred as a result of the social interaction inherent in the rearing process. Emotionally stable mothers that had mothered emotionally stable infants scored highest in the open-field tests, as did their infants, whereas emotionally disturbed mothers that had mothered emotionally disturbed infants scored lowest, along with their infants. Scores for the other two combinations ranged between these two extremes [see bottom illustration on page 48].

In the course of these experiments we noted that pups from litters of eight or more tended to score significantly lower on their activity tests than pups from smaller litters. This suggested that the effects of interaction with siblings might be as deserving of study as the effects of the mother-infant relation. Experiments with a variety of animals conducted by other investigators had, in fact, indicated that early social relations among infants are decisive in the development of normal peer relations in adult life [see "Social Deprivation in Monkeys," by Harry F. and Margaret Kuenne Harlow; SCIENTIFIC AMERICAN Offprint 473]. We proceeded, therefore, to test the effects of association with agemates after weaning on the later emotional behavior of our rats.

For these experiments we divided the infants into two groups, regularly subjecting the members of one group to handling during the first three weeks of life and leaving the other group unhandled, thereby ensuring that the former would later show a low emotionality score and the latter a high one. The usual laboratory procedure is to put the infants into laboratory cages after weaning and keep them there until they are tested in adulthood. It occurred to us, however, that the barren environment of the laboratory cage does little to improve the emotional state of the disturbed rats and probably contributes to their maladaptive stress-response in later life. We suspected that a "richer," or more complex, environment might have a therapeutic effect on some of the emotionally disturbed pups. Accordingly we constructed several large "free environment" boxes about four feet square that contained ramps, tunnels, platforms and plenty of food and water. In one box we placed 16 just-weaned pups that had

REARING IN FREE ENVIRONMENT tended to increase the adult activity scores of both normal (top) and emotionally disturbed rats (bottom), although the latter continued to show many of the characteristic signs of unstable behavior while in the free-environment box. Play behavior among the normal pups included much wrestling and nipping, whereas disturbed pups generally crowded into the corners of the box. Each box contains eight rats.

| INFANT EXPERIENCE (FIRST 25 DAYS) | POSTWEANING ENVIRONMENT (26 TO 50 DAYS) | DEFECATION SCORE ON OPEN-FIELD TEST (180 DAYS) |
|---|---|---|
| HANDLED | CAGE | 8.83 |
| UNHANDLED | CAGE | 27 |
| HANDLED | FREE ENVIRONMENT | .83 |
| UNHANDLED | FREE ENVIRONMENT | 21.38 |

THERAPEUTIC EFFECT of rearing in the free environment compared with rearing in laboratory cages shows up in the lowered defecation scores of both normal and emotionally disturbed rats at maturity. Normal rats were produced by a routine of daily handling in early infancy; unhandled infants tended to be less stable emotionally even as adults.

been handled in infancy; in another we placed 16 pups that had not received this treatment. An equal number of handled and unhandled pups were put into laboratory cages as controls. At 50 days all the animals were placed in laboratory cages and at 180 days they were tested in the open field. As expected, the rats handled during infancy adapted more normally to the unfamiliar test situation, registering higher activity scores and lower defecation scores than the unhandled rats. Within each group, however, the animals that had been reared in the free environment showed lower defecation scores than their controls, which had spent the same 25-day period in laboratory cages [see illustration above]. Apparently the free-environment experience provided an effective form of group psychotherapy in which the opportunity for a broader range of social interaction had combined with the greater diversity of perceptual and motor experience to stimulate and develop the adaptive capacities of the emotionally disturbed rats.

In our next experiment we set out to determine whether or not exposure to emotionally more stable rats in the free-environment situation would have a still greater therapeutic effect on the emotionally disturbed individuals. At weaning we put groups composed of equal numbers of handled and unhandled infants into the boxes; at 50 days we placed all these rats in laboratory cages with littermates of the same sex and at 76 days began open-field testing. To our surprise the effect of mixing the handled and unhandled infants was to make their emotional behavior even more dissimilar at maturity. The activity scores of the unhandled adults were lower and those

of the handled adults were higher than the scores of the rats that had been subjected to the same experience with fellows of like emotional background [see bottom illustration on this page].

Our observations of these animals during their stay in the free-environment box indicated that the accentuation of differences was probably influenced by the play behavior of the pups. The handled animals moved about freely and explored the whole box, whereas the unhandled animals huddled in one or two corners. The emotionally freer pups also played more vigorously, wrestling and nipping each other and attempting to engage the emotionally disturbed rats in their activity. This undoubtedly increased the emotional disturbance of the latter. At the same time, the passivity of the unhandled rats apparently served to reduce the inhibitions of the handled rats and to encourage them to even greater activity.

The results of this experiment seem to run counter to the suggestion advanced by several child psychologists that putting an emotionally upset child in with a group of well-adjusted children will improve the emotional development of the upset child. The contrasting situation we observed with our rats was remarkably similar to the results of an experiment conducted by the child psychologist Fritz Redl several years ago. Redl introduced a boy with considerable masochistic tendencies into a group of normal boys who were all friends and members of the same club. In a short time the disturbed boy had stirred up in the others "more sadistic-pleasure temptations than they could cope with" and their violent aggressions disrupted the social relations of the entire group. We have conducted further experiments in which we have varied the ratio of handled to nonhandled rats, thus duplicating more closely the conditions of this analogous human situation. The results are as yet inconclusive.

With considerable confidence, however, we can say that social experience immediately after weaning will bring about relatively permanent changes in the emotional behavior of the adult animal. What is more, such experience can be deliberately designed in experiments to enhance or depress the capacity for adaptive emotional behavior. The combined techniques of stimulation in infancy and open-field testing provide the basis for many more experiments dealing with the function of social variables in the emotional development of individual animals.

| INFANT EXPERIENCE (FIRST 25 DAYS) | POSTWEANING FREE-ENVIRONMENT GROUPINGS (26 TO 50 DAYS) | ACTIVITY SCORE ON OPEN-FIELD TEST (76 DAYS) |
|---|---|---|
| HANDLED | ALL HANDLED | 270 |
| UNHANDLED | ALL UNHANDLED | 210 |
| HANDLED | EQUALLY MIXED | 382 |
| UNHANDLED | EQUALLY MIXED | 169 |

DISPARITY between the adult activity scores of handled and unhandled infants increased when both were reared together in the free-environment box. The passivity of the unhandled pups during this period apparently stimulated the handled pups to even greater activity, which in turn served to further increase the emotional disturbance of the former.

# Commentary

I am haunted by the thought that the author of the fascinating article "Early Experience and Emotional Development" may have indulged in circular reasoning. In gauging emotional development in rats Victor H. Denenberg used as a measure of flexible and adaptive behavior a rat's activity in running about and exploring the floor of an empty four-foot-square box. Per contra, he took as being emotionally disturbed those rats that cowered in corners of the box, crept about timidly and defecated frequently.

Can we be certain those standards are correct and not reversed? Presumably there is in the experiment and its findings some implicit analogy with other mammals and their behavior. So the question arises whether, in a world ever more dangerous and uncertain for both mice and men, the mother and her pups that proceed with caution and take what meager protection the environment affords are emotionally disturbed or, alternately, are really adapting to their conditions as well as could be expected.

Keeping the bowels open and the flanks covered in these hostile times may well be as good a way to adapt to unfamiliar situations as any other—better, perhaps, than hyperactive, constipated scuttling in aimless search for an unknown goal.

ALFRED FRIENDLY

*The Washington Post*
Washington, D.C.
August, 1963.

Alfred Friendly's question concerning the validity of the measure of emotional reactivity is quite cogent. Both observation and experimental data help to supply an answer. A common observation with laboratory rodents is that they "freeze" and defecate when placed in a strange and frightening situation.

The experimental data relating stimulation in infancy to open-field performance and other patterns of behavior, and to physiological measurements, are too lengthy to detail here. Briefly, animals who are stimulated in infancy are more active and defecate less in the open-field test than nonstimulated controls; such animals also learn an avoidance response more rapidly, have a lower ratio of adrenal weight to body weight following chronic stress and are better able to survive a severe environmental stress.

The patterns of behavior suggested by Friendly in his last paragraph may be appropriate for humans, but it is doubtful that they would be of any adaptive value to rats. And whether or not these patterns would be adaptive for humans would be determined in the long run by the adequacy of the plumbing and the efficiency of the sanitation system.

VICTOR H. DENENBERG

Department of Psychology
Purdue University
Lafayette, Ind.
August, 1963.

# 10

# Opinions and Social Pressure

SOLOMON E. ASCH

*November 1955*

That social influences shape every person's practices, judgments and beliefs is a truism to which anyone will readily assent. A child masters his "native" dialect down to the finest nuances; a member of a tribe of cannibals accepts cannibalism as altogether fitting and proper. All the social sciences take their departure from the observation of the profound effects that groups exert on their members. For psychologists, group pressure upon the minds of individuals raises a host of questions they would like to investigate in detail.

How, and to what extent, do social forces constrain people's opinions and attitudes? This question is especially pertinent in our day. The same epoch that has witnessed the unprecedented technical extension of communication has also brought into existence the deliberate manipulation of opinion and the "engineering of consent." There are many good reasons why, as citizens and as scientists, we should be concerned with studying the ways in which human beings form their opinions and the role that social conditions play.

Studies of these questions began with the interest in hypnosis aroused by the French physician Jean Martin Charcot (a teacher of Sigmund Freud) toward the end of the 19th century. Charcot believed that only hysterical patients could be fully hypnotized, but this view was soon challenged by two other physicians, Hyppolyte Bernheim and A. A. Liébault, who demonstrated that they could put most people under the hypnotic spell. Bernheim proposed that hypnosis was but an extreme form of a normal psychological process which became known as "suggestibility." It was shown that monotonous reiteration of instructions could induce in normal persons in the waking state involuntary bodily changes such as swaying or rigidity of the arms, and sensations such as warmth and odor.

It was not long before social thinkers seized upon these discoveries as a basis for explaining numerous social phenomena, from the spread of opinion to the formation of crowds and the following of leaders. The sociologist Gabriel Tarde summed it all up in the aphorism: "Social man is a somnambulist."

When the new discipline of social psychology was born at the beginning of this century, its first experiments were

EXPERIMENT IS REPEATED in the Laboratory of Social Relations at Harvard University. Seven student subjects are asked by the experimenter (*right*) to compare the length of lines (*see diagram on the next page*). Six of the subjects have been coached beforehand to give unanimously wrong answers. The seventh (*sixth from the left*) has merely been told that it is an experiment in perception.

essentially adaptations of the suggestion demonstration. The technique generally followed a simple plan. The subjects, usually college students, were asked to give their opinions or preferences concerning various matters; some time later they were again asked to state their choices, but now they were also informed of the opinions held by authorities or large groups of their peers on the same matters. (Often the alleged consensus was fictitious.) Most of these studies had substantially the same result: confronted with opinions contrary to their own, many subjects apparently shifted their judgments in the direction of the views of the majorities or the experts. The late psychologist Edward L. Thorndike reported that he had succeeded in modifying the esthetic preferences of adults by this procedure. Other psychologists reported that people's evaluations of the merit of a literary passage could be raised or lowered by ascribing the passage to different authors. Apparently the sheer weight of numbers or authority sufficed to change opinions, even when no arguments for the opinions themselves were provided.

Now the very ease of success in these experiments arouses suspicion. Did the subjects actually change their opinions, or were the experimental victories scored only on paper? On grounds of common sense, one must question whether opinions are generally as watery as these studies indicate. There is some reason to wonder whether it was not the investigators who, in their enthusiasm for a theory, were suggestible, and whether the ostensibly gullible subjects were not providing answers which they thought good subjects were expected to give.

The investigations were guided by certain underlying assumptions, which today are common currency and account for much that is thought and said about the operations of propaganda and public opinion. The assumptions are that peo-

ple submit uncritically and painlessly to external manipulation by suggestion or prestige, and that any given idea or value can be "sold" or "unsold" without reference to its merits. We should be skeptical, however, of the supposition that the power of social pressure necessarily implies uncritical submission to it: independence and the capacity to rise above group passion are also open to human beings. Further, one may question on psychological grounds whether it is possible as a rule to change a person's judgment of a situation or an object without first changing his knowledge or assumptions about it.

In what follows I shall describe some experiments in an investigation of the effects of group pressure which was carried out recently with the help of a number of my associates. The tests not only demonstrate the operations of group pressure upon individuals but also illustrate a new kind of attack on the problem and some of the more subtle questions that it raises.

A group of seven to nine young men, all college students, are assembled in a classroom for a "psychological experiment" in visual judgment. The experimenter informs them that they will be comparing the lengths of lines. He shows two large white cards. On one is a single vertical black line—the standard whose length is to be matched. On the other card are three vertical lines of various lengths. The subjects are to choose the one that is of the same length as the line on the other card. One of the three actually is of the same length; the other two are substantially different, the difference ranging from three quarters of an inch to an inch and three quarters.

The experiment opens uneventfully. The subjects announce their answers in the order in which they have been seated in the room, and on the first round every person chooses the same matching line.

Then a second set of cards is exposed; again the group is unanimous. The members appear ready to endure politely another boring experiment. On the third trial there is an unexpected disturbance. One person near the end of the group disagrees with all the others in his selection of the matching line. He looks surprised, indeed incredulous, about the disagreement. On the following trial he disagrees again, while the others remain unanimous in their choice. The dissenter becomes more and more worried and hesitant as the disagreement continues in succeeding trials; he may pause before announcing his answer and speak in a low voice, or he may smile in an embarrassed way.

What the dissenter does not know is that all the other members of the group were instructed by the experimenter beforehand to give incorrect answers in unanimity at certain points. The single individual who is not a party to this prearrangement is the focal subject of our experiment. He is placed in a position in which, while he is actually giving the correct answers, he finds himself unexpectedly in a minority of one, opposed by a unanimous and arbitrary majority with respect to a clear and simple fact. Upon him we have brought to bear two opposed forces: the evidence of his senses and the unanimous opinion of a group of his peers. Also, he must declare his judgments in public, before a majority which has also stated its position publicly.

The instructed majority occasionally reports correctly in order to reduce the possibility that the naive subject will suspect collusion against him. (In only a few cases did the subject actually show suspicion; when this happened, the experiment was stopped and the results were not counted.) There are 18 trials in each series, and on 12 of these the majority responds erroneously.

How do people respond to group pressure in this situation? I shall report first the statistical results of a series in which a total of 123 subjects from three institutions of higher learning (not including my own, Swarthmore College) were placed in the minority situation described above.

Two alternatives were open to the subject: he could act independently, repudiating the majority, or he could go along with the majority, repudiating the evidence of his senses. Of the 123 put to the test, a considerable percentage yielded to the majority. Whereas in ordinary circumstances individuals matching the lines will make mistakes less than 1

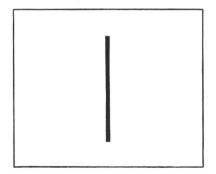
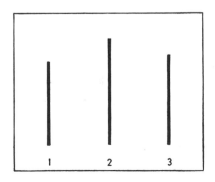

SUBJECTS WERE SHOWN two cards. One bore a standard line. The other bore three lines, one of which was the same length as the standard. The subjects were asked to choose this line.

**EXPERIMENT PROCEEDS** as follows. In the top picture the subject (*center*) hears rules of experiment for the first time. In the second picture he makes his first judgment of a pair of cards, disagreeing with the unanimous judgment of the others. In the third he leans forward to look at another pair of cards. In the fourth he shows the strain of repeatedly disagreeing with the majority. In the fifth, after 12 pairs of cards have been shown, he explains that "he has to call them as he sees them." This subject disagreed with the majority on all 12 trials. Seventy-five per cent of experimental subjects agree with the majority in varying degrees.

per cent of the time, under group pressure the minority subjects swung to acceptance of the misleading majority's wrong judgments in 36.8 per cent of the selections.

Of course individuals differed in response. At one extreme, about one quarter of the subjects were completely independent and never agreed with the erroneous judgments of the majority. At the other extreme, some individuals went with the majority nearly all the time. The performances of individuals in this experiment tend to be highly consistent. Those who strike out on the path of independence do not, as a rule, succumb to the majority even over an extended series of trials, while those who choose the path of compliance are unable to free themselves as the ordeal is prolonged.

The reasons for the startling individual differences have not yet been investigated in detail. At this point we can only report some tentative generalizations from talks with the subjects, each of whom was interviewed at the end of the experiment. Among the independent individuals were many who held fast because of staunch confidence in their own judgment. The most significant fact about them was not absence of responsiveness to the majority but a capacity to recover from doubt and to re-establish their equilibrium. Others who acted independently came to believe that the majority was correct in its answers, but they continued their dissent on the simple ground that it was their obligation to call the play as they saw it.

Among the extremely yielding persons we found a group who quickly reached the conclusion: "I am wrong, they are right." Others yielded in order "not to spoil your results." Many of the individuals who went along suspected that the majority were "sheep" following the first responder, or that the majority were victims of an optical illusion; nevertheless, these suspicions failed to free them at the moment of decision. More disquieting were the reactions of subjects who construed their difference from the majority as a sign of some general deficiency in themselves, which at all costs they must hide. On this basis they desperately tried to merge with the majority, not realizing the longer-range consequences to themselves. All the yielding subjects underestimated the frequency with which they conformed.

Which aspect of the influence of a majority is more important—the size of the majority or its unanimity? The experiment was modified to examine this

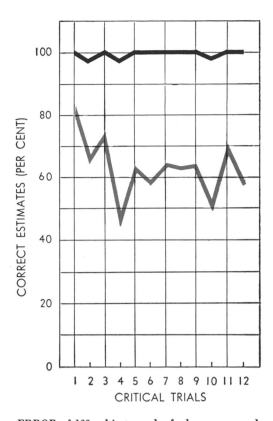

ERROR of 123 subjects, each of whom compared lines in the presence of six to eight opponents, is plotted in the colored curve. The accuracy of judgments not under pressure is indicated in black.

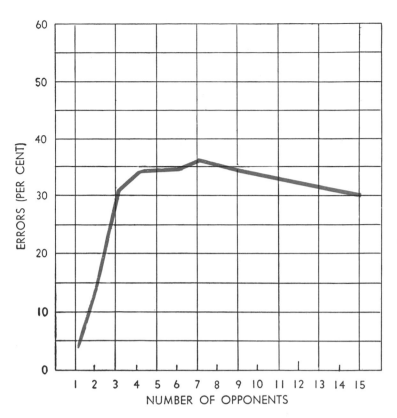

SIZE OF MAJORITY which opposed them had an effect on the subjects. With a single opponent the subject erred only 3.6 per cent of the time; with two opponents he erred 13.6 per cent; three, 31.8 per cent; four, 35.1 per cent; six, 35.2 per cent; seven, 37.1 per cent; nine, 35.1 per cent; 15, 31.2 per cent.

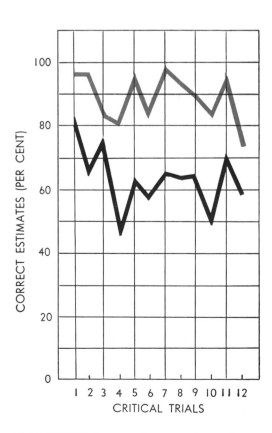

TWO SUBJECTS supporting each other against a majority made fewer errors (colored curve) than one subject did against a majority (black curve).

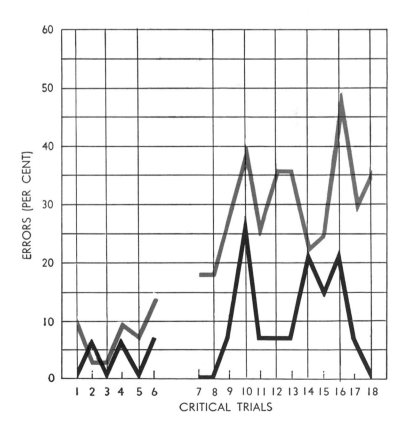

PARTNER LEFT SUBJECT after six trials in a single experiment. The colored curve shows the error of the subject when the partner "deserted" to the majority. Black curve shows error when partner merely left the room.

question. In one series the size of the opposition was varied from one to 15 persons. The results showed a clear trend. When a subject was confronted with only a single individual who contradicted his answers, he was swayed little: he continued to answer independently and correctly in nearly all trials. When the opposition was increased to two, the pressure became substantial: minority subjects now accepted the wrong answer 13.6 per cent of the time. Under the pressure of a majority of three, the subjects' errors jumped to 31.8 per cent. But further increases in the size of the majority apparently did not increase the weight of the pressure substantially. Clearly the size of the opposition is important only up to a point.

Disturbance of the majority's unanimity had a striking effect. In this experiment the subject was given the support of a truthful partner—either another individual who did not know of the prearranged agreement among the rest of the group, or a person who was instructed to give correct answers throughout.

The presence of a supporting partner depleted the majority of much of its power. Its pressure on the dissenting individual was reduced to one fourth: that is, subjects answered incorrectly only one fourth as often as under the pressure of a unanimous majority [see chart at lower left on preceding page]. The weakest persons did not yield as readily. Most interesting were the reactions to the partner. Generally the feeling toward him was one of warmth and closeness; he was credited with inspiring confidence. However, the subjects repudiated the suggestion that the partner decided them to be independent.

Was the partner's effect a consequence of his dissent, or was it related to his accuracy? We now introduced into the experimental group a person who was instructed to dissent from the majority but also to disagree with the subject. In some experiments the majority was always to choose the worst of the comparison lines and the instructed dissenter to pick the line that was closer to the length of the standard one; in others the majority was consistently intermediate and the dissenter most in error. In this manner we were able to study the relative influence of "compromising" and "extremist" dissenters.

Again the results are clear. When a moderate dissenter is present, the effect of the majority on the subject decreases by approximately one third, and extremes of yielding disappear. Moreover, most of the errors the subjects do make

are moderate, rather than flagrant. In short, the dissenter largely controls the choice of errors. To this extent the subjects broke away from the majority even while bending to it.

On the other hand, when the dissenter always chose the line that was more flagrantly different from the standard, the results were of quite a different kind. The extremist dissenter produced a remarkable freeing of the subjects; their errors dropped to only 9 per cent. Furthermore, all the errors were of the moderate variety. We were able to conclude that dissent *per se* increased independence and moderated the errors that occurred, and that the direction of dissent exerted consistent effects.

In all the foregoing experiments each subject was observed only in a single setting. We now turned to studying the effects upon a given individual of a change in the situation to which he was exposed. The first experiment examined the consequences of losing or gaining a partner. The instructed partner began by answering correctly on the first six trials. With his support the subject usually resisted pressure from the majority: 18 of 27 subjects were completely independent. But after six trials the partner joined the majority. As soon as he did so, there was an abrupt rise in the subjects' errors. Their submission to the majority was just about as frequent as when the minority subject was opposed by a unanimous majority throughout.

It was surprising to find that the experience of having had a partner and of having braved the majority opposition with him had failed to strengthen the individuals' independence. Questioning at the conclusion of the experiment suggested that we had overlooked an important circumstance; namely, the strong specific effect of "desertion" by the partner to the other side. We therefore changed the conditions so that the partner would simply leave the group at the proper point. (To allay suspicion it was announced in advance that he had an appointment with the dean.) In this form of the experiment, the partner's effect outlasted his presence. The errors increased after his departure, but less markedly than after a partner switched to the majority.

In a variant of this procedure the trials began with the majority unanimously giving correct answers. Then they gradually broke away until on the sixth trial the naive subject was alone and the group unanimously against him. As long as the subject had anyone on his side, he

was almost invariably independent, but as soon as he found himself alone, the tendency to conform to the majority rose abruptly.

As might be expected, an individual's resistance to group pressure in these experiments depends to a considerable degree on how wrong the majority is. We varied the discrepancy between the standard line and the other lines systematically, with the hope of reaching a point where the error of the majority would be so glaring that every subject would repudiate it and choose independently. In this we regretfully did not succeed. Even when the difference between the lines was seven inches, there were still some who yielded to the error of the majority.

The study provides clear answers to a few relatively simple questions, and it raises many others that await investigation. We would like to know the degree of consistency of persons in situations which differ in content and structure. If consistency of independence or conformity in behavior is shown to be a fact, how is it functionally related to qualities of character and personality? In what ways is independence related to sociological or cultural conditions? Are leaders more independent than other people, or are they adept at following their followers? These and many other questions may perhaps be answerable by investigations of the type described here.

Life in society requires consensus as an indispensable condition. But consensus, to be productive, requires that each individual contribute independently out of his experience and insight. When consensus comes under the dominance of conformity, the social process is polluted and the individual at the same time surrenders the powers on which his functioning as a feeling and thinking being depends. That we have found the tendency to conformity in our society so strong that reasonably intelligent and well-meaning young people are willing to call white black is a matter of concern. It raises questions about our ways of education and about the values that guide our conduct.

Yet anyone inclined to draw too pessimistic conclusions from this report would do well to remind himself that the capacities for independence are not to be underestimated. He may also draw some consolation from a further observation: those who participated in this challenging experiment agreed nearly without exception that independence was preferable to conformity.

# Commentary

I feel impelled to take issue with the conclusions reached by Solomon E. Asch in his article "Opinions and Social Pressure." The constant readiness to be proven wrong constitutes the most indispensable prerequisite for a scientist. On all matters of sense perception, capable of objective measurement and verification, a true scientist will invariably question and reject his own subjective impressions when confronted with objective evidence to the contrary, such as actual physical measurement. Even more striking results could have been obtained by actual measurement with two unequal, "fixed" measuring rods. Under the conditions of the stated experiments, the closest approach to objective verification available to the subject was the near unanimous judgment of his equals. One could thus make a strong case for a thesis contrary to that implied by the author: That the highest trait favoring the scientific progress of our society is the unflinching readiness to be proven wrong as adjudged by the best available methods of verification.

Stubborn attachment to disproven pet theories (which the article would dub "capacity for independence" or "individualism") is unworthy of intelligent men. Furthermore, the unselfish willingness to accept the best available objective evidence in preference to one's own subjective impressions by no means implies conformist behavior in the fields of ethical, moral, political and social principles and convictions.

A well-rounded person cannot do without a blend of both character ingredients: some measure of reliance on the common achievements of his fellow men, as well as some measure of critical independence. The results of these experiments thus prove nothing whatever to a natural scientist, and point to no conclusion or lesson. Their value, if any, lies exclusively in the determination of quantitative orders of magnitude for very specialized sets of conditions.

KURT EISEMANN

New York, N. Y.
January, 1956

It comes as a surprise that the investigation of social pressure could be interpreted as confusing independence with dogmatic attachment to one's views. The comments of Mr. Eisemann neglect a modest but vital fact about the experimental situation: It was the task of the person who served as observer to report what *he* saw, not what others were seeing. His role was that of a person testifying to a fact within his experience, and analogous to that of a member of a jury who is expected to consider the views of others but who may not delegate his responsibility. This was clearly understood by all.

Indeed, our observations show that the problem of the observer was not restricted to deciding whether the majority was accurate. There were persons who, although convinced that the majority was judging correctly and that their own judgments were wrong, remained completely independent. There were others who complied with the majority although convinced that it was wrong. These observations cannot be reconciled with the view that independence under the present conditions marked a failure to show a decent respect for the opinion of others, and that conformity was the sign of an unselfish scientific temper. The evidence points in a different direction: independence required a measure of strength, while failure of independence was connected with self-distrust and fear.

Mr. Eisemann's letter raises a wider and a humanly more important issue. If I understand him rightly, he comes close at some points to saying that agreement among persons is at times tantamount to proof and the best criterion of truth, placing upon other individuals the obligation to acknowledge it. Had he limited his claim to suggesting that we ought to take into account and examine the views of others, one could hardly quarrel with it. But the history of human affairs—and of science itself—establishes sufficiently that error and distortion can have quite a following. Apparently we need to discriminate between valid and baseless consensus. And what other means of doing so have we than our own understanding? I prefer to say that agreement is not proof, that consensus is baseless unless it is independently confirmed in the experience and insights of each person. Our individual "subjective impressions" may be all too frail, but they are all we have.

SOLOMON E. ASCH

January, 1956

# 11

# Cognitive Dissonance

LEON FESTINGER

*October 1962*

There is an experiment in psychology that you can perform easily in your own home if you have a child three or four years old. Buy two toys that you are fairly sure will be equally attractive to the child. Show them both to him and say: "Here are two nice toys. This one is for you to keep. The other I must give back to the store." You then hand the child the toy that is his to keep and ask: "Which of the two toys do you like better?" Studies have shown that in such a situation most children will tell you they prefer the toy they are to keep.

This response of children seems to conflict with the old saying that the grass is always greener on the other side of the fence. Do adults respond in the same way under similar circumstances or does the adage indeed become true as we grow older? The question is of considerable interest because the adult world is filled with choices and alternative courses of action that are often about equally attractive. When they make a choice of a college or a car or a spouse or a home or a political candidate, do most people remain satisfied with their choice or do they tend to wish they had made a different one? Naturally any choice may turn out to be a bad one on the basis of some objective measurement, but the question is: Does some psychological process come into play immediately after the making of a choice that colors one's attitude, either favorably or unfavorably, toward the decision?

To illuminate this question there is another experiment one can do at home, this time using an adult as a subject rather than a child. Buy two presents for your wife, again choosing things you are reasonably sure she will find about equally attractive. Find some plausible excuse for having both of them in your possession, show them to your wife and ask her to tell you how attractive each one is to her. After you have obtained a good measurement of attractiveness, tell her that she can have one of them, whichever she chooses. The other you will return to the store. After she has made her choice, ask her once more to evaluate the attractiveness of each of them. If you compare the evaluations of attractiveness before and after the choice, you will probably find that the chosen present has increased in attractiveness and the rejected one decreased.

Such behavior can be explained by a new theory concerning "cognitive dissonance." This theory centers around the idea that if a person knows various things that are not psychologically consistent with one another, he will, in a variety of ways, try to make them more consistent. Two items of information that psychologically do not fit together are said to be in a dissonant relation to each other. The items of information may be about behavior, feelings, opinions, things in the environment and so on. The word "cognitive" simply emphasizes that the theory deals with relations among items of information.

Such items can of course be changed. A person can change his opinion; he can change his behavior, thereby changing the information he has about it; he can even distort his perception and his information about the world around him. Changes in items of information that produce or restore consistency are referred to as dissonance-reducing changes.

Cognitive dissonance is a motivating state of affairs. Just as hunger impels a person to eat, so does dissonance impel a person to change his opinions or his behavior. The world, however, is much

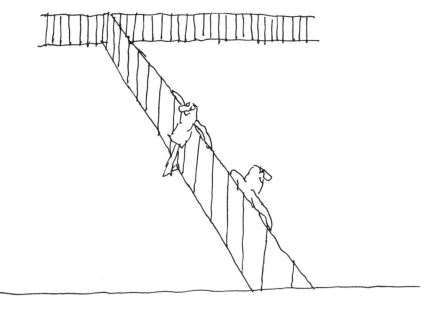

*The grass is not always greener on the other side of the fence*

*Consequences of making a decision between two reasonably attractive alternatives*

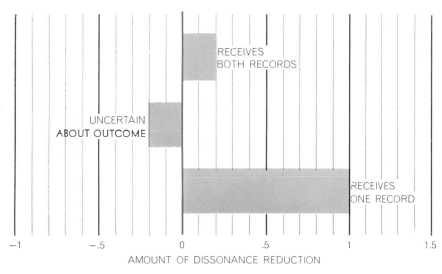

RECEIVES BOTH RECORDS

UNCERTAIN ABOUT OUTCOME

RECEIVES ONE RECORD

−1          −.5          0          .5          1          1.5

AMOUNT OF DISSONANCE REDUCTION

**DISSONANCE REDUCTION is a psychological phenomenon found to occur after a person has made a choice between two approximately equal alternatives. The effect of the phenomenon is to enhance the attractiveness of the chosen object or chosen course of action. The chart summarizes the results of an experiment in which high school girls rated the attractiveness of 12 "hit" records before and after choosing one of them as a gift. Substantial dissonance reduction occurred under only one of three experimental conditions described in the text. Under two other conditions no systematic reduction was observed.**

more effectively arranged for hunger reduction than it is for dissonance reduction. It is almost always possible to find something to eat. It is not always easy to reduce dissonance. Sometimes it may be very difficult or even impossible to change behavior or opinions that are involved in dissonant relations. Consequently there are circumstances in which appreciable dissonance may persist for long periods.

To understand cognitive dissonance as a motivating state, it is necessary to have a clearer conception of the conditions that produce it. The simplest definition of dissonance can, perhaps, be given in terms of a person's expectations. In the course of our lives we have all accumulated a large number of expectations about what things go together and what things do not. When such an expectation is not fulfilled, dissonance occurs.

For example, a person standing unprotected in the rain would expect to get wet. If he found himself in the rain and he was not getting wet, there would exist dissonance between these two pieces of information. This unlikely example is one where the expectations of different people would all be uniform. There are obviously many instances where different people would not share the same expectations. Someone who is very self-confident might expect to succeed at whatever he tried, whereas someone who had a low opinion of himself might normally expect to fail. Under these circumstances what would produce dissonance for one person might produce consonance for another. In experimental investigations, of course, an effort is made to provide situations in which expectations are rather uniform.

Perhaps the best way to explain the theory of cognitive dissonance is to show its application to specific situations. The rest of this article, therefore, will be devoted to a discussion of three examples of cognitive dissonance. I shall discuss the effects of making a decision, of lying and of temptation. These three examples by no means cover all the situations in which dissonance can be created. Indeed, it seldom happens that everything a person knows about an action he has taken is perfectly consistent with his having taken it. The three examples, however, may serve to illustrate the range of situations in which dissonance can be expected to occur. They will also serve to show the kinds of dissonance-reduction effects that are obtained under a special circumstance: when dissonance involves the person's behavior and the action in question is difficult to change.

Let us consider first the consequences of making a decision. Imagine the situation of a person who has carefully weighed two reasonably attractive alternatives and then chosen one of them—a decision that, for our purposes, can be regarded as irrevocable. All the information this person has concerning the attractive features of the rejected alternative (and the possible unattractive features of the chosen alternative) are now inconsistent, or dissonant, with the knowledge that he has made the given choice. It is true that the person also knows many things that are consistent or consonant with the choice he has made, which is to say all the attractive features of the chosen alternative and unattractive features of the rejected one. Nevertheless, some dissonance exists and after the decision the individual will try to reduce the dissonance.

There are two major ways in which the individual can reduce dissonance in this situation. He can persuade himself that the attractive features of the rejected alternative are not really so attractive as he had originally thought, and that the unattractive features of the chosen alternative are not really unattractive. He can also provide additional justification for his choice by exaggerating the attractive features of the chosen alternative and the unattractive features of the rejected alternative. In other words, according to the theory the process of dissonance reduction should lead, after the decision, to an increase in the desirability of the chosen alternative and a decrease in the desirability of the rejected alternative.

This phenomenon has been demonstrated in a variety of experiments. A brief description of one of these will suffice to illustrate the precise nature of the effect. In an experiment performed by Jon Jecker of Stanford University, high school girls were asked to rate the attractiveness of each of 12 "hit" records. For each girl two records that she had rated as being only moderately attractive were selected and she was asked which of the two she would like as a gift. After having made her choice, the girl again rated the attractiveness of all the records. The dissonance created by the decision could be reduced by increasing the attractiveness of the chosen record and decreasing the attractiveness of the rejected record. Consequently a measurement of dissonance reduction could be obtained by summing both of these kinds of changes in ratings made before and after the decision.

Different experimental variations were employed in this experiment in order to examine the dynamics of the process of dissonance reduction. Let us look at three of these experimental variations. In all three conditions the girls, when they were making their choice, were given to understand there was a slight possibility that they might actually be given both records. In one condition they were asked to rerate the records after they had made their choice but before they knew definitely whether they would receive both records or only the one they chose. The results for this condition should indicate whether dissonance reduction begins with having made the choice or whether it is suspended until the uncertainty is resolved. In a second condition the girls were actually given both records after their choice and were then asked to rerate

*Further consequences of making a difficult decision*

all the records. Since they had received both records and therefore no dissonance existed following the decision, there should be no evidence of dissonance reduction in this condition. In a third condition the girls were given only the record they chose and were then asked to do the rerating. This, of course, resembles the normal outcome of a decision and the usual dissonance reduction should occur.

The chart on page 59 shows the results for these three conditions. When the girls are uncertain as to the outcome, or when they receive both records, there is no dissonance reduction—that is, no systematic change in attractiveness of the chosen and rejected records. The results in both conditions are very close to zero—one slightly positive, the other slightly negative. When they receive only the record they chose, however, there is a large systematic change in rating to reduce dissonance. Since dissonance reduction is only observed in this last experimental condition, it is evident that dissonance reduction does not occur during the process of making a decision but only after the decision is made and the outcome is clear.

Let us turn now to the consequences of lying. There are many circumstances in which, for one reason or another, an individual publicly states something that is at variance with his private belief. Here again one can expect dissonance to arise. There is an inconsistency between knowing that one really believes one thing and knowing tnat one has publicly stated something quite different. Again, to be sure, the individual knows things that are consonant with his overt, public behavior. All the reasons that induced him to make the public statement are consonant with his having made it and provide him with some justification for his behavior. Nevertheless, some dissonance exists and, according to the theory, there will be attempts to reduce it. The degree to which the dissonance is bothersome for the individual will depend on two things. The more deviant his public statement is from his private belief, the greater will be the dissonance. The greater the amount of justification the person has for having made the public statement, the less bothersome the dissonance will be.

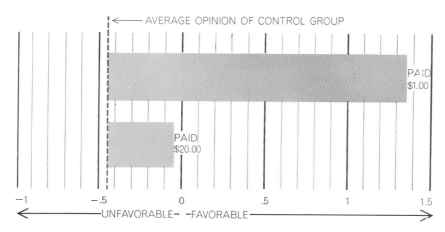

CONSEQUENCES OF LYING are found to vary, depending on whether the justification for the lie is large or small. In this experiment students were persuaded to tell others that a boring experience was really fun. Those in one group were paid only $1 for their co-operation; in a second group, $20. The low-paid students, having least justification for lying, experienced most dissonance and reduced it by coming to regard the experience favorably.

How can the dissonance be reduced? One method is obvious. The individual can remove the dissonance by retracting his public statement. But let us consider only those instances in which the public statement, once made, cannot be changed or withdrawn; in other words, in which the behavior is irrevocable. Under such circumstances the major avenue for reduction of the dissonance is change of private opinion. That is, if the private opinion were changed so that it agreed with what was publicly stated, obviously the dissonance would be gone. The theory thus leads us to expect that after having made an irrevocable public statement at variance with his private belief, a person will tend to change his private belief to bring it into line with his public statement. Furthermore, the degree to which he changes his private belief will depend on the amount of justification or the amount of pressure for making the public statement initially. The less the original justification or pressure, the greater the dissonance and the more the person's private belief can be expected to change.

An experiment recently conducted at Stanford University by James M. Carlsmith and me illustrates the nature of this effect. In the experiment, college students were induced to make a statement at variance with their own belief. It was done by using students who had

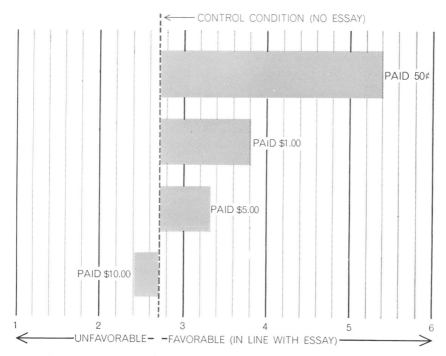

GRADED CHANGE OF OPINION was produced by paying subjects various sums for writing essays advocating opinions contrary to their beliefs. When examined later, students paid the least had changed their opinion the most to agree with what they had written. Only the highest paid group held to their original opinion more strongly than did a control group.

volunteered to participate in an experiment to measure "motor performance." The purported experiment lasted an hour and was a boring and fatiguing session. At the end of the hour the experimenter thanked the subject for his participation, indicating that the experiment was over. The real purpose of the hour-long session, however, was to provide each subject with an identical experience about which he would have an unfavorable opinion.

At the end of the fatiguing hour the experimenter enlisted the subject's aid in preparing the next person for the experiment. The subject was led to believe that, for experimental purposes, the next person was supposed to be given the impression that the hour's session was going to be very interesting and lots of fun. The subject was persuaded to help in this deception by telling the next subject, who was waiting in an adjoining room, that he himself had just finished the hour and that it had indeed been very interesting and lots of fun. The first subject was then interviewed by someone else to determine his actual private opinion of the experiment.

Two experimental conditions were run that differed only in the amount of pressure, or justification given the subject for stating a public opinion at variance with his private belief. All subjects, of course, had the justification of helping to conduct a scientific experiment. In addition to this, half of the subjects were paid $1 for their help—a relatively small amount of money; the other subjects were paid $20—a rather large sum for the work involved. From the theory we would expect that the subjects who were paid only $1, having less justification for their action, would have more dissonance and would change their private beliefs more in order to reduce the dissonance. In other words, we would expect the greatest change in private opinion among the subjects given the least tangible incentive for changing.

The upper illustration on the preceding page shows the results of the experiment. The broken line in the chart shows the results for a control group of subjects. These subjects participated in the hour-long session and then were asked to give their private opinion of it. Their generally unfavorable views are to be expected when no dissonance is induced between private belief and public statement. It is clear from the chart that introducing such dissonance produced a change of opinion so that the subjects who were asked to take part in a deception finally came to think better of the session than did the control subjects. It

*The effect of rewards on lying*

is also clear that only in the condition where they were paid a dollar is this opinion change appreciable. When they were paid a lot of money, the justification for misrepresenting private belief is high and there is correspondingly less change of opinion to reduce dissonance.

Another way to summarize the result is to say that those who are highly rewarded for doing something that involves dissonance change their opinion less in the direction of agreeing with what they did than those who are given very little reward. This result may seem surprising, since we are used to thinking that reward is effective in creating change. It must be remembered, however, that the critical factor here is that the reward is being used to induce a behavior that is dissonant with private opinion.

To show that this result is valid and not just a function of the particular situation or the particular sums of money used for reward, Arthur R. Cohen of New York University conducted a similar experiment in a different context. Cohen paid subjects to write essays advocating an opinion contrary to what

they really believed. Subjects were paid either $10, $5, $1 or 50 cents to do this. To measure the extent to which dissonance was reduced by their changing their opinion, each subject was then given a questionnaire, which he left unsigned, to determine his private opinion on the issue. The extent to which the subjects reduced dissonance by changing their opinion to agree with what they wrote in the essay is shown in the lower illustration on the preceding page. Once again it is clear that the smaller the original justification for engaging in the dissonance-producing action, the greater the subsequent change in private opinion to bring it into line with the action.

The final set of experiments I shall discuss deals with the consequences of resisting temptation. What happens when a person wants something and discovers that he cannot have it? Does he now want it even more or does he persuade himself that it is really not worth having? Sometimes our common general understanding of human behavior can provide at least crude answers to such questions. In this case, however, our common understanding is ambiguous, because it supplies two contradictory answers. Everyone knows the meaning of the term "sour grapes"; it is the attitude taken by a person who persuades himself that he really does not want what he cannot have. But we are also familiar with the opposite reaction. The child who is not allowed to eat candy and hence loves candy passionately; the woman who adores expensive clothes even though she cannot afford to own them; the man who has a hopeless obsession for a woman who spurns his attentions. Everyone "understands" the behavior of the person who longs for what he cannot have.

Obviously one cannot say one of these reactions is wrong and the other is right; they both occur. One might at least, however, try to answer the question: Under what circumstances does one reaction take place and not the other? If we examine the question from the point of view of the theory of dissonance, a partial answer begins to emerge.

Imagine the psychological situation that exists for an individual who is tempted to engage in a certain action but for one reason or another refrains. An analysis of the situation here reveals its similarity to the other dissonance-producing situations. An individual's knowledge concerning the attractive aspects of the activity toward which he was tempted is dissonant with the knowledge that he has refrained from engaging in the activity. Once more, of course, the individual has some knowledge that is consonant with his behavior in the situation. All the pressures, reasons and justifications for refraining are consonant with his actual behavior. Nevertheless, the dissonance does exist, and there will be psychological activity oriented toward reducing this dissonance.

As we have already seen in connection with other illustrations, one major way to reduce dissonance is to change one's opinions and evaluations in order to bring them closer in line with one's actual behavior. Therefore when there is

dissonance produced by resisting temptation, it can be reduced by derogating or devaluing the activity toward which one was tempted. This derivation from the theory clearly implies the sour-grapes attitude, but both theory and experiment tell us that such dissonance-reducing effects will occur only when there was insufficient original justification for the behavior. Where the original justification for refraining from the action was great, little dissonance would have occurred and there would have been correspondingly little change of opinion in order to reduce dissonance. Therefore one might expect that if a person had resisted temptation in a situation of strong prohibition or strong threatened punishment, little dissonance would have been created and one would not observe the sour-grapes effect. One would expect this effect only if the person resisted temptation under conditions of weak deterrent.

This line of reasoning leaves open the question of when the reverse effect occurs—that is, the situation in which desire for the "unattainable" object is increased. Experimentally it is possible to look at both effects. This was done by Elliot Aronson and Carlsmith, at Stanford University, in an experiment that sheds considerable light on the problem. The experiment was performed with children who were about four years old. Each child was individually brought into a large playroom in which there were five toys on a table. After the child had had an opportunity to play briefly with each toy, he was asked to rank the five in order of attractiveness. The toy that the child liked second best was then left on the table and the other four toys were spread around on the floor. The experimenter told the child that he had to leave for a few minutes to do an errand but would be back soon. The experimenter then left the room for 10 minutes. Various techniques were employed to "prohibit" the child from playing with the particular toy that he liked second best while the experimenter was out of the room.

For different children this prohibition was instituted in three different ways. In one condition there was no temptation at all; the experimenter told the child he could play with any of the toys in the room and then took the second-best toy with him when he left. In the other two conditions temptation was present: the second-best toy was left on the table in the experimenter's absence. The children were told they could play with any of the toys in the room except the one on the table. The children in one

*Temptation accompanied by a severe threat*

*Temptation accompanied by a mild threat*

group were threatened with mild punishment if they violated the prohibition, whereas those in the other group were threatened with more severe punishment. (The actual nature of the punishment was left unspecified.)

During his absence from the room the experimenter observed each child through a one-way mirror. None of the children in the temptation conditions played with the prohibited toy. After 10 minutes were up the experimenter returned to the playroom and each child was again allowed to play briefly with each of the five toys. The attractiveness of each toy for the child was again measured. By comparing the before and after measurements of the attractiveness of the toy the child originally liked second best, one can assess the effects of the prohibition. The results are shown in the chart at the bottom of this page.

When there was no temptation—that is, when the prohibited toy was not physically present—there was of course no dissonance, and the preponderant result is an increase in the attractiveness of the prohibited toy. When the temptation is present but the prohibition is enforced by means of a severe threat of punishment, there is likewise little dissonance created by refraining, and again the preponderant result is an increase in the attractiveness of the prohibited toy. In other words, it seems clear that a prohibition that is enforced in such a way as not to introduce dissonance results in a greater desire for the prohibited activity.

The results are quite different, however, when the prohibition is enforced by only a mild threat of punishment. Here we see the result to be expected from the theory of dissonance. Because the justification for refraining from playing with the toy is relatively weak, there is appreciable dissonance between the child's knowledge that the toy is attractive and his actual behavior. The tendency to reduce this dissonance is strong enough to more than overcome the effect apparent in the other two conditions. Here, as a result of dissonance reduction, we see an appreciable sour-grapes phenomenon.

The theory of cognitive dissonance obviously has many implications for everyday life. In addition to throwing light on one's own behavior, it would seem to carry useful lessons for everyone concerned with understanding human behavior in a world where everything is not black and white.

*Consequences of resisting temptation when deterrence varies*

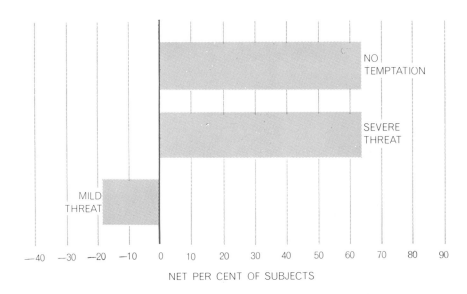

NET PER CENT OF SUBJECTS

CONSEQUENCES OF TEMPTATION were explored by prohibiting children from playing with a desirable toy. Later the children were asked to re-evaluate the attractiveness of the forbidden toy. In one case the prohibition was enforced by removing the toy from the child's presence. In the second case the prohibition took the form of a threat of severe punishment; in the third case, a threat of mild punishment. The chart shows the net per cent of children who thought the forbidden toy more attractive after the experiment than before. ("Net per cent" means the per cent who found the toy more attractive minus the per cent who found it less so.) Evidently only those threatened mildly experienced much dissonance, and they reduced it by downgrading toy's desirability. Others thought the toy more desirable.

# 12

## The Anthropology of Manners

EDWARD T. HALL, JR.

*April 1955*

*The Goops they lick their fingers*
*    and the Goops they lick their knives;*
*They spill their broth on the table cloth—*
*    Oh, they lead disgusting lives.*
*The Goops they talk while eating,*
*    and loud and fast they chew;*
*And that is why I'm glad that I*
*    am not a Goop—are you?*

In Gelett Burgess' classic on the Goops we have an example of what anthropologists call "an enculturating device"—a means of conditioning the young to life in our society. Having been taught the lesson of the goops from childhood (with or without the aid of Mr. Burgess) Americans are shocked when they go abroad and discover whole groups of people behaving like goops—eating with their fingers, making noises and talking while eating. When this happens, we may (1) remark on the barbarousness or quaintness of the "natives" (a term cordially disliked all over the world) or (2) try to discover the nature and meaning of the differences in behavior. One rather quickly discovers that what is good manners in one context may be bad in the next. It is to this point that I would like to address myself.

The subject of manners is complex; if it were not, there would not be so many injured feelings and so much misunderstanding in international circles everywhere. In any society the code of manners tends to sum up the culture—to be a frame of reference for all behavior. Emily Post goes so far as to say: "There is not a single thing that we do, or say, or choose, or use, or even think, that does not follow or break one of the exactions of taste, or tact, or ethics of good manners, or etiquette—call it what you will." Unfortunately many of the most important standards of acceptable behavior in different cultures are elusive: they are intangible, undefined and unwritten.

An Arab diplomat who recently arrived in the U. S. from the Middle East attended a banquet which lasted several hours. When it was over, he met a fellow countryman outside and suggested they go get something to eat, as he was starving. His friend, who had been in this country for some time, laughed and said: "But, Habib, didn't you know that if you say, 'No, thank you,' they think you really don't want any?" In an Arab country etiquette dictates that the person being served must refuse the proffered dish several times, while his host urges him repeatedly to partake. The other side of the coin is that Americans in the Middle East, until they learn better, stagger away from banquets having eaten more than they want or is good for them.

When a public-health movie of a baby being bathed in a bathinette was shown in India recently, the Indian women who saw it were visibly offended. They wondered how people could be so inhuman as to bathe a child in stagnant (not running) water. Americans in Iran soon learn not to indulge themselves in their penchant for chucking infants under the chin and remarking on the color of their eyes, for the mother has to pay to have the "evil eye" removed. We also learn that in the Middle East you don't hand people things with your left hand, because it is unclean. In India we learn not to touch another person, and in Southeast Asia we learn that the head is sacred.

In the interest of intercultural understanding various U. S. Government agencies have hired anthropologists from time to time as technical experts. The State Department especially has pioneered in the attempt to bring science to bear on this difficult and complex problem. It began by offering at the Foreign Service Institute an intensive four-week course for Point 4 technicians. Later these facilities were expanded to include other foreign service personnel.

The anthropologist's job here is not merely to call attention to obvious taboos or to coach people about types of thoughtless behavior that have very little to do with culture. One should not need an anthropologist to point out, for instance, that it is insulting to ask a foreigner: "How much is this in real money?" Where technical advice is most needed is in the interpretation of the unconscious aspects of a culture—the things people do automatically without being aware of the full implications of what they have done. For example, an ambassador who has been kept waiting for more than half an hour by a foreign visitor needs to understand that if his visitor "just mutters an apology" this is not necessarily an insult. The time system in the foreign country may be composed of different basic units, so that the visitor is not as late as he may appear to us. You must know the time system of the country to know at what point apologies are really due.

Twenty years of experience in working with Americans in foreign lands convinces me that the real problem in preparing them to work overseas is not with taboos, which they catch on to rather quickly, but rather with whole congeries of habits and attitudes which anthropologists have only recently begun to describe systematically.

Can you remember tying your shoes this morning? Could you give the rules for when it is proper to call another person by his first name? Could you describe the gestures you make in conversation? These examples illustrate how much of our behavior is "out of awareness," and how easy it is to get into trouble in another culture.

Nobody is continually aware of the quality of his own voice, the subtleties of stress and intonation that color the meaning of his words or the posture and distance he assumes in talking to another person. Yet all these are taken as cues to the real nature of an utterance, regardless of what the words say. A simple illustration is the meaning in the tone of voice. In the U. S. we raise our voices not only when we are angry but also when we want to emphasize a point, when we are more than a certain distance from another person, when we are concluding a meeting and so on. But to the Chinese, for instance, overloudness of the voice is most characteristically associated with anger and loss of self-control. Whenever we become really interested in something, they are apt to have the feeling we are angry, in spite

of many years' experience with us. Very likely most of their interviews with us, however cordial, seem to end on a sour note when we exclaim heartily: "WELL, I'M CERTAINLY GLAD YOU DROPPED IN, MR. WONG."

The Latin Americans, who as a rule take business seriously, do not understand our mixing business with informality and recreation. We like to put our feet up on the desk. If a stranger enters the office, we take our feet down. If it turns out that the stranger and we have a lot in common, up go the feet again—a cue to the other fellow that we feel at ease. If the office boy enters, the feet stay up; if the boss enters and our relationship with him is a little strained at the moment, they go down. To a Latin American this whole behavior is shocking. All he sees in it is insult or just plain rudeness.

Differences in attitudes toward space —what would be territoriality in lower forms of life—raise a number of other in-

teresting points. U. S. women who go to live in Latin America all complain about the "waste" of space in the houses. On the other hand, U. S. visitors to the Middle East complain about crowding, in the houses and on the street cars and buses. Everywhere we go space seems to be distorted. When we see a gardener in the mountains of Italy planting a single row on each of six separate terraces, we wonder why he spreads out his crop so that he has to spend half his time climbing up and down. We overlook the complex chain of communication that would be broken if he didn't cultivate alongside his brothers and his cousin and if he didn't pass his neighbors and talk to them as he moves from one terrace to the next.

A colleague of mine was caught in a snowstorm while traveling with companions in the mountains of Lebanon. They stopped at the next house and asked to be put up for the night. The

*The American in Latin America*

house had only one room. Instead of distributing the guests around the room, their host placed them next to the pallet where he slept with his wife—so close that they almost touched the couple. To have done otherwise in that country would have been unnatural and unfriendly. In the U. S. we distribute ourselves more evenly than many other people. We have strong feelings about touching and being crowded; in a streetcar, bus or elevator we draw ourselves in. Toward a person who relaxes and lets himself come into full contact with others in a crowded place we usually feel reactions that could not be printed on this page. It takes years for us to train our children not to crowd and lean on us. We tell them to stand up, that it is rude to slouch, not to sit so close or not to "breathe down our necks." After a while they get the point. By the time we Americans are in our teens we can tell what relationship exists between a man and woman by how they walk or sit together.

In Latin America, where touching is more common and the basic units of space seem to be smaller, the wide automobiles made in the U. S. pose problems.

People don't know where to sit. North Americans are disturbed by how close the Latin Americans stand when they converse. "Why do they have to get so close when they talk to you?" "They're so pushy." "I don't know what it is, but it's something in the way they stand next to you." And so on. The Latin Americans, for their part, complain that people in the U. S. are distant and cold—*retraídos* (withdrawing and uncommunicative).

An analysis of the handling of space during conversations shows the following: A U. S. male brought up in the Northeast stands 18 to 20 inches away when talking face to face to a man he does not know very well; talking to a woman under similar circumstances, he increases the distance about four inches. A distance of only eight to 13 inches between males is considered either very aggressive or indicative of a closeness of a type we do not ordinarily want to think about. Yet in many parts of Latin America and the Middle East distances which are almost sexual in connotation are the only ones at which people can talk comfortably. In Cuba, for instance, there is nothing suggestive in a man's talking to

an educated woman at a distance of 13 inches. If you are a Latin American, talking to a North American at the distance he insists on maintaining is like trying to talk across a room.

To get a more vivid idea of this problem of the comfortable distance, try starting a conversation with a person eight or 10 feet away or one separated from you by a wide obstruction in a store or other public place. Any normally enculturated person can't help trying to close up the space, even to the extent of climbing over benches or walking around tables to arrive within comfortable distance. U. S. businessmen working in Latin America try to prevent people from getting uncomfortably close by barricading themselves behind desks, typewriters or the like, but their Latin American office visitors will often climb up on desks or over chairs and put up with loss of dignity in order to establish a spatial context in which interaction can take place for them.

The interesting thing is that neither party is specifically aware of what is wrong when the distance is not right. They merely have vague feelings of discomfort or anxiety. As the Latin American approaches and the North American backs away, both parties take offense without knowing why. When a North American, having had the problem pointed out to him, permits the Latin American to get close enough, he will immediately notice that the latter seems much more at ease.

My own studies of space and time have engendered considerable cooperation and interest on the part of friends and colleagues. One case recently reported to me had to do with a group of seven-year-olds in a crowded Sunday-school classroom. The children kept fighting. Without knowing quite what was involved, the teacher had them moved to a larger room. The fighting stopped. It is interesting to speculate as to what would have happened had the children been moved to a smaller room.

The embarrassment about intimacy in space applies also to the matter of addressing people by name. Finding the proper distance in the use of names is even more difficult than in space, because the rules for first-naming are unbelievably complex. As a rule we tend to stay on the "mister" level too long with Latins and some others, but very often we swing into first naming too quickly, which amounts to talking down to them. Whereas in the U. S. we use Mr. with the surname, in Latin America the first and last names are used together

*The diplomat and the "native'*

and señor (Sr.) is a title. Thus when one says, "My name is Sr. So-and-So," it is interpreted to mean, "I am the Honorable, his Excellency So-and-So." It is no wonder that when we stand away, barricade ourselves behind our desks (usually a reflection of status) and call ourselves mister, our friends to the south wonder about our so-called "good-neighbor" policy and think of us as either high-hat or unbelievably rude. Fortunately most North Americans learn some of these things after living in Latin America for a while, but the aversion to being touched and to touching sometimes persists after 15 or more years of residence and even under such conditions as intermarriage.

The difference in sense of time is another thing of which we are not aware. An Iranian, for instance, is not taught that it is rude to be late in the same way that we in the U. S. are. In a general way we are conscious of this, but we fail to realize that their time system is structured differently from ours. The different cultures simply place different values on the time units.

Thus let us take as a typical case of the North European time system (which has regional variations) the situation in the urban eastern U. S. A middle-class business man meeting another of equivalent rank will ordinarily be aware of being two minutes early or late. If he is three minutes late, it will be noted as significant but usually neither will say anything. If four minutes late, he will mutter something by way of apology; at five minutes he will utter a full sentence of apology. In other words, the major unit is a five-minute block. Fifteen minutes is the smallest significant period for all sorts of arrangements and it is used very commonly. A half hour of course is very significant, and if you spend three quarters of an hour or an hour, either the business you transact or the relationship must be important. Normally it is an insult to keep a public figure or a person of significantly higher status than yourself waiting even two or three minutes, though the person of higher position can keep you waiting or even break an appointment.

Now among urban Arabs in the Eastern Mediterranean, to take an illustrative case of another time system, the unit that corresponds to our five-minute period is 15 minutes. Thus when an Arab arrives nearly 30 minutes after the set time, by his reckoning he isn't even "10 minutes" late yet (in our time units). Stated differently, the Arab's tardiness will not amount to one significant period (15 minutes in our system). An American

can normally will wait no longer than 30 minutes (two significant periods) for another person to turn up in the middle of the day. Thereby he often unwittingly insults people in the Middle East who want to be his friends.

How long is one expected to stay when making a duty call at a friend's house in the U. S.? While there are regional variations, I have observed that the minimum is very close to 45 minutes, even in the face of pressing commitments elsewhere, such as a roast in the oven. We may think we can get away in 30 minutes by saying something about only stopping for "a minute," but usually we discover that we don't feel comfortable about leaving until 45 minutes have elapsed. I am referring to afternoon social calls; evening calls last much longer and operate according to a different system. In Arab countries an American pay-

ing a duty call at the house of a desert sheik causes consternation if he gets up to leave after half a day. There a duty call lasts three days—the first day to prepare the feast, the second for the feast itself and the third to taper off and say farewell. In the first half day the sheik has barely had time to slaughter the sheep for the feast. The guest's departure would leave the host frustrated.

There is a well-known story of a tribesman who came to Kabul, the capital of Afghanistan, to meet his brother. Failing to find him, he asked the merchants in the market place to tell his brother where he could be found if the brother showed up. A year later the tribesman returned and looked again. It developed that he and his brother had agreed to meet in Kabul but had failed to specify what year! If the Afghan time system were structured similarly to our own, which it apparently is not, the brother would

*Western culture in the Middle East*

not offer a full sentence of apology until he was five years late.

Informal units of time such as "just a minute," ."a while," "later," "a long time," "a spell," "a long, long time," "years" and so on provide us with the culturological equivalent of Evil-Eye Fleegle's "double-whammy" (in *Li'l Abner*). Yet these expressions are not as imprecise as they seem. Any American who has worked in an office with someone else for six months can usually tell within five minutes when that person will be back if he says, "I'll be gone for a while." It is simply a matter of learning from experience the individual's system of time indicators. A reader who is interested in communications theory can fruitfully speculate for a while on the very wonderful way in which culture provides the means whereby the receiver puts back all the redundant material that

was stripped from such a message. Spelled out, the message might go somewhat as follows: "I am going downtown to see So-and-So about the Such-and-Such contract, but I don't know what the traffic conditions will be like or how long it will take me to get a place to park nor do I know what shape So-and-So will be in today, but taking all this into account I think I will be out of the office about an hour but don't like to commit myself, so if anyone calls you can say I'm not sure how long I will be; in any event I expect to be back before 4 o'clock."

Few of us realize how much we rely on built-in patterns to interpret messages of this sort. An Iranian friend of mine who came to live in the U. S. was hurt and puzzled for the first few years. The new friends he met and liked would say on parting: "Well, I'll see you later." He mournfully complained: "I kept ex-

pecting to see them, but the 'later' never came." Strangely enough we ourselves are exasperated when a Mexican can't tell us precisely what he means when he uses the expression *mañana*.

The role of the anthropologist in preparing people for service overseas is to open their eyes and sensitize them to the subtle qualities of behavior—tone of voice, gestures, space and time relationships—that so often build up feelings of frustration and hostility in other people with a different culture. Whether we are going to live in a particular foreign country or travel in many, we need a frame of reference that will enable us to observe and learn the significance of differences in manners. Progress is being made in this anthropological study, but it is also showing us how little is known about human behavior.

# 13

# Experiments in Group Conflict

MUZAFER SHERIF

*November 1956*

Conflict between groups—whether between boys' gangs, social classes, "races" or nations—has no simple cause, nor is mankind yet in sight of a cure. It is often rooted deep in personal, social, economic, religious and historical forces. Nevertheless it is possible to identify certain general factors which have a crucial influence on the attitude of any group toward others. Social scientists have long sought to bring these factors to light by studying what might be called the "natural history" of groups and group relations. Intergroup conflict and harmony is not a subject that lends itself easily to laboratory experiments. But in recent years there has been a beginning of attempts to investigate the problem under controlled yet lifelike conditions, and I shall report here the results of a program of experimental studies of groups which I started in 1948. Among the persons working with me

were Marvin B. Sussman, Robert Huntington, O. J. Harvey, B. Jack White, William R. Hood and Carolyn W. Sherif. The experiments were conducted in 1949, 1953 and 1954; this article gives a composite of the findings.

We wanted to conduct our study with groups of the informal type, where group organization and attitudes would evolve naturally and spontaneously, without formal direction or external pressures. For this purpose we conceived that an isolated summer camp would make a good experimental setting, and that decision led us to choose as subjects boys about 11 or 12 years old, who would find camping natural and fascinating. Since our aim was to study the development of group relations among these boys under carefully controlled conditions, with as little interference as possible from personal neuroses, background influences or prior experiences, we selected

normal boys of homogeneous background who did not know one another before they came to the camp.

They were picked by a long and thorough procedure. We interviewed each boy's family, teachers and school officials, studied his school and medical records, obtained his scores on personality tests and observed him in his classes and at play with his schoolmates. With all this information we were able to assure ourselves that the boys chosen were of like kind and background: all were healthy, socially well-adjusted, somewhat above average in intelligence and from stable, white, Protestant, middle-class homes.

None of the boys was aware that he was part of an experiment on group relations. The investigators appeared as a regular camp staff—camp directors, counselors and so on. The boys met one another for the first time in buses that

**MEMBERS OF ONE GROUP** of boys raid the bunkhouse of another group during the first experiment of the author and his associates, performed at a summer camp in Connecticut. The rivalry of the groups was intensified by the artificial separation of their goals.

took them to the camp, and so far as they knew it was a normal summer of camping. To keep the situation as lifelike as possible, we conducted all our experiments within the framework of regular camp activities and games. We set up projects which were so interesting and attractive that the boys plunged into them enthusiastically without suspecting that they might be test situations. Unobtrusively we made records of their behavior, even using "candid" cameras and microphones when feasible.

We began by observing how the boys became a coherent group. The first of our camps was conducted in the hills of northern Connecticut in the summer of 1949. When the boys arrived, they were all housed at first in one large bunkhouse. As was to be expected, they quickly formed particular friendships and chose buddies. We had deliberately put all the boys together in this expectation, because we wanted to see what would happen later after the boys were separated into different groups. Our object was to reduce the factor of personal attraction in the formation of groups. In a few days we divided the boys into two groups and put them in different cabins. Before doing so, we asked each boy informally who his best friends were, and then took pains to place the "best friends" in different groups so far as possible. (The pain of separation was assuaged by allowing each group to go at once on a hike and camp-out.)

As everyone knows, a group of strangers brought together in some common activity soon acquires an informal and spontaneous kind of organization. It comes to look upon some members as leaders, divides up duties, adopts unwritten norms of behavior, develops an *esprit de corps*. Our boys followed this pattern as they shared a series of experiences. In each group the boys pooled their efforts, organized duties and divided up tasks in work and play. Different individuals assumed different responsibilities. One boy excelled in cooking. Another led in athletics. Others, though not outstanding in any one skill, could be counted on to pitch in and do their level best in anything the group attempted. One or two seemed to disrupt activities, to start teasing at the wrong moment or offer useless suggestions. A few boys consistently had good suggestions and showed ability to coordinate the efforts of others in carrying them through. Within a few days one person had proved himself more resourceful and skillful than the rest. Thus, rather quick-

MEMBERS OF BOTH GROUPS collaborate in common enterprises during the second experiment, performed at a summer camp in Oklahoma. At the top the boys of the two groups prepare a meal. In the middle the two groups surround a water tank while trying to solve a water-shortage problem. At the bottom the members of one group entertain the other.

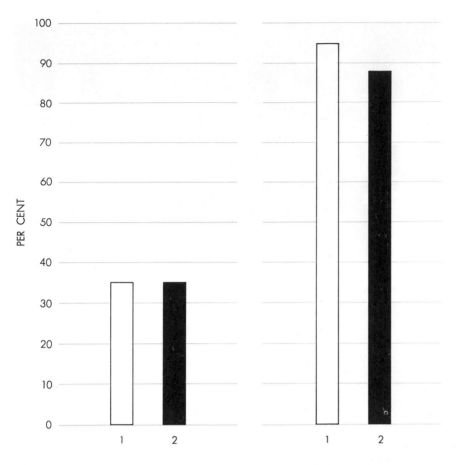

**FRIENDSHIP CHOICES** of campers for others in their own cabin are shown for Red Devils (*white*) and Bulldogs (*black*). At first a low percentage of friendships were in the cabin group (*left*). After five days, most friendship choices were within the group (*right*).

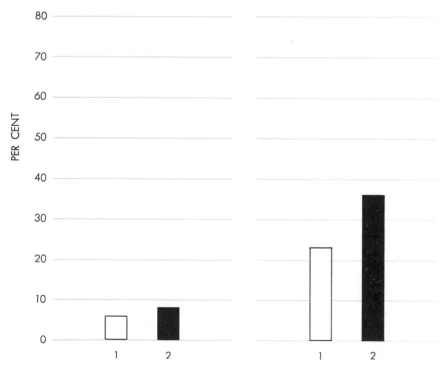

**DURING CONFLICT** between the two groups in the Robber's Cave experiment there were few friendships between cabins (*left*). After cooperation toward common goals had restored good feelings, the number of friendships between groups rose significantly (*right*).

ly, a leader and lieutenants emerged. Some boys sifted toward the bottom of the heap, while others jockeyed for higher positions.

We watched these developments closely and rated the boys' relative positions in the group, not only on the basis of our own observations but also by informal sounding of the boys' opinions as to who got things started, who got things done, who could be counted on to support group activities.

As the group became an organization, the boys coined nicknames. The big, blond, hardy leader of one group was dubbed "Baby Face" by his admiring followers. A boy with a rather long head became "Lemon Head." Each group developed its own jargon, special jokes, secrets and special ways of performing tasks. One group, after killing a snake near a place where it had gone to swim, named the place "Moccasin Creek" and thereafter preferred this swimming hole to any other, though there were better ones nearby.

Wayward members who failed to do things "right" or who did not contribute their bit to the common effort found themselves receiving the "silent treatment," ridicule or even threats. Each group selected symbols and a name, and they had these put on their caps and T-shirts. The 1954 camp was conducted in Oklahoma, near a famous hideaway of Jesse James called Robber's Cave. The two groups of boys at this camp named themselves the Rattlers and the Eagles.

Our conclusions on every phase of the study were based on a variety of observations, rather than on any single method. For example, we devised a game to test the boys' evaluations of one another. Before an important baseball game, we set up a target board for the boys to throw at, on the pretense of making practice for the game more interesting. There were no marks on the front of the board for the boys to judge objectively how close the ball came to a bull's-eye, but, unknown to them, the board was wired to flashing lights behind so that an observer could see exactly where the ball hit. We found that the boys consistently overestimated the performances by the most highly regarded members of their group and underestimated the scores of those of low social standing.

The attitudes of group members were even more dramatically illustrated during a cook-out in the woods. The staff supplied the boys with unprepared food and let them cook it themselves. One boy promptly started to build a fire, asking

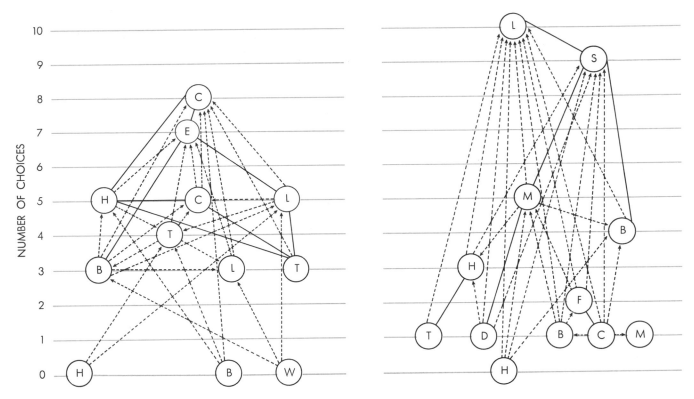

SOCIOGRAMS represent patterns of friendship choices within the fully developed groups. One-way friendships are indicated by broken arrows; reciprocated friendships, by solid lines. Leaders were among those highest in the popularity scale. Bulldogs (*left*) had a close-knit organization with good group spirit. Low-ranking members participated less in the life of the group but were not rejected. Red Devils (*right*) lost the tournament of games between the groups. They had less group unity and were sharply stratified.

for help in getting wood. Another attacked the raw hamburger to make patties. Others prepared a place to put buns, relishes and the like. Two mixed soft drinks from flavoring and sugar. One boy who stood around without helping was told by the others to "get to it." Shortly the fire was blazing and the cook had hamburgers sizzling. Two boys distributed them as rapidly as they became edible. Soon it was time for the watermelon. A low-ranking member of the group took a knife and started toward the melon. Some of the boys protested. The most highly regarded boy in the group took over the knife, saying, "You guys who yell the loudest get yours last."

When the two groups in the camp had developed group organization and spirit, we proceeded to the experimental studies of intergroup relations. The groups had had no previous encounters; indeed, in the 1954 camp at Robber's Cave the two groups came in separate buses and were kept apart while each acquired a group feeling.

Our working hypothesis was that when two groups have conflicting aims— *i.e.*, when one can achieve its ends only at the expense of the other—their members will become hostile to each other

even though the groups are composed of normal well-adjusted individuals. There is a corollary to this assumption which we shall consider later. To produce friction between the groups of boys we arranged a tournament of games: baseball, touch football, a tug-of-war, a treasure hunt and so on. The tournament started in a spirit of good sportsmanship. But as it progressed good feeling soon evaporated. The members of each group began to call their rivals "stinkers," "sneaks" and "cheaters." They refused to have anything more to do with individuals in the opposing group. The boys in the 1949 camp turned against buddies whom they had chosen as "best friends" when they first arrived at the camp. A large proportion of the boys in each group gave negative ratings to all the boys in the other. The rival groups made threatening posters and planned raids, collecting secret hoards of green apples for ammunition. In the Robber's Cave camp the Eagles, after a defeat in a tournament game, burned a banner left behind by the Rattlers; the next morning the Rattlers seized the Eagles' flag when they arrived on the athletic field. From that time on name-calling, scuffles and raids were the rule of the day.

Within each group, of course, solidar-

ity increased. There were changes: one group deposed its leader because he could not "take it" in the contests with the adversary; another group overnight made something of a hero of a big boy who had previously been regarded as a bully. But morale and cooperativeness within the group became stronger. It is noteworthy that this heightening of cooperativeness and generally democratic behavior did not carry over to the group's relations with other groups.

We now turned to the other side of the problem: How can two groups in conflict be brought into harmony? We first undertook to test the theory that pleasant social contacts between members of conflicting groups will reduce friction between them. In the 1954 camp we brought the hostile Rattlers and Eagles together for social events: going to the movies, eating in the same dining room and so on. But far from reducing conflict, these situations only served as opportunities for the rival groups to berate and attack each other. In the dining-hall line they shoved each other aside, and the group that lost the contest for the head of the line shouted "Ladies first!" at the winner. They threw paper, food and vile names at each other at the tables. An Eagle bumped by a Rattler was admonished by his fellow Eagles to brush "the dirt" off his clothes.

We then returned to the corollary of our assumption about the creation of conflict. Just as competition generates friction, working in a common endeavor should promote harmony. It seemed to us, considering group relations in the everyday world, that where harmony between groups is established, the most decisive factor is the existence of "superordinate" goals which have a compelling appeal for both but which neither could achieve without the other. To test this hypothesis experimentally, we created a series of urgent, and natural, situations which challenged our boys.

One was a breakdown in the water supply. Water came to our camp in pipes from a tank about a mile away. We arranged to interrupt it and then called the boys together to inform them of the crisis. Both groups promptly volunteered to search the water line for the trouble. They worked together harmoniously, and before the end of the afternoon they had located and corrected the difficulty.

A similar opportunity offered itself when the boys requested a movie. We told them that the camp could not afford to rent one. The two groups then got together, figured out how much each group would have to contribute, chose the film by a vote and enjoyed the showing together.

One day the two groups went on an outing at a lake some distance away. A large truck was to go to town for food. But when everyone was hungry and ready to eat, it developed that the truck would not start (we had taken care of that). The boys got a rope—the same rope they had used in their acrimonious tug-of-war—and all pulled together to start the truck.

These joint efforts did not immediately dispel hostility. At first the groups returned to the old bickering and name-calling as soon as the job in hand was finished. But gradually the series of cooperative acts reduced friction and conflict. The members of the two groups began to feel more friendly to each other. For example, a Rattler whom the Eagles disliked for his sharp tongue and skill in defeating them became a "good egg."

The boys stopped shoving in the meal line. They no longer called each other names, and sat together at the table. New friendships developed between individuals in the two groups.

In the end the groups were actively seeking opportunities to mingle, to entertain and "treat" each other. They decided to hold a joint campfire. They took turns presenting skits and songs. Members of both groups requested that they go home together on the same bus, rather than on the separate buses in which they had come. On the way the bus stopped for refreshments. One group still had five dollars which they had won as a prize in a contest. They decided to spend this sum on refreshments. On their own initiative they invited their former rivals to be their guests for malted milks.

Our interviews with the boys confirmed this change. From choosing their "best friends" almost exclusively in their own group, many of them shifted to listing boys in the other group as best friends [see lower chart on page 72]. They were glad to have a second chance to rate boys in the other group, some of them remarking that they had changed their minds since the first rating made after the tournament. Indeed they had. The new ratings were largely favorable [see chart on this page].

E fforts to reduce friction and prejudice between groups in our society have usually followed rather different methods. Much attention has been given to bringing members of hostile groups together socially, to communicating accurate and favorable information about one group to the other, and to bringing the leaders of groups together to enlist their influence. But as everyone knows, such measures sometimes reduce intergroup tensions and sometimes do not. Social contacts, as our experiments demonstrated, may only serve as occasions for intensifying conflict. Favorable information about a disliked group may be ignored or reinterpreted to fit stereotyped notions about the group. Leaders cannot act without regard for the prevailing temper in their own groups.

What our limited experiments have shown is that the possibilities for achieving harmony are greatly enhanced when groups are brought together to work toward common ends. Then favorable information about a disliked group is seen in a new light, and leaders are in a position to take bolder steps toward cooperation. In short, hostility gives way when groups pull together to achieve overriding goals which are real and compelling to all concerned.

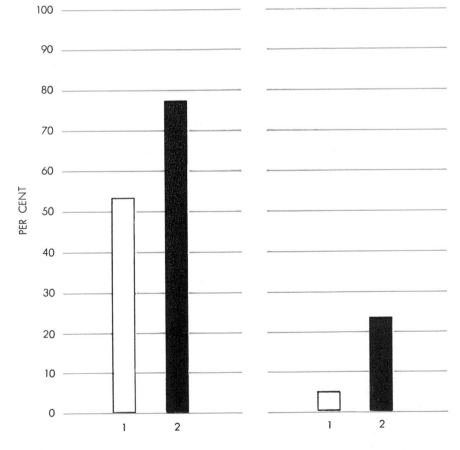

NEGATIVE RATINGS of each group by the other were common during the period of conflict (*left*) but decreased when harmony was restored (*right*). The graphs show percent who thought that *all* (rather than *some* or *none*) of the other group were cheaters, sneaks, etc.

# 14

# The Effects of Observing Violence

LEONARD BERKOWITZ

*February 1964*

An ancient view of drama is that the action on the stage provides the spectators with an opportunity to release their own strong emotions harmlessly through identification with the people and events depicted in the play. This idea dates back at least as far as Aristotle, who wrote in *The Art of Poetry* that drama is "a representation . . . in the form of actions directly presented, not narrated; with incidents arousing pity and fear in such a way as to accomplish a purgation of such emotions."

Aristotle's concept of catharsis, a term derived from the Greek word for purgation, has survived in modern times. It can be heard on one side of the running debate over whether or not scenes of violence in motion pictures and television programs can instigate violent deeds, sooner or later, by people who observe such scenes. Eminent authorities contend that filmed violence, far from leading to real violence, can actually have beneficial results in that the viewer may purge himself of hostile impulses by watching other people behave aggressively, even if these people are merely actors appearing on a screen. On the other hand, authorities of equal stature contend that, as one psychiatrist told a Senate subcommittee, filmed violence is a "preparatory school for delinquency." In this view emotionally immature individuals can be seriously affected by fighting or brutality in films, and disturbed young people in particular can be led into the habit of expressing their aggressive energies by socially destructive actions.

Until recently neither of these arguments had the support of data obtained by controlled experimentation; they had to be regarded, therefore, as hypotheses, supported at best by unsystematic observation. Lately, however, several psychologists have undertaken laboratory tests of the effects of filmed aggression. The greater control obtained in these tests, some of which were done in my laboratory at the University of Wisconsin with the support of the National Science Foundation, provides a basis for some statements that have a fair probability of standing up under continued testing.

First, it is possible to suggest that the observation of aggression is more likely to induce hostile behavior than to drain off aggressive inclinations; that, in fact, motion picture or television violence can stimulate aggressive actions by normal people as well as by those who are emotionally disturbed. I would add an important qualification: such actions by normal people will occur only under appropriate conditions. The experiments point to some of the conditions that might result in aggressive actions by people in an audience who had observed filmed violence.

Second, these findings have obvious social significance. Third, the laboratory tests provide some important information about aggressive behavior in general. I shall discuss these three statements in turn.

Catharsis appeared to have occurred in one of the first experiments, conducted by Seymour Feshbach of the University of Colorado. Feshbach deliberately angered a group of college men; then he showed part of the group a filmed prizefight and the other students a more neutral film. He found that the students who saw the prizefight exhibited less hostility than the other students on two tests of aggressiveness administered after the film showings. The findings may indicate that the students who had watched the prizefight had vented their anger vicariously.

That, of course, is not the only possible explanation of the results. The men who saw the filmed violence could have become uneasy about their own aggressive tendencies. Watching someone being hurt may have made them think that aggressive behavior was wrong; as a result they may have inhibited their hostile responses. Clearly there was scope for further experimentation, particularly studies varying the attitude of the subjects toward the filmed aggression.

Suppose the audience were put in a frame of mind to regard the film violence as justified—for instance because a villain got a beating he deserved. The concept of symbolic catharsis would predict in such a case that an angered person might enter vicariously into the scene and work off his anger by thinking of himself as the winning fighter, who was inflicting injury on the man who had provoked him. Instead of accepting

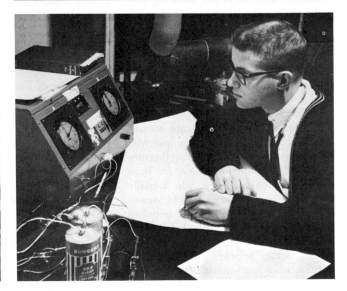

TYPICAL EXPERIMENT tests reaction of angered man to filmed violence. Experiment begins with introduction of subject (*white shirt*) to a man he believes is a co-worker but who actually is a confederate of the author's. In keeping with pretense that experiment is to test physiological reactions, student conducting the experiment takes blood-pressure readings. He assigns the men a task and leaves; during the task the confederate insults the subject. Experimenter returns and shows filmed prizefight. Confederate leaves; experimenter tells subject to judge a floor plan drawn by confederate and to record opinion by giving confederate electric shocks. Shocks actually go to recording apparatus. The fight film appeared to stimulate the aggressiveness of angered men.

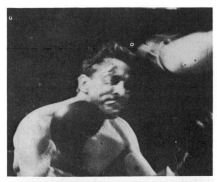

FILMED AGGRESSION shown in author's experiments was from the motion picture *Champion* and included these scenes in which Kirk Douglas receives a bad beating. Watchers had been variously prepared; after showing they were tested for aggressive tendencies.

this thesis, my associates and I predicted that justified film aggression would lead to stronger rather than weaker manifestations of hostility. We believed that the rather low volume of open hostility in the Feshbach experiment was attributable to film-induced inhibitions. If this were so, an angered person who saw what appeared to be warranted aggression might well think he was justified in expressing his own hostile desires.

To test this hypothesis we conducted three experiments. Since they resulted in essentially similar findings and employed comparable procedures, I shall describe only the latest. In this experiment we brought together two male college students at a time. One of them was the subject; the other was a confederate of the experimenter and had been coached on how to act, although of course none of this was known to the subject. Sometimes we introduced the confederate to the subject as a college boxer and at other times we identified him as a speech major. After the introduction the experimenter announced that the purpose of the experiment was to study physiological reactions to various tasks. In keeping with that motif he took blood-pressure readings from each man. Then he set the pair to work on the first task: a simple intelligence test. During this task the confederate either deliberately insulted the subject—

for example, by remarks to the effect that "You're certainly taking a long time with that" and references to "cow-college students" at Wisconsin—or, in the conditions where we were not trying to anger the subject, behaved in a neutral manner toward him. On the completion of the task the experimenter took more blood-pressure readings (again only to keep up the pretense that the experiment had a physiological purpose) and then informed the men that their next assignment was to watch a brief motion picture scene. He added that he would give them a synopsis of the plot so that they would have a better understanding of the scene. Actually he was equipped with two different synopses.

To half of the subjects he portrayed the protagonist of the film, who was to receive a serious beating, as an unprincipled scoundrel. Our idea was that the subjects told this story would regard the beating as retribution for the protagonist's misdeeds; some tests we administered in connection with the experiment showed that the subjects indeed had little sympathy for the protagonist. We called the situation we had created with this synopsis of the seven-minute fight scene the "justified fantasy aggression."

The other subjects were given a more favorable description of the protagonist. He had behaved badly, they were told, but this was because he had been vic-

timized when he was young; at any rate, he was now about to turn over a new leaf. Our idea was that the men in this group would feel sympathetic toward the protagonist; again tests indicated that they did. We called this situation the "less justified fantasy aggression."

Then we presented the film, which was from the movie *Champion;* the seven-minute section we used showed Kirk Douglas, as the champion, apparently losing his title. Thereafter, in order to measure the effects of the film, we provided the subjects with an opportunity to show aggression in circumstances where that would be a socially acceptable response. We separated each subject and accomplice and told the subject that his co-worker (the confederate) was to devise a "creative" floor plan for a dwelling, which the subject would judge. If the subject thought the floor plan was highly creative, he was to give the co-worker one electric shock by depressing a telegraph key. If he thought the floor plan was poor, he was to administer more than one shock; the worse the floor plan, the greater the number of shocks. Actually each subject received the same floor plan.

The results consistently showed a greater volume of aggression directed against the anger-arousing confederate by the men who had seen the "bad guy" take a beating than by the men who had been led to feel sympathy for the protagonist in the film [*see the illustration on page* 78]. It was clear that the people who saw the justified movie violence had not discharged their anger through vicarious participation in the aggression but instead had felt freer to attack their tormentor in the next room. The motion picture scene had apparently influenced their judgment of the propriety of aggression. If it was all right for the movie villain to be injured aggressively, they seemed to think, then perhaps it was all right for them to attack the villain in their own lives—the person who had insulted them.

Another of our experiments similarly demonstrated that observed aggression has little if any effectiveness in reducing aggressive tendencies on the part of an observer. In this experiment some angered men were told by another student how many shocks they should give the person, supposedly in the next room, who had provoked them. Another group of angered men, instead of delivering the shocks themselves, watched the other student do it. Later the members of both groups had an opportunity to deliver the shocks personally. Consistently the men who had watched in the first part of the experiment now displayed stronger aggression than did the people who had been able to administer shocks earlier. Witnessed aggression appeared to have been less satisfying than self-performed aggression.

Our experiments thus cast considerable doubt on the possibility of a cathartic purge of anger through the observation of filmed violence. At the very least, the findings indicated that such a catharsis does not occur as readily as many authorities have thought.

Yet what about the undoubted fact that aggressive motion pictures and violent athletic contests provide relaxation and enjoyment for some people? A person who was tense with anger sometimes comes away from them feeling calmer. It seems to me that what happens here is quite simple: He calms down not because he has discharged his anger vicariously but because he was carried away by the events he witnessed. Not thinking of his troubles, he ceased to stir himself up and his anger dissipated. In addition, the enjoyable motion

picture or game could have cast a pleasant glow over his whole outlook, at least temporarily.

The social implications of our experiments have to do primarily with the moral usually taught by films. Supervising agencies in the motion picture and television industries generally insist that films convey the idea that "crime does not pay." If there is any consistent principle used by these agencies to regulate how punishment should be administered to the screen villain, it would seem to be the talion law: an eye for an eye, a tooth for a tooth.

Presumably the audience finds this concept of retaliation emotionally satisfying. Indeed, we based our "justified fantasy aggression" situation on the concept that people seem to approve of hurting a scoundrel who has hurt others. But however satisfying the talion principle may be, screenplays based on it can lead to socially harmful consequences. If the criminal or "bad guy" is punished aggressively, so that others do to him what he has done to them,

the violence appears justified. Inherent in the likelihood that the audience will regard it as justified is the danger that some angered person in the audience will attack someone who has frustrated *him*, or perhaps even some innocent person he happens to associate with the source of his anger.

Several experiments have lent support to this hypothesis. O. Ivar Lövaas of the University of Washington found in an experiment with nursery school children that the youngsters who had been exposed to an aggressive cartoon film displayed more aggressive responses with a toy immediately afterward than a control group shown a less aggressive film did. In another study Albert Bandura and his colleagues at Stanford University noted that preschool children who witnessed the actions of an aggressive adult in a motion picture tended later, after they had been subjected to mild frustrations, to imitate the kind of hostile behavior they had seen.

This tendency of filmed violence to stimulate aggression is not limited to children. Richard H. Walters of the Uni-

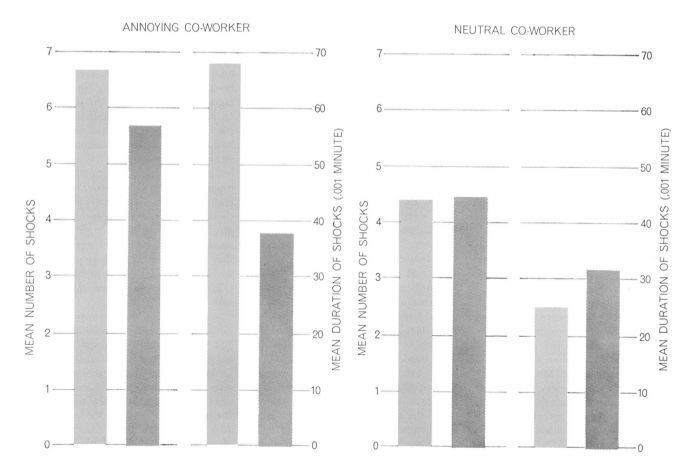

**RESPONSES OF SUBJECTS** invited to commit aggression after seeing prizefight film varied according to synopsis they heard beforehand. One (*colored bars*) called Douglas' beating deserved; the other (*gray bars*) said it was undeserved. After the film the subjects were told they could give electric shocks to an annoying or neutral co-worker based on his "creativeness" in doing a task. Seeing a man receive what had been described as a well-deserved beating apparently lowered restraints against aggressive behavior.

versity of Waterloo in Ontario found experimentally that male hospital attendants who had been shown a movie of a knife fight generally administered more severe punishment to another person soon afterward than did other attendants who had seen a more innocuous movie. The men in this experiment were shown one of the two movie scenes and then served for what was supposedly a study of the effects of punishment. They were to give an electric shock to someone else in the room with them each time the person made a mistake on a learning task. The intensity of the electric shocks could be varied. This other person, who was actually the experimenter's confederate, made a constant number of mistakes, but the people who had seen the knife fight gave him more intense punishment than the men who had witnessed the nonaggressive film. The filmed violence had apparently aroused aggressive tendencies in the men and, since the situation allowed the expression of aggression, their tendencies were readily translated into severe aggressive actions.

These experiments, taken together with our findings, suggest a change in approach to the manner in which screenplays make their moral point. Although it may be socially desirable for a villain to receive his just deserts at the end of a motion picture, it would seem equally desirable that this retribution should not take the form of physical aggression.

The key point to be made about aggressiveness on the basis of experimentation in this area is that a person's hostile tendencies will persist, in spite of any satisfaction he may derive from filmed violence, to the extent that his frustrations and aggressive habits persist. There is no free-floating aggressive energy that can be released through a wide range of different activities. A drive to hurt cannot be reduced through attempts to master other drives, as Freud proposed, or by observing others as they act aggressively.

In fact, there have been studies suggesting that even if the angered person performs the aggression himself, his hostile inclinations are not satisfied unless he believes he has attacked his tormentor and not someone else. J. E. Hokanson of Florida State University has shown that angered subjects permitted to commit aggression against the person who had annoyed them often display a drop in systolic blood pressure. They seem to have experienced a physiological relaxation, as if they had satisfied their aggressive urges. Systolic pressure declines less, however, when the angered people carry out the identical motor activity involved in the aggression but without believing they have attacked the source of their frustration.

I must now qualify some of the observations I have made. Many aggressive motion pictures and television programs have been presented to the public, but the number of aggressive incidents demonstrably attributable to such shows is quite low. One explanation for this is that most social situations, unlike the conditions in the experiments I have described, impose constraints on aggression. People are usually aware of the social norms prohibiting attacks on others, consequently they inhibit whatever hostile inclinations might have been aroused by the violent films they have just seen.

Another important factor is the attributes of the people encountered by a person after he has viewed filmed violence. A man who is emotionally aroused does not necessarily attack just anyone. Rather, his aggression is directed toward specific objectives. In other words, only certain people are capable of drawing aggressive responses from him. In my theoretical analyses of the sources of aggressive behavior I have suggested that the arousal of anger only creates a readiness for aggression. The theory holds that whether or not this predisposition is translated into actual aggression depends on the presence of appropriate cues: stimuli associated with the present or previous instigators of anger. Thus if someone has been insulted, the sight or the thought of others who have provoked him, whether then or earlier, may evoke hostile responses from him.

An experiment I conducted in conjunction with a graduate student provides some support for this train of thought. People who had been deliberately provoked by the experimenter were put to work with two other people, one a person who had angered them earlier and the other a neutral person. The subjects showed the greatest hostility, following their frustration by the experimenter, to the co-worker they disliked. He, by having thwarted them previously, had acquired the stimulus quality that caused him to draw aggression from them after they had been aroused by the experimenter.

My general line of reasoning leads me to some predictions about aggressive behavior. In the absence of any strong inhibitions against aggression, people who have recently been angered and have then seen filmed aggression will be more likely to act aggressively than people who have not had those experiences. Moreover, their strongest attacks will be directed at those who are most directly connected with the provocation or at others who either have close associations with the aggressive motion picture or are disliked for any reason.

One of our experiments showed results consistent with this analysis. In this study male college students, taken separately, were first either angered or not angered by A, one of the two graduate

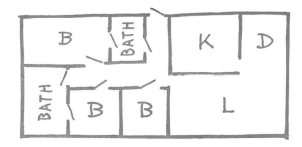

| SUBJECT | A | B | C | D | E | F | G | H |
|---|---|---|---|---|---|---|---|---|
| NUMBER OF SHOCKS | 6 | 3 | 8 | 3 | 6 | 7 | 5 | 4 |
| DURATION (.001 MINUTE) | 46 | 38 | 76 | 10 | 120 | 49 | 60 | 28 |

**TASK BY ANNOYING CO-WORKER** supposedly was to draw a floor plan. Actually each subject saw the floor plan shown here. The subject was asked to judge the creativeness of the plan and to record his opinion by pressing a telegraph key that he thought would give electric shocks to the co-worker: one shock for a good job and more for poor work. Responses of eight subjects who saw prizefight film are shown; those in color represent men told that Douglas deserved his beating; those in black, men informed it was undeserved.

ANGERED SUBJECTS

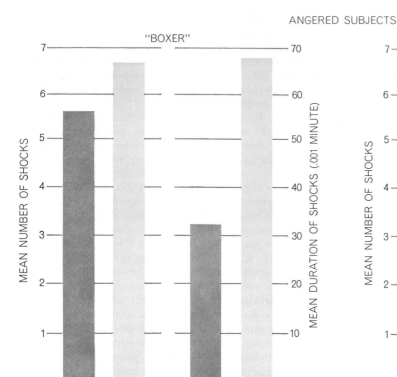

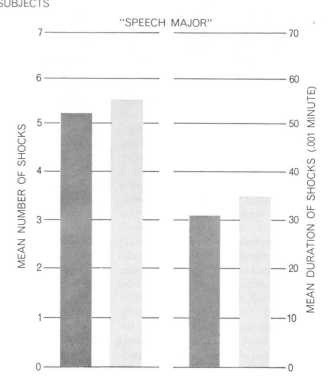

**CO-WORKER'S INTRODUCTION** also produced variations in aggressiveness of subjects. Co-worker was introduced as boxer or as speech major; reactions shown here are of men who were angered by co-worker and then saw either a fight film (*colored bars*) or a neutral film (*gray bars*). Co-worker received strongest attacks when subjects presumably associated him with fight film.

UNANGERED SUBJECTS

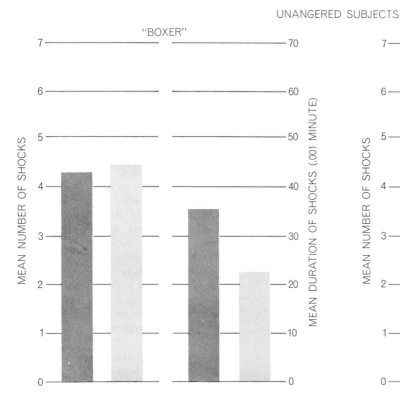

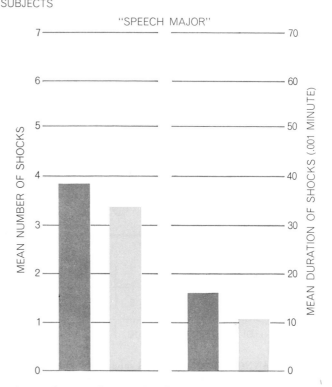

**SIMILAR TEST,** varied by the fact that the co-worker behaved neutrally toward the subjects and therefore presumably did not anger them, produced these reactions. The greater number of shocks given to the co-worker introduced as a boxer than to the one introduced as a speech major apparently reflected a tendency to take a generally negative attitude toward persons identified as boxers.

students acting as experimenters. *A* had been introduced earlier either as a college boxer or as a speech major. After *A* had had his session with the subject, *B*, the second experimenter, showed the subject a motion picture: either the prizefight scene mentioned earlier or a neutral film. (One that we used was about canal boats in England; the other, about the travels of Marco Polo.)

We hypothesized that the label "college boxer" applied to *A* in some of the cases would produce a strong association in the subject's mind between *A* and the boxing film. In other words, any aggressive tendencies aroused in the subject would be more likely to be directed at *A* the college boxer than at *A* the speech major. The experiment bore out this hypothesis. Using questionnaires at the end of the session as the measures of hostility, we found that the deliberately angered subjects directed more hostility at *A*, the source of their anger, when they had seen the fight film and he had been identified as a boxer. Angered men who had seen the neutral film showed no particular hostility to *A* the boxer. In short, the insulting experimenter received the strongest verbal attacks when he was also associated with the aggressive film. It is also noteworthy that in this study the boxing film did not influence the amount of hostility shown toward *A* when he had not provoked the subjects.

A somewhat inconsistent note was introduced by our experiments, described previously, in "physiological reactions." Here the nonangered groups, regardless of which film they saw, gave the confederate more and longer shocks when

they thought he was a boxer than when they understood him to be a speech major [*see bottom illustration on opposite page*]. To explain this finding I assume that our subjects had a negative attitude toward boxers in general. This attitude may have given the confederate playing the role of boxer the stimulus quality that caused him to draw aggression from the angered subjects. But it could only have been partially responsible, since the insulted subjects who saw the neutral film gave fewer shocks to the boxer than did the insulted subjects who saw the prizefight film.

Associations between the screen and the real world are important. People seem to be emotionally affected by a screenplay to the extent that they associate the events of the drama with their own life experiences. Probably adults are less strongly influenced than children because they are aware that the film is make-believe and so can dissociate it from their own lives. Still, it seems clear from the experiments I have described that an aggressive film can induce aggressive actions by anyone in the audience. In most instances I would expect that effect to be short-lived. The emotional reaction produced by filmed violence probably dies away rather rapidly as the viewer enters new situations and encounters new stimuli. Subjected to different influences, he becomes less and less ready to attack other people.

Television and motion pictures, however, may also have some persistent effects. If a young child sees repeatedly that screen heroes gain their ends

through aggressive actions, he may conclude that aggression is desirable behavior. Fortunately screenplays do not consistently convey that message, and in any event the child is exposed to many other cultural norms that discourage aggression.

As I see it, the major social danger inherent in filmed violence has to do with the temporary effects produced in a fairly short period immediately following the film. For that period, at least, a person—whether an adult or a child —who had just seen filmed violence might conclude that he was warranted in attacking those people in his own life who had recently frustrated him. Further, the film might activate his aggressive habits so that for the period of which I speak he would be primed to act aggressively. Should he then encounter people with appropriate stimulus qualities, people he dislikes or connects psychologically with the film, this predisposition could lead to open aggression.

What, then, of catharsis? I would not deny that it exists. Nor would I reject the argument that a frustrated person can enjoy fantasy aggression because he sees characters doing things he wishes he could do, although in most cases his inhibitions restrain him. I believe, however, that effective catharsis occurs only when an angered person perceives that his frustrater has been aggressively injured. From this I argue that filmed violence is potentially dangerous. The motion picture aggression has increased the chance that an angry person, and possibly other people as well, will attack someone else.

# Commentary

Leonard Berkowitz' reference to Aristotle's theory of catharsis ["The Effects of Observing Violence," by Leonard Berkowitz] is entirely unnecessary for his discussion of the effects of the spectacle of violence, and as he uses it, it makes no sense historically or psychologically.

Aristotle is talking precisely about "tragic" feelings, which are constrictive states of the organism: to open out foreboding and sadness to felt fear and grief.

He is not referring at all to expansive feelings such as anger or lust; on the contrary, these would be *excited* by battle songs or comedy. In the context of Greek civic therapy it is inconceivable that public poets would want to reduce the aggressive or the lustful, although they might well want to diminish guilt and mourning....

The most important current context for Professor Berkowitz' violence hypoth-

esis is in designing pacifist films, and I agree with him that showing violence, the brutality and horrors of war, probably has an exacerbating rather than a pacifist effect. (For the psychology, see "Designing Pacifist Films" in my *Utopian Essays*.)

Regarded simply as abreaction, letting off steam, a spectacle arousing fear and grief is different from one arousing rage, since a spectator in a theater can actually physically be chilled to the bone or have a good cry but he cannot actually kick or punch anybody. There is stimulation without discharge; no wonder Professor Berkowitz' subjects are left touchy. He should set up his experiment to allow a physical outlet—punching a bag—then he might find a cathartic effect.

But certainly Professor Berkowitz must know that Aristotle's catharsis is not a doctrine of mere abreaction. Nearly the entire *Poetics* is devoted to the management of the plot to the proper resolution so that the aroused feelings can be reintegrated. In the history of interpretation of the text, catharsis has been called "idealization," "transcendence," "poetic justice," "reconciliation" and so on. The emotions are neutralized when there is some new awareness. From this point of view Aristotle himself would agree with Professor Berkowitz' empirical observation of diminishing of aggression against the "good guy"; but this need not be merely superego inhibition. The difference is between Professor Berkowitz' stereotyped fictions, which can result only in conformist inhibition, and genuine poetry that gives insight and compassion, for example the much-remarked saddening and pacifist effect of the *Iliad,* in contrast to most war stories.

Allow me a last observation. We must remember that with regard to psychological, pedagogic and political matters, the ancients had as much empirical evidence as we do—they too lived in cities, brought up children, quarreled, practiced rhetoric. Generally too they had a simpler vision than we do. It is extremely unlikely that, on matters such as the effects of a good cry or of showing violence, they would be as far off base as Professor Berkowitz thinks.

PAUL GOODMAN

Institute for Policy Studies
Washington, D.C.
June, 1964

---

A number of readers have taken me to task for oversimplifying Aristotle's discussion of catharsis. My apologies if I did him an injustice. But when it comes to the matter of considering the possible effects of media violence, Aristotle's exact analysis is somewhat less important than the view generally attributed to him; authorities *have* quoted him in asserting the presumed social benefits of aggressive scenes, and it is this assertion (oversimplified though it may be) to which part of the article was addressed.

One other point about Aristotle's analysis. I believe that the "new awareness" resulting from the use of "genuine poetry" or "art" is not more likely to lead to a "drainage" of anger than is ordinary melodrama or mere reportage (as some have put it). Here we have to be careful in distinguishing between catharsis, a reduction in some emotional state, and a decreased probability of *overt* aggression due to inhibitions. Insight or self-awareness can arouse inhibitions ("I should not attack this person; I would behave the same way if I were in his place"). Compassion can arouse inhibitions ("He is acting in that awful manner only because of the way he has been treated or because he is suffering"). Indeed, we could say that the less justified movie-aggression condition produced compassion toward Kirk Douglas in just this way, and independent evidence indicates that inhibitions were aroused in the subjects feeling the compassion. And similarly, poetry or art that makes the audience conscious of the ideals of our society may also arouse inhibitions; our ideals proscribe certain forms of behavior as "bad" and thus not to be carried out. Contrary to what Mr. Goodman says, then, the diminishing of aggression against the "good guy" because of compassion for him, or because the audience is poetically reminded that all men are brothers, may well be the result of inhibitions....

To get on with the discussion, I am not certain what sort of film Mr. Goodman has in mind when he talks about movies showing the brutality and horrors of war. If the audience regards the violence as brutal and horrible, however, its effect probably would be to diminish the likelihood of aggression rather than have an exacerbating effect. The audience, in a sense, is made extremely aware of the harmful consequences of this behavior and inhibitions against this behavior are aroused....

Finally, as for the matter of movies providing stimulation without an opportunity for discharge, this is beside the point. The adherents of the symbolic-catharsis notion have contended that aggressive movies *in themselves* produce the release, not some activity afterward....

LEONARD BERKOWITZ

University of Wisconsin
Madison, Wis.
June, 1964

# III

## POPULATION AND HETEROGENEITY

# III

## Population and Heterogeneity

### *Introduction*

T. R. Malthus, who started it all with his *Essay on Population* in 1798, said that all the controls of population can be subsumed under two heads: misery and vice. By misery he meant principally disease, starvation, and the ills resulting from homelessness. Vice included war, social conflict, and (as he later concluded) contraception. His population analysis was, of course, based on the human situation.

Students of animal populations have come to much the same conclusions, with certain necessary modifications. As is characteristic of controversial matters, there has been a tendency for many students to take a somewhat unbalanced view and emphasize one controlling factor to the near exclusion of others. Special pleading is all too common. However, by reading widely, it is easy to arrive at a balance that individual proponents may not have. It is clear that not only starvation, but also predators and parasites are controllers of population. In different populations, at different times, one or another may be predominant. The essential aspect of a controller of population is that it is *density-dependent*. That is, the effect of such a factor must increase disproportionately with increase in size of the population.

V. C. Wynne-Edwards, in his "Population Control in Animals," describes the working of a population-dependent factor that was not sufficiently emphasized until recently. As the author says, "the threat of starvation tomorrow, not hunger itself today," is often the effective agent in triggering control measures. Simple starvation is a retroactive (and painful) control agent. Many species, however, effectively avoid such retroactive control by a control that can be called anticipatory. Among human beings anticipatory control can be, and should be, completely conscious.

Among nonhuman animals, however, there is no reason to think that a population control measure of this sort is conscious; it is merely a logical consequence of instinctive behavior. Territoriality is the best known of the behavioral characteristics that keep a population in check. Male song birds, by being programmed to demand a minimum acreage of territory, so divide the resources of their world as to insure that none of the breeders is in serious danger of starving to death. They do this not by bloody battles fought to the finish, but by the symbolic aggression of their songs. In the words of Wynne-Edwards, they compete for conventional prizes (acreage) by conventional methods (bird song). The key word is "conventional"; like any word that is too much used, the immediate content of this word has been degraded; we must refurbish this word, restoring its rich primitive meaning, if we are to understand the significance of conventional maneuvers in population control.

The conventions of all animals but man are instinctually determined. If human populations are to be controlled by awarding conventional prizes through conventional methods, we must ask ourselves what conventions we want to create. How will we validate them—that is, make their acceptance obligatory on all members of our society? When there are infractions—that is, when individual members of society seek to change the rules of the game—how shall we meet this challenge?

And, in general, what *should* happen to the losers in a conventional conflict anyway?

In his article on "Population" the sociologist Kingsley Davis brings us to the controversial human area. The problem is dangerous for various reasons, one of which is increasing diversity in the world. The differences between poor countries and rich countries are now greater than they ever were at any other period in man's history, and they are still increasing. This is surely a threat to international stability.

One evidence of incipient instability can be found in the purely verbal sphere. Because of a decent, though possibly misplaced, consideration for the feelings of the citizens of poor countries we have become trapped in an infinite regress of euphemisms. In an earlier day the inhabitants of the wealthy countries did not hesitate to speak of their poor neighbors as "savage," "primitive," "backward," or "poor." Now such terms are considered insulting. Over the past two decades they have been replaced by a succession of unstable but well-meant euphemisms: "underdeveloped countries," "less well developed," "emerging," "less privileged," "have-not," "catch-up," "low-income," "needy," "poorest third," "recipient," "expectant," and "developing." The average half-life of such euphemisms seems to be about two years. Perhaps it is time to use a simple descriptive term, like "poor countries," and let the citizens of those countries take care of their psychotherapy in their own way.

Putting nomenclature to one side, a most disturbing fact about our efforts to help the poor countries has been that we have had to learn all over again the ancient wisdom that "the road to Hell is paved with good intentions." We have learned that, of all "good" actions, exporting public health techniques to poor countries is the most productive of evil. In the past we comforted ourselves and dispelled our doubts about these activities by assuming that the poor countries of the world would simply repeat the history of Western Europe, though at a faster pace. The graph on page 109 reveals that this comforting assumption cannot be true. With the best intentions in the world we have laid the groundwork for massive tragedy in the poor countries, a tragedy the like of which was never experienced in Europe. Bitterly, we remember Oscar Wilde's words: "The people who do most harm are the people who try to do most good."

In the past century we have made the world somewhat nearer to the heart's desire by moving ever closer to a complete conquest of disease. But in doing so we neglected to take into account that disease is a marvelously efficient density-dependent population controller. If we rid ourselves of this agent, what new density-dependent agent must we accept? One possibility is given in John B. Calhoun's "Population Density and Social Pathology." A particular meaning of Malthus' general term "vice" is shown in the particular context of a rat population. Some critics have rightly protested the too-facile extrapolation of these results to the human situation. People are not rats, and no doubt have different psychological reactions to overcrowding. Granted: but if human populations are not to be governed by the particular corrective feedback devices found in Calhoun's rat population, by what density-dependent feedbacks are they to be controlled? If we want to control our destiny at all, we cannot avoid facing this question.

In the course of the past two centuries scientists have been preoccupied, successively, with: *matter, energy,* and now *information.* When Malthus wrote his essay, scientists were concerned largely with determining the properties of matter; it is no wonder, then, that Malthus saw the limit of population in terms of "means of subsistence," by which he generally meant food. By the latter part of the nineteenth century scientific interest had shifted to energy, and nutritionists and students of population began to view the population problem in terms of the capture and the distribution of energy in the form of calorie-rich food. By the middle of the twentieth century the dissemination of information had become the great frontier of science, and a careful reading of Calhoun's paper shows that it is an overload of information, rather than a shortage of matter or energy, that is involved in limiting Calhoun's population.

Information is a function of communication between organisms. The matter or energy that is transferred from one to another is of trivial importance; it is the meaning of that which is transferred that is of overriding significance. A squeal, a look, an odor, or a shove can all convey information—information that means something to the receiving organism, and sets in train a series of innate or learned reactions. Once we enter the realm of information theory an important and novel principle alters our view of the significance of population density, a principle that illustrates the danger of assuming that what is true in one field of study is true in another.

It is axiomatic in economic theory that almost any manufacturing process enjoys the benefits of "economies of scale." The unit cost of automobiles goes down markedly as we increase the rate of production from a thousand units per year to a million. The more complex the machine, the greater the economies of scale. So well has Western man learned this lesson that he unthinkingly assumes that all growth is good. Quite unconsciously, he considers population problems almost solely in terms of matter and energy. But when we are dealing with information a different principle arises. We can think of information as being the product of a network of elements joined by lines of relationship. With two elements, there is one relationship; with three elements, there are three relationships; with four, there are six; with five, there are ten, and so on. The general formula for the number of relationships is given thus:

$$1 + 2 + 3 + \ldots + (n-1) + n = \frac{n(n-1)}{2}.$$

When $n$ becomes very large we can (with only a little loss of accuracy) say that the number of relationships is proportionate to $n^2$. Therefore, when we are confronted with problems of communication and information we encounter *dis*economies of scale. The larger the group communicating, the greater the need for communication. Attempting to know the same relative proportion of what goes on in an expanding world that was known about what went on in a static one leads to a very real threat of "information overload." In the most general terms, this is what Calhoun's rats suffered from.

Though the reactions to information overload are no doubt different in humans (and no doubt different at different times in history) it is hard to find any reason to doubt that the ultimate limit of human population will be determined more by the obscure phenomenon of information overload than it will be by the more easily definable problems posed by matter and energy.

Admitting this brings us to the world of "Cybernetics," here described by the originator of the term and a pioneer in the development of the discipline, Norbert Wiener. The basic ideas are described by Wiener, and in the following paper Arnold Tustin elaborates on the meanings of the concept "Feedback." Tustin's article describes the basic differences between feedback that produces normal results and feedback that produces pathological results. In the control of populations in nature we can find examples of both, the lemmings being a classical instance of the latter. As we move toward a day in which human populations will be deliberately controlled to achieve a zero growth rate, we need to keep the general characteristics of feedback control in mind if we are to avoid setting up pathological and painful oscillations.

There are two contrasting ways of viewing persistent evils:

As isolated events, each an effect of a cause.
As the stable state of a system with negative (corrective) feedbacks that restore the system to its setpoint after each disturbance.

The first view has almost universally characterized the attitude of the administrators of our poverty programs until very recently. As one sociologist said: "There is nothing wrong with poverty that money won't cure." This sounds self-evidently true; but according to Oscar Lewis it is subtly false. "The Culture of Poverty" constitutes a poorly delineated, but tolerable, cybernetic system. Disturb such a system by pouring in money and you merely create prosperous poor people (momentarily prosperous, that is).

Lewis' viewpoint is a controversial one; but there is much anecdotal material, at least, to back it up. Hundreds of millions of dollars have been poured into urban slums in the United States in the belief that slums were the cause of poverty. The failure of these attempts is now undisputed. Something more is required. It is by no means true that there is a perfect correlation between the amount of poverty and the evils of poverty. The poverty of primitive people is quite commonly far greater than that found in an American slum—but it is generally unaccompanied by a culture of poverty. The *ghetto*—that popular and much misused word of our day—in its historical origins had a high esprit de corps, much poverty but no culture of poverty.

What are we to do about our communities that maintain such a culture? This is not an easy question to answer. While acknowledging that "it is easier to praise poverty than to live in it," Lewis insists that the culture of poverty has positive values. Perhaps a more convincing testimony is that of the movie actress Sophia Loren. Recalling her childhood days, she said:

The roots of my life are in Naples and in the neighboring town of Pozzuoli where I grew up. A guidebook describes this area as being one of the most squalid slums in all Italy, but it never seemed that way to me. Life is open in Naples. There is vitality in Naples—warmth and comfort in the streets. In Naples you have a feeling that all humanity is very close and loving.

The bittersweet root-paradox of reform is that to be successful, reform must (as Lewis says) "effectively destroy the psychological and social core of the culture of poverty." Bluntly put: to save a people (in one sense) the reformer must be willing to destroy them (in another). The reformer, like the successful surgeon, must have the moral strength to cut decisively. This means he must know he is right in his values. (Heaven grant that he is.) A civilization that is in doubt about its cultural values is not likely to produce successful reformers.

Lewis estimates that of the 50 million Americans who could be called poor about 20 per cent live in a culture of poverty. This raises an interesting question for which there is not, at present, even the ghost of an answer: comparing those who are "culturally" poor with those who are *merely* poor, which group is reproducing at a greater rate? Numerous variables are involved: birth rates and death rates in both groups as well as rates of out-migration. This is clearly a crucial question, if you believe that the culture of poverty exists. If you do not, it doesn't matter.

One of the many ways in which society witlessly works to keep the poor impoverished is by maintaining class discrimination in birth control. No single method of birth control is perfect; when it fails, a woman may seek abortion as a back-stop method. A contraceptive failure rate of only one per cent would produce 250,000 unwanted pregnancies annually in the United States. Under good medical conditions an early abortion is only one-sixth as dangerous as a normal childbirth, so there is no medical reason for forbidding it. For historical reasons, however, legislative prohibition of abortion has been almost universal in the United States. In spite of the laws, middle- and upper-class women can get competently performed abortions. Poor women have to find hack operators, or do the operation themselves—or bear more unwanted children, to their sorrow and society's expense. The poor are made poorer by compulsory pregnancy, a particularly clearcut example of the destructive effects of positive feedback.

Abortion, now the subject of much concern throughout the world, was first effectively brought to the American public's attention by the account of the 1955 Arden House Conference, published in 1958. *Scientific America's* review of this book, by James R. Newman, was the only review to appear in any newsstand magazine, so strong was the taboo against the word "abortion" in the '50s. Within six years every major magazine in the country was discussing the problem.

In 1969, some ten years after Newman's review, the first full-length article on abortion appeared in *Scientific American.* The medical statisticians Christopher Tietze and Sarah Lewit reviewed the known facts and indicated the gaps in our knowledge. Although a considerable change in public attitudes in the United States took place in the short span of two years, 1965 to 1967, only 25 percent of the population approved abortion if the couple could not afford another child. Since this is the major reason for wanting abortion, such an attitude bodes ill for poor women for some time to come.

Although many positive feedback mechanisms tend to increase the poverty of the poor, their plight is not, in principle, beyond remedy. William Mangin, in "Squatter Settlements," presents a case study that illustrates Lewis' point that poverty can exist without creating an enervating culture of poverty. Under certain conditions poverty and external opposition can even create a magnificent esprit de corps, bringing about the development of men and women who are capable of climbing the socio-economic ladder. The North American tradition has always said this could happen, but we seem lately to have forgotten the tradition. Can the lesson we relearn from Latin America be applied to contemporary problems in the United States? Does the deliberate fostering of a sense of being abused diminish the ability of the abler poor to rise above their origins? Tough question!

In "The Renewal of Cities," Nathan Glazer recounts some horrendous tales of past failures brought about by unintelligent interference with the metabolism of cities. It is axiomatic in cybernetics that ill-informed interference with a feedback system may actually aggravate the disease one is trying to cure. The social system of the city in a free enterprise system is a sort of cybernetic system. In an attempt to change it only a little bit, a federally financed program of "urban renewal" was set up in the United States in the 1950's. It was particularly aimed at improving the lot of the urban Negro—which it spectacularly failed to do. "Urban renewal" was soon cynically referred to as "Negro removal."

An important influence in creating our misconceptions of the city is what we might call "the tyranny of the camera"—the belief that the problem can be photographed. The photographs on facing pages 164 and 165 will serve as a suitable text to illustrate the point. A minor aspect of the tyranny of the camera is simple lying. Notice that the "before" picture was taken on a day when smog was heavy; the "after" picture was judiciously photographed on a smog-free day. No one has asserted that the building program altered the smog level of the city; but the viewer of the paired photographs may unconsciously be led to infer this.

More fundamental than the outright lying represented by biased photography is the fact that what is really important about living cannot be photographed. We can easily photograph the broken bricks and patched roofs of condemned homes before their destruction, but the camera cannot record the broken lines of mutual responsibility and community feeling resulting from the destruction of a long-existing web of social relationships among the dispossessed citizenry. The "after" buildings look neat on the outside—but what about the lives of the people inside? Cultural relations are complex and easily damaged; to these, the camera is blind. Choices in government housing are overwhelmingly dictated by what takes a good photograph.

Certain fundamentally important aspects of the evolution of cities are criminally neglected in housing programs. As Glazer points out, some of the housing projects have spent as much as a million dollars per acre (of the taxpayer's money) to secure "suitable" land. There is an unspoken assumption that the price of land must never go down, but always up. Yet it is precisely cheap land that is required to house the poor, the aged, the pensioners, the Bohemians, the unambitious, and the over-large families. Since the people who stand to gain by a rise in the price of land are the influential members of society, and those who would gain by a fall in its price are the already dispossessed all the forces of urban renewal work in the same dysfunctional direction.

No one ever considers the possibility that a city has a natural optimum size, beyond which growth is pathological and cancer-like. Recent studies indicate that smoke and fumes cost the average Manhattan family an added $850 a year in cleaning bills, hair-dos, and medical expenses. In 1968 it took more than $10,000 a year for a family of four to live a "moderate" existence in New York, as compared with less than $6,000 in 1961. The rise in cost was distinctly greater than that for the country as a whole. To a biologist, such facts suggest pathology. If a city like New York has already passed the optimum size, enabling it to grow still larger by pouring in federal funds to do what cannot be done by private enterprise is sending good money after bad. If there is some fundamental reason for wanting all people to live in cities perhaps this can be defended. But no one has answered, or even asked, this fundamental question.

The problem of unlimited city growth, bad as it is everywhere in the world, is made worse in the United States by being interlocked with another problem, the unresolved problem of Negroes and their role in society. As several students have pointed out, the urban problem is predominantly a Negro problem. Negroes, once almost entirely rural, have now moved into the cities, where, economically weak and discriminated against, they have concentrated in the decaying centers. Wealthier whites, seeking to preserve their way of life, have migrated steadily outward toward the periphery, to the suburbs. This evolution of the city seems good to neither blacks nor whites. But how can it be stopped? What mechanisms can we devise to oppose the present evolution without introducing some other processes that may be worse?

In "The Social Power of the Negro" James P. Comer describes how the problem looks from inside the black community. Though money is needed to combat the present situation, money is not enough and does not by itself get to the heart of the problem. It is the "life style" of the blacks that needs changing. To be more powerful, blacks need to develop a greater sense of independence. For whites who would like to help, this raises a curious problem, one that is related to problems of parenthood. As his child grows up, a wise parent wants to encourage him to be independent of his parents. In what terms shall this encouragement be couched? Suppose a father says: "Don't do as I say—make up your own mind. Be independent!" In speaking explicitly in this fashion a parent unwittingly threatens his child with what Gregory Bateson has called the "double bind." After hearing this instruction, the child cannot win. If he continues taking his cues from his father, he is being dependent. On the other hand, if he makes up his own mind, then by virtue of the fact that he is obeying his father's instruction to do so, he is still being dependent. He's damned if he does and damned if he doesn't. This paradox is part of the Oedipus situation.

Something like this holds for the racial situation. By virtue of their possession of most of the social and economic power, whites stand in a quasi-parental relation to the blacks. Those who want to help the blacks are tempted to draw up blueprints to show them how to achieve independence. But this introduces the double bind, of which sensitive blacks are keenly aware. Avoiding the double bind in race relationships will take patience, subtlety, and understanding, even as it does in a parental relationship.

Any attempt to do something about the plight of the inner city runs head-on into the problem of human variety. The confrontation is generally not explicit—and therein lies a significant portion of the trouble. Because the subject of human differences operates under a taboo, different people with different unexamined assumptions reach different conclusions. Since the assumptions and arguments are unexamined and implicit, each of the antagonists views the practical conclusions of his opponents as being based on irrational grounds. No early resolution of the difficulties is in sight. The best that we can hope to do here is to present two papers which bear on the problem from strikingly different angles, leaving the reader to try to reconcile the contradictory trains of thought thereby started.

Before discussing Anthony Allison's paper the stage needs to be set. What are we to say about human differences? What do they imply about the social order? Several stages in the history of thought on these questions can be distinguished. The earliest was that which we know by the name "eugenics," as it was conceived around the year 1900. In this view, people could be arranged along a continuum, from inferior to superior. With the discovery of the mechanism of heredity, there was a strong tendency to attribute the greater part of the differences to heredity, that is, to genes. The eugenic ideal held that we should reward the able and ignore the failures—or at least try to see to it that the failures did not breed as much as the able. Genes were either good or bad, and we should discourage the bad from increasing.

The blood groups and their genetic basis were discovered about 1900. By about 1930 it seemed clear that the blood groups were without adaptive significance. There was no evidence that it made any difference whether you were of blood group A, B, AB, or O. The various races, ethnic groups, tribes—call them what you will—had different proportions of the blood group genes. Since there was no evidence that one allele was any better than its alternative, it was widely accepted that blood group genes were "neutral genes." Mankind was said to be polymorphic with respect to these genes—to exhibit *polymorphism.*

Then the sickle cell gene was discovered. Polymorphism existed here too: a person could be either homozygous for the normal gene, heterozygous for the sickle cell gene, or homozygous for the sickle cell gene. Homozygosity for the sickle cell gene had a deleterious effect: it produced the near-lethal condition of sickle cell anemia. The sufferer from this seldom survived to maturity, and, in any case, almost never reproduced. Clearly, such a gene should be rapidly eliminated from all populations—but it wasn't. In some places in Africa it was actually a very common gene. This was a puzzle. At first, it was suggested that the gene enjoyed a high mutation rate. This seemed unlikely, and in fact is not so. The true explanation is given in Allison's "Sickle Cells and Evolution." In a malarial environment the heterozygote is superior to either homozygote. Such heterozygote superiority necessarily leads to a quasi-stable situation, with selective losses to both forms of homozygosity in every generation.

The sickle predicament is a very special one but it has general significance. As Allison later said in another publication:

> The great value of abnormal haemoglobin studies has been that they have changed the whole climate of opinion about genetically-determined characters in man. We now expect polymorphic characters to have selective value, and are often disappointed when there is no indication as to the nature of the selective agencies operating on them.

In the light of this finding the so-called neutral genes were examined again, and much more closely. By using very sensitive statistical tests it was possible to show that a person of blood group A was more likely to suffer from stomach cancer than were members of the other blood groups, and that people of blood group O were more likely to have peptic ulcers. The statistical differences are slight, and not enough to affect the practical decisions of anyone belonging to either of these groups. But they are enough to allow natural selection to exert its effect.

A striking change of professional opinion rapidly took place. William C. Boyd, one of the leaders in the study of the "neutral" genes, in 1959 introduced a panel discussion on "Selective Factors in the ABO Polymorphism" with the following words.

> Ladies and gentlemen, you see before you a reformed character. I formerly believed, and even used it as an argument for the use of blood groups in physical anthropology, that the blood group genes were selectively neutral. In taking this attitude I was doubtless influenced by certain physical anthropologists who had been arguing that human classification should be based on "non-adaptive" characters, as being the least likely to be changed by the action of evolutionary forces. I have since come to believe that there probably are no neutral genes and that classification has to be based on genes which have been affected by the forces which bring about racial differentiation, mainly mutation and selection. (In fairness it should be remarked that some of the anthropologists also changed their minds.) I am, therefore, a former believer in neutral genes who has recanted . . .

With the disappearance of the assumption of neutral genes, every polymorphism is seen as a problem crying for a solution. (To summarize our position at the present time it is enough to say that we still have a great sufficiency of unsolved problems!) The analysis of the sickle cell case re-emphasized a truth that had long been known, but which is constantly in danger of being forgotten, namely: *"fitness" is a function of the environment.* The sickle cell trait, miserable as it seems to us, is actually superior in a malarious environment where the benefits of modern medicines are unknown. By analogy, the cancer-predisposing gene A is presumably superior in some environment, as is also the ulcer-predisposing gene O. The wildly polymorphic nature of human populations would seem, then, to be indicative of the great variety of environments in which human beings live. That the fine-scale environment of humans does exhibit tremendous variety is something we have always known—but which we are inclined to forget unless reminded of it periodically.

The complementary relationship of genetic variety and environmental variety leads to difficult problems that we have not yet tackled on the theoretical level, much less the practical level. When a particular variant of the environment disappears, owing to the "march of history," what happens—what *should* happen— to the genetic variant that was best suited to that environment? If we were as unconscious of ourselves, and as little given to planning and interference as other animals, the inappropriate genetic variant would simply reproduce at a slower rate than the others and would finally disappear. Put bluntly, Nature's response to the complaints of the loser in the competitive battle is simply this: "Drop dead!"

That we are unwilling to accept so brutal an answer is not surprising. Under the accelerating drive of historical change we seem to be creating new environments at a rate that is much faster than the rate at which new genotypes can be assembled in quantity. In a sense, we are all misfits. It is no wonder that we take seriously the Biblical advice, "Judge not, that ye be not judged." It is by no means clear what we should do about the lack of coupling of genetic variance and environmental variance. If, following the advice of eugenicists, we were to control the breeding of human beings deliberately and rigidly, how should we control it? Should we try to determine the number of individuals needed to fill each human environment, and then breed precisely that number of individuals of the proper genotype? Should we allow *laissez-faire* in breeding, and then expand each environment to match the number of each relevant genotype that happened to be produced? Should the design of society be determined by a "battle of the cradles"?

Or should the quantitative mix of the various environments be determined by some other criteria, and breeding be controlled to produce the necessary variants in the proper numbers?

Or should we just say, "To Hell with it all"?

Perhaps we would like to do nothing—but that is never one of our options. Things will happen, though we close our eyes ever so tight. People of similar genotypes will come together, either because they are forced together by others, or because they seek togetherness for the psychological comfort that it gives. There are veritable libraries of books dealing with the problems associated with human variety, but there is not even the beginning of a fundamental approach to their solution.

Genetic complexity and environmental complexity are interrelated in subtle and poorly understood ways, and it is for this reason that most professional geneticists have very little to say in public about racial questions. They would rather hold their counsel until they know more—perhaps for several decades. Of particular advice they have little to give at the moment. In a very general way, they would like to remind others of these principles:

1. There are no neutral genes.
2. Selection is inescapable.
3. Different environments select for different genotypes.
4. For human beings, the total environment comprises not only the physical environment but also the social and psychological environments, both of which may be much the more important under present-day conditions.
5. Every change in social and political structure undoubtedly alters selective conditions, though generally in unknown ways.
6. Rapid evolution is possible.
7. "Man makes himself," by setting up his own selective criteria.

The items listed above hardly constitute answers to our pressing human problems. They are at best broad generalizations posed in a global, and not very helpful, way. But they must not be forgotten as we continue to struggle with the fantastically difficult problems posed by human diversity and our psychological, social, and political reactions to the facts of diversity.

Differences between individuals have a hereditary component. This broad generalization is certainly true, but it is so broad as to be almost useless. To sharpen its application to particular cases we would need to test different individuals under identical conditions. Any differences in performance that might then develop we could attribute to heredity. Unfortunately, "identical conditions" are more of a dream than a reality. Decade by decade we become increasingly more impressed with the subtlety of environmental influences, and with the difficulty of controlling them in such a way as to create the genuinely "identical conditions" that are so desirable for definitive testing.

Robert Rosenthal and Lenore F. Jacobson tell how "Teacher Expectations for the Disadvantaged" are themselves part of the environment of each student. Under rigorously controlled experimental conditions they determined that the "self-fulfilling prophecy" is a reality and constitutes an important factor in education. This demonstration is rich in practical implications, so rich as to be frightening to those whose ideal it is to give every child the best possible start in life. It is obvious that we will fall far short of this goal under the crowded conditions of overpopulation. It is also probable that falling short will activate a positive feedback system in which the disadvantage of one generation will carry on to augment the disadvantage of the next.

The study by Rosenthal and Jacobson is a splendid one which validates assumptions we had long suspected to be correct but for which we did not have "hard data." For this, we must all be grateful. And yet there is an aspect of this study that disturbs many people. This is the question of the ethics of experimentation in the behavioral sciences. Because of the self-fulfilling prophecy an experimenter has a heavy responsibility for what he does to the subjects of his experiments. Falsifying the records of students in order to create the impression that they were better students than they actually were, raised no serious ethical issue. No parent would object to his child being "given a leg up."

But what of the children whose records were not so falsified and who consequently were put at a competitive disadvantage in the experimental classroom? What if some parent maintained that his child's future had been permanently damaged by the experiment, and cited the experimental results in evidence? How would the experimenters answer this charge in a court of law?

Furthermore, to mention a matter that is not quite so serious, what of the teachers involved in such an experiment? For the experiment to work, the teachers had to be deceived. What was their reaction when they later learned of the deceit? What will be their response to a future behavioral scientist who seeks to obtain their cooperation in another experiment?

These are hard questions. The behavioral sciences labor under difficulties that are unknown to the natural sciences.

# 15

# Population Control in Animals

V. C. WYNNE-EDWARDS

*August 1964*

In population growth the human species is conspicuously out of line with the rest of the animal kingdom. Man is almost alone in showing a long-term upward trend in numbers; most other animals maintain their population size at a fairly constant level. To be sure, many of them fluctuate in number from season to season, from year to year or from decade to decade; notable examples are arctic lemmings, migratory locusts living in the subtropical dry belt, many northern game birds and certain fur-bearing animals. Such fluctuations, however, tend to swing erratically around a constant average value. More commonly animal populations maintain a steady state year after year and even century after century. If and when the population does rise or fall permanently, because of some change in the environment, it generally stabilizes again at a new level.

This well-established fact of population dynamics deserves to be studied with close attention, because the growth of human populations has become in recent years a matter of increasing concern. What sort of mechanism is responsible for such strict control of the size of populations? Each animal population, apart from man's, seems to be regulated in a homeostatic manner by some system that tends to keep it within not too wide limits of a set average density. Ecologists have been seeking to discover the nature of this system for many years. I shall outline here a new hypothesis that I set forth in full detail in a recently published book, *Animal Dispersion in Relation to Social Behaviour.*

The prevailing hypothesis has been that population is regulated by a set of negative natural controls. It is assumed that animals will produce young as fast as they efficiently can, and that the main factors that keep population

density within fixed limits are predators, starvation, accidents and parasites causing disease. On the face of it this assumption seems entirely reasonable; overcrowding should increase the death toll from most of these factors and thus act to cut back the population when it rises to a high density. On close examination, however, these ideas do not stand up.

The notions that predators or disease are essential controllers of population density can be dismissed at once. There are animals that effectively have no predators and are not readily subject to disease and yet are limited to a stable level of population; among notable examples are the lion, the eagle and the skua [see "The Antarctic Skua," by Carl R. Eklund; SCIENTIFIC AMERICAN, February, 1964]. Disease per se does not act on a large scale to control population growth in the animal world. This leaves starvation as the possible control. The question of whether starvation itself acts directly to remove a population surplus calls for careful analysis.

Even a casual examination makes it clear that in most animal communities starvation is rare. Normally all the individuals in the habitat get enough food to survive. Occasionally a period of drought or severe cold may starve out a population, but that is an accident of weather—a disaster that does not arise from the density of population. We must therefore conclude that death from hunger is not an important density-dependent factor in controlling population size except in certain unusual cases.

Yet the density of population in the majority of habitats does depend directly on the size of the food supply; the close relation of one to the other is clear in representative situations where both variables have been measured [see

*illustration on page 97*]. We have, then, the situation that no individual starves but the population does not outgrow the food supply available in its habitat under normal conditions.

For many of the higher animals one can see therefore that neither predators, disease nor starvation can account for the regulation of numbers. There is of course accidental mortality, but it strikes in unpredictable and haphazard ways, independently of population density, and so must be ruled out as a stabilizer of population. All these considerations point to the possibility that the animals themselves must exercise the necessary restraint!

Man's own history provides some vivid examples of what is entailed here. By overgrazing he has converted once rich pastures into deserts; by overhunting he has exterminated the passenger pigeon and all but eliminated animals such as the right whale, the southern fur seal and, in many of their former breeding places, sea turtles; he is now threatening to exterminate all five species of rhinoceros inhabiting tropical Africa and Asia because the horns of those animals are valued for their alleged aphrodisiac powers. Exploiting the riches of today can exhaust and destroy the resources of tomorrow. The point is that animals face precisely this danger with respect to their food supply, and they generally handle it more prudently than man does.

Birds feeding on seeds and berries in the fall or chickadees living on hibernating insects in winter are in such a situation. The stock of food to begin with is so abundant that it could feed an enormous population. Then, however, it would be gone in hours or days, and the birds must depend on this food supply for weeks or months. To make it

UNEMPLOYED BIRDS are visible at a gannetry on Cape St. Mary in Newfoundland. They are the ones on the slope at left; the main colony is on the large adjacent slope. The unemployed gannets are excluded from breeding, apparently as part of the colony's automatic mechanisms for controlling the population level. These birds do, however, constitute a reserve for raising the population level.

MASSED MANEUVERS by starling flocks occur frequently on fine evenings, particularly in the fall. The maneuvers are an example of communal activity that appears to have the purpose of provid- ing the flock with an indication of population density. If the density is too high or too low in relation to the food supply, the flock automatically increases the activities that will improve the balance.

last through the season the birds must restrict the size of their population in advance. The same necessity holds in situations where unlimited feeding would wipe out the sources that replenish the food supply. Thus the threat of starvation tomorrow, not hunger itself today, seems to be the factor that decides what the density of a population ought to be. Long before starvation would otherwise occur, the population must limit its growth in order to avoid disastrous overexploitation of its food resources.

All this implies that animals restrict their population density by some artificial device that is closely correlated with the food supply. What is required is some sort of automatic restrictive mechanism analogous to the deliberate conventions or agreements by which nations limit the exploitation of fishing grounds.

One does not need to look far to realize that animals do indeed possess conventions of this kind. The best-known is the territorial system of birds. The practice of staking out a territory for nesting and rearing a family is common among many species of birds. In the breeding season each male lays claim to an area of not less than a certain minimum size and keeps out all other males of the species; in this way a group of males will parcel out the available ground as individual territories and put a limit on crowding. It is a perfect example of an artificial mechanism geared to adjusting the density of population to the food resources. Instead of competing directly for the food itself the members compete furiously for pieces of ground, each of which then becomes the exclusive food preserve of its owner. If the standard territory is large enough to feed a family, the entire group is safe from the danger of overtaxing the food supply.

The territorial convention is just one example of a convention that takes many other forms, some of them much more sophisticated or abstract. Seabirds, for instance, being unable to stake out a territory or nest on the sea itself,

PLACE IN HIERARCHY is at stake in this contest between male black bucks in India. Many mammal and bird groups have a hierarchical system or a system of defended territories. Successful individuals acquire food and breeding rights; the others leave, or perhaps stay as a reserve available for breeding if needed. By such means the group correlates its population with food resources.

adopt instead a token nesting place on the shore that represents their fishing rights. Each nesting site occupies only a few square feet, but the birds' behavior also limits the overall size of their colony, thereby restricting the number that will fish in the vicinity. Any adults that have not succeeded in winning a site within the perimeter of the colony are usually inhibited from nesting or starting another colony nearby.

Other restrictive conventions practiced by animals are still more abstract. Often the animals compete not for actual property, such as a nesting site, but merely for membership in the group, and only a certain number are accepted. In all cases the effect is to limit the density of the group living in the given habitat and unload any surplus population to a safe distance.

Not the least interesting fact is that the competition itself tends to take an abstract or conventional form. In their contest for a territory birds seldom actually draw blood or kill each other. Instead they merely threaten with aggressive postures, vigorous singing or displays of plumage. The forms of intimidation of rivals by birds range all the way from the naked display of weapons to the triumph of splendor revealed in the peacock's train.

This hypothesis about the mechanism of population control in animals leads to a generalization of broader scope, namely that this was the origin or root of all social behavior in animals, including man. Surprisingly there has been no generally acceptable theory of how the first social organizations arose. One can now argue logically, however, that the kind of competition under conventional rules that is typified by the territorial system of birds was the earliest form of social organization. Indeed, a society can be defined as a group of individuals competing for conventional prizes by conventional methods. To put it another way, it is a brotherhood tempered by rivalry. One does not need to ponder very deeply to see how closely this cap fits even human societies.

A group of birds occupying an area divided into individual territories is plainly a social organization, and it exhibits a considerable range of characteristically social behavior. This is well illustrated by the red grouse of Scotland—a bird that is being studied intensively in a long-term research project near Aberdeen.

The grouse population on a heather moor consists of individuals known to one another and differing among themselves in social standing. The dominant males hold territories almost all year round, the most aggressive claiming on the average the largest territories. Their individual domains cover the moor like a mosaic [*see top illustration on page 99*]. The community admits as members some socially subordinate males and unmated hens that have no territories of their own, but with the onset of winter, or with a decline in the food supply for some other reason, these supernumeraries at the bottom of the social ladder get squeezed out. Only as many as can be supported by the lowered food level are allowed to stay. Thus the social hierarchy of the red grouse works as a safety valve or overflow mechanism, getting rid of any excess that would overtax the food resources. The existence of the peck-order system among birds has been known for some time, but its functional reason for being has been unclear; it now appears that the lowest members of the order serve as a dispensable reserve that can fill in as replacements for casualties among the established members or be dropped as circumstances require.

Certain definite rules mark the competition of the red grouse males for territory and status. One is that, at least in the fall, they crow and threaten only on fine mornings between first light and two or three hours later. So aggressive is this struggle that the stress forces some of the losers to make a break away from the moor; on unfamiliar ground and without their usual food they soon weaken and are killed by predators or disease. Once the early-morning contest is over, however, those birds that remain in the habitat flock together amicably and feed side by side for the rest of the day.

The convention of competing at dawn or at dusk and leaving the rest of the day free for feeding and other peaceable activities is exceedingly common among animals of various kinds. The changes of light at dawn and dusk are, of course, the most conspicuous recurrent events of the day, and this no doubt explains why they serve so often as a signal for joint or communal activities. There are many familiar manifestations of this timing: the dawn chorus of songbirds and crowing cocks, the flight of ducks at dusk, the massed maneuvers of starlings and blackbirds at their roosts as darkness falls; the evening choruses of almost innumerable other birds, various tropical bats, frogs, cicadas and fishes such as the croaker, and the morning concerts of howler monkeys.

All these synchronized outbursts give an indication of the numbers present

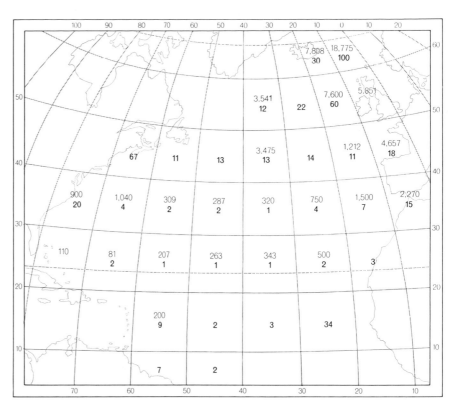

POPULATION AND FOOD SUPPLY show a correlation in the North Atlantic Ocean. The figures in light type give the average volume of plankton found per cubic centimeter of water; the darker figures show the average daily count of ocean birds that feed on plankton.

in the respective populations. They provide an index of the population density in the habitat from day to day, and so feed to the group information that causes it, not deliberately but automatically, to step up those activities that may be necessary to restore the balance between the density and the food supply.

The daily community display puts a changing pressure on the members taking part. If the stress is great enough, a reduction in the population can be triggered off; if it is felt lightly or not at all, there is room for new recruits. Overcrowding will lead to expulsion of the population surplus, as in the case of the red grouse. In the breeding season the density index, in the form of the daily display, can influence the proportion of adults that mate and breed; likewise the number of young can be restricted in a variety of other ways to the quota that the habitat will allow.

In the light of this hypothesis one would expect these "epideictic" displays (that is, population-pressure dem-onstrations) to be particularly prom-inent at the outset of the breeding season. That is actually the case. In birds the demonstrators are usually the males; they can be called the epideic-tic sex. They may swarm and dance in the air (as many flying insects do) or engage in ritual tournaments, gymnas-tics or parades (characteristic of sage grouse, prairie chickens, tropical hum-mingbirds, manakins and birds-of-para-dise). The intensity of these activities depends on the density of the popula-tion: the more males there are, the keener the competition. The new hy-pothesis suggests that this will result in greater stress among the males and sharper restriction of the size of the population.

In many animals the males have vocal abilities the females lack; this is true of songbirds, cicadas, most crickets and katydids, frogs, drumfishes, howler monkeys and others. Contrary to what was once thought, these males use their voices primarily not to woo females but in the contest with their fellow males for real estate and status. The same ap-plies to many of the males' adornments and scent glands, as well as to their weapons. This newly recognized fact calls for some rethinking of the whole vexed subject of sexual selection.

Epideictic displays rise to a height not only as a prelude to the breeding season but also at the time of animal migrations. They show the scale of the impending change in the population density of the habitat and, during the migration, give an indication of the size of the flocks that have gathered at the stopping places, thereby enabling the migrants to avoid dangerous congestion at any one place. Locusts build up for a great flight with spectacular massed maneuvers, and comparable excitement marks the nightly roosting of migratory chimney swifts and other big gatherings of birds, fruit bats and insects.

Altogether the hypothesis that animal populations regulate themselves through the agency of social conventions of this kind seems to answer satisfactorily sev-eral of the major questions that have concerned ecologists. Basically the aver-age population level is set by the long-term food resources of the habitat. A system of behavioral conventions acts as homeostatic machinery that prevents the growth of the population from de-parting too far from the optimal density. Fluctuations from this average can be explained as being due partly to tempo-rary accidents (such as climatic ex-tremes) and partly to the working of the homeostatic machinery itself, which allows the population density to build up when the food yields are good and thins it down when the yields fall below average. At any particular time the availability of food in relation to the number of mouths to be fed—in other words, the standard of living at the moment—determines the response of the regulating mechanism. The mecha-nism acts by controlling the rate of re-cruitment, by creating a pressure to emigrate or sometimes by producing stresses that result in large-scale mor-tality.

It has been particularly gratifying to find that the hypothesis offers ex-planations of several social enigmas on which there has been no good theory, such as the biological origin of social behavior; the function of the social hierarchy, or peck-order system, among birds; the chorus of birds and similar social events synchronized at dawn and dusk.

The theory has wide ramifications, which I have discussed at length in my book. The one that interests us most,

POPULATION-CONTROL DEVICES include the territory, of which the four basic types are depicted. Birds or mammals with territories have an established right to the available food; they also are the ones that breed. The others are in effect squeezed out. At top are two types of territory occupied by single males and their mates. At bottom are the types occupied by animals that live in colonies. One is virtually exclusive. The other is overlapping; shown here are islands from which five seabird colonies fan out within a maximum radius.

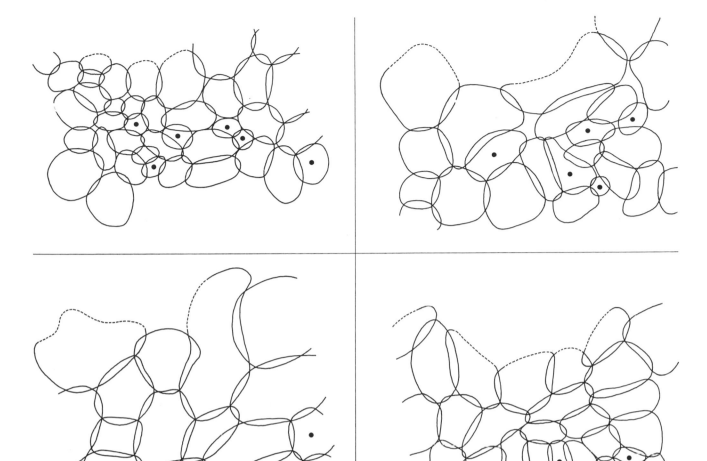

**TERRITORIAL VARIATIONS** of Scottish red grouse males reflect a form of population control. The drawings show the territorial holdings of individual cocks in four successive springs (1958–1961) on the same 140 acres of moorland. Some of the smaller territories, marked by dots, were held by males who remained unmated. Average territory size varies from year to year, thus affecting the density of breeding; in these four years the number of territories ranged between 40 in 1958 (*top left*) and 16 in 1960 (*bottom left*). The density of breeding is correlated with the food supply, which is to say with the quantity and quality of the heather.

**BLACK GROUSE MALES** are depicted in an "epideictic display," or ceremonial demonstration, that appears to be a form of population control. It evidently provides a measure of the population density within the area, because many males participate simultaneously on a communal strutting ground. It also serves as a means of excluding some less prominent males, who seldom display and often are chased away by the dominant birds. Epideictic displays also occur among many other bird and mammal species.

of course, is its bearing on the problem of the unchecked growth of the human population. The hypothesis opens up to clearer view the differences between man's demographic history and that of other animals.

There are two outstanding differences. In the first place, the homeostatic control of animal populations is strictly automatic: even the social conventions of behavior are innate rather than deliberately arrived at. In part the density-dependent control in many animals, including some of the mammals, is exercised by means of a biological reaction—either reduction of the rate of ovulation through a change in the output of hormones, or resorption of the embryos in the uterus as a result of stress (as occurs in rabbits, foxes and deer). Man's fertility and population growth, on the other hand, are subject only to his conscious and deliberate behavior. The second important difference is that modern man has progressively and enormously increased the food productivity of his habitat.

Primitive man, limited to the food he could get by hunting, had evolved a system for restricting his numbers by tribal traditions and taboos, such as prohibiting sexual intercourse for mothers while they were still nursing a baby, practicing compulsory abortion and infanticide, offering human sacrifices, conducting headhunting expeditions against rival tribes and so forth. These customs, consciously or not, kept the population density nicely balanced against the feeding capacity of the hunting range. Then, some 8,000 to 10,000 years ago, the agricultural revolution removed that limitation. There was no longer any reason to hold down the size of the tribe; on the contrary, power and wealth accrued to those tribes that allowed their populations to multiply, to develop farms, villages and even towns. The old checks on population growth were gradually discarded and forgotten. The rate of reproduction became a matter of individual choice rather than of tribal or community control. It has remained so ever since.

Given opportunity for procreation and a low death rate, the human population, whether well fed or hungry, now shows a tendency to expand without limit. Lacking the built-in homeostatic system that regulates the density of animal populations, man cannot look to any natural process to restrain his rapid growth. If the growth is to be slowed down, it must be by his own deliberate and socially applied efforts.

# 16

# Population

KINGSLEY DAVIS

*September 1963*

Just as the nation-state is a modern phenomenon, so is the explosive increase of the human population. For hundreds of millenniums *Homo sapiens* was a sparsely distributed animal. As long as this held true man could enjoy a low mortality in comparison to other species and could thus breed slowly in relation to his size. Under primitive conditions, however, crowding tended to raise the death rates from famine, disease and warfare. Yet man's fellow mammals even then might well have voted him the animal most likely to succeed. He had certain traits that portended future dominance: a wide global dispersion, a tolerance for a large variety of foods (assisted by his early adoption of cooking) and a reliance on group co-operation and socially transmitted techniques. It was only a matter of time before he and his kind would learn how to live together in communities without paying the penalty of high death rates.

Man remained sparsely distributed during the neolithic revolution, in spite of such advances as the domestication of plants and animals and the invention of textiles and pottery. Epidemics and pillage still held him back, and new kinds of man-made disasters arose from erosion, flooding and crop failure. Indeed, the rate of growth of the world population remained low right up to the 16th and 17th centuries.

Then came a spectacular quickening of the earth's human increase. Between 1650 and 1850 the annual rate of increase doubled, and by the 1920's it had doubled again. After World War II, in the decade from 1950 to 1960, it took another big jump [*see middle illustration on page 104*]. The human population is now growing at a rate that is impossible to sustain for more than a moment of geologic time.

Since 1940 the world population has grown from about 2.5 billion to 3.2 billion. This increase, within 23 years, is more than the *total* estimated population of the earth in 1800. If the human population were to continue to grow at the rate of the past decade, within 100 years it would be multiplied sixfold.

Projections indicate that in the next four decades the growth will be even more rapid. The United Nations' "medium" projections give a rate during the closing decades of this century high enough, if continued, to multiply the world population sevenfold in 100 years. These projections are based on the assumption that the changes in mortality and fertility in regions in various stages of development will be roughly like those of the recent past. They do not, of course, forecast the actual population, which may turn out to be a billion or two greater than that projected for the year 2000 or to be virtually nil. So far the UN projections, like most others in recent decades, are proving conservative. In 1960 the world population was 75 million greater than the figure given by the UN's "high" projection (published in 1958 and based on data up to 1955).

In order to understand why the revolutionary rise of world population has occurred, we cannot confine ourselves to the global trend, because this trend is a summation of what is happening in regions that are at any one time quite different with respect to their stage of development. For instance, the first step in the demographic evolution of modern nations—a decline in the death rate—began in northwestern Europe long before it started elsewhere. As a result, although population growth is now slower in this area than in the rest of the world, it was here that the unprecedented upsurge in human numbers began. Being most advanced in demographic development, northwestern Europe is a good place to start in our analysis of modern population dynamics.

In the late medieval period the average life expectancy in England, according to life tables compiled by the historian J. C. Russell, was about 27 years. At the end of the 17th century and during most of the 18th it was about 31 in England, France and Sweden, and in the first half of the 19th century it advanced to 41.

The old but reliable vital statistics from Denmark, Norway and Sweden show that the death rate declined erratically up to 1790, then steadily and more rapidly. Meanwhile the birth rate remained remarkably stable (until the latter part of the 19th century). The result was a marked increase in the excess of births over deaths, or what demographers call "natural increase" [*see illustration on page 107*]. In the century from about 1815 until World War I the average annual increase in the three Scandinavian countries was 11.8 per 1,000—nearly five times what it had been in the middle of the 18th century, and sufficient to triple the population in 100 years.

For a long time the population of northwestern Europe showed little reaction to this rapid natural increase. But when it came, the reaction was emphatic; a wide variety of responses occurred, all of which tended to reduce the growth of the population. For example, in the latter part of the 19th century people began to emigrate from Europe by the millions, mainly to America, Australia and South Africa. Between 1846 and 1932 an estimated 27 million people emigrated overseas from Europe's 10 most advanced countries. The three Scandinavian countries alone sent out 2.4 million, so that in 1915 their combined population was 11.1 million instead of the 14.2 million it would otherwise have been.

In addition to this unprecedented exodus there were other responses, all of which tended to reduce the birth rate. In spite of opposition from church and state, agitation for birth control began and induced abortions became common. The age at marriage rose. Childlessness became frequent. The result was a decline in the birth rate that eventually overtook the continuing decline in the death rate. By the 1930's most of the industrial European countries had age-specific fertility rates so low that, if the rates had continued at that level, the population would eventually have ceased to replace itself.

In explaining this vigorous reaction one gets little help from two popular clichés. One of these—that population growth is good for business—would hardly explain why Europeans were so bent on stopping population growth. The other—that numerical limitation comes from the threat of poverty because "population always presses on the means of subsistence"—is factually untrue. In every one of the industrializing countries of Europe economic growth outpaced population growth. In the United Kingdom, for example, the real per capita income increased 2.3 times between the periods 1855–1859 and 1910–1914. In Denmark from 1770 to 1914 the rise of the net domestic product in constant prices was two and a half times the natural increase rate; in Norway and Sweden from the 1860's to 1914 it was respectively 1.4 and 2.7 times the natural increase rate. Clearly the strenuous efforts to lessen population growth were due to some stimulus other than poverty.

The stimulus, in my view, arose from the clash between new opportunities on the one hand and larger families on the other. The modernizing society of northwestern Europe necessarily offered new opportunities to people of all classes: new ways of gaining wealth, new means of rising socially, new symbols of status. In order to take advantage of those opportunities, however, the individual and his children required education, special skills, capital and mobility—none of which was facilitated by an improvident marriage or a large family. Yet because mortality was being reduced (and reduced more successfully in the childhood than in the adult ages) the size of families had become potentially larger than before. In Sweden, for instance, the mortality of the period 1755–1775 allowed only 6.1 out of every 10 children born to reach the age of 10, whereas the mor-

tality of 1901–1910 allowed 8.5 to survive to that age. In order to avoid the threat of a large family to his own and his children's socioeconomic position, the individual tended to postpone or avoid marriage and to limit reproduction within marriage by every means available. Urban residents had to contend particularly with the cost and inconvenience of young children in the city. Rural families had to adjust to the lack of enough land to provide for new marriages when the children reached marriageable age. Land had become less available not only because of the plethora of families with numerous youths but also because, with modernization, more capital was needed per farm and because the old folks, living longer, held on to the property. As a result farm youths postponed marriage, flocked to the cities or went overseas.

In such terms we can account for the paradox that, as the progressive European nations became richer, their population growth slowed down. The process of economic development itself provided the motives for curtailment of reproduction, as the British sociologist J. A. Banks has made clear in his book *Prosperity and Parenthood*. We can see now that in all modern nations the long-run trend is one of low mortality, a relatively modest rate of reproduction and slow population growth. This is an efficient demographic system that allows such countries, in spite of their "maturity," to continue to advance economically at an impressive speed.

Naturally the countries of northwestern Europe did not all follow an identical pattern. Their stages differed somewhat in timing and in the pattern of preference among the various means of population control. France, for example, never attained as high a natural increase as Britain or Scandinavia did. This was not due solely to an earlier decline in the birth rate, as is often assumed, but also to a slower decline in the death rate. If we historically substitute the Swedish death rate for the French, we revise the natural increase upward by almost the same amount as we do by substituting the Swedish birth rate. In accounting for the early and easy drop in French fertility one recalls that France, already crowded in the 18th century and in the van of intellectual radicalism and sophistication, was likely to have a low threshold for the adoption of abortion and contraception. The death rate, however, remained comparatively high because France did not keep economic pace with her more rapidly industrializ-

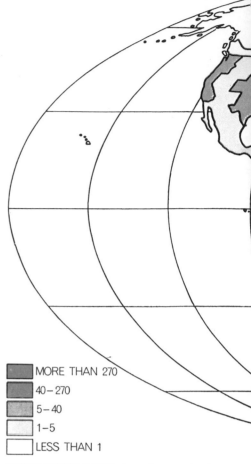

MORE THAN 270

40–270

5–40

1–5

LESS THAN 1

**POPULATION MAPS** show density (*top*) as of 1961 and per cent increase per year

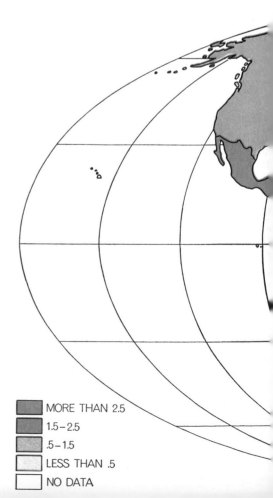

MORE THAN 2.5

1.5–2.5

.5–1.5

LESS THAN .5

NO DATA

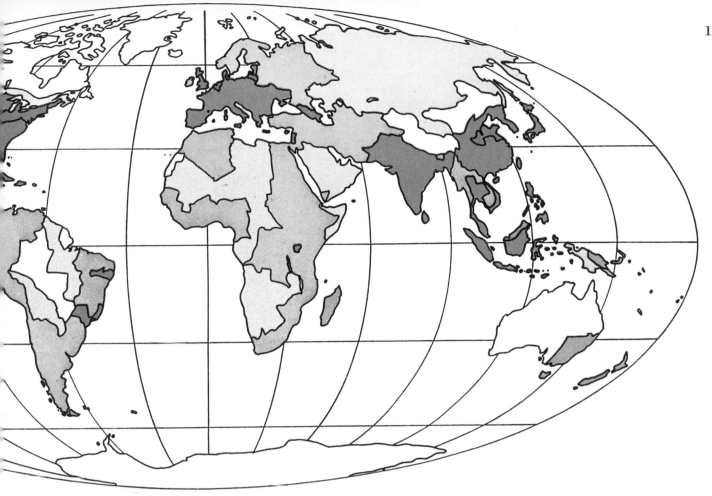

between 1958 and 1961 (*bottom*). Except for the countries of largest area the density has been averaged within the boundaries of each nation. The densities are given in terms of the number of people per square kilometer. The data are primarily from UN publications.

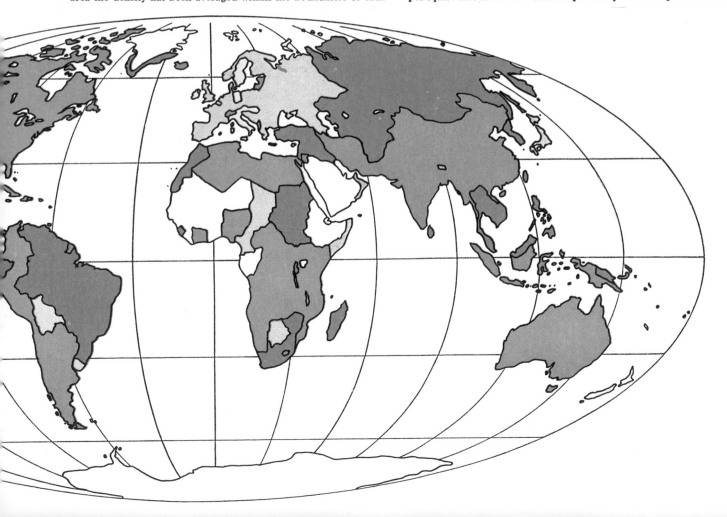

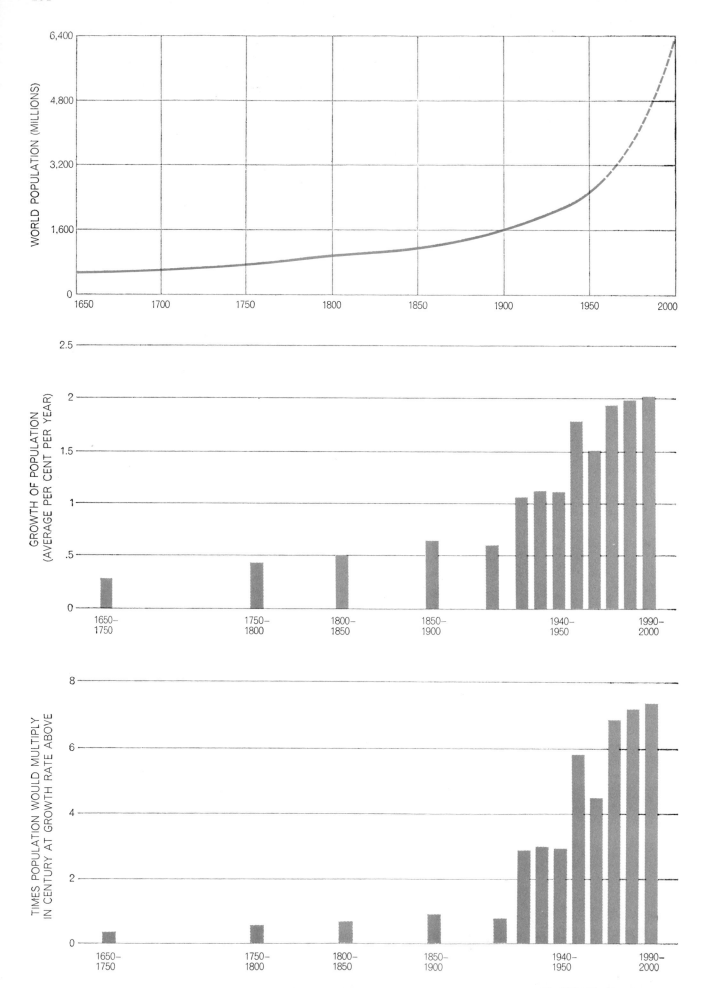

ing neighbors. As a result the relatively small gap between births and deaths gave France a slower growth in population and a lesser rate of emigration.

Ireland also has its own demographic history, but like France it differs from the other countries in emphasis rather than in kind. The emphasis in Ireland's escape from human inflation was on emigration, late marriage and permanent celibacy. By 1891 the median age at which Irish girls married was 28 (compared with 22 in the U.S. at that date); nearly a fourth of the Irish women did not marry at all, and approximately a third of all Irish-born people lived outside of Ireland. These adjustments, begun with the famine of the 1840's and continuing with slight modifications until today, were so drastic that they made Ireland the only modern nation to experience an absolute decline in population. The total of 8.2 million in 1841 was reduced by 1901 to 4.5 million.

The Irish preferences among the means of population limitation seem to come from the island's position as a rural region participating only indirectly in the industrial revolution. For most of the Irish, land remained the basis for respectable matrimony. As land became inaccessible to young people they postponed marriage. In doing so they were not discouraged by their parents, who wished to keep control of the land, or by their religion. Their Catholicism, which they embraced with exceptional vigor both because they were rural and because it was a rallying point for Irish nationalism as against the Protestant English, placed a high value on celibacy. The clergy, furthermore, were powerful enough to exercise strict control over courtship and thus to curtail illicit pregnancy and romance as factors leading to marriage. They were also able to exercise exceptional restraint on abortion and contraception. Although birth control was practiced to some extent, as evidenced by a decline of fertility within marriage, its influence was so small as to make early marriage synonymous with a large family and therefore to be avoided. Marriage was also discouraged by the ban on divorce and by the lowest participation of married women in the labor force to be found in Europe. The

POPULATION GROWTH of world from 1650 to 1960 is shown by curve at top of opposite page, projected to the year 2000. The middle chart shows the rate of growth, and the bottom chart the number of times the population would multiply in 100 years at that growth rate for various periods.

country's failure to industrialize meant that the normal exodus from farms to cities was at the same time an exodus from Ireland itself.

Ireland and France illustrate contrasting variations on a common theme. Throughout northwestern Europe the population upsurge resulting from the fall in death rates brought about a multiphasic reaction that eventually reduced the population growth to a modest pace. The main force behind this response was not poverty or hunger but the desire of the people involved to preserve or improve their social standing by grasping the opportunities offered by the newly emerging industrial society.

Is this an interpretation applicable to the history of any industrialized country, regardless of traditional culture? According to the evidence the answer is yes. We might expect it to be true, as it currently is, of the countries of southern and eastern Europe that are finally industrializing. The crucial test is offered by the only nation outside the European tradition to become industrialized: Japan. How closely does Japan's demographic evolution parallel that of northwestern Europe?

If we superpose Japan's vital-rate curves on those of Scandinavia half a century earlier [see illustration on page 107], we see a basically similar, although more rapid, development. The reported statistics, questionable up to 1920 but good after that, show a rapidly declining death rate as industrialization took hold after World War I. The rate of natural increase during the period from 1900 to 1940 was almost exactly the same as Scandinavia's between 1850 and 1920, averaging 12.1 per 1,000 population per year compared with Scandinavia's 12.3. And Japan's birth rate, like Europe's, began to dip until it was falling faster than the death rate, as it did in Europe. After the usual baby boom following World War II the decline in births was precipitous, amounting to 50 per cent from 1948 to 1960—perhaps the swiftest drop in reproduction that has ever occured in an entire nation. The rates of childbearing for women in various ages are so low that, if they continued indefinitely, they would not enable the Japanese population to replace itself.

In thus slowing their population growth have the Japanese used the same means as the peoples of northwestern Europe did? Again, yes. Taboo-ridden Westerners have given disproportionate attention to two features of the change— the active role played by the Japanese government and the widespread resort to

abortion—but neither of these disproves the similarity. It is true that since the war the Japanese government has pursued a birth-control policy more energetically than any government ever has before. It is also clear, however, that the Japanese people would have reduced their childbearing of their own accord. A marked decline in the reproduction rate had already set in by 1920, long before there was a government policy favoring this trend.

As for abortion, the Japanese are unusual only in admitting its extent. Less superstitious than Europeans about this subject, they keep reasonably good records of abortions, whereas most of the other countries have no accurate data. According to the Japanese records, registered abortions rose from 11.8 per 1,000 women of childbearing age in 1949 to a peak of 50.2 per 1,000 in 1955. We have no reliable historical information from Western countries, but we do know from many indirect indications that induced abortion played a tremendous role in the reduction of the birth rate in western Europe from 1900 to 1940, and that it still plays a considerable role. Furthermore, Christopher Tietze of the National Committee for Maternal Health has assembled records that show that in five eastern European countries where abortion has been legal for some time the rate has shot up recently in a manner strikingly similar to Japan's experience. In 1960–1961 there were 139 abortions for every 100 births in Hungary, 58 per 100 births in Bulgaria, 54 in Czechoslovakia and 34 in Poland. The countries of eastern Europe are in a developmental stage comparable to that of northwestern Europe earlier in the century.

Abortion is by no means the sole factor in the decline of Japan's birth rate. Surveys made since 1950 show the use of contraception before that date, and increasing use thereafter. There is also a rising frequency of sterilization. Furthermore, as in Europe earlier, the Japanese are postponing marriage. The proportion of girls under 20 who have ever married fell from 17.7 per cent in 1920 to 1.8 per cent in 1955. In 1959 only about 5 per cent of the Japanese girls marrying for the first time were under 20, whereas in the U.S. almost half the new brides (48.5 per cent in the registration area) were that young.

Finally, Japan went through the same experience as western Europe in another respect—massive emigration. Up to World War II Japan sent millions of emigrants to various regions of Asia, Oceania and the Americas.

In short, in response to a high rate of

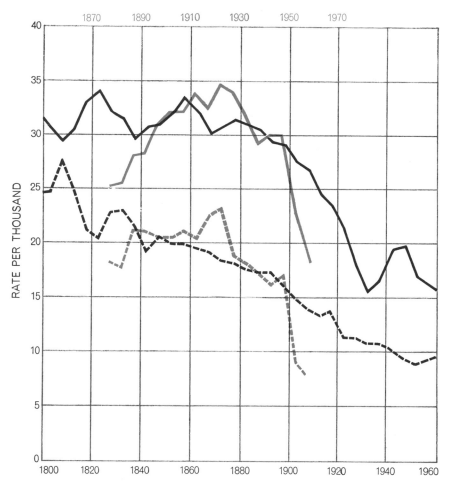

**BIRTH AND DEATH RATES** for Denmark, Norway and Sweden combined (*black lines and dates*) are compared with Japanese rates (*colored lines and dates*) of 50 years later. Japan has been passing through a population change similar to that which occurred earlier in Scandinavia. Area between respective birth-rate curves (*solid lines*) and death-rate curves (*broken lines*) shows natural increase, or population growth that would have occurred without migration. In past few years both Japanese rates have dropped extremely rapidly.

natural increase brought by declining mortality, Japan reacted in the same ways as the countries of northwestern Europe did at a similar stage. Like the Europeans, the Japanese limited their population growth in their own private interest and that of their children in a developing society, rather than from any fear of absolute privation or any concern with overpopulation in their homeland. The nation's average 5.4 per cent annual growth in industrial output from 1913 to 1958 exceeded the performance of European countries at a similar stage.

As our final class of industrialized

**MEXICO CITY, a part of which is shown in the vertical aerial photograph on the opposite page, displays the rapid population growth and the urbanization that characterize much of the world today. In 1940 the population of the metropolitan area of the city was 1,758,000, in 1950 it reached 3,050,000 and by 1960 it had jumped to 4,830,000.**

countries we must now consider the frontier group—the U.S., Canada, Australia, New Zealand, South Africa and Russia. These countries are distinguished from those of northwestern Europe and Japan by their vast wealth of natural resources in relation to their populations; they are the genuinely affluent nations. They might be expected to show a demographic history somewhat different from that of Europe. In certain particulars they do, yet the general pattern is still much the same.

One of the differences is that the riches offered by their untapped resources invited immigration. All the frontier industrial countries except Russia received massive waves of emigrants from Europe. They therefore had a more rapid population growth than their industrializing predecessors had experienced. As frontier countries with great room for expansion, however, they were also characterized by considerable internal migration and continuing new opportunities.

As a result their birth rates remained comparatively high. In the decade from 1950 to 1960, with continued immigration, these countries grew in population at an average rate of 2.13 per cent a year, compared with 1.76 per cent for the rest of the world. It was the four countries with the sparsest settlement (Canada, Australia, New Zealand and South Africa), however, that accounted for this high rate; in the U.S. and the U.S.S.R. the growth rate was lower—1.67 per cent per year.

Apparently, then, in pioneer industrial countries with an abundance of resources population growth holds up at a higher level than in Japan or northwestern Europe because the average individual feels it is easier for himself and his children to achieve a respectable place in the social scale. The immigrants attracted by the various opportunities normally begin at a low level and thus make the status of natives relatively better. People marry earlier and have slightly larger families. But this departure from the general pattern for industrial countries appears to be only temporary.

In the advanced frontier nations, as in northwestern Europe, the birth rate began to fall sharply after 1880, and during the depression of the 1930's it was only about 10 per cent higher than in Europe. Although the postwar baby boom has lasted longer than in other advanced countries, it is evidently starting to subside now, and the rate of immigration has diminished. There are factors at work in these affluent nations that will likely limit their population growth. They are among the most urbanized countries in the world, in spite of their low average population density. Their birth rates are extremely sensitive to business fluctuations and social changes. Furthermore, having in general the world's highest living standards, their demand for resources, already staggering, will become fantastic if both population and per capita consumption continue to rise rapidly, and their privileged position in the world may become less tolerated.

Let us shift now to the other side of the population picture: the nonindustrial, or underdeveloped, countries.

As a class the nonindustrial nations since 1930 have been growing in population about twice as fast as the industrial ones. This fact is so familiar and so taken for granted that its irony tends to escape us. When we think of it, it is astonishing that the world's most impoverished nations, many of them already overcrowded by any standard, should be generating additions to the population at the highest rate.

The underdeveloped countries have about 69 per cent of the earth's adults—and some 80 per cent of the world's children. Hence the demographic situation itself tends to make the world constantly more underdeveloped, or impoverished, a fact that makes economic growth doubly difficult.

How can we account for the paradox that the world's poorest regions are producing the most people? One is tempted to believe that the underdeveloped countries are simply repeating history: that they are in the same phase of rapid growth the West experienced when it began to industrialize and its death rates fell. If that is so, then sooner or later the developing areas will limit their population growth as the West did.

It is possible that this may prove to be true in the long run. But before we accept the comforting thought we should take a close look at the facts as they are.

In actuality the demography of the nonindustrial countries today differs in essential respects from the early history of the present industrial nations. Most striking is the fact that their rate of human multiplication is far higher than the West's ever was. The peak of the industrial nations' natural increase rarely rose above 15 per 1,000 population per year; the highest rate in Scandinavia was 13, in England and Wales 14, and even in Japan it was slightly less than 15. True, the U.S. may have hit a figure of 30 per 1,000 in the early 19th century, but if so it was with the help of heavy immigration of young people (who swelled the births but not the deaths) and with the encouragement of an empty continent waiting for exploitation.

In contrast, in the present underdeveloped but often crowded countries the natural increase per 1,000 population is everywhere extreme. In the decade from 1950 to 1960 it averaged 31.4 per year in Taiwan, 26.8 in Ceylon, 32.1 in Malaya, 26.7 in Mauritius, 27.7 in Albania, 31.8 in Mexico, 33.9 in El Salvador and 37.3 in Costa Rica. These are not birth rates; they are the *excess* of births over deaths! At an annual natural increase of 30 per 1,000 a population will double itself in 23 years.

The population upsurge in the backward nations is apparently taking place at an earlier stage of development—or perhaps we should say *un*development—than it did in the now industrialized nations. In Britain, for instance, the peak of human multiplication came when the country was already highly industrialized and urbanized, with only a fifth

of its working males in agriculture. Comparing four industrial countries at the peak of their natural increase in the 19th century (14.1 per 1,000 per year) with five nonindustrial countries during their rapid growth in the 1950's (32.2 per 1,000 per year), I find that the industrial countries were 38.5 per cent urbanized and had 27.9 per cent of their labor force in manufacturing, whereas now the nonindustrial countries are 29.4 per cent urbanized and have only 15.1 per cent of their people in manufacturing. In short, today's nonindustrial populations are growing faster and at an earlier stage than was the case in the demographic cycle that accompanied industrialization in the 19th century.

As in the industrial nations, the main generator of the population upsurge in the underdeveloped countries has been a fall in the death rate. But their resulting excess of births over deaths has proceeded faster and farther, as a comparison of Ceylon in recent decades with Sweden in the 1800's shows [*see illustration on next page*].

In most of the underdeveloped nations the death rate has dropped with record speed. For example, the sugar-growing island of Mauritius in the Indian Ocean within an eight-year period after the war raised its average life expectancy from

33 to 51—a gain that took Sweden 130 years to achieve. Taiwan within two decades has increased its life expectancy from 43 to 63; it took the U.S. some 80 years to make this improvement for its white population. According to the records in 18 underdeveloped countries, the crude death rate has dropped substantially in each decade since 1930; it fell some 6 per cent in the 1930's and nearly 20 per cent in the 1950's, and according to the most recent available figures the decline in deaths is still accelerating.

The reasons for this sharp drop in mortality are in much dispute. There are two opposing theories. Many give the credit to modern medicine and public health measures. On the other hand, the public health spokesmen, rejecting the accusation of complicity in the world's population crisis, belittle their own role and maintain that the chief factor in the improvement of the death rate has been economic progress.

Those in the latter camp point out that the decline in the death rate in northwestern Europe followed a steadily rising standard of living. Improvements in diet, clothing, housing and working conditions raised the population's resistance to disease. As a result many dangerous ailments disappeared or subsided without specific medical attack. The same process, say the public health peo-

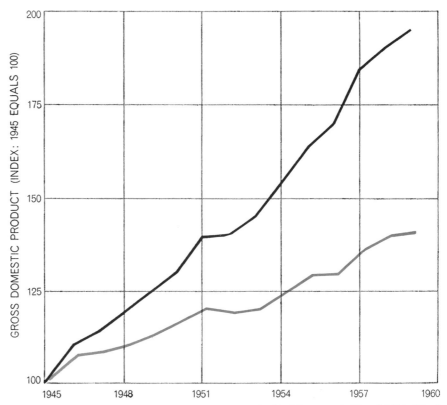

**GROSS DOMESTIC PRODUCT** of Latin America doubled between 1945 and 1959 (*black line*) but population growth held down the increase in per capita product (*colored line*).

8

ple, is now at work in the developing countries.

On the other side, most demographers and economists believe that economic conditions are no longer as important as they once were in strengthening a community's health. The development of medical science has provided lifesaving techniques and medicines that can be transported overnight to the most backward areas. A Stone Age people can be endowed with a low 20th-century death rate within a few years, without waiting for the slow process of economic development or social change. International agencies and the governments of the affluent nations have been delighted to act as good Samaritans and send out public health missionaries to push disease-fighting programs for the less developed countries.

The debate between the two views is hard to settle. Such evidence as we have indicates that there is truth on both sides. Certainly the newly evolving countries have made economic progress. Their economic advance, however, is not nearly rapid enough to account for the very swift decline in their death rates, nor do they show any clear correlation between economic growth and improve-

ment in life expectancy. For example, in Mauritius during the five-year period from 1953 to 1958 the per capita income fell by 13 per cent, yet notwithstanding this there was a 36 per cent drop in the death rate. On the other hand, in the period between 1945 and 1960 Costa Rica had a 64 per cent increase in the per capita gross national product and a 55 per cent decline in the death rate. There seems to be no consistency—no significant correlation between the two trends when we look at the figures country by country. In 15 underdeveloped countries for which such figures are available we find that the decline in death rate in the 1950's was strikingly uniform (about 4 per cent per year), although the nations varied greatly in economic progress— from no improvement to a 6 per cent annual growth in per capita income.

Our tentative conclusion must be, therefore, that the public health people are more efficient than they admit. The billions of dollars spent in public health work for underdeveloped areas has brought down death rates, irrespective of local economic conditions in these areas. The programs instituted by outsiders to control cholera, malaria, plague

and other diseases in these countries have succeeded. This does not mean that death control in underdeveloped countries has become wholly or permanently independent of economic development but that it has become temporarily so to an amazing degree.

Accordingly the unprecedented population growth in these countries bears little relation to their economic condition. The British economist Colin G. Clark has contended that rapid population growth stimulates economic progress. This idea acquires plausibility from the association between human increase and industrialization in the past and from the fact that in advanced countries today the birth rate (but not the death rate) tends to fluctuate with business conditions. In today's underdeveloped countries, however, there seems to be little or no visible connection between economics and demography.

In these countries neither births nor deaths have been particularly responsive to economic change. Some of the highest rates of population growth ever known are occurring in areas that show no commensurate economic advance. In 34 such countries for which we have data, the correlation between population growth and economic gain during the 1950's was negligible, and the slight edge was on the negative side: − .2. In 20 Latin-American countries during the period from 1954 to 1959, while the annual gain in per capita gross domestic product fell from an average of 2 per cent to 1.3 per cent, the population growth rate *rose* from 2.5 to 2.7 per cent per year.

All the evidence indicates that the population upsurge in the underdeveloped countries is not helping them to advance economically. On the contrary, it may well be interfering with their economic growth. A surplus of labor on the farms holds back the mechanization of agriculture. A rapid rise in the number of people to be maintained uses up income that might otherwise be utilized for long-term investment in education, equipment and other capital needs. To put it in concrete terms, it is difficult to give a child the basic education he needs to become an engineer when he is one of eight children of an illiterate farmer who must support the family with the produce of two acres of ground.

By definition economic advance means an increase in the amount of product per unit of human labor. This calls for investment in technology, in improvement of the skills of the labor force and in administrative organization and planning. An economy that must spend a dispro-

NEW DEMOGRAPHIC PATTERN is appearing in the nonindustrialized nations. The birth rate (*solid line*) has not been falling significantly, whereas the death rate (*broken line*) has dropped precipitously, as illustrated by Ceylon (*color*). The spread between the two rates has widened. In nations such as Sweden (*black*), however, birth rate dropped during development long before death rate was as low as in most underdeveloped countries today.

portionate share of its income in support- ing the consumption needs of a growing population—and at a low level of con- sumption at that—finds growth difficult because it lacks capital for improve- ments.

A further complication lies in the process of urbanization. The shifts from villages and farmsteads to cities is seem- ingly an unavoidable and at best a pain- ful part of economic development. It is most painful when the total popula- tion is skyrocketing; then the cities are bursting both from their own multipli- cation and from the stream of migrants from the villages. The latter do not move to cities because of the opportunities there. The opportunities are few and unemployment is prevalent. The mi- grants come, rather, because they are impelled by the lack of opportunity in the crowded rural areas. In the cities they hope to get something—a menial job, government relief, charities of the rich. I have recently estimated that if the population of India increases at the rate projected for it by the UN, the net num- ber of migrants to cities between 1960 and 2000 will be of the order of 99 to 201 million, and in 2000 the largest city will contain between 36 and 66 million inhabitants. One of the greatest problems now facing the governments of under- developed countries is what to do with these millions of penniless refugees from the excessively populated countryside.

Economic growth is not easy to achieve. So far, in spite of all the talk and the earnest efforts of under- developed nations, only one country out- side the northwestern European tradi- tion has done so: Japan. The others are struggling with the handicap of a popu- lation growth greater than any indus- trializing country had to contend with in the past. A number of them now realize that this is a primary problem, and their governments are pursuing or contemplating large-scale programs of birth-limitation. They are receiving little help in this matter, however, from the industrial nations, which have so willing- ly helped them to lower their death rates.

The Christian nations withhold this help because of their official taboos against some of the means of birth-limita- tion (although their own people private- ly use all these means). The Communist nations withhold it because limitation of population growth conflicts with official Marxist dogma (but Soviet citizens con- trol births just as capitalist citizens do, and China is officially pursuing policies calculated to reduce the birth rate).

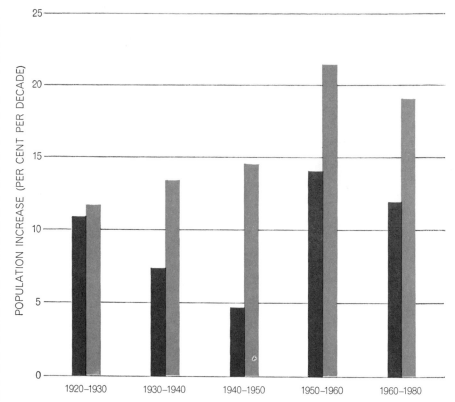

DIFFERENTIAL POPULATION GROWTH in underdeveloped regions (*colored bars*) and developed regions (*gray bars*) is plotted. The 1960–1980 projections may turn out to be low.

The West's preoccupation with the technology of contraception seems un- justified in view of its own history. The peoples of northwestern Europe utilized all the available means of birth limitation once they had strong motives for such limitation. The main question, then, is whether or not the peoples of the present underdeveloped countries are likely to acquire such motivation in the near fu- ture. There are signs that they will. Sur- veys in India, Jamaica and certain other areas give evidence of a growing desire among the people to reduce the size of their families. Furthermore, circum- stances in the underdeveloped nations today are working more strongly in this direction than they did in northwestern Europe in the 19th century.

As in that earlier day, poverty and deprivation alone are not likely to gen- erate a slowdown of the birth rate. But personal aspirations are. The agrarian peoples of the backward countries now look to the industrialized, affluent fourth of the world. They nourish aspirations that come directly from New York, Paris and Moscow. No more inclined to be satisfied with a bare subsistence than their wealthier fellows would be, they are demanding more goods, education, opportunity and influence. And they are beginning to see that many of their desires are incompatible with the en-

larged families that low mortality and customary reproduction are giving them.

They live amid a population density far greater than existed in 19th-century Europe. They have no place to which to emigrate, no beckoning continents to colonize. They have rich utopias to look at and industrial models to emulate, whereas the Europeans of the early 1800's did not know where they were going. The peoples of the underdevel- oped, overpopulated countries therefore seem likely to start soon a multiphasic limitation of births such as began to sweep through Europe a century ago. Their governments appear ready to help them. Government policy in these coun- tries is not quibbling over means or con- fining itself to birth-control technology; its primary task is to strengthen and accelerate the peoples' motivation for reproductive restraint.

Meanwhile the industrial countries also seem destined to apply brakes to their population growth. The steadily rising level of living, multiplied by the still growing numbers of people, is en- gendering a dizzying rate of consump- tion. It is beginning to produce painful scarcities of space, of clean water, of clean air and of quietness. All of this may prompt more demographic modera- tion than these countries have already exercised.

# 17

# Population Density
# and Social Pathology

JOHN B. CALHOUN

*February 1962*

In the celebrated thesis of Thomas Malthus, vice and misery impose the ultimate natural limit on the growth of populations. Students of the subject have given most of their attention to misery, that is, to predation, disease and food supply as forces that operate to adjust the size of a population to its environment. But what of vice? Setting aside the moral burden of this word, what are the effects of the social behavior of a species on population growth—and of population density on social behavior?

Some years ago I attempted to submit this question to experimental inquiry. I confined a population of wild Norway rats in a quarter-acre enclosure. With an abundance of food and places to live and with predation and disease eliminated or minimized, only the animals' behavior with respect to one another remained as a factor that might affect the increase in their number. There could be no escape from the behavioral consequences of rising population density. By the end of 27 months the population had become stabilized at 150 adults Yet adult mortality was so low that 5,000 adults might have been expected from the observed reproductive rate. The reason this larger population did not materialize was that infant mortality was extremely high. Even with only 150 adults in the enclosure, stress from social interaction led to such disruption of maternal behavior that few young survived.

With this background in mind I turned to observation of a domesticated albino strain of the Norway rat under more controlled circumstances indoors. The data for the present discussion come from the histories of six different populations. Each was permitted to increase to approximately twice the number that my experience had indicated could occupy the available space with only moderate stress from social interaction. In each

case my associates and I maintained close surveillance of the colonies for 16 months in order to obtain detailed records of the modifications of behavior induced by population density.

The consequences of the behavioral pathology we observed were most apparent among the females. Many were unable to carry pregnancy to full term or to survive delivery of their litters if they did. An even greater number, after successfully giving birth, fell short in their maternal functions. Among the males the behavior disturbances ranged from sexual deviation to cannibalism and from frenetic overactivity to a pathological withdrawal from which individuals would emerge to eat, drink and move about only when other members of the community were asleep. The social organization of the animals showed equal disruption. Each of the experimental populations divided itself into several groups, in each of which the sex ratios were drastically modified. One group might consist of six or seven females and one male, whereas another would have 20 males and only 10 females.

The common source of these disturbances became most dramatically apparent in the populations of our first series of three experiments, in which we observed the development of what we called a behavioral sink. The animals would crowd together in greatest number in one of the four interconnecting pens in which the colony was maintained. As many as 60 of the 80 rats in each experimental population would assemble in one pen during periods of feeding. Individual rats would rarely eat except in the company of other rats. As a result extreme population densities developed in the pen adopted for eating, leaving the others with sparse populations.

Eating and other biological activities were thereby transformed into social ac-

tivities in which the principal satisfaction was interaction with other rats. In the case of eating, this transformation of behavior did not keep the animals from securing adequate nutrition. But the same pathological "togetherness" tended to disrupt the ordered sequences of activity involved in other vital modes of behavior such as the courting of sex partners, the building of nests and the nursing and care of the young. In the experiments in which the behavioral sink developed, infant mortality ran as high as 96 per cent among the most disoriented groups in the population. Even in the absence of the behavioral sink, in the second series of three experiments, infant mortality reached 80 per cent among the corresponding members of the experimental populations.

The design of the experiments was relatively simple. The three populations of the first series each began with 32 rats; each population of the second series began with 56 rats. In all cases the animals were just past weaning and were evenly divided between males and females. By the 12th month all the populations had multiplied and each comprised 80 adults. Thereafter removal of the infants that survived birth and weaning held the populations steady. Although the destructive effects of population density increased during the course of the experiments, and the mortality rate among the females and among the young was much higher in the 16th month than it was earlier, the number of young that survived to weaning was always large enough to offset the effects of adult mortality and actually to increase the population. The evidence indicates, however, that in time failures of reproductive function would have caused the colonies to die out. At the end of the first series of experiments eight rats—the four healthi-

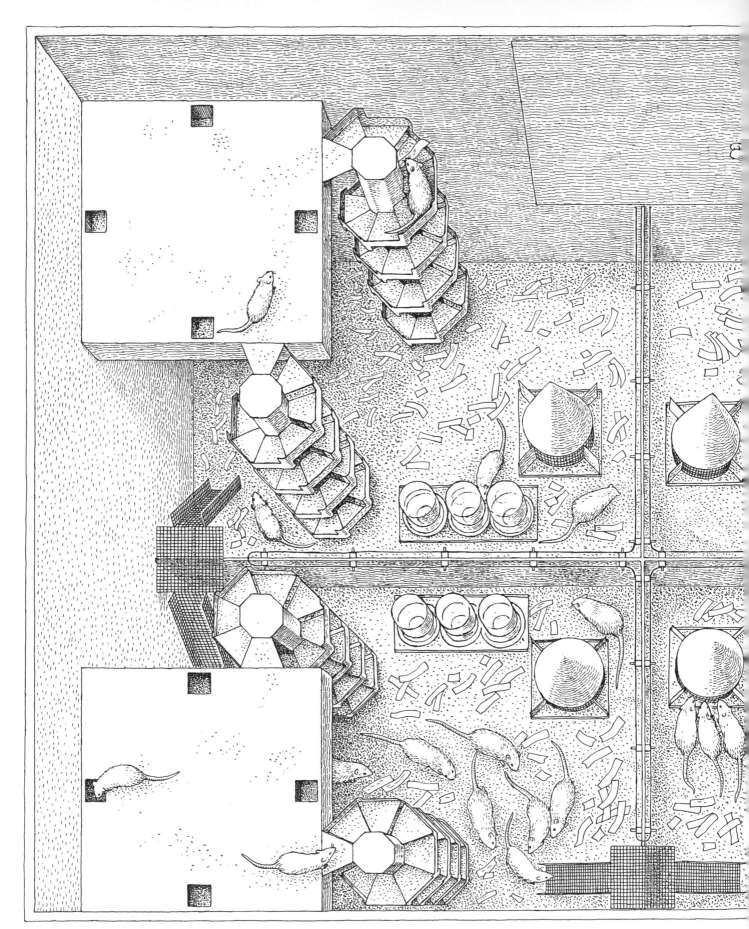

EFFECT OF POPULATION DENSITY on the behavior and social organization of rats was studied by confining groups of 80 animals in a 10-by-14-foot room divided into four pens by an electrified fence. All pens (numbered 1, 2, 3 and 4 clockwise from door) were complete dwelling units. Conical objects are food hoppers; trays with three bottles are drinking troughs. Elevated burrows, reached by winding staircases, each had five nest boxes, seen in pen 1, where top of burrow has been removed. Ramps connected all pens but 1 and 4. Rats therefore tended to concentrate in pens 2 and 3. Development of a "behavioral sink," which further increased population in one pen, is reflected in pen 2, where three rats are eating simultaneously. Rat approaching ramp in pen 3 is an estrous female

est males and the four healthiest females in each of two populations—were permitted to survive. These animals were six months old at the time, in the prime of life. Yet in spite of the fact that they no longer lived in overpopulated environments, they produced fewer litters in the next six months than would normally have been expected. Nor did any of the offspring that were born survive to maturity.

The males and females that initiated each experiment were placed, in groups of the same size and sex composition, in each of the four pens that partitioned a 10-by-14-foot observation room. The pens were complete dwelling units; each contained a drinking fountain, a food hopper and an elevated artificial burrow, reached by a winding staircase and holding five nest boxes. A window in the ceiling of the room permitted observation, and there was a door in one wall. With space for a colony of 12 adults in each pen—the size of the groups in which rats are normally found—this setup should have been able to support 48 rats comfortably. At the stabilized number of 80, an equal distribution of the animals would have found 20 adult rats in each pen. But the animals did not dispose themselves in this way.

Biasing factors were introduced in the physical design of the environment to encourage differential use of the four pens. The partitions separating the pens were electrified so that the rats could not climb them. Ramps across three of the partitions enabled the animals to get from one pen to another and so traverse the entire room. With no ramps to permit crossing of the fourth partition, however, the pens on each side of it became the end pens of what was topologically a row of four. The rats had to make a complete circuit of the room to go from the pen we designated 1 to the pen designated 4 on the other side of the partition separating the two. This arrangement of ramps immediately skewed the mathematical probabilities in favor of a higher population density in pens 2 and 3 than in pens 1 and 4. Pens 2 and 3 could be reached by two ramps, whereas pens 1 and 4 had only one each.

The use of pen 4 was further discouraged by the elevation of its burrow to a height greater than that of the burrow in the other end pen. The two middle pens were similarly distinguished from each other, the burrow in pen 3 being higher than that in pen 2. But here the differential appears to have played a smaller role, although pen 2 was used somewhat more often than pen 3.

With the distribution of the rats

pursued by a pack of males. In pens 2 and 3, where population density was **highest,** males outnumbered females. In pens 1 and 4, a dominant male was usually able to expel all other males and possess a harem of females. Dominant males are sleeping at the base of the ramps in pens 1 and 4. They wake when other males approach, preventing incursions into their territories. The three rats peering down from a ramp are probers, one of the deviant behavioral types produced by the pressures of a high population density.

biased by these physical arrangements, the sizes of the groups in each pen could have been expected to range from as few as 13 to as many as 27. With the passage of time, however, changes in behavior tended to skew the distribution of the rats among the pens even more. Of the 100 distinct sleeping groups counted in the 10th to 12th month of each experiment, only 37 fell within the expected size range. In 33 groups there were fewer than 13 rats, and in 30 groups the count exceeded 27. The sex ratio approximated equality only in those groups that fell within the expected size range. In the smaller groups, generally composed of eight adults, there were seldom more than two males. In the larger groups, on the other hand, there were many more males than females. As might be expected, the smaller groups established themselves in the end pens, whereas the larger groups were usually observed to form in the middle pens. The female members of the population distributed themselves about equally in the four pens, but the male population was concentrated almost overwhelmingly in the middle pens.

One major factor in the creation of this state of affairs was the struggle for status that took place among the males. Shortly after male rats reach maturity, at about six months of age, they enter into a round robin of fights that eventually fixes their position in the social hierarchy. In our experiments such fights took place among the males in all the pens, both middle and end. In the end pens, however, it became possible for a single dominant male to take over the area as his territory. During the period when the social hierarchy was being established, the subordinate males in all pens adopted the habit of arising early. This enabled them to eat and drink in peace. Since rats generally eat in the course of their normal wanderings, the subordinate residents of the end pens were likely to feed in one of the middle pens. When, after feeding, they wanted to

FOOD HOPPER used in first series of experiments is seen at the left in this drawing. Water tray is at the right. The hopper, covered with wire grating and holding hard pellets of food, made eating a lengthy activity during which one rat was likely to meet another.

Thus it fostered the development of a behavioral sink: the animals would eat only in the presence of others, and they preferred one of the four hoppers in the room to all the others. In time 75 per cent of the animals crowded into the pen containing this hopper to eat.

WATER FOUNTAIN used in second series of experiments is seen at the right in this drawing. Food hopper is at the left. The fountain was operated by pressing a lever. Thus it made drinking a lengthy activity, associated with the presence of others. But it

did not create a behavioral sink. Although the rats would drink only if other animals were present, they engaged in this activity in their home pens, immediately after awakening. The fountain therefore acted to produce an even distribution of the population.

return to their original quarters, they would find it very difficult. By this time the most dominant male in the pen would probably have awakened, and he would engage the subordinates in fights as they tried to come down the one ramp to the pen. For a while the subordinate would continue its efforts to return to what had been its home pen, but after a succession of defeats it would become so conditioned that it would not even make the attempt. In essence the dominant male established his territorial dominion and his control over a harem of females not by driving the other males out but by preventing their return.

Once a male had established his dominion over an end pen and the harem it contained, he was usually able to maintain it. Although he slept a good deal of the time, he made his sleeping quarters at the base of the ramp. He was, therefore, on perpetual guard. Awakening as soon as another male appeared at the head of the ramp, he had only to open his eyes for the invader to wheel around and return to the adjoining pen. On the other hand, he would sleep calmly through all the comings and goings of his harem; seemingly he did not even hear their clatterings up and down the wire ramp. His conduct during his waking hours reflected his dominant status. He would move about in a casual and deliberate fashion, occasionally inspecting the burrow and nests of his harem. But he would rarely enter a burrow, as some other males did, merely to ferret out the females.

A territorial male might tolerate other males in his domain provided they respected his status. Such subordinate males inhabited the end pens in several of the experiments. Phlegmatic animals, they spent most of their time hidden in the burrow with the adult females, and their excursions to the floor lasted only as long as it took them to obtain food and water. Although they never attempted to engage in sexual activity with any of the females, they were likely, on those rare occasions when they encountered the dominant male, to make repeated attempts to mount him. Generally the dominant male tolerated these advances.

In these end pens, where population density was lowest, the mortality rate among infants and females was also low. Of the various social environments that developed during the course of the experiments, the brood pens, as we called them, appeared to be the only healthy ones, at least in terms of the survival of the group. The harem females generally made good mothers. They nursed their young, built nests for them and protected them from harm. If any situation arose that a mother considered a danger to her pups, she would pick the infants up one at a time and carry them in her mouth to a safer place. Nothing would distract her from this task until the entire litter had been moved. Half the infants born in the brood pens survived.

The pregnancy rates recorded among the females in the middle pens were no lower than those recorded in the end pens. But a smaller percentage of these pregnancies terminated in live births. In the second series of experiments 80 per cent of the infants born in the middle pens died before weaning. In the first series 96 per cent perished before this time. The males in the middle pens were no less affected than the females by the pressures of population density. In both series of experiments the social pathology among the males was high. In the first series, however, it was more aggravated than it was in the second.

This increase in disturbance among the middle-pen occupants of the first series of experiments was directly related to the development of the phenomenon of the behavioral sink—the outcome of any behavioral process that collects animals together in unusually great numbers. The unhealthy connotations of the term are not accidental: a behavioral sink does act to aggravate all forms of pathology that can be found within a group.

The emergence of a behavioral sink was fostered by the arrangements that were made for feeding the animals. In these experiments the food consisted of small, hard pellets that were kept in a circular hopper formed by wire mesh. In consequence satisfaction of hunger required a continuous effort lasting several minutes. The chances therefore were good that while one rat was eating another would join it at the hopper. As was mentioned earlier, rats usually eat intermittently throughout their waking hours, whenever they are hungry and food is available. Since the arrangement of the ramps drew more rats into the middle pens than into the end ones, it was in these pens that individuals were most likely to find other individuals eating. As the population increased, the association of eating with the presence of other animals was further reinforced. Gradually the social aspect of the activity became determinant: the rats would rarely eat except at hoppers already in use by other animals.

At this point the process became a vicious circle. As more and more of the rats tended to collect at the hopper in one of the middle pens, the other hoppers became less desirable as eating places. The rats that were eating at these undesirable locations, finding themselves deserted by their groupmates, would transfer their feeding to the more crowded pen. By the time the three experiments in the first series drew to a close half or more of the populations were sleeping as well as eating in that pen. As a result there was a decided increase in the number of social adjustments each rat had to make every day. Regardless of which pen a rat slept in, it would go to one particular middle pen several times a day to eat. Therefore it was compelled daily to make some sort of adjustment to virtually every other rat in the experimental population.

No behavioral sinks developed in the second series of experiments, because we offered the rats their diet in a different way. A powdered food was set out in an open hopper. Since it took the animals only a little while to eat, the probability that two animals would be eating simultaneously was considerably reduced. In order to foster the emergence of a behavioral sink I supplied the pens with drinking fountains designed to prolong the drinking activity. The effect of this arrangement was unquestionably to make the animals social drinkers; they used the fountain mainly when other animals lined up at it. But the effect was also to discourage them from wandering and to prevent the development of a behavioral sink. Since rats generally drink immediately on arising, drinking and the social interaction it occasioned tended to keep them in the pens in which they slept. For this reason all social pathology in the second series of experiments, although severe, was less extreme than it was in the first series.

Females that lived in the densely populated middle pens became progressively less adept at building adequate nests and eventually stopped building nests at all. Normally rats of both sexes build nests, but females do so most vigorously around the time of parturition. It is an undertaking that involves repeated periods of sustained activity, searching out appropriate materials (in our experiments strips of paper supplied an abundance), transporting them bit by bit to the nest and there arranging them to form a cuplike depression, frequently sheltered by a hood. In a crowded middle pen, however, the ability of females to persist in this biologically essential activity became markedly impaired. The first sign of disruption was a failure to build the nest to normal specifications.

These females simply piled the strips of paper in a heap, sometimes trampling them into a pad that showed little sign of cup formation. Later in the experiment they would bring fewer and fewer strips to the nesting site. In the midst of transporting a bit of material they would drop it to engage in some other activity occasioned by contact and interaction with other individuals met on the way. In the extreme disruption of their behavior during the later months of the population's history they would build no nests at all but would bear their litters on the sawdust in the burrow box.

SLEEPING FEMALES    DRINKING
SLEEPING MALES    EATING

The middle-pen females similarly lost the ability to transport their litters from one place to another. They would move only part of their litters and would scatter them by depositing the infants in different places or simply dropping them on the floor of the pen. The infants thus abandoned throughout the pen were seldom nursed. They would die where they were dropped and were thereupon generally eaten by the adults.

The social stresses that brought about this disorganization in the behavior of the middle-pen females were imposed with special weight on them when they came into heat. An estrous female would be pursued relentlessly by a pack of males, unable to escape from their soon

unwanted attentions. Even when she retired to a burrow, some males would follow her. Among these females there was a correspondingly high rate of mortality from disorders in pregnancy and parturition. Nearly half of the first- and second-generation females that lived in the behavioral-sink situation had died of these causes by the end of the 16th month. Even in the absence of the extreme stresses of the behavioral sink, 25 per cent of the females died. In contrast, only 15 per cent of the adult males in both series of experiments died.

A female that lived in a brood pen was sheltered from these stresses even though during her periods of estrus she would leave her pen to mate with males

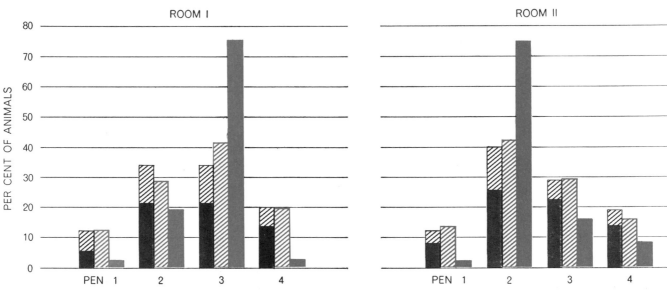

BEHAVIORAL SINK developed in the first series of three experiments, drawing half the rats either into pen 2 or pen 3 of each room to drink and sleep, and even more into that pen to eat. Chart describes the situation in the 13th month of the experiment. By then the population distributions were fairly stable and many females in the densely populated pens had died. One male in room

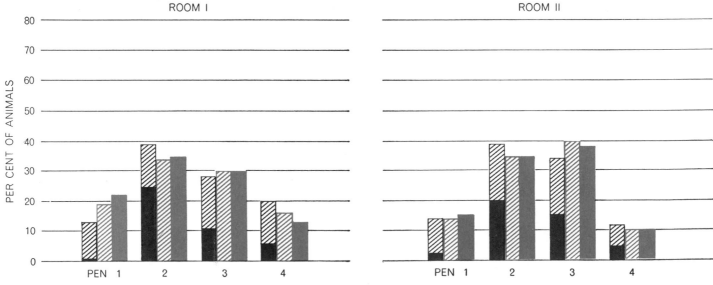

POPULATION DISTRIBUTIONS in the second series of three experiments, in which no behavioral sink developed, were more even than they were in the first series, and the death rate among females and infants was lower. Chart shows the situation in the 13th month, when one male had established pens 3 and 4 of room III as his territory, and another was taking over pen 2, thus

in the other pens of the room. Once she was satiated, however, she could return to the brood pen. There she was protected from the excessive attention of other males by the territorial male.

For the effect of population density on the males there is no index as explicit and objective as the infant and maternal mortality rates. We have attempted a first approximation of such an index, however, by scoring the behavior of the males on two scales: that of dominance and that of physical activity. The first index proved particularly effective in the early period of the experiments, when the males were approaching adulthood and beginning the fights that eventually fixed their status in the social hierarchy.

ROOM III

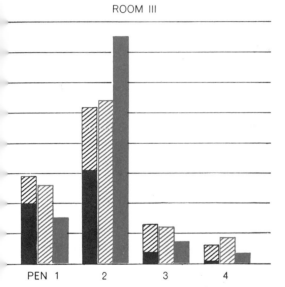

III had established pens 3 and 4 as his territory. Subsequently a male in room I took over pen 1, expelling all the other males.

ROOM III

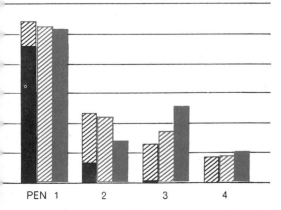

PEN 1     2     3     4

forcing most of the males into pen 1. Pen 1 in rooms I and II had also become territories; later pen 4 in room II became a territory.

The more fights a male initiated and the more fights he won, the more likely he was to establish a position of dominance. More than half the animals in each experiment gave up the struggle for status after a while, but among those that persisted a clear-cut hierarchy developed.

In the crowded middle pens no one individual occupied the top position in this hierarchy permanently. In every group of 12 or more males one was the most aggressive and most often the victor in fights. Nevertheless, this rat was periodically ousted from his position. At regular intervals during the course of their waking hours the top-ranking males engaged in free-for-alls that culminated in the transfer of dominance from one male to another. In between these tumultuous changings of the guard relative calm prevailed.

The aggressive, dominant animals were the most normal males in our populations. They seldom bothered either the females or the juveniles. Yet even they exhibited occasional signs of pathology, going berserk, attacking females, juveniles and the less active males, and showing a particular predilection —which rats do not normally display— for biting other animals on the tail.

Below the dominant males both on the status scale and in their level of activity were the homosexuals—a group perhaps better described as pansexual. These animals apparently could not discriminate between appropriate and inappropriate sex partners. They made sexual advances to males, juveniles and females that were not in estrus. The males, including the dominants as well as the others of the pansexuals' own group, usually accepted their attentions. The general level of activity of these animals was only moderate. They were frequently attacked by their dominant associates, but they very rarely contended for status.

Two other types of male emerged, both of which had resigned entirely from the struggle for dominance. They were, however, at exactly opposite poles as far as their levels of activity were concerned. The first were completely passive and moved through the community like somnambulists. They ignored all the other rats of both sexes, and all the other rats ignored them. Even when the females were in estrus, these passive animals made no advances to them. And only very rarely did other males attack them or approach them for any kind of play. To the casual observer the passive animals would have appeared to be the healthiest and most attractive

members of the community. They were fat and sleek, and their fur showed none of the breaks and bare spots left by the fighting in which males usually engage. But their social disorientation was nearly complete.

Perhaps the strangest of all the types that emerged among the males was the group I have called the probers. These animals, which always lived in the middle pens, took no part at all in the status struggle. Nevertheless, they were the most active of all the males in the experimental populations, and they persisted in their activity in spite of attacks by the dominant animals. In addition to being hyperactive, the probers were both hypersexual and homosexual, and in time many of them became cannibalistic. They were always on the alert for estrous females. If there were none in their own pens, they would lie in wait for long periods at the tops of the ramps that gave on the brood pens and peer down into them. They always turned and fled as soon as the territorial rat caught sight of them. Even if they did not manage to escape unhurt, they would soon return to their vantage point.

The probers conducted their pursuit of estrous females in an abnormal manner. Mating among rats usually involves a distinct courtship ritual. In the first phase of this ritual the male pursues the female. She thereupon retires for a while into the burrow, and the male lies quietly in wait outside, occasionally poking his head into the burrow for a moment but never entering it. (In the wild forms of the Norway rat this phase usually involves a courtship dance on the mound at the mouth of the burrow.) The female at last emerges from the burrow and accepts the male's advances. Even in the disordered community of the middle pens this pattern was observed by all the males who engaged in normal heterosexual behavior. But the probers would not tolerate even a short period of waiting at the burrows in the pens where accessible females lived. As soon as a female retired to a burrow, a prober would follow her inside. On these expeditions the probers often found dead young lying in the nests; as a result they tended to become cannibalistic in the later months of a population's history.

Although the behavioral sink did not develop in the second series of experiments, the pathology exhibited by the populations in both sets of experiments, and in all pens, was severe. Even

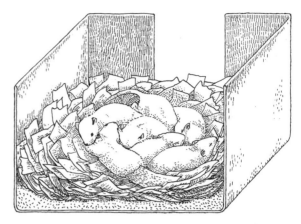

**NORMAL MATERNAL BEHAVIOR** among rats includes building a fluffy, well-shaped nest for the young. The drawing at the left shows such a nest, holding a recently born litter. The drawing at the right shows this same nest about two weeks later. It has been flattened by the weight of the animals' bodies but it still offers ample protection and warmth, and the remaining pups can still rest comfortably. In these experiments half the offspring of normal mothers survived infancy and were successfully weaned.

**ABNORMAL MATERNAL BEHAVIOR,** shown by females exposed to the pressures of population density, includes failure to build adequate nests. The drawing at the left shows the recently born young of a disturbed female. She started to make a nest but never finished it. The drawing at the right shows her young about two weeks later. One pup has already left and another is leaving. Neither can survive alone. In these experiments the mortality rate among infants of disturbed mothers was as high as 96 per cent.

in the brood pens females could raise only half their young to weaning. Nor does the difference in infant mortality between the middle pens of the first and second series—96 per cent in the first as opposed to 80 per cent in the second—represent a biologically significant improvement.

provement. It is obvious that the behavioral repertoire with which the Norway rat has emerged from the trials of evolution and domestication must break down under the social pressures generated by population density. In time, refinement of experimental procedures

and of the interpretation of these studies may advance our understanding to the point where they may contribute to the making of value judgments about analogous problems confronting the human species.

# 18

# Cybernetics

NORBERT WIENER
*November 1948*

Cybernetics is a word invented to define a new field in science. It combines under one heading the study of what in a human context is sometimes loosely described as thinking and in engineering is known as control and communication. In other words, cybernetics attempts to find the common elements in the functioning of automatic machines and of the human nervous system, and to develop a theory which will cover the entire field of control and communication in machines and in living organisms.

It is well known that between the most complex activities of the human brain and the operations of a simple adding machine there is a wide area where brain and machine overlap. In their more elaborate forms, modern computing machines are capable of memory, association, choice and many other brain functions. Indeed, the experts have gone so far in the elaboration of such machines that we can say the human brain behaves very much like the machines. The construction of more and more complex mechanisms actually is bringing us closer to an understanding of how the brain itself operates.

The word cybernetics is taken from the Greek *kybernetes,* meaning steersman. From the same Greek word, through the Latin corruption *gubernator,* came the term governor, which has been used for a long time to designate a certain type of control mechanism, and was the title of a brilliant study written by the Scottish physicist James Clerk Maxwell 80 years ago. The basic concept which both Maxwell and the investigators of cybernetics mean to describe by the choice of this term is that of a feedback mechanism, which is especially well represented by the steering engine of a ship. Its meaning is made clear by the following example.

Suppose that I pick up a pencil. To do this I have to move certain muscles. Only an expert anatomist knows what all these muscles are, and even an anatomist could hardly perform the act by a conscious exertion of the will to contract each muscle concerned in succession. Actually what we will is not to move individual muscles but to pick up the pencil. Once we have determined on this, the motion of the arm and hand proceeds in such a way that we may say that the amount by which the pencil is not yet picked up is decreased at each stage. This part of the action is not in full consciousness.

To perform an action in such a manner, there must be a report to the nervous system, conscious or unconscious, of the amount by which we have failed to pick up the pencil at each instant. The report may be visual, at least in part, but it is more generally kinesthetic, or to use a term now in vogue, proprioceptive. If the proprioceptive sensations are wanting, and we do not replace them by a visual or other substitute, we are unable to perform the act of picking up the pencil, and find ourselves in a state known as ataxia. On the other hand, an excessive feedback is likely to be just as serious a handicap. In the latter case the muscles overshoot the mark and go into an uncontrollable oscillation. This condition, often associated with injury to the cerebellum, is known as purpose tremor.

Here, then, is a significant parallel between the workings of the nervous system and of certain machines. The feedback principle introduces an important new idea in nerve physiology. The central nervous system no longer appears to be a self-contained organ receiving signals from the senses and discharging into the muscles. On the contrary, some of its most characteristic activities are explainable only as circular processes, traveling from the nervous system into the muscles and re-entering the nervous system through the sense organs. This finding seems to mark a step forward in the study of the nervous system as an integrated whole.

The new approach represented by cybernetics—an integration of studies which is not strictly biological or strictly physical, but a combination of the two—has already given evidence that it may help to solve many problems in engineering, in physiology and very likely in psychiatry.

This work represents the outcome of a program undertaken jointly several years ago by the writer and Arturo Rosenblueth, then of the Harvard Medical School and now of the National Institute of Cardiology of Mexico. Dr. Rosenblueth is a physiologist; I am a mathematician. For many years Dr. Rosenblueth and I had shared the conviction that the most fruitful areas for the growth of the sciences were those which had been neglected as no-man's lands between the various established fields. Dr. Rosenblueth always insisted that a proper exploration of these blank spaces on the map of science could be made only by a team of scientists, each a specialist but each possessing a thoroughly sound acquaintance with the fields of his fellows.

Our collaboration began as the result of a wartime project. I had been assigned, with a partner, Julian H. Bigelow, to the problem of working out a fire-control apparatus for anti-aircraft artillery which would be capable of tracking the curving course of a plane and predicting its future position. We soon came to the conclusion that any solution of the problem must depend heavily on the feedback principle, as it operated not only in the apparatus but in the human operators of the gun and of the plane. We approached

Dr. Rosenblueth with a specific question concerning oscillations in the nervous system, and his reply, which cited the phenomenon of purpose tremor, confirmed our hypothesis about the importance of feedback in voluntary activity.

The ideas suggested by this discussion led to several joint experiments, one of which was a study of feedback in the muscles of cats. The scope of our investigations steadily widened, and as it did so scientists from widely diverse fields joined our group. Among them were the mathematicians John von Neumann of the Institute for Advanced Study and Walter Pitts of Massachusetts Institute of Technology; the physiologists Warren McCulloch of the University of Pennsylvania and Lorente de No of the Rockefeller Institute; the late Kurt Lewin, psychologist, of M.I.T.; the anthropologists Gregory Bateson and Margaret Mead; the economist Oskar Morgenstern of the Institute for Advanced Study; and others in psychology, sociology, engineering, anatomy, neurophysiology, physics, and so on.

The study of cybernetics is likely to have fruitful applications in many fields, from the design of control mechanisms for artificial limbs to the almost complete mechanization of industry. But in our view it encompasses much wider horizons. If the 17th and early 18th centuries were the age of clocks, and the latter 18th and 19th centuries the age of steam engines, the present time is the age of communication and control. There is in electrical engineering a division which is known as the split between the technique of strong currents and the technique of weak currents; it is this split which separates the age just passed from that in which we are living. What distinguishes communication engineering from power engineering is that the main interest of the former is not the economy of energy but the accurate reproduction of a signal.

At every stage of technique since Daedalus, the ability of the artificer to produce a working simulacrum of a living organism has always intrigued people. In the days of magic, there was the bizarre and sinister concept of the Golem, that figure of clay into which the rabbi of Prague breathed life. In Isaac Newton's time the automaton became the clockwork music box. In the 19th century, the automaton was a glorified heat engine, burning a combustible fuel instead of the glycogen of human muscles. The automaton of our day opens doors by means of photocells, or points guns to the place at which a radar beam picks up a hostile airplane, or computes the solution of a differential equation.

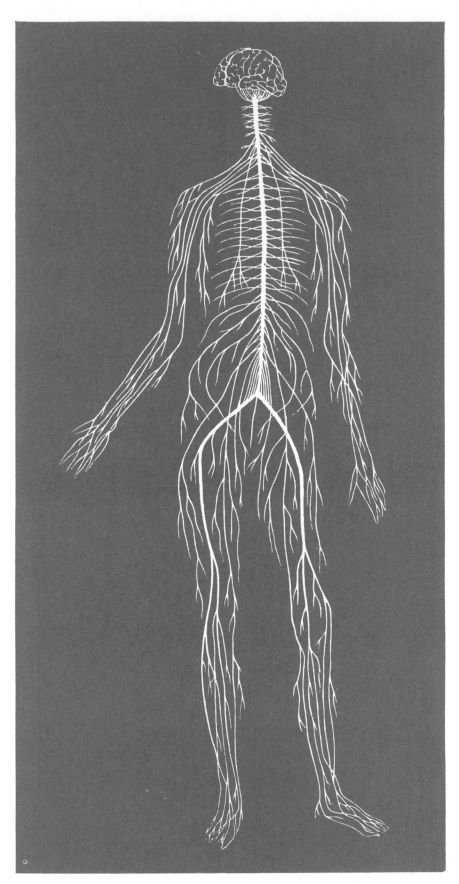

**THE NERVOUS SYSTEM,** in the cybernetic view, is more than a self-contained apparatus for receiving and transmitting signals. It is a circuit in which a feedback principle operates as certain impulses enter muscles and re-enter the nervous system through the sense organs.

Under the influence of the prevailing view in the science of the 19th century, the engineering of the body was naturally considered to be a branch of power engineering. Even today this is the predominant point of view among classically minded, conservative physiologists. But we are now coming to realize that the body is very far from a conservative system, and that the power available to it is much less limited than was formerly believed. We are beginning to see that such important elements as the neurones—the units of the nervous complex of our bodies —do their work under much the same conditions as vacuum tubes, their relatively

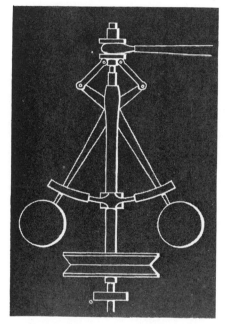

GOVERNOR of a steam engine is an example of feedback, one of the most important fundamental concepts in cybernetics.

small power being supplied from outside by the body's circulation, and that the bookkeeping which is most essential to describe their function is not one of energy.

In short, the newer study of automata, whether in the metal or in the flesh, is a branch of communications engineering, and its cardinal ideas are those of the message, of the amount of disturbance or "noise" (a term taken from the telephone engineer), of the quantity of information to be transmitted, of coding technique, and so on.

This view obviously has implications which affect many branches of science. Let us consider here the application of cybernetics to the problem of mental disorders. The realization that the brain and computing machines have much in common may suggest new and valid approaches to psychopathology, and even to psychiatry.

These begin with perhaps the simplest question of all: how the brain avoids gross blunders or gross miscarriages of activity due to the malfunction of individual parts. Similar questions referring to the computing machine are of great practical importance, for here a chain of operations, each of which covers only a fraction of a millionth of a second, may last a matter of hours or days. It is quite possible for a chain of computational operations to involve a billion separate steps. Under these circumstances, the chance that at least one operation will go amiss is far from negligible, even though the reliability of modern electronic apparatus has exceeded the most sanguine expectations.

In ordinary computational practice by hand or by desk machines, it is the custom to check every step of the computation and, when an error is found, to localize it by a backward process starting from the first point where the error is noted. To do this with a high-speed machine, the check must proceed at the pace of the original machine, or the whole effective order of speed of the machine will conform to that of the slower process of checking.

A much better method of checking, and in fact the one generally used in practice, is to refer every operation simultaneously to two or three separate mechanisms. When two such mechanisms are used, their answers are automatically collated against each other; and if there is a discrepancy, all data are transferred to permanent storage, the machine stops and a signal is sent to the operator that something is wrong. The operator then compares the results, and is guided by them in his search for the malfunctioning part, perhaps a tube which has burned out and needs replacement. If three separate mechanisms are used for each stage, there will practically always be agreement between two of the three mechanisms, and this agreement will give the required result. In this case the collation mechanism accepts the majority report, and the machine need not stop. There is a signal, however, indicating where and how the minority report differs from the majority report. If this occurs at the first moment of discrepancy, the indication of the position of the error may be very precise.

It is conceivable, and not implausible, that at least two of the elements of this process are also represented in the nervous system. It is hardly to be expected that any important message is entrusted for transmission to a single neurone, or that an important operation is entrusted to a single neuronal mechanism. Like the computing machine, the brain probably

works on a variant of the famous principle expounded by Lewis Carroll in *The Hunting of the Snark:* "What I tell you three times is true."

It is also improbable that the various channels available for the transfer of information generally go from one end of their course to the other without connecting with one another. It is much more probable that when a message reaches a certain level of the nervous system, it may leave that point and proceed to the next by one or more alternative routes. There may be parts of the nervous system, especially in the cortex, where this interchangeability is much limited or abolished. Still, the principle holds, and it probably holds most clearly for the relatively unspecialized cortical areas which serve the purpose of association and of what we call the higher mental functions.

So far we have been considering errors in performance that are normal, and pathological only in an extended sense. Let us now turn to those that are much more clearly pathological. Psychopathology has been rather a disappointment to the instinctive materialism of the doctors, who have taken the view that every disorder must be accompanied by actual lesions of some specific tissue involved. It is true that specific brain lesions, such as injuries, tumors, clots and the like, may be accompanied by psychic symptoms, and

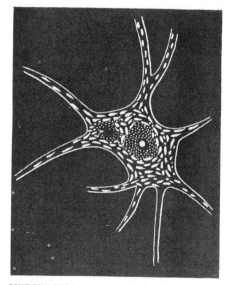

NERVE CELL performs its functions under much the same conditions as a vacuum tube, obtaining its power from outside.

that certain mental diseases, such as paresis, are the sequelae of general bodily disease and show a pathological condition of the brain tissue. But there is no way of identifying the brain of a schizophrenic of one of the strict Kraepelin types, nor of a manic-depressive patient, nor of a paranoiac. These we call functional disorders.

This distinction between functional and organic disorders is illuminated by the consideration of the computing machine. It is not the empty physical structure of the computing machine that corresponds to the brain—to the adult brain, at least— but the combination of this structure with the instructions given it at the beginning of a chain of operations and with all the additional information stored and gained from outside in the course of its operation.

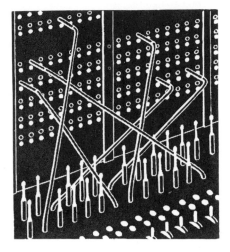

TELEPHONE EXCHANGE, when it is over- loaded, has breakdowns rather similar to the kind that occur in human beings.

This information is stored in some physical form — in the form of memory. But part of it is in the form of circulating memories, with a physical basis that vanishes when the machine is shut down or the brain dies, and part is in the form of long-time memories, which are stored in a way at which we can only guess, but probably also in a form with a physical basis that vanishes at death.

There is therefore nothing surprising in considering the functional mental disorders fundamentally as diseases of memory, of the circulating information kept by the brain in active state and of the long-time permeability of synapses. Even the grosser disorders such as paresis may produce a large part of their effects not so much by the destruction of tissue which they involve and the alteration of synaptic thresholds as by the secondary disturbances of traffic, the overload of what remains of the nervous system and the rerouting of messages which must follow such primary injuries.

In a system containing a large number of neurones, circular processes can hardly be stable for long periods of time. Either they run their course, dissipate themselves and die out, as in the case of memories belonging to the specious present, or they embrace more and more neurones in their system, until they occupy an inordinate part of the neurone pool. This is what we should expect to be the case in the malignant worry that accompanies anxiety neuroses. In such a case, it is possible that the patient simply does not have the room —i.e., a sufficient number of neurones— to carry out his normal processes of thought. Under such conditions, there may be less going on in the brain to occupy the neurones not yet affected, so that they are all the more readily involved in the expanding process. Furthermore, the permanent memory becomes more and more deeply involved, and the pathological process which began at the level of the circulating memories may repeat itself in a more intractable form at the level of the permanent memories. Thus what started as a relatively trivial and accidental disturbance of stability may build itself up into a process totally destructive to the normal mental life.

Pathological processes of a somewhat similar nature are not unknown in the case of mechanical or electrical computing machines. A tooth of a wheel may slip under such conditions that no tooth with which it engages can pull it back into its normal relations, or a high-speed electrical computing machine may go into a circular process that seems impossible to stop.

How do we deal with these accidents in the case of the machine? We first try to clear the machine of all information, in the hope that when it starts again with different data the difficulty will not recur. If this fails and the difficulty is inaccessible to the clearing mechanism, we shake the machine or, if it is electrical, subject it to an abnormally large electrical impulse in the hope that we may jolt the inaccessible part into a position where the false cycle of its activities will be interrupted. If even this fails, we may disconnect an erring part of the apparatus, for it is possible that what remains may be adequate for our purpose.

In the case of the brain, there is no normal process, except death, that can clear it of all past impressions. Of the normal non-fatal processes, sleep comes closest to clearing the brain. How often we find that the best way to handle a complicated worry or an intellectual muddle is to sleep on it! Sleep, however, does not clear away the deeper memories, nor indeed is a malignant state of worry compatible with adequate sleep.

Thus we are often forced to resort to more violent types of intervention in the memory cycle. The most violent of these involve surgery on the brain, leaving behind permanent damage, mutilation and the abridgement of the powers of the victim, for the mammalian central nervous system seems to possess no power of regeneration. The principal type of surgical intervention that has been practiced is known as prefrontal lobotomy, or leucotomy. It consists in the removal or isolation of a portion of the prefrontal lobe of the cortex. It is currently having a certain vogue, probably not unconnected with the fact that it makes the custodial care of many patients easier. (Let me remark in passing that killing them makes their custodial care still easier.) Prefrontal lobotomy does seem to have a genuine effect on malignant worry, not by bringing the patient nearer to a solution of his problem, but by damaging or destroying the capac-

AUTOMATON of the 15th century was one of a long series of attempts to produce a working simulacrum of a living organism.

ity for maintained worry, known in the terminology of another profession as the conscience. It appears to impair the circulating memory, i.e., the ability to keep in mind a situation not actually presented.

The various forms of shock treatment— electric, insulin, metrazol — are less drastic methods of doing a very similar thing. They do not destroy brain tissue, or at least are not intended to destroy it, but they do have a decidedly damaging effect on the memory. In so far as the shock treatment affects recent disordered memories, which are probably scarcely worth preserving anyhow, it has something to recommend it as against lobotomy, but it is sometimes followed by deleterious effects on the permanent memory and the personality. As it is used at present, it is another violent, imperfectly understood, imperfectly controlled method to interrupt a mental vicious circle.

In long-established cases of mental disorder, the permanent memory is as badly deranged as the circulating memory. We do not seem to possess any purely pharmaceutical or surgical weapon for intervening selectively in the permanent memory. This is where psychoanalysis and the other psychotherapeutic measures come in.

Whether psychoanalysis is taken in the orthodox Freudian sense or in the modified senses of Jung and of Adler, or whether the psychotherapy is not strictly psychoanalytic at all, the treatment is clearly based on the concept that the stored information of the mind lies on many levels of accessibility. The effect and accessibility of this stored information are vitally conditioned by affective experiences that we cannot always uncover by introspection. The technique of the psychoanalyst consists in a series of means to discover and interpret these hidden memories, to make the patient accept them for what they are, and thus to modify, if not their content, at least the affective tone they carry, and make them less harmful.

All this is perfectly consistent with the cybernetic point of view. Our theory perhaps explains, too, why there are circumstances in which a joint use of shock treatment and psychotherapy is indicated, combining a physical or pharmacological therapy for the malignant reverberations in the nervous system and a psychological therapy for the damaging long-time memories which might re-establish the vicious circle broken up by the shock treatments.

We have already mentioned the traffic problem of the nervous system. It has been noted by many writers that each form of organization has an upper limit of size beyond which it will not function. Thus insect organization is limited by the length of tubing over which the spiracle method of bringing air by diffusion directly to the breathing tissues will function; a land animal cannot be so big that the legs or other portions in contact with the ground will be crushed by its weight, and so on. The same sort of thing is observed in engineering structures. Skyscrapers are limited in size by the fact that when they exceed a certain height, the elevator space needed for the upper stories consumes an excessive part of the cross section of the lower floors. Beyond a certain span, the best pos-

sible suspension bridge will collapse under its own weight. Similarly, the size of a single telephone exchange is limited.

In a telephone system, the important limiting factor is the fraction of the time during which a subscriber will find it impossible to put a call through. A 90 per cent chance of completing calls is probably good enough to permit business to be carried on with reasonable facility. A success of 75 per cent is annoying but will permit business to be carried on after a fashion; if half the calls are not completed, subscribers will begin to ask to have their telephones taken out. Now, these represent all-over figures. If the calls go through a number of distinct stages of switching, and the probability of failure is independent and equal for each stage, in order to get a high probability of final success the probability of success at each stage must be higher than the final one. Thus to obtain a 75 per cent chance for the completion of the call after five stages, we must have about 95 per cent chance of success at each stage. The more stages there are, the more rapidly the service becomes extremely bad when a critical level of failure for the individual call is exceeded, and extremely good when this critical level of failure is not quite reached. Thus a switching service involving many stages and designed for a certain level of failure shows no obvious signs of failure until the traffic comes up to the edge of the critical point, when it goes completely to pieces and we have a catastrophic traffic jam.

So man, with the best developed nervous system of all the animals, probably involving the longest chains of effectively operated neurones, is likely to perform a complicated type of behavior efficiently very close to the edge of an overload, when he will give way in a serious and catastrophic manner. This overload may take place in several ways: by an excess in the amount of traffic to be carried; by a physical removal of channels for the carrying of traffic; or by the excessive occupation of such channels by undesirable systems of traffic, such as circulating memories that have accumulated to the extent of becoming pathological worries. In all these cases, a point is reached—quite suddenly—when the normal traffic does not have space enough allotted to it, and we have a form of mental breakdown, very possibly amounting to insanity.

This will first affect the faculties or operations involving the longest chains of neurones. There is appreciable evidence, of various kinds, that these are precisely the processes recognized as the highest in our ordinary scale of valuation.

If we compare the human brain with that of a lower mammal, we find that it is much more convoluted. The relative thickness of the gray matter is much the same, but it is spread over a far more involved system of grooves and ridges. The effect of this is to increase the amount of gray matter at the expense of the amount of white matter. Within a ridge, this decrease of the white matter is largely a decrease in length rather than in number of fibers, as the opposing folds are nearer together than the same areas would be on a smooth-surfaced brain of the same size. On the other hand, when it comes to the connectors between different ridges, the distance they have to run is increased by the convolution of the brain.

Thus the human brain would seem to be fairly efficient in the matter of the short-distance connectors, but defective in the matter of long-distance trunk lines. This means that in the case of a traffic jam, the processes involving parts of the brain quite remote from one another should suffer first. That is, processes involving several centers, a number of different motor processes and a considerable number of association areas should be among the least stable in cases of insanity. These are precisely the processes which we should normally class as higher, thereby confirming our theory, as experience does also, that the higher processes deteriorate first in insanity.

The phenomena of handedness and of hemispheric dominance suggest other interesting speculations. Right-handedness, as is well known, is generally associated with left-brainedness, and left-handedness with right-brainedness. The dominant hemisphere has the lion's share of the higher cerebral functions. In the adult, the effect of an extensive injury in the secondary hemisphere is far less serious than the effect of a similar injury in the dominant hemisphere. At a relatively early stage in his career, Louis Pasteur suffered a cerebral hemorrhage on the right side which left him with a moderate degree of one-sided paralysis. When he died, his brain was examined and the damage to its right side was found to be so extensive that it has been said that after his injury "he had only half a brain." Nevertheless, after his injury he did some of his best work. A similar injury to the left side of the brain in a right-handed adult would almost certainly have been fatal; at the least it would have reduced the patient to an animal condition.

In the first six months of life, an extensive injury to the dominant hemisphere may compel the normally secondary hemi-

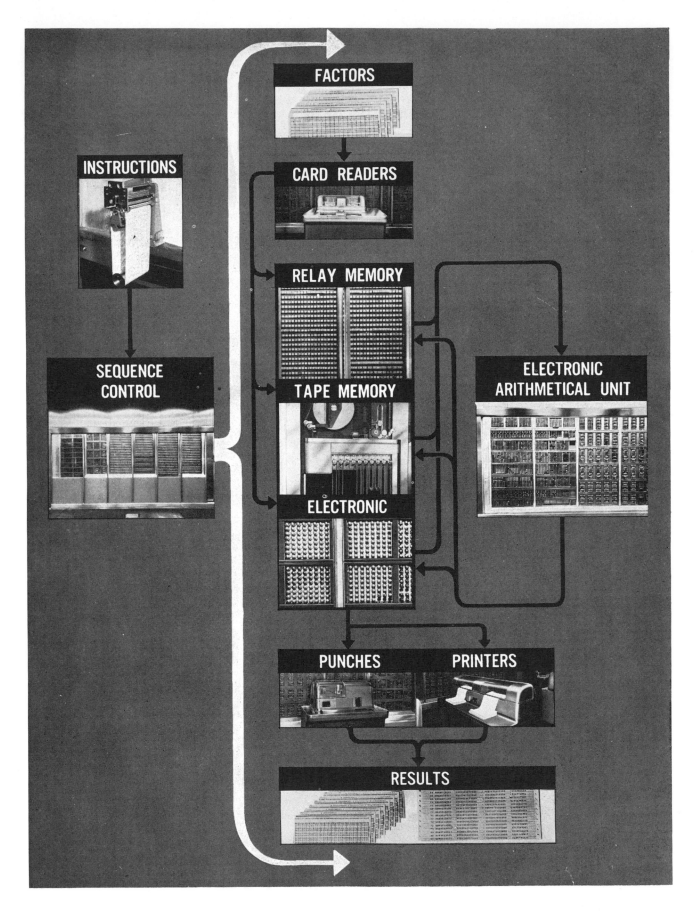

DIAGRAM of the Selective Sequence Electronic Calculator built by the International Business Machines Corporation, provides another cybernetic comparison. Physical structure of the machine is not analogous to the brain. The structure plus instructions and stored memories is analogous. The machine has electronic and relay circuits for temporary memory, punched cards for permanent memory.

sphere to take its place, so that the patient appears far more nearly normal than he would have been had the injury occurred at a later stage. This is quite in accordance with the great flexibility shown by the nervous system in the early weeks of life. It is possible that, short of very serious injuries, handedness is reasonably flexible in the very young child. Long before the child is of school age, however, the natural handedness and cerebral dominance are established for life. Many people have changed the handedness of their children by education, though of course they could not change its physiological basis in hemispheric dominance. These hemispheric changelings often become stutterers and develop other defects of speech, reading and writing.

We now see at least one possible explanation for this phenomenon. With the education of the secondary hand, there has been a partial education of that part of the secondary hemisphere which deals with skilled motions such as writing. Since these motions are carried out in the closest possible association with reading, and with speech and other activities which are inseparably connected with the dominant hemisphere, the neurone chains involved in these processes must cross over from hemisphere to hemisphere, and in any complex activity they must do this again and again. But the direct connectors between the hemispheres in a brain as large as that of man are so few in number that they are of very little help. Consequently the interhemispheric traffic must go by roundabout routes through the brain stem. We know little about these routes, but they are certainly long, scanty and subject to interruption. As a consequence, the processes associated with speech and writing are very likely to be involved in a traffic jam, and stuttering is the most natural thing in the world.

The human brain is probably too large already to use in an efficient manner all the facilities which seem to be present. In a cat, the destruction of the dominant hemisphere seems to produce relatively less damage than in man, while the destruction of the secondary hemisphere probably produces more damage. At any rate, the apportionment of function in the two hemispheres is more nearly equal. In man, the gain achieved by the increase in the size and complexity of the brain is partly nullified by the fact that less of the organ can be used effectively at one time.

It is interesting to reflect that we may be facing one of those limitations of nature in which highly specialized organs reach a level of declining efficiency and ultimately lead to the extinction of the species. The human brain may be as far along on its road to destructive specialization as the great nose horns of the last of the titanotheres.

# 19

# Feedback

ARNOLD TUSTIN

*September 1952*

FOR hundreds of years a few examples of true automatic control systems have been known. A very early one was the arrangement on windmills of a device to keep their sails always facing into the wind. It consisted simply of a miniature windmill which could rotate the whole mill to face in any direction. The small mill's sails were at right angles to the main ones, and whenever the latter faced in the wrong direction, the wind caught the small sails and rotated the mill to the correct position. With steam power came other automatic mechanisms: the engine-governor, and then the steering servo-engine on ships, which operated the rudder in correspondence with movements of the helm. These devices, and a few others such as simple voltage regulators, constituted man's achievement in automatic control up to about 20 years ago.

In the past two decades necessity, in the form of increasingly acute problems arising in our ever more complex technology, has given birth to new families of such devices. Chemical plants needed regulators of temperature and flow; air warfare called for rapid and precise control of searchlights and anti-aircraft guns; radio required circuits which would give accurate amplification of signals.

Thus the modern science of automatic control has been fed by streams from many sources. At first, it now seems surprising to recall, no connection between these various developments was recognized. Yet all control and regulating systems depend on common principles. As soon as this was realized, progress became much more rapid. Today the design of controls for a modern boiler or a guided missile, for example, is based largely on principles first developed in the design of radio amplifiers.

Indeed, studies of the behavior of automatic control systems give us new insight into a wide variety of happenings in nature and in human affairs. The notions that engineers have evolved from these studies are useful aids in understanding how a man stands upright without toppling over, how the human heart beats, why our economic system suffers from slumps and booms, why the rabbit population in parts of Canada regularly fluctuates between scarcity and abundance.

The chief purpose of this article is to make clear the common pattern that underlies all these and many other varied phenomena. This common pattern is the existence of feedback, or—to express the same thing rather more generally—interdependence.

We should not be able to live at all, still less to design complex control systems, if we did not recognize that there are regularities in the relationship between events—what we call "cause and effect." When the room is warmer, the thermometer on the wall reads higher. We do not expect to make the room warmer by pushing up the mercury in the thermometer. But now consider the case when the instrument on the wall is not a simple thermometer but a thermostat, contrived so that as its reading goes above a chosen setting, the fuel supply to the furnace is progressively reduced, and, conversely, as its reading falls below that setting, the fuel flow is increased. This is an example of a familiar control system. Not only does the reading of the thermometer depend on the warmth of the room, but the warmth of the room also depends on the reading of the thermometer. The two quantities are interdependent. Each is a cause, and each an effect, of the other. In such cases we have a closed chain or sequence—what engineers call a "closed loop" (*see diagram on the opposite page*).

In analyzing engineering and scientific problems it is very illuminating to sketch out first the scheme of dependence and see how the various quantities involved in the problem are determined by one another and by disturbances from outside the system. Such a diagram enables one to tell at a glance whether a system is an open or a closed one. This is an important distinction, because a closed system possesses several significant properties. Not only can it act as a regulator, but it is capable of

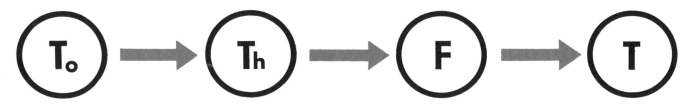

**OPEN SEQUENCE** of control is illustrated by a system for regulating the temperature of a room. $T_o$ is a variation in the temperature outdoors. Th is the variation of a thermometer. F is the fuel control of a furnace. T is the variation of the temperature in the room. In such a system of control there is no feedback.

various "self-excitatory" types of behavior—like a kitten chasing its own tail.

The now-popular name for this process is "feedback." In the case of the thermostat, the thermometer's information about the room temperature is fed back to open or close the valve, which in turn controls the temperature. Not all automatic control systems are of the closed-loop type. For example, one might put the thermometer outside in the open air, and connect it to work the fuel valve through a specially shaped cam, so that the outside temperature regulates the fuel flow. In this open-sequence system the room temperature has no effect; there is no feedback. The control compensates only that disturbance of room temperature caused by variation of the outdoor temperature. Such a system is not necessarily a bad or useless system; it might work very well under some circumstances. But it has two obvious shortcomings. Firstly, it is a "calibrated" system; that is to say, its correct working would require careful preliminary testing and special shaping of the cam to suit each particular application. Secondly, it could not deal with any but standard conditions. A day that was windy as well as cold would not get more fuel on that account.

The feedback type of control avoids these shortcomings. It goes directly to the quantity to be controlled, and it corrects indiscriminately for all kinds of disturbance. Nor does it require calibration for each special condition.

Feedback control, unlike open-sequence control, can never work without *some* error, for the error is depended upon to bring about the correction. The objective is to make the error as small as possible. This is subject to certain limitations, which we must now consider.

The principle of control by feedback is quite general. The quantities that it may control are of the most varied kinds, ranging from the frequency of a national electric-power grid to the degree of anesthesia of a patient under surgical operation. Control is exercised by negative feedback, which is to say that the information fed back is the amount of departure from the desired condition.

ANY QUANTITY may be subjected to control if three conditions are met. First, the required changes must be controllable by some physical means, a regulating organ. Second, the controlled quantity must be measurable, or at least comparable with some standard; in other words, there must be a measuring device. Third, both regulation and measurement must be rapid enough for the job in hand.

As an example, take one of the simplest and commonest of industrial requirements: to control the rate of flow of liquid along a pipe. As the regulating organ we can use a throttle valve, and as the measuring device, some form of flowmeter. A signal from the flowmeter, telling the actual rate of flow through the pipe, goes to the "controller"; there it is compared with a setting giving the required rate of flow. The amount and direction of "error," *i.e.*, deviation from this setting, is then transmitted to the throttle valve as an operating signal to bring about adjustment in the required direction (*see diagram at the top of page 131*).

In flow-control systems the signals are usually in the form of variations in air pressure, by which the flowmeter measures the rate of flow of the liquid. The pressure is transmitted through a small-bore pipe to the controller, which is essentially a balance piston. The difference between this received pressure and the setting regulates the air pressure in another pipeline that goes to the regulating valve.

Signals of this kind are slow, and difficulties arise as the system becomes complex. When many controls are concentrated at a central point, as is often the case, the air-pipes that transmit the signals may have to be hundreds of feet long, and pressure changes at one end reach the other only after delays of some seconds. Meanwhile the error may have become large. The time-delay often creates another problem: overcorrection of the error, which causes the system to oscillate about the required value instead of settling down.

For further light on the principles involved in control systems let us consider the example of the automatic gun-director. In this problem a massive gun must be turned with great precision to angles indicated by a fly-power pointer on a clock-dial some hundreds of feet away. When the pointer moves, the gun must turn correspondingly. The quantity to be controlled is the angle of the gun. The reference quantity is the angle of the clock-dial pointer. What is needed is a feedback loop which constantly compares the gun angle with the pointer angle and arranges matters so that if the gun angle is too small, the gun is driven forward, and if it is too large, the gun is driven back.

The key element in this case is some device which will detect the error of angular alignment between two shafts remote from each other, and which does not require more force than is available at the fly-power transmitter shaft. There are several kinds of electrical elements that will serve such a purpose. The one usually selected is a pair of the miniature alternating-current machines known as selsyns. The two selsyns, connected respectively to the transmitter shaft and the gun, provide an electrical signal proportional to the error of alignment. The signal is amplified and fed to a generator which in turn feeds a motor that

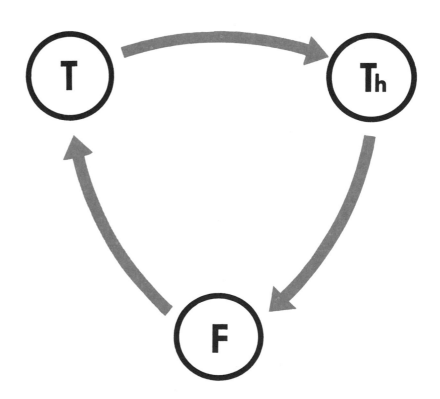

**CLOSED SEQUENCE** of control is illustrated by a system for regulating the temperature of a room by means of a thermostat. Here Th is a thermostat rather than a thermometer. In such a system there is feedback.

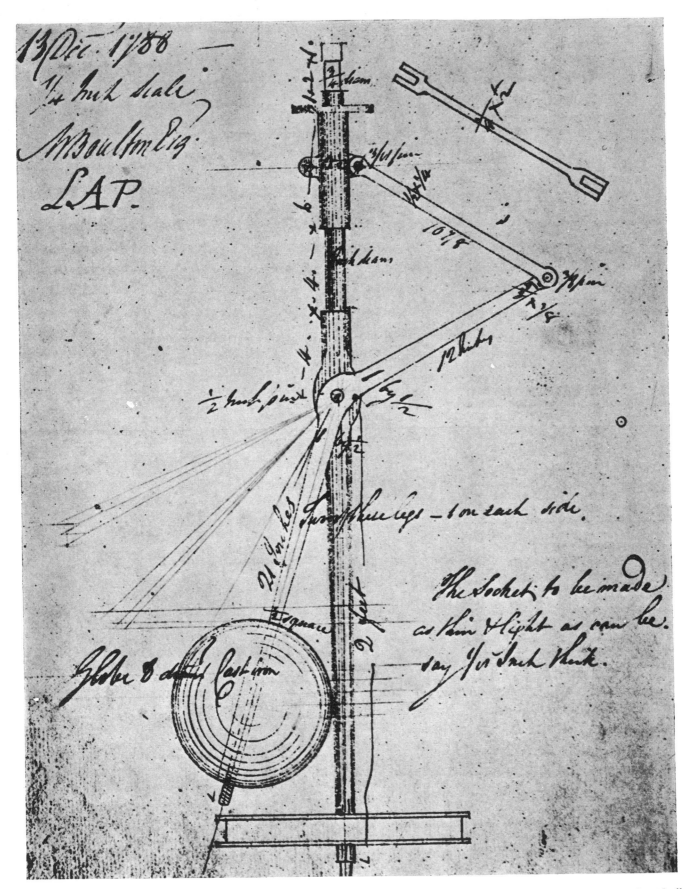

**EARLIEST KNOWN DRAWING** of the flyball-governor was made in 1788. The governor was invented by James Watt. At the upper left appear the date and the name of Watt's associate Matthew Boulton. Only half of a governor is shown in the drawing, but below the center are the words: "Two of these legs—1 on each side." Later Watt attempted to prevent the oscillation of the governor by fitting it with stops for the balls, one to keep them from coming too close together and the other to prevent them from "opening too wide asunder."

drives the gun (*see diagram on the next page*).

THIS GIVES the main lines of a practicable scheme, but if a system were built as just described, it would fail. The gun's inertia would carry it past the position of correct alignment; the new error would then cause the controller to swing it back, and the gun would hunt back and forth without ever settling down.

This oscillatory behavior, maintained by "self-excitation," is one of the principal limitations of feedback control. It is the chief enemy of the control-system designer, and the key to progress has been the finding of various simple means to prevent oscillation. Since oscillation is a very general phenomenon, it is worth while to look at the mechanism in detail, for what we learn about oscillation in man-made control systems may suggest means of inhibiting oscillations of other kinds—such as economic booms and slumps, or periodic swarms of locusts.

Consider any case in which a quantity that we shall call the output depends on another quantity we shall call the input. If the input quantity oscillates in value, then the output quantity also will oscillate, not simultaneously or necessarily in the same way, but with the same frequency. Usually in physical systems the output oscillation lags behind the input. For example, if one is boiling water and turns the gas slowly up and down, the amount of steam increases and decreases the same number of times per minute, but the maximum amount of steam in each cycle must come rather later than the maximum application of heat, because of the time required for heating. If the first output quantity in turn affects some further quantity, the variation of this second quantity in the sequence will usually lag still more, and so on. The lag (as a proportion of one oscillation) also usually increases with frequency—the faster the input is varied, the farther behind the output falls.

Now suppose that in a feedback system some quantity in the closed loop is oscillating. This causes the successive quantities around the loop to oscillate also. But the loop comes around to the original quantity, and we have here the mechanism by which an oscillation may maintain itself. To see how this can happen, we must remember that with the feedback negative, the motion it causes would be opposite to the original motion, if it were not for the lags. It is only when the lags add up to just half a cycle that the feedback maintains the assumed motion. Thus any system with negative feedback will maintain a continuous oscillation when disturbed if (a) the time-delays in response at some frequency add up to half a period of oscillation, and (b) the feedback ef-

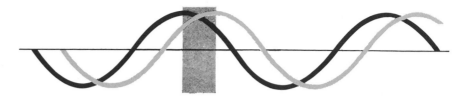

**REGULAR OSCILLATORY VARIATION** of a quantity put into a feedback system (*black curve*) is followed by a similar variation in the output quantity (*gray curve*). The gray rectangle indicates the time-delay.

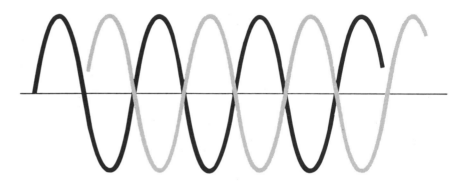

**ONE TYPE OF OSCILLATION** occurs when the feedback (*gray curve*) of a system is equal and opposed to its error (*black curve*). Here the term error is used to mean any departure of the system from its desired state.

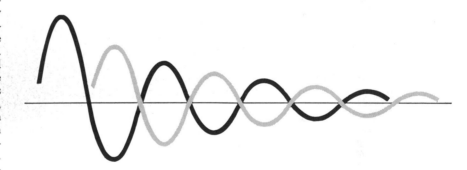

**SECOND TYPE OF OSCILLATION** occurs when the feedback (*black curve*) of a system is less than and opposed to the error (*gray curve*). This set of conditions tends to damp the disturbance in the system.

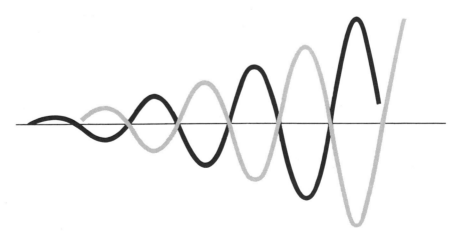

**THIRD TYPE OF OSCILLATION** occurs when the feedback (*gray curve*) of a system is greater than and opposed to the error (*black curve*). This set of conditions tends to amplify the disturbance in the system

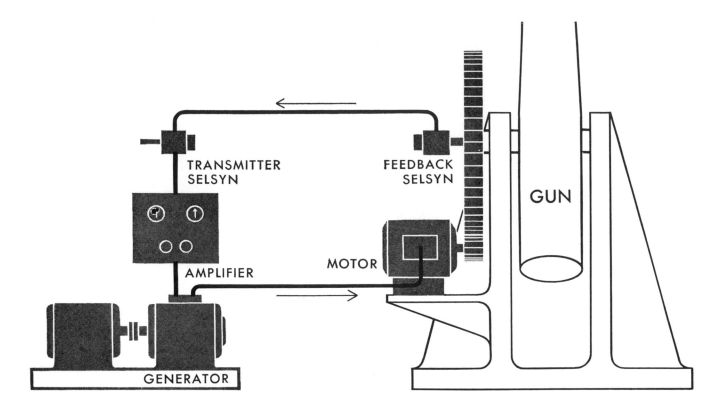

**ELEVATION OF A GUN** is controlled by an electrical feedback system. The closed sequence, or closed loop, runs from the position of the gun through the feedback selsyn, the transmitter selsyn, the amplifier and the generator to the motor. The selsyn is an electrical device which transmits position or speed of rotation.

fect is sufficiently large at this frequency.

In a linear system, that is, roughly speaking, a system in which effects are directly proportional to causes, there are three possible results. If the feedback, at the frequency for which the lag is half a period, is equal in strength to the original oscillation, there will be a continuous steady oscillation which just sustains itself. If the feedback is greater than the oscillation at that frequency, the oscillation builds up; if it is smaller, the oscillation will die away.

This situation is of critical importance for the designer of control systems. On the one hand, to make the control accurate, one must increase the feedback; on the other, such an increase may accentuate any small oscillation. The control breaks into an increasing oscillation and becomes useless.

TO ESCAPE from the dilemma the designer can do several things. Firstly, he may minimize the time-lag by using electronic tubes or, at higher power levels, the new varieties of quick-response direct-current machines. By dividing the power amplification among a multiplicity of stages, these special generators have a smaller lag than conventional generators. The lag is by no means negligible, however.

Secondly, and this was a major advance in the development of control systems, the designer can use special elements that introduce a time-lead, anticipating the time-lag. Such devices, called phase-advancers, are often based on the properties of electric capacitors, because alternating current in a capacitor circuit leads the voltage applied to it.

Thirdly, the designer can introduce other feedbacks besides the main one, so designed as to reduce time-lag. Modern achievements in automatic control are based on the use of combinations of such devices to obtain both accuracy and stability.

So far we have been treating these systems as if they were entirely linear. A system is said to be linear when all effects are strictly proportional to causes. For example, the current through a resistor is proportional to the voltage applied to it; the resistor is therefore a linear element. The same does not apply to a rectifier or electronic tube. These are non-linear elements.

None of the elements used in control systems gives proportional or linear dependence over all ranges. Even a resistor will burn out if the current is too high. Many elements, however, are linear over the range in which they are required to work. And when the range of variation is small enough, most elements will behave in an approximately linear fashion, simply because a very small bit of a curved graph does not differ significantly from a straight line.

We have seen that linear closed-sequence systems are delightfully simple to understand and—even more important—very easy to handle in exact mathematical terms. Because of this, most introductory accounts of control systems either brazenly or furtively assume that all such systems are linear. This gives the rather wrong impression that the principles so deduced may have little application to real, non-linear, systems. In practice, however, most of the characteristic behavior of control systems is affected only in detail by the non-linear nature of the dependences. It is essential to be clear that non-linear systems are not excluded from feedback control. Unless the departures from linearity are large or of special kinds, most of what has been said applies with minor changes to non-linear systems.

LONG BEFORE man existed, evolution hit upon the need for anti-oscillating features in feedback control and incorporated them in the body mechanisms of the animal world. Signals in the animal body are transmitted by trains of pulses along nerve fibers. When a sensory organ is stimulated, the stimulus will produce pulses at a greater rate if it is increasing than if it is decreasing. The pattern of nerve response to an oscillatory stimulus is shown in the diagram on page 133. The maximum response, or output signal, occurs before the maximum of the stimulus. This is just the anticipatory type of effect (the time-lead) that is required for high-

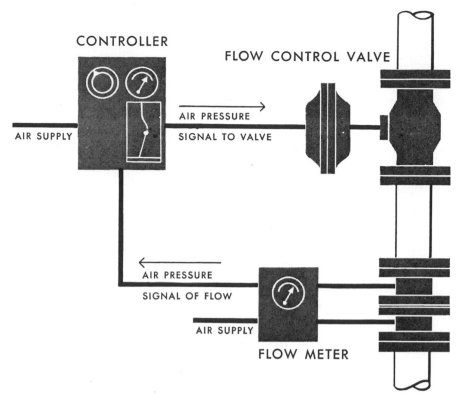

CONTROLLER

FLOW CONTROL VALVE

AIR SUPPLY

AIR PRESSURE
SIGNAL TO VALVE

AIR PRESSURE
SIGNAL OF FLOW

AIR SUPPLY

FLOW METER

**RATE OF FLOW IN A PIPE** is controlled by a pneumatic feedback system. Here the closed loop runs from the flow of fluid in the pipe through the flow meter and the recording controller to the flow-control valve.

accuracy control. Physiologists now believe that the anticipatory response has evolved in the nervous system for, at least in part, the same reason that man wants it in his control mechanisms—to avoid overshooting and oscillation. Precisely what feature of the structure of the nerve mechanism gives this remarkable property is not yet fully understood.

Fascinating examples of the consequences of interdependence arise in the fluctuations of animal populations in a given territory. These interactions are sometimes extremely complicated.

Charles Darwin invoked such a scheme to explain why there are more bumblebees near towns. His explanation was that near towns there are more cats; this means fewer field mice, and field mice are the chief ravagers of bees' nests. Hence near towns bees enjoy more safety.

The interdependence of animal species sometimes produces a periodic oscillation. Just to show how this can happen, and leaving out complications that are always present in an actual situation, consider a territory inhabited by rabbits and lynxes, the rabbits being the chief

food of the lynxes. When rabbits are abundant, the lynx population will increase. But as the lynxes become abundant, the rabbit population falls, because more rabbits are caught. Then as the rabbits diminish, the lynxes go hungry and decline. The result is a self-maintaining oscillation, sustained by negative feedback with a time-delay (*see diagram below*).

Curves of variation such that when R is large L is rising, but when L becomes large R is falling must have the periodic oscillatory character indicated. This is not, of course, the complete picture of such phenomena as the well-known "fur cycle" of Canada, but it illustrates an important element in the mechanisms that cause it.

THE PERIODIC booms and slumps in economic activity stand out as a major example of oscillatory behavior due to feedback. In 1935 the economist John Maynard Keynes gave the first adequate and satisfying account of the essential mechanisms on which the general level of economic activity depends. Although Keynes did not use the terminology of control-system theory, his account fits precisely the same now-familiar pattern.

Keynes' starting point was the simple notion that the level of economic activity depends on the rate at which goods are bought. He took the essential further step of distinguishing two kinds of buying—of consumption goods and of capital goods. The latter is the same thing as the rate of investment. The money available to buy all these goods is not automatically provided by the wages and profits disbursed in making them, because normally some of this money is saved. The system would therefore run down and stop if it were not for the constant injection of extra demand in the form of new investment. Therefore the level of economic activity and employment depends on the rate of invest-

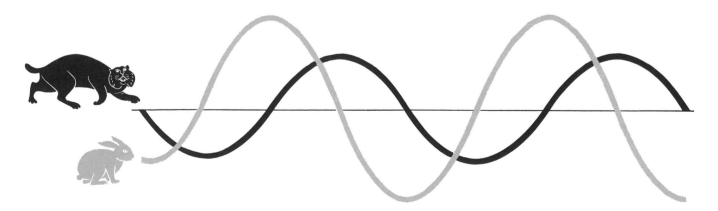

**RABBIT AND LYNX** population cycles are an example of a feedback system in nature. Here a fall in the relatively small population of lynxes (*black curve*) is followed by a rise in the large population of rabbits (*gray curve*). This is followed by a rise in the lynx population, a fall in the rabbit population and so on.

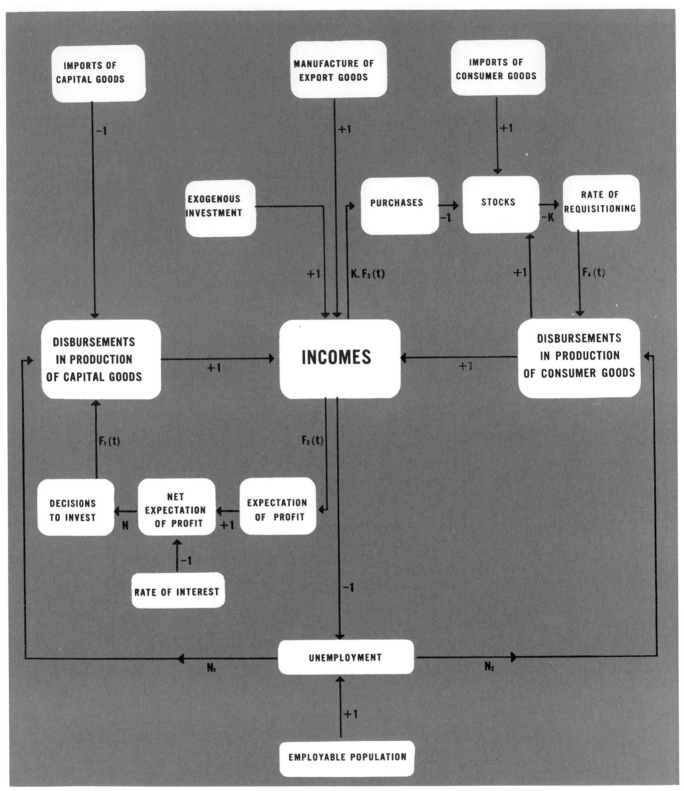

**FEEDBACK IN AN ECONOMIC SYSTEM** is blocked out in this diagram by the author. The scheme of dependence is based upon the ideas advanced by the late J. M. Keynes. Total incomes arise from disbursements for consumer goods on the one hand and capital goods on the other. But each of these is dependent in turn, *via* its subsidiary closed loop, upon total incomes. Keynes was especially concerned with the factors which determine the relationship between the two loops and the relative flow of money into them from total incomes. He showed this to be a highly sensitive relationship, since a comparatively small increase in the flow of money around the capital-goods loop is amplified *via* the consumer-goods loop into a much larger change in total incomes. This is precisely analogous to the behavior of similarly coupled electrical feedback circuits. Pursuing the analogy, the author has entered on the diagram some symbols for values that would have to be defined to design a complete electrical analogue for the economic system. K, for example, is Keynes' "propensity to consume"; $F_1$ (t) represents the time-lag between the decision to invest and the purchase of capital goods; $N_1$ and $N_2$ stand for non-linear functions which curtail increase in production as unemployment approaches zero.

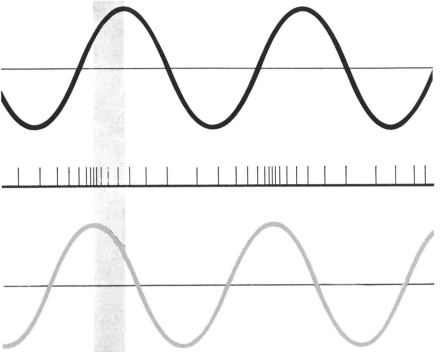

**FEEDBACK IN THE NERVOUS SYSTEM** has a sophisticated feature: the anticipation of the input signal by the output. The black curve at the top of this diagram represents the input signal. The row of vertical lines in the middle indicates the number of nerve pulses in a given time. The gray curve at the bottom translates these pulses into an output signal. The gray rectangle indicates how much the output signal leads the input.

ment. This is the first dependence. The rate of investment itself, however, depends on the expectation of profit, and this in turn depends on the trend, present and expected, of economic activity. Thus not only does economic activity depend on the rate of investment, but the rate of investment depends on economic activity.

Modern theories of the business cycle aim to explain in detail the nature of these dependences and their characteristic non-linearities. This clarification of the mechanisms at work immediately suggests many ways in which, by proper timing of investment expenditure, by more rational business forecasting, and so on, a stable level of optimal economic activity may be achieved in the near future. The day when it can unequivocally be said that slumps belong to the past will certainly be the beginning of a brighter chapter in human history.

THE EXAMPLES of feedback given here are merely a few selected to illustrate general principles. Many more will be described in other articles in this issue. In this article on "theory" I should like to touch on a further point: some ways in which the properties of automatic control systems or other complex feedback systems may be investigated in detail, and their performance perfected.

Purely mathematical methods are remarkably powerful when the system happens to be linear. Sets of linear dif-

ferential equations are the happy hunting ground of mathematicians. They can turn the equations into a variety of equivalent forms, and generally play tunes on them. For the more general class of non-linear systems, the situation is quite different. There exact determination of the types of motion implied by a set of dependences is usually very laborious or practically impossible.

To determine the behavior of such complex systems two principal kinds of machines are being used. The first is the "analogue" computer. The forms of this type of computer are varied, but they all share a common principle: some system of physical elements is set up with relationships analogous to those existing in the system to be investigated, and the interdependence among them is then worked out in proportional terms. The second kind of aid is the new high-speed digital computer. In this type of machine the quantities are represented by numbers rather than by physical equivalents. The implications of the equations involved are explored by means of arithmetical operations on these numbers. The great speed of operation of these modern machines makes possible calculations that could not be attempted by human computers because of the time required.

The theory of control systems is now so well understood that, with such modern aids, the behavior of even extremely complex systems can be largely pre-

dicted in advance. Although this is a new branch of science, it is already in a state that ensures rapid further progress.

AT THE commencement of this account of control systems it was necessary to assume that the human mind can distinguish "cause" and "effect" and describe the regularities of nature in these terms. It may be fitting to conclude by suggesting that the concepts reviewed are not without relevance to the grandest of all problems of science and philosophy: the nature of the human mind and the significance of our forms of perception of what we call reality.

In much of the animal world, behavior is controlled by reflexes and instinct-mechanisms in direct response to the stimulus of the immediate situation. In man and the higher animals the operation of what we are subjectively aware of as the "mind" provides a more flexible and effective control of behavior. It is not at present known whether these conscious phenomena involve potentialities of matter other than those we study in physics. They may well do so, and we must not beg this question in the absence of evidence.

Whatever the nature of the means or medium involved, the function of the central nervous system in the higher animals is clear. It is to provide a biologically more effective control of behavior under a combination of inner and environmental stimuli. An inner analogue or simulation of relevant aspects of the external world, which we are aware of as our idea of the environment, controls our responses, superseding mere instinct or reflex reaction. The world is still with us when we shut our eyes, and we use the "play of ideas" to predict the consequences of action. Thus our activity is adjusted more elaborately and advantageously to the circumstances in which we find ourselves.

This situation is strikingly similar in principle (though immensely more complex) to the introduction of a predictor in the control of a gun, for all predictors are essentially analogues of the external situation. The function of mind is to predict, and to adjust behavior accordingly. It operates like an analogue computer fed by sensory clues.

It is not surprising, therefore, that man sees the external world in terms of cause and effect. The distinction is largely subjective. "Cause" is what might conceivably be manipulated. "Effect" is what might conceivably be purposed.

Man is far from understanding himself, but it may turn out that his understanding of automatic control is one small further step toward that end.

# 20

# The Culture of Poverty

OSCAR LEWIS

*October 1966*

Poverty and the so-called war against it provide a principal theme for the domestic program of the present Administration. In the midst of a population that enjoys unexampled material well-being—with the average annual family income exceeding $7,000—it is officially acknowledged that some 18 million families, numbering more than 50 million individuals, live below the $3,000 "poverty line." Toward the improvement of the lot of these people some $1,600 million of Federal funds are directly allocated through the Office of Economic Opportunity, and many hundreds of millions of additional dollars flow indirectly through expanded Federal expenditures in the fields of health, education, welfare and urban affairs.

Along with the increase in activity on behalf of the poor indicated by these figures there has come a parallel expansion of publication in the social sciences on the subject of poverty. The new writings advance the same two opposed evaluations of the poor that are to be found in literature, in proverbs and in popular sayings throughout recorded history. Just as the poor have been pronounced blessed, virtuous, upright, serene, independent, honest, kind and happy, so contemporary students stress their great and neglected capacity for self-help, leadership and community organization. Conversely, as the poor have been characterized as shiftless, mean, sordid, violent, evil and criminal,

so other students point to the irreversibly destructive effects of poverty on individual character and emphasize the corresponding need to keep guidance and control of poverty projects in the hands of duly constituted authorities. This clash of viewpoints reflects in part the infighting for political control of the program between Federal and local officials. The confusion results also from the tendency to focus study and attention on the personality of the individual victim of poverty rather than on the slum community and family and from the consequent failure to distinguish between poverty and what I have called the culture of poverty.

The phrase is a catchy one and is used and misused with some frequency in the current literature. In my writings it is the label for a specific conceptual model that describes in positive terms a subculture of Western society with its own structure and rationale, a way of life handed on from generation to generation along family lines. The culture of poverty is not just a matter of deprivation or disorganization, a term signifying the absence of something. It is a culture in the traditional anthropological sense in that it provides human beings with a design for living, with a ready-made set of solutions for human problems, and so serves a significant adaptive function. This style of life transcends national boundaries and regional and rural-urban differences

within nations. Wherever it occurs, its practitioners exhibit remarkable similarity in the structure of their families, in interpersonal relations, in spending habits, in their value systems and in their orientation in time.

Not nearly enough is known about this important complex of human behavior. My own concept of it has evolved as my work has progressed and remains subject to amendment by my own further work and that of others. The scarcity of literature on the culture of poverty is a measure of the gap in communication that exists between the very poor and the middle-class personnel—social scientists, social workers, teachers, physicians, priests and others—who bear the major responsibility for carrying out the antipoverty programs. Much of the behavior accepted in the culture of poverty goes counter to cherished ideals of the larger society. In writing about "multiproblem" families social scientists thus often stress their instability, their lack of order, direction and organization. Yet, as I have observed them, their behavior seems clearly patterned and reasonably predictable. I am more often struck by the inexorable repetitiousness and the iron entrenchment of their lifeways.

The concept of the culture of poverty may help to correct misapprehensions that have ascribed some behavior patterns of ethnic, national or regional groups as distinctive characteristics. For

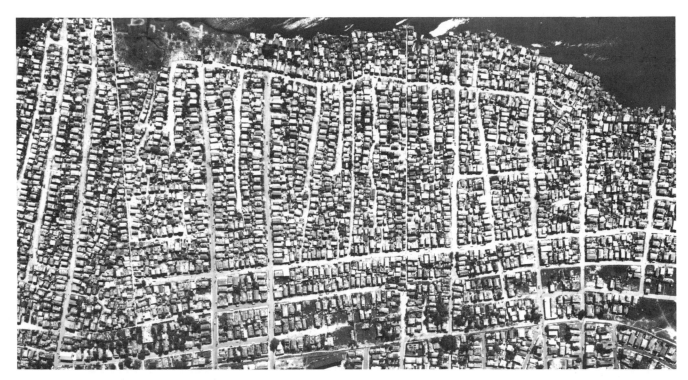

SAN JUAN SLUM AREA in the Santurce district sprawls along the edge of the tidal inlet (*top*) that connects the city's harbor with San José Lake. Rickety buildings have been erected on stilts beyond the high-water line and narrow alleyways crisscross the district. Compared to this area, many of New York's worst slum areas, such as the ones that appear below, are nearly middle-class.

EL BARRIO, the original nuclear Latin-American slum area of Manhattan, occupies the greater part of this aerial photograph. Lying roughly between Central Park and the East River north of 99th Street and south of 125th Street in Manhattan, this is the area that received the pioneer Puerto Rican immigrants to New York in the early years of this century. Photograph was made in 1961.

example, a high incidence of common-law marriage and of households headed by women has been thought to be distinctive of Negro family life in this country and has been attributed to the Negro's historical experience of slavery. In actuality it turns out that such households express essential traits of the culture of poverty and are found among diverse peoples in many parts of the world and among peoples that have had no history of slavery. Although it is now possible to assert such generalizations, there is still much to be learned about this difficult and affecting subject. The absence of intensive anthropological studies of poor families in a wide variety of national contexts—particularly the lack of such studies in socialist countries—remains a serious handicap to the formulation of dependable cross-cultural constants of the culture of poverty.

My studies of poverty and family life have centered largely in Mexico. On occasion some of my Mexican friends have suggested delicately that I turn to a study of poverty in my own country. As a first step in this direction I am currently engaged in a study of Puerto Rican families. Over the past three years my staff and I have been assembling data on 100 representative families in four slums of Greater San Juan and some 50 families of their relatives in New York City.

Our methods combine the traditional techniques of sociology, anthropology and psychology. This includes a battery of 19 questionnaires, the administration of which requires 12 hours per informant. They cover the residence and employment history of each adult; family relations; income and expenditure; complete inventory of household and personal possessions; friendship patterns, particularly the *compadrazgo*, or godparent, relationship that serves as a kind of informal social security for the children of these families and establishes special obligations among the adults; recreational patterns; health and medical history; politics; religion; world view and "cosmopolitanism." Open-end interviews and psychological tests (such as the thematic apperception test, the Rorschach test and the sentence-completion test) are administered to a sampling of this population.

All this work serves to establish the context for close-range study of a selected few families. Because the family is a small social system, it lends itself to

WATERFRONT SHACKS of a Puerto Rican slum provide a sharp contrast to the modern construction that characterizes the prosperous parts of San Juan's Santurce district (*rear*). The author has found that residents in clearly delineated slum neighborhoods such as this one often have a community sense similar to that characteristic of villagers in rural areas. Such *esprit de corps* is

the holistic approach of anthropology. Whole family studies bridge the gap between the conceptual extremes of the culture at one pole and of the individual at the other, making possible observation of both culture and personality as they are interrelated in real life. In a large metropolis such as San Juan or New York the family is the natural unit of study.

Ideally our objective is the naturalistic observation of the life of "our" families, with a minimum of intervention. Such intensive study, however, necessarily involves the establishment of deep personal ties. My assistants include two Mexicans whose families I had studied; their "Mexican's-eye view" of the Puerto Rican slum has helped to point up the similarities and differences between the Mexican and Puerto

uncommon among participants in the culture of poverty; although gregarious, they seldom manage to become well organized.

Rican subcultures. We have spent many hours attending family parties, wakes and baptisms, responding to emergency calls, taking people to the hospital, getting them out of jail, filling out applications for them, hunting apartments with them, helping them to get jobs or to get on relief. With each member of these families we conduct tape-recorded interviews, taking down their life stories and their answers to questions on a wide variety of topics. For the ordering of our material we undertake to reconstruct, by close interrogation, the history of a week or more of consecutive days in the lives of each family, and we observe and record complete days as they unfold. The first volume to issue from this study is to be published next month under the title of *La Vida, a Puerto Rican Family in the Culture of Poverty—San Juan and New York* (Random House).

There are many poor people in the world. Indeed, the poverty of the two-thirds of the world's population who live in the underdeveloped countries has been rightly called "the problem of problems." But not all of them by any means live in the culture of poverty. For this way of life to come into being and flourish it seems clear that certain preconditions must be met.

The setting is a cash economy, with wage labor and production for profit and with a persistently high rate of unemployment and underemployment, at low wages, for unskilled labor. The society fails to provide social, political and economic organization, on either a voluntary basis or by government imposition, for the low-income population. There is a bilateral kinship system centered on the nuclear progenitive family, as distinguished from the unilateral extended kinship system of lineage and clan. The dominant class asserts a set of values that prizes thrift and the accumulation of wealth and property, stresses the possibility of upward mobility and explains low economic status as the result of individual personal inadequacy and inferiority.

Where these conditions prevail the way of life that develops among some of the poor is the culture of poverty. That is why I have described it as a subculture of the Western social order. It is both an adaptation and a reaction of the poor to their marginal position in a class-stratified, highly individuated, capitalistic society. It represents an effort to cope with feelings of hopelessness and despair that arise from the

realization by the members of the marginal communities in these societies of the improbability of their achieving success in terms of the prevailing values and goals. Many of the traits of the culture of poverty can be viewed as local, spontaneous attempts to meet needs not served in the case of the poor by the institutions and agencies of the larger society because the poor are not eligible for such service, cannot afford it or are ignorant and suspicious.

Once the culture of poverty has come into existence it tends to perpetuate itself. By the time slum children are six or seven they have usually absorbed the basic attitudes and values of their subculture. Thereafter they are psychologically unready to take full advantage of changing conditions or improving opportunities that may develop in their lifetime.

My studies have identified some 70 traits that characterize the culture of poverty. The principal ones may be described in four dimensions of the system: the relationship between the subculture and the larger society; the nature of the slum community; the nature of the family, and the attitudes, values and character structure of the individual.

The disengagement, the nonintegration, of the poor with respect to the major institutions of society is a crucial element in the culture of poverty. It reflects the combined effect of a variety of factors including poverty, to begin with, but also segregation and discrimination, fear, suspicion and apathy and the development of alternative institutions and procedures in the slum community. The people do not belong to labor unions or political parties and make little use of banks, hospitals, department stores or museums. Such involvement as there is in the institutions of the larger society—in the jails, the army and the public welfare system— does little to suppress the traits of the culture of poverty. A relief system that barely keeps people alive perpetuates rather than eliminates poverty and the pervading sense of hopelessness.

People in a culture of poverty produce little wealth and receive little in return. Chronic unemployment and underemployment, low wages, lack of property, lack of savings, absence of food reserves in the home and chronic shortage of cash imprison the family and the individual in a vicious circle. Thus for lack of cash the slum householder makes frequent purchases of small quantities of food at higher prices. The slum economy turns inward; it shows a high incidence of pawning of personal goods, borrowing at usurious rates of interest, informal credit arrangements among neighbors, use of secondhand clothing and furniture.

There is awareness of middle-class values. People talk about them and even claim some of them as their own. On the whole, however, they do not live by them. They will declare that marriage by law, by the church or by both is the ideal form of marriage, but few will marry. For men who have no steady jobs, no property and no prospect of wealth to pass on to their children, who live in the present without expectations of the future, who want to avoid the expense and legal difficulties involved in marriage and divorce, a free union or consensual marriage makes good sense. The women, for their part, will turn down offers of marriage from men who are likely to be immature, punishing and generally unreliable. They feel that a consensual union gives them some of the freedom and flexibility men have. By not giving the fathers of their children legal status as husbands, the women have a stronger claim on the children. They also maintain exclusive rights to their own property.

Along with disengagement from the larger society, there is a hostility to the basic institutions of what are regarded as the dominant classes. There is hatred of the police, mistrust of government and of those in high positions and a cynicism that extends to the church. The culture of poverty thus holds a certain potential for protest and for entrainment in political movements aimed against the existing order.

With its poor housing and overcrowding, the community of the culture of poverty is high in gregariousness, but it has a minimum of organization beyond the nuclear and extended family. Occasionally slum dwellers come together in temporary informal groupings; neighborhood gangs that cut across slum settlements represent a considerable advance beyond the zero point of the continuum I have in mind. It is the low level of organization that gives the culture of poverty its marginal and anomalous quality in our highly organized society. Most primitive peoples have achieved a higher degree of sociocultural organization than contemporary urban slum dwellers. This is not to say that there may not be a sense of community and *esprit de corps* in a slum neighborhood. In fact, where slums are

isolated from their surroundings by enclosing walls or other physical barriers, where rents are low and residence is stable and where the population constitutes a distinct ethnic, racial or language group, the sense of community may approach that of a village. In Mexico City and San Juan such territoriality is engendered by the scarcity of low-cost housing outside of established slum areas. In South Africa it is actively enforced by the *apartheid* that confines rural migrants to prescribed locations.

The family in the culture of poverty does not cherish childhood as a specially prolonged and protected stage in the life cycle. Initiation into sex comes early. With the instability of consensual marriage the family tends to be mother-centered and tied more closely to the mother's extended family. The female head of the house is given to authoritarian rule. In spite of much verbal emphasis on family solidarity, sibling rivalry for the limited supply of goods and maternal affection is intense. There is little privacy.

The individual who grows up in this culture has a strong feeling of fatalism, helplessness, dependence and inferiority. These traits, so often remarked in the current literature as characteristic of the American Negro, I found equally strong in slum dwellers of Mexico City and San Juan, who are not segregated or discriminated against as a distinct ethnic or racial group. Other traits include a high incidence of weak ego structure, orality and confusion of sexual identification, all reflecting maternal deprivation; a strong present-time orientation with relatively little disposition to defer gratification and plan for the future, and a high tolerance for psychological pathology of all kinds. There is widespread belief in male superiority and among the men a strong preoccupation with *machismo,* their masculinity.

Provincial and local in outlook, with little sense of history, these people know only their own neighborhood and their own way of life. Usually they do not have the knowledge, the vision or the ideology to see the similarities between their troubles and those of their counterparts elsewhere in the world. They are not class-conscious, although they are sensitive indeed to symbols of status.

The distinction between poverty and the culture of poverty is basic to the model described here. There are numerous examples of poor people whose way of life I would not characterize as be-

longing to this subculture. Many primitive and preliterate peoples that have been studied by anthropologists suffer dire poverty attributable to low technology or thin resources or both. Yet even the simplest of these peoples have a high degree of social organization and a relatively integrated, satisfying and self-sufficient culture.

In India the destitute lower-caste peoples—such as the Chamars, the leatherworkers, and the Bhangis, the sweepers—remain integrated in the larger society and have their own panchayat institutions of self-government. Their panchayats and their extended unilateral kinship systems, or clans, cut across village lines, giving them a strong sense of identity and continuity. In my studies of these peoples I found no culture of poverty to go with their poverty.

The Jews of eastern Europe were a poor urban people, often confined to ghettos. Yet they did not have many traits of the culture of poverty. They had a tradition of literacy that placed great value on learning; they formed many voluntary associations and adhered with devotion to the central community organization around the rabbi, and they had a religion that taught them they were the chosen people.

I would cite also a fourth, somewhat speculative example of poverty dissociated from the culture of poverty. On the basis of limited direct observation in one country—Cuba—and from indirect evidence, I am inclined to believe the culture of poverty does not exist in socialist countries. In 1947 I undertook a study of a slum in Havana. Recently I had an opportunity to revisit the same slum and some of the same families. The physical aspect of the place had changed little, except for a beautiful new nursery school. The people were as poor as before, but I was impressed to find much less of the feelings of despair and apathy, so symptomatic of the culture of poverty in the urban slums of the U.S. The slum was now highly organized, with block committees, educational committees, party committees. The people had found a new sense of power and importance in a doctrine that glorified the lower class as the hope of humanity, and they were armed. I was told by one Cuban official that the Castro government had practically eliminated delinquency by giving arms to the delinquents!

Evidently the Castro regime—revising Marx and Engels—did not write off the so-called *lumpenproletariat* as an inherently reactionary and antirevolutionary

**PUERTO RICAN BOYS** in a Manhattan upper East Side neighborhood use the fenced-off yard of a deserted school as an impromptu playground. The culture of poverty does not cherish childhood as a protected and especially prolonged part of the life cycle. Sexual maturity is early, the male is accepted as a superior and men are strongly preoccupied with *machismo*, or masculinity.

force but rather found in them a revolutionary potential and utilized it. Frantz Fanon, in his book *The Wretched of the Earth*, makes a similar evaluation of their role in the Algerian revolution: "It is within this mass of humanity, this people of the shantytowns, at the core of the *lumpenproletariat*, that the rebellion will find its urban spearhead. For the *lumpenproletariat*, that horde of starving men, uprooted from their tribe and from their clan, constitutes one of the most spontaneous and most radically revolutionary forces of a colonized people."

It is true that I have found little revolutionary spirit or radical ideology among low-income Puerto Ricans. Most of the families I studied were politically conservative, about half of them favoring the Statehood Republican Party, which provides opposition on the right to the Popular Democratic Party that dominates the politics of the commonwealth. It seems to me, therefore, that disposition for protest among people living in the culture of poverty will vary considerably according to the national context and historical circumstances. In contrast to Algeria, the independence movement in Puerto Rico has found little popular support. In Mexico, where the cause of independence carried long ago, there is no longer any such movement to stir the dwellers in the new and old slums of the capital city.

Yet it would seem that any movement—be it religious, pacifist or revolutionary—that organizes and gives hope to the poor and effectively promotes a sense of solidarity with larger groups must effectively destroy the psychological and social core of the culture of poverty. In this connection, I suspect that the civil rights movement among American Negroes has of itself done more to improve their self-image and self-respect than such economic gains as it has won although, without doubt, the two kinds of progress are mutually reinforcing. In the culture of poverty of the American Negro the additional disadvantage of racial discrimination has generated a potential for revolutionary protest and organization that is absent in the slums of San Juan and Mexico City and, for that matter, among the poor whites in the South.

If it is true, as I suspect, that the culture of poverty flourishes and is endemic to the free-enterprise, pre-welfare-state stage of capitalism, then it is also endemic in colonial societies. The most likely candidates for the culture of poverty would be the people who come from the lower strata of a rapidly changing society and who are already partially alienated from it. Accordingly the subculture is likely to be found where imperial conquest has smashed the native social and economic structure and held the natives, perhaps for generations, in servile status, or where feudalism is yielding to capitalism in the later evolution of a colonial economy. Landless rural workers who migrate to the cities, as in Latin America, can be expected to fall into this way of life more readily than migrants from stable peasant villages with a well-organized traditional culture, as in India. It remains to be seen, however, whether the culture of poverty has not already begun to develop in the slums of Bombay and Calcutta. Compared with Latin America also, the strong corporate nature of many African tribal societies may tend to inhibit or delay the formation of a full-blown culture of poverty in the new towns and cities of that continent. In South Africa the institutionalization of repression and discrimination under *apartheid* may also have begun to promote an immunizing sense of identity and group consciousness among the African Negroes.

One must therefore keep the dynamic aspects of human institutions forward in observing and assessing the evidence for the presence, the waxing or the waning of this subculture. Measured on the dimension of relationship to the larger society, some slum dwellers may have a warmer identification with their national tradition even though they suffer deeper poverty than members of a similar community in another country. In Mexico City a high percentage of our respondents, including those with little or no formal schooling, knew of Cuauhtémoc,

**MOTHER AND DAUGHTER** stand together by the door of a run-down apartment building on upper Park Avenue. Because common-law marriage offers the female participant in the culture of poverty more protection of her property rights and surer custody of her children than formal marriage does, the mother is usually the head of the household and family ties are to her kin and not the father's.

Hidalgo, Father Morelos, Juárez, Díaz, Zapata, Carranza and Cárdenas. In San Juan the names of Rámon Power, José de Diego, Baldorioty de Castro, Rámon Betances, Nemesio Canales, Lloréns Torres rang no bell; a few could tell about the late Albizu Campos. For the lower-income Puerto Rican, however, history begins with Muñoz Rivera and ends with his son Muñoz Marín.

The national context can make a big difference in the play of the crucial traits of fatalism and hopelessness. Given the advanced technology, the high level of literacy, the all-pervasive reach of the media of mass communications and the relatively high aspirations of all sectors of the population, even the poorest and most marginal communities of the U.S. must aspire to a larger future than the slum dwellers of Ecuador and Peru, where the actual possibilities are more limited and where an authoritarian social order persists in city and country. Among the 50 million U.S. citizens now more or less officially certified as poor, I would guess that about 20 percent live in a culture of poverty. The largest numbers in this group are made up of Negroes, Puerto Ricans, Mexicans, American Indians and Southern poor whites. In these figures there is some reassurance for those concerned, because it is much more difficult to undo the culture of poverty than to cure poverty itself.

Middle-class people—this would cer-tainly include most social scientists—tend to concentrate on the negative aspects of the culture of poverty. They attach a minus sign to such traits as present-time orientation and readiness to indulge impulses. I do not intend to idealize or romanticize the culture of poverty—"it is easier to praise poverty than to live in it." Yet the positive aspects of these traits must not be over-looked. Living in the present may develop a capacity for spontaneity, for the enjoyment of the sensual, which is often blunted in the middle-class, future-oriented man. Indeed, I am often struck by the analogies that can be drawn between the mores of the very rich—of the "jet set" and "café society" —and the culture of the very poor. Yet it is, on the whole, a comparatively superficial culture. There is in it much pathos, suffering and emptiness. It does not provide much support or satisfaction; its pervading mistrust magnifies individual helplessness and isolation. Indeed, poverty of culture is one of the crucial traits of the culture of poverty.

The concept of the culture of poverty provides a generalization that may help to unify and explain a number of phenomena hitherto viewed as peculiar to certain racial, national or regional groups. Problems we think of as being distinctively our own or distinctively Negro (or as typifying any other ethnic group) prove to be endemic in countries where there are no segregated ethnic minority groups. If it follows that the elimination of physical poverty may not by itself eliminate the culture of poverty, then an understanding of the subculture may contribute to the design of measures specific to that purpose.

What is the future of the culture of poverty? In considering this question one must distinguish between those countries in which it represents a relatively small segment of the population and those in which it constitutes a large one. In the U.S. the major solution proposed by social workers dealing with the "hard core" poor has been slowly to raise their level of living and incorporate them in the middle class. Wherever possible psychiatric treatment is prescribed.

In underdeveloped countries where great masses of people live in the culture of poverty, such a social-work solution does not seem feasible. The local psychiatrists have all they can do to care for their own growing middle class. In those countries the people with a culture of poverty may seek a more revolutionary solution. By creating basic structural changes in society, by redistributing wealth, by organizing the poor and giving them a sense of belonging, of power and of leadership, revolutions frequently succeed in abolishing some of the basic characteristics of the culture of poverty even when they do not succeed in curing poverty itself.

# 21

# Abortion as a Disease of Societies

JAMES R. NEWMAN

*January 1959*

A review of *Abortion in the United States*, edited by Mary Steichen Calderone. New York: Hoeber-Harper and Row, 1958.

Abortion is an ancient practice, but even in antiquity it provoked sharp differences of opinion. Plato, in the *Republic*, approved abortion to prevent the birth of incestuous offspring; Aristotle, always a practical fellow, looked upon it as a useful Malthusian governor. The Hippocratic oath, on the other hand, contains the words "I will not give to a woman a pessary to produce abortion"; Seneca and Cicero condemned abortion on ethical grounds; and the Justinian Code prohibited it. There seems little doubt, however, that in the Roman Empire and the Hellenistic world abortion was, as one authority has stated, "very common among the upper classes." The Christian Church took a stern stand against this "pagan attitude," and pronounced abortion a sin. In many states the law followed church doctrine and made the sin a crime. But in Anglo-Saxon law abortion was considered "an ecclesiastical offense only."

Abortion is today a world problem. Surveys and studies by individuals and by UNESCO show the practice to be widespread in, among others, the Scandinavian countries, Finland, Germany, the U.S.S.R., Japan, Mexico, Puerto Rico, Latin America and the U. S. George Devereux's book, *A Study of Abortion in Primitive Societies*, which covers some 400 preindustrial societies as well as 20 historical and modern nations, concludes that abortion "is an absolutely universal phenomenon."

The problem has grown with the growth of population. Poverty, ill health, vanity, social customs, aversion to pain are familiar reasons for resorting to abortion; other factors are the "urban middle-class attitude toward illegitimacy," inadequate sex education, prohibitions against birth control. In most societies the problem has usually been mishandled. Ignorance, hypocrisy and inhumanity have presided in judgment over it. Because a frank admission of the dimensions of the problem might force remedial action which would provoke intense religious and social opposition, politicians, physicians, public health officials and others who make rules and opinion are wont to pretend that the practice is negligible. They are aided in this courageous stand by the general lack of knowledge in the matter. It is true, to be sure, that except for limited samples reliable statistics do not exist. Nevertheless it is clear, on the basis of a mass of data, that an enormous number of abortions—running into the millions—are performed annually the world over. We are faced, therefore, with a matter of the highest medical and social importance: a disease of society, the more serious because many communities refuse to recognize it and do nothing to eliminate its causes and mitigate its effects.

This challenging and engrossing book is evidence that there are men and women who recognize the nature and scope of the problem and are working to bring it into the open. In 1955 a conference on abortion, sponsored by the Planned Parenthood Federation of America, was held at Arden House in Harriman, N.Y., and at the New York Academy of Medicine. The participants, including leading gynecologists, psychiatrists, social workers, lawyers and public health officials, discussed such topics as the incidence of abortion, methods used in illegal abortion, causes of death due to abortion, the psychiatric and legal aspects of abortion. A report on the discussions, together with a concluding statement which summarizes the facts and makes recommendations, is given in this volume. I cannot emphasize too strongly the value of the material as a social document.

The primary concern of the conference was with abortion in the U. S. It is a shocking picture. One has only to look at it from the legal side to understand why.

Of the 49 states, all but six prohibit abortion except when it is necessary "to preserve the life of the woman." Six states permit a therapeutic abortion "to save the life of the unborn child" (whatever this absurd phrase may mean); in only two states is abortion permitted "to preserve the health of the woman." Under the hodgepodge of statutes the physician's position is perilous and uncertain. It is not easy in most cases to prove conclusively that the life of the woman depends on the abortion; moreover, in some jurisdictions a physician on trial for having performed an illegal abortion has to plead necessity as an affirmative defense, and thus the burden of proof is on him and not on the prosecution. Powerful social disapproval operates further to narrow the statutory restrictions. Some hospitals are unwilling to permit use of their facilities even for abortions sanctioned by law. In many states a woman with serious heart disease is not entitled to have her pregnancy interrupted. Though the child of a woman who in early pregnancy contracts German measles has about a 20-per-cent chance of being born blind, mentally retarded or otherwise affected, the woman cannot claim the right to a therapeutic abortion.

The law is not merely, as Mr. Bumble said, "a ass"; it is much worse. In the U. S. neither rape nor incest, even if the victim is a very young girl, is ground for abortion. Social disgrace or poverty or any other humanitarian reason receives no consideration whatever. And as regards the extraordinary attitude of the law toward the illegitimate child, nothing is more cogent than the view of Iwan Bloch, expressed in his famous book *The Sexual Life of Our Time.* The State, he said, considers as sacred the life of the child before it is born and punishes anyone who interferes with its preservation, but then considers the same child a bastard as soon as it is born and for the rest of its life.

Since it is extremely difficult to obtain a "therapeutic abortion," the illegal practice flourishes. When people want something badly enough they will have it, regardless of the law. The unmarried pregnant woman has the desperate choice of suicide, of bearing and raising an illegitimate child, or of going to an abortionist. The married pregnant woman who does not want to bear her child may be less desperate, but she too is usually beset with anxieties and confused as to what course to follow. One of the participants in the conference, G. Lottrell Timanus, who for years practised in Maryland as an abortionist, vividly described the plight of the average unwilling pregnant woman: "She does not know what to do or where to turn. There is no place available where she can air her situation comfortably and quietly and confidentially. Her only resource at present is to go to a local physician, and, under present standards, he is afraid even to look at her. He has no place to send her. He has no recommendations to make to her. So, consequently, she goes to an abortionist." And that, as Ashley Montagu says in his introduction to this volume, "is how our society deals with the problem of abortion in its most elementary form." It abandons to the abortionist the woman in her greatest need.

The statistics of illegal abortion are, as I have said, not easy to assemble. But consider one or two bits and pieces of relevant data. The late Alfred Kinsey told the conference that of some 5,000 women in the sample interviewed by his Institute, 10 per cent of those married had had induced abortions by 20 years of age; 22 per cent had had at least one induced abortion by 45 years of age; more than 90 per cent of the single women who got pregnant had undergone abortion. And among all the single white females in the Kinsey sample who had had sexual intercourse, 20 per cent had had abortions. The significance of these figures can be judged by examining a single pertinent record of New York City's Board of Health, which shows a ratio of well under five therapeutic abortions per 1,000 live births. No less interesting is the information provided by Dr. Timanus. In 20 years of practice in Baltimore he alone performed 5,210 abortions, all but a tiny fraction of them illegal. He kept careful records of his cases, and the compilations of their various features are given in this report. (Timanus is now retired, having fallen into disagreement with the law. In this connection he made an observation which epitomizes the hypocrisy of the medical profession's stand on abortion. In the 20-year period during which he had operated as an abortionist, he had served 353 doctors; yet of this entire group of estimable men, not one was willing to come forward and testify for him, or even to admit that he had sent Timanus a single patient.) Many other personal estimates of illegal abortions were given by individual participants in the conference. The statistics committee did not feel that these furnished the basis for an exact figure for the total population; however, the members were prepared to subscribe to a range of "plausible estimates," from a minimum of 200,000 to a maximum of 1,200,000 abortions per year in the U. S. Even accepting the lower figure, the consequences of illegal abortion are appalling: incalculable human suffering and misery, illness and death, heavy economic loss, the corruption of social and ethical standards.

What is to be done? Knowledgeable persons with a humane outlook are prepared to act but find it hard to agree on a program. Speakers at the conference presented evidence of the magnitude of the abortion practice in the U. S., discussed its causes and made plain its tragic effects. Yet in the final session on conclusions and recommendations there was a tendency to split hairs, to argue over false issues and irrelevancies and to gloss over essential questions. The participants recognized that they were dealing with an explosive problem and that it was as difficult to draft specific reforms as it would be risky to advocate them. The cake of social custom has a hard crust; prejudices die hard. Moreover, as several speakers pointed out, since the Catholic Church is opposed to abortion on any ground, it was safe to anticipate vehement criticism of any recommendations for modifying existing abortion statutes which the conference might adopt. A statement of conclusions was finally prepared, which more than three quarters of the participants signed; they were careful to state, however, that their signatures represented personal agreement and in no sense committed the hospitals, universities, health boards or other organizations with which they were connected. The statement deserves close attention.

Present laws and mores, the signatories firmly assert, have failed entirely to control illegal abortion. "Rather this has continued to an extent ignored or, perhaps, condoned by a large proportion of the general public and even of the medical and legal professions." Since one cannot legislate the practice out of existence, it is folly to keep on the books laws which do not receive public sanction and observance. The constitutional amendment that prohibited liquor is an example of an unpopular and unenforceable law which led to evils vastly greater than those it was designed to meet. The signatories make a striking comparison between the high incidence of abortion as a disease of society and the high incidence of venereal disease three decades ago. Both involve health, mores and morals; both are problems in epidemiology. Until recently the venereal-disease problem was kept in the cupboard; prejudice and prudery barred bringing it into the open. Then at last physicians and public health agencies decided to smash the barriers and ventilate the problem. This led to salutary controls even before antibiotics finally broke the back of venereal disease. The majority of those attending the conference felt that "the same type of frontal assault should now be made on the problem of intentional abortion."

The motives for abortion, the report emphasizes, are often complex. Ill health may play a part, but more usually "poor social and economic environment, disturbed marital relations, psychiatric or neurotic disturbance within the family, or, quite simply, a need to keep her family at its present size" induce a woman to seek this way out. Abortion is always a damaging experience, the conferees declare, and while it meets an immediate crisis it frequently does not solve the basic difficulties that created the crisis. To recognize this fact is to recognize that the vast number of illegal abortions each year is many times the number consistent with sound medical or social practice. But to reduce the number is a goal not to be achieved within the framework of present attitudes and laws. Abortion must therefore receive the attention not only of professionals and specialists in medicine, sociology, psychology, education, religion and law, but also of all responsible citizens.

Several remedial measures are recommended.

First, medical, psychological and social studies of women seeking abortion are needed to furnish reliable information on background, motivation, mechanism and results. More must be known about the problem before it can be dealt with intelligently.

Second, consultation centers for women seeking abortion should be set up. These would operate under joint medical and social auspices and, if possible, through sponsorship of state health and welfare departments. Models for such centers are to be found in the Scandinavian countries. Clinics in Norway, Denmark and Sweden give instruction on the use of contraceptives to anyone who asks for it, regardless of economic or marital status. They provide facilities for sex education, and furnish advice on abortion. An enlightened and realistic outlook informs these programs. Bard Brekke, a Norwegian public health official participating in the conference, gave an excellent account of the basic principles that guide the maternal health centers. "We believe," he said, "that when a woman wants an abortion, there must be something wrong either with herself or with her life situation or both, and that frequently she represents not an individual medical and social problem only, but that of a whole family in need of some social or sociomedical treatment." Sometimes an abortion must be part of a "total treatment plan" of social, psychiatric and medical services, and then the center recommends it. But the attempt is always made to correct the underlying situation and to help the woman and her family without resorting to abortion. "In not a few cases," Dr. Brekke added, "we have found it possible to do what we would like to call rehabilitative or reconstructive sociomedical work, but in many cases these women and their families represent chronic social disease; they are in need of continuous social and psychiatric care and, as in other forms of chronic disease, these records can rarely be closed."

The third measure is to extend, under medical supervision, the practice of giving free advice on contraception. There is admittedly little evidence to support the claim that increased availability of contraceptive services will reduce the illegal abortion rate. But until dependable statistics can decide the question there is every reason to adopt and act on the assumption that such services will help. Under present circumstances there is a lamentable inequality of access to contraceptive information. "The law in its majestic equality," Anatole France once reminded us, "forbids the rich as well as the poor to sleep under bridges, to beg in the streets, and to steal bread." But even in the U. S. there are disadvantages to being poor. In most communities contraceptive information can be obtained only through a private physician. Hospitals serving the lower income groups do not furnish routine contraceptive facilities. One has to pay to get what one needs; moreover, one has to know what one needs. It was noted at the conference that among the uneducated it is still not rare to find women who do not connect pregnancy with sexual intercourse. It is not surprising, therefore, that there are many women who know nothing about contraceptives. Free clinics and a vigorous education program are essential if this approach to illegal abortion is to make any headway.

The same inequality marks the availability of legal abortions. Reports made at the conference show a much higher therapeutic abortion rate in private than in voluntary hospitals, and similar differences are reported for private and ward patients in the same hospital. There are more illegal-abortion deaths among poor people than among those better off; more abortion deaths among Negroes than among whites. A participant observed that "the difference between an illegal and a therapeutic abortion is $300 and knowing the right person." This remark is wildly wrong only in respect to the sum mentioned. In most cities it costs more these days to get an abortion.

The one formidable obstacle to reform is, of course, the law itself. Existing statutes are so fanatically narrow and backward that the only possible way to live with them is to break them. If they are interpreted in their plain meaning, there is almost no justification for therapeutic abortion, for with improvements in modern medicine it rarely becomes necessary to perform an abortion to save life. From 1951 to 1953, 37.8 per cent of all therapeutic abortions in New York City were based on "psychiatric indications." This means that in each of these 642 cases psychiatrists certified that abortion was necessary to save the woman's life, i.e., to prevent suicide. Obviously this is nonsense. In most instances the certifying psychiatrist knew that he was in a position to predict suicide, yet he knew also that it was dangerous to force a child upon a woman suffering from serious mental illness. The crime then, if there was one, was the legislators', not his.

In their final recommendation the signatories urged that authoritative bodies such as the National Conference of Commissioners on Uniform State Laws, the American Law Institute and the Council of State Governments should frame a model law to replace existing statutes. Again it is to the Scandinavian countries that one must turn for instruction on sensible abortion laws. The statute adopted by the Swedish Parliament accepts "humanitarian" and "eugenic" as well as medical indications for the lawful interruption of pregnancy. The weak, worn-out or exhausted mother who cannot take another child without serious consequences to herself and her family can get relief. A pregnant girl under 15 years of age, a victim of rape, incest or other criminal coercion, a mental defective—each of these is considered entitled to a legal abortion. The eugenic indication applies when it seems probable that the mother or father of the expected child would transmit to the offspring a hereditary mental disease or deficiency or a serious illness or physical defect. In such cases sterilization is usually required to accompany the interruption of pregnancy. And in 1946 Sweden added a paragraph to the law in which an indication of a "sociomedical nature" was established. According to this provision, abortion should be granted when, taking into consideration "the conditions of life of the woman, and other circumstances," it is fair to infer that her physical or mental strength would be damaged by the birth and care of the child.

It must not be supposed that this enlightened law is administered loosely. A committee of the Royal Medical Board exercises strict supervision. Members of the committee are a physician, a layman (usually a woman) and the Chief of the Bureau of Social Psychiatry. A geneticist joins the committee when cases of eugenic indications are to be decided. The applicant is never seen in person; written material and other evidence are the bases for the findings. The effect of the law has been greatly to increase during the last 15 years the number of legally induced abortions; the present rate seems to be about five per 100 term deliveries. Compare this with a rate of about five therapeutic abortions per 1,000 live births in New York City. The difference between these rates is a measure of the difference between the rate of illegal abortion in the U. S. and in Sweden.

# 22

# Abortion

CHRISTOPHER TIETZE and SARAH LEWIT
*January 1969*

The practice of abortion goes back to human traditions far older than the earliest written history. Abortion is still the most widespread, and the most clandestine, method of fertility control in the modern world. In recent years several nations have legalized the practice, and as a consequence induced abortion is emerging from the shadows and has become a topic of worldwide discussion and controversy. The debate ranges over a wide spectrum of considerations: moral, ethical, medical, social, economic, legal, political and humanitarian. The experience of countries that have made abortion legally permissible is now beginning to provide a body of reliable data with which to evaluate the pros and cons of the practice.

That induced abortion was probably common among prehistoric peoples is suggested by anthropological studies of primitive tribes still living in isolated parts of the world today. The motivation is much the same as it is in civilized societies: the illness or extreme youth or advanced age of the pregnant woman, economic hardship, the desire of unmarried women to avoid disgrace and a variety of other biological and social considerations. The methods employed by women in primitive societies to terminate unwanted pregnancy vary from hard physical exertion to the application of heat or skin irritants to various parts of the body or the insertion of a variety of instruments into the uterus.

Techniques for abortion are mentioned in some of the oldest medical texts known to man. An ancient Chinese work, said to have been written by the Emperor Shen Nung 4,600 years ago, contains a recipe for the induction of abortion by the use of mercury. In ancient Greece, Hippocrates recommended violent exercise as the best method, and Aristotle and Plato advocated abortion to limit the size of the population and maintain an economically healthy society. With the rise of the Jewish and Christian religions abortion fell under moral condemnation. During the Renaissance, however, popular disapproval of abortion was relaxed considerably. In English common law abortion was not regarded as a punishable offense unless it was committed after "quickening," or the first felt movements of the fetus in the womb. Even then it was classed only as a misdemeanor. Abortion did not become a statutory crime in England until 1803; it did not become one in the U.S. until about 1830. Today it is still prohibited in most states of the U.S. except in cases of serious hazard to the mother's life.

The arguments for and against abortion are widely known. The moral objection is based primarily on the contention that human life begins with the union of the egg and the sperm, so that destruction of the fertilized ovum is an act of homicide, even murder. Oppo-

nents of abortion hold that it opens the door to the brutalization of society, encouraging "mercy deaths," infanticide, use of the gas chamber and other violations of the sanctity of life. Abortion is said to undermine the social structure by encouraging promiscuous sexual relations, by weakening family ties and by raising legal problems with respect to property ownership and inheritance. In the medical area many physicians find the abortion operation distasteful because of their training as guardians of human life, and some believe it may cause sterility, menstrual disorders, ectopic pregnancy (implantation of the fertilized ovum in the Fallopian tube), miscarriage, abnormal delivery or guilt feelings that may lead to neurotic or even psychotic symptoms.

The defenders of abortion take the position that the embryo does not become a human being until after a certain time, fixed variously from 12 to 28 weeks of gestation. Their arguments for the legalization of abortion are based on the premise that the overriding consideration should be the health (mental and physical) of the prospective mother and child. If there is a risk that the child may be malformed because of genetic factors, illness (for example German measles during the first three months of pregnancy) or injury (such as ingestion of thalidomide by the pregnant woman), she and her physician should be allowed to decide whether or not the pregnancy

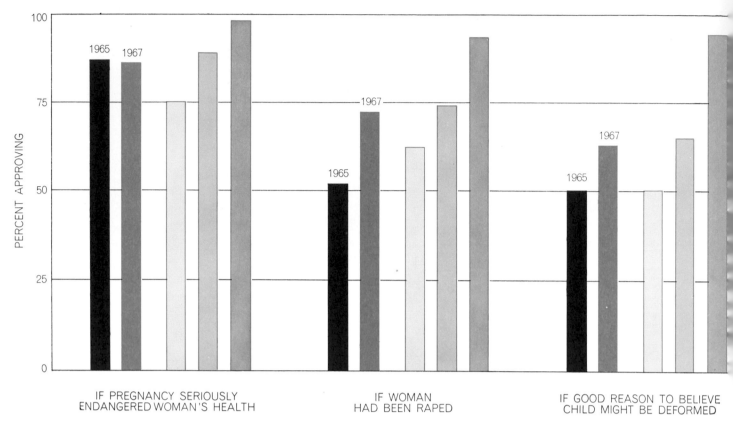

U.S. OPINION on abortion has become somewhat more permissive in recent years. The chart shows what percentage of respondents answered yes when asked if they thought it would be all right for a woman to have a pregnancy interrupted for each of several

should continue. The birth and rearing of a deformed or mentally defective child may have profoundly disturbing effects on a family. It is contended further that the outlawing of abortion, with the consequent lack of medical regulation and safeguards, has made it a leading public health problem in many countries because large numbers of women are maimed or killed by unskilled or irresponsible operators. The defenders of legalized abortion point out that present techniques for the operation, when it is performed by a competent physician, are so safe that the risk to life is well below the risk of childbearing itself. To the moral objections they reply that most abortions are performed on married women with the consent of their husbands, and that a woman who has been the victim of a criminal rape should not be forced to suffer the additional indignity of having to bear the child. At the philosophical level it is argued that no woman should be forced to bear a child she does not want, and that the implied rights of every child include the right to be born wanted and loved.

Clinically we can divide induced abortions into two broad categories: those performed clandestinely by lay abortionists and those performed by physicians using techniques that minimize the health hazard. The methods employed by lay abortionists are generally either ineffective or potentially dangerous to the pregnant woman. Over the centuries many kinds of internal medication have been tried, ranging from the bush tea of preliterate tribes to white phosphorus to hormones; almost invariably these treatments fail to induce abortion. It does not seem likely that a safe and reliable "abortion pill" will be realized in the near future. The most common technique employed by lay abortionists in the U.S. is the insertion of a foreign body, usually a rubber tube, into the cervix. This procedure brings on contractions, bleeding and the expulsion of the fetus. Part of the placenta may remain in the uterus, however, so that bleeding persists and the patient must go to a physician or a hospital to have the placental fragment curetted (scraped out). In any case, the lay operation usually carries a high risk of infection, owing to the lack of aseptic precautions.

In the medical profession the favored technique for early abortion (that is, during the first 12 weeks of pregnancy) has been dilation and curettage, or "D & C." A series of metal dilators, each slightly larger than the preceding one, is inserted into the cervical canal to stretch it, and when the passage has been opened wide enough, the contents of the uterus are scraped out with a curette. The entire procedure usually takes only a few minutes. In recent years a suction method has become popular, particularly in eastern Europe. After dilation of the cervix a metal, glass or plastic tube with a lateral opening near the end is inserted into the uterus and moved about to dislodge the embryo from the uterine wall; the fragments are then drawn out by means of a vacuum pump connected to the tube. This technique is easier, quicker and less traumatic than D & C.

Abortion is seldom performed after the 12th week of pregnancy, even in countries where it is legal with few restrictions. Late abortion may, however, be necessary to save the life of the pregnant woman. The method usually employed is to remove the fetus by a Cesarean-like operation or to inject into the uterus a highly concentrated solution of salt or sugar, which ends the development of the fetus and causes it to be expelled within a few days.

In nearly two-thirds of the world (in terms of population) abortion is prohibited entirely or is allowed only for narrowly defined medical reasons. The U.S. is one of the restrictive countries.

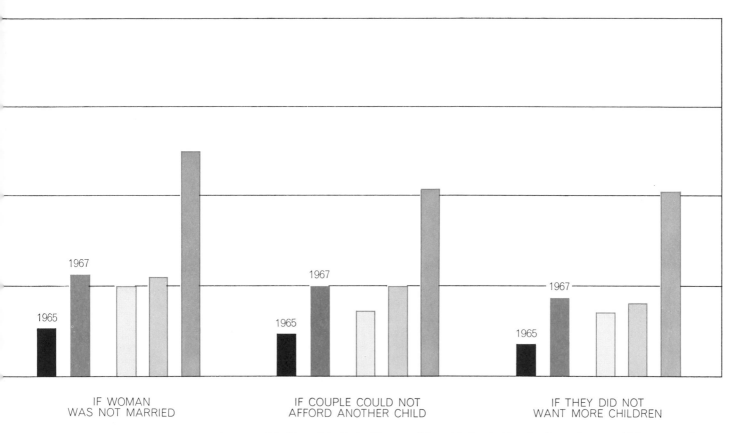

1965    1967    1967    1965    1967    1965

IF WOMAN
WAS NOT MARRIED

IF COUPLE COULD NOT
AFFORD ANOTHER CHILD

IF THEY DID NOT
WANT MORE CHILDREN

reasons. The bars give results for 1965 (*black*) and 1967 (*solid color*) and in 1967 for Catholics (*light tint*), Protestants (*darker*) and Jews (*darkest*). Abortion is much more widely approved as an emergency measure than as an elective method of birth regulation.

In 43 states abortion is permitted only if the pregnancy threatens the mother's life. The other seven states and the District of Columbia are somewhat more permissive. In 1967 and 1968 five states —California, Colorado, Georgia, Maryland and North Carolina—adopted liberalized abortion laws based on the model code recommended by the American Law Institute. The code allows abortion in cases where there is a substantial risk that the mother's physical or mental health may be gravely impaired or that the child would be born with a serious physical or mental defect, or when the pregnancy is the result of rape or incest.

The medical profession in the U.S. has generally interpreted the abortion laws conservatively. Most of the large hospitals require a careful review by a board of physicians of each application for an abortion. It is estimated that the number of therapeutic abortions performed in U.S. hospitals from 1963 through 1965 was only about 8,000 per year, a ratio of about two abortions to 1,000 births. About two out of five abortions were performed for psychiatric reasons, about a fourth because the mother had German measles during the first three months of pregnancy and the rest for a variety of other health reasons. In New York City in 1960–1962 the

abortion ratios ranged from 3.9 per 1,000 births in proprietary hospitals to only .1 per 1,000 in municipal hospitals, indicating that legal abortion was less readily available to low-income families than to those with higher incomes. There was also a marked ethnic differential: the ratio of therapeutic abortions per 1,000 deliveries was 2.6 for white women, .5 for Negro women and .1 for Puerto Rican women.

It is extremely difficult to obtain a reliable estimate of the number of illegal abortions in the U.S. In 1955 a Conference on Abortion, sponsored by the Planned Parenthood Federation and the New York Academy of Medicine, was held at Arden House; a committee of the conference could only estimate that the number of illegal abortions might be as low as 200,000 or as high as 1,200,000 per year. No information on which to base a better estimate has been obtained since that time. Nor do we have reliable data for determining the number of deaths from illegal abortions in the U.S. Some 30 years ago it was judged that such deaths might number 5,000 to 10,000 per year, but this rate, even if it was approximately correct at the time, cannot be anywhere near the true rate now. The total number of deaths from

all causes among women of reproductive age in the U.S. is not more than about 50,000 per year. The National Center for Health Statistics listed 235 deaths from abortion in 1965. Total mortality from illegal abortions was undoubtedly larger than that figure, but in all likelihood it was under 1,000.

The late Alfred C. Kinsey and his associates, in their famous studies of sexual behavior more than a decade ago, found that among the women they interviewed abortions were performed in most cases (about 90 percent) by physicians, although not necessarily physicians in good standing. It is suspected that many American women seeking abortions go abroad to countries where the laws are not as restrictive or are not enforced. Among the less privileged women who do not have access to a skilled physician the abortion risk is substantial compared with other risks of pregnancy. According to a study by Edwin M. Gold of deaths of women from complications of pregnancy or childbirth in New York City in 1960–1962, abortion was the cause of death for 25 percent of the white women, 49 percent of the nonwhite women and 56 percent of the Puerto Rican women.

Since the Arden House Conference of 1955 the subject of abortion has been discussed by many professional and civic groups. In 1967 the House of Delegates of the American Medical Association endorsed liberalization of the abortion laws, and in 1968 the American Public Health Association urged repeal of all restrictive statutes.

Public opinion polls also support liberalization. In the National Fertility Study conducted in the fall of 1965 a considerable majority favored legalization of abortions to protect the health of the mother, and a smaller majority approved abortion to prevent the birth of a defective child or in cases of rape. On the other hand, only 13 percent at that time approved of abortion for unmarried women and only 8 percent for any woman who wanted it. Two years later a Gallup poll conducted for the Population Council showed a "substantial liberalization of attitudes," with a rise to 28 percent approving abortions for the unmarried and 21 percent for any applicant.

The two-thirds of the world where prohibitions like those in the U.S. are in effect include countries in the Western Hemisphere, southern and western Europe, Africa and most of Asia, with the exception of Japan and China. In Britain and the Scandinavian countries,

constituting 2 percent of the world population, the laws are considerably more liberal, allowing abortion on broad medical, eugenic and humanitarian indications. Japan, China and most of the countries of eastern Europe, which account for the remaining nearly one-third of the world population, have gone further in liberalization, permitting abortion at the request of the pregnant woman or on broadly interpreted social indications.

The countries with the longest uninterrupted experience are Sweden and Denmark, which began to liberalize their abortion laws in the 1930's and have gradually relaxed the conditions for abortion over the years. In both countries the criteria for permission are primarily based on the physical and mental hazards for the mother, taking into account "the conditions under which the woman will have to live." Indeed, most of the abortions have been approved on psychiatric grounds. In Sweden, where most applications originally were referred to the Royal Medical Board, a large proportion of the procedures are now performed on the recommendation of two physicians. In Denmark certification by the chief of the obstetrics and gynecology department of a hospital is sufficient authority for an abortion on

purely medical considerations; in cases involving other factors the application must be approved by a committee of the Mother's Aid Institution consisting of a psychiatrist, a gynecologist and a social worker.

The annual number of legal abortions in Sweden rose from 400 in 1939 to more than 6,300 in 1951 and then declined, and in the past few years it has risen again to an estimated 9,600 in 1967. The ratio of abortions to live births was five per 1,000 in 1939 and 79 per 1,000 in 1967. Denmark's experience has paralleled Sweden's [see illustration on opposite page]. As legal abortions increased, the death rate associated with the operation declined. In Sweden, for example, mortality fell from 257 per 100,000 legal abortions in the period 1946–1948 to 39 per 100,000 in the period 1960–1966.

In Britain abortion to protect the health of the mother has been legal since a 1938 court decision in a famous rape case in which the judge held that abortion was justified "if the doctor is of the opinion...that the probable consequences [of the birth] will be to make the woman a physical and mental wreck." The British medical profession, however, took a conservative attitude toward making use of this ruling. In 1967, after a long and bitter legislative struggle, Parliament enacted an abortion law similar to the Scandinavian model and going beyond it in certain important respects. The new British law authorizes abortion on the certification by two physicians that continuance of the pregnancy would involve risk to the life or the physical or mental health of the pregnant woman, even if the risk is very slight: it need only be greater than that of abortion, which, if performed under the proper conditions, involves less danger to life than pregnancy and childbearing. Furthermore, the British law permits abortion if the prospective birth would imperil the physical or mental health of "existing children" of the pregnant woman's family or if there is a substantial risk that the child would be born defective. Since the new British abortion law came into force only in April, 1968, no significant information has yet become available concerning its effects.

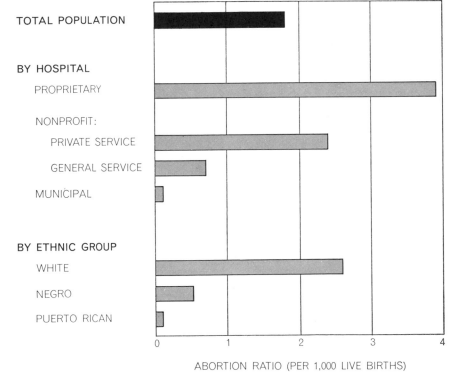

INDUCED ABORTION is generally illegal in the U.S. but may be performed in hospitals under certain circumstances. It is more readily available to people with money, as indicated by this chart giving the abortion ratio in New York City from 1960 through 1962.

What has been the experience of the countries with the most permissive abortion laws, that is, the countries of eastern Europe and Japan and China? No information is available on China, but fairly detailed data have been re-

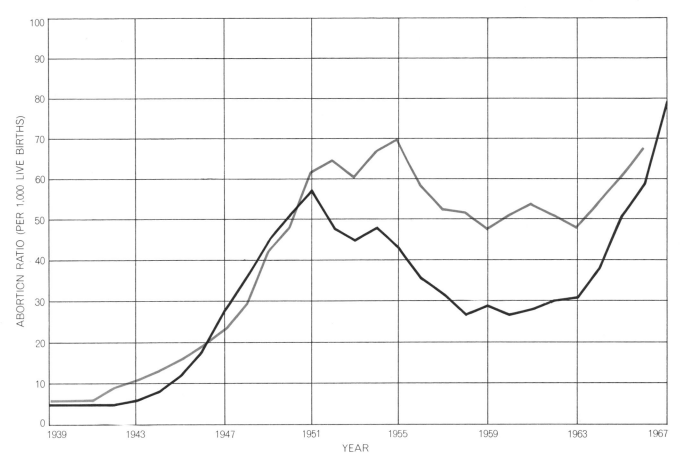

SCANDINAVIAN COUNTRIES have been liberalizing their abortion laws since the 1930's. The chart shows the trend of the abortion ratio in Sweden (*black*) and Denmark (*color*) since 1939. The ratios rose to peaks in the 1950's, fell and are now rising again.

ported from such countries as Hungary and Czechoslovakia, which (along with the U.S.S.R.) enacted liberal abortion statutes in the 1950's, and from Japan, which has had a "Eugenic Protection Law" since 1948.

In most of these countries the laws permit abortion on broadly interpreted social indications, if not at the request of the pregnant woman. Poland, for example, lists any "difficult social situation" as an acceptable reason for a legal abortion. Czechoslovakia allows abortion on grounds of the woman's advanced age, the presence of three or more children in the family, the loss or disability of the husband, a broken home, a threat to the family's living standard when the woman bears the main economic responsibility, "difficult circumstances" in the case of an unmarried woman and pregnancy resulting from rape. Similarly, in Yugoslavia abortion is permissible whenever the birth of the child would unavoidably create a serious personal, familial or economic hardship for the mother. As a general rule, in the eastern European countries abortions for medical reasons are performed in hospitals without charge;

in other cases the applicant pays part of the costs.

Not surprisingly, where the abortion laws are as permissive as they are in eastern Europe the number of legal abortions has risen sharply. In Hungary, for example, the increase has been so great that the number of legal abortions now exceeds the number of births. In 1967 there were 187,500 legal abortions in that country as against 148,900 live births [*see top illustration on next page*]. The contrasting curves, showing that as the abortion rate has gone up the birthrate has gone down, suggest that for many women in Hungary abortion has replaced contraception as the means of birth control. Similarly, the statistics in Czechoslovakia indicate that legalized abortion has reduced the birthrate, although it is still considerably higher than the abortion rate [*see middle illustration on next page*].

The drop in the birthrate has caused two countries of eastern Europe to revise their abortion laws. In October, 1966, Romania substituted a restrictive statute for one allowing abortion on re-

quest. Abortions on medical grounds are permitted now only if the pregnancy threatens the woman's life, and the other acceptable indications are limited to cases in which the woman is over 45 years of age, or has four or more children, or may give birth to a malformed child or has become pregnant as a result of rape or incest. Following the enactment of the new Romanian law the birthrate rose from 13.7 (per 1,000 of population) in the fourth quarter of 1966 to 38.4 in the third quarter of 1967. This rise indicates that before the repeal of the permissive law the annual rate of legal abortions in Romania had been at least 24.7 per 1,000 of population, which is considerably higher than the high rate of 18.4 that was reported for Hungary in 1967.

In 1968 Bulgaria also imposed restrictions—somewhat less drastic—on abortion. Women with three or more children may still have an abortion on request, but those with one or two children must apply to a board that "shall explain the harmfulness and dangers of abortion, the need to take the pregnancy to full term, the financial support that the family

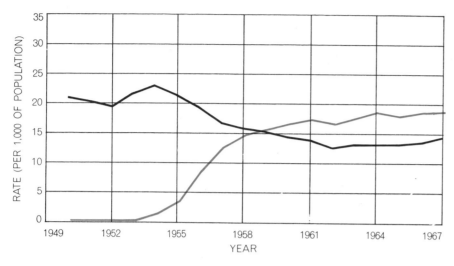

IN HUNGARY, where abortion is allowed essentially on request, abortion rate (*color*) now exceeds birthrate (*black*) by one-fourth; legal abortion is a routine means of birth control.

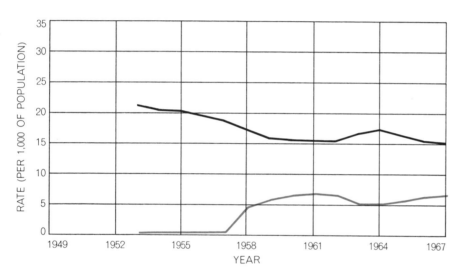

IN CZECHOSLOVAKIA grounds for abortion recognize social and economic, as well as medical, factors. In recent years the abortion rate (*color*) is up, birthrate (*black*) down.

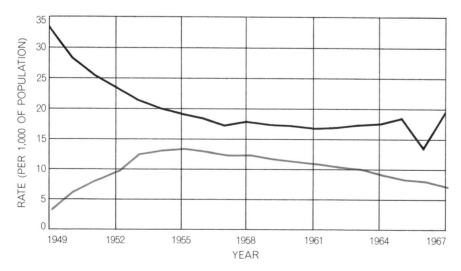

IN JAPAN, where abortion is freely available, abortion rate (*color*) may be understated. This is indicated by its failure to rise when the birthrate (*black*) dropped sharply in 1966.

will receive after the birth of a child and, in general, shall make every effort to dissuade every woman who expresses the desire to have her pregnancy interrupted from doing so. If, nevertheless, the woman concerned persists in asking for her pregnancy to be interrupted, the board shall give its approval to this effect." Childless women may be aborted on medical grounds only.

In Japan the Eugenic Protection Law authorizes abortion if the pregnancy will seriously affect the woman's health "from the physical or economic viewpoint." This provision has been interpreted so broadly that any Japanese woman can have a legal abortion on request. The number of legal abortions reported in Japan rose from 246,000 in 1949, the year after passage of the law, to 1,170,000 in 1955. After that year the reported number declined, and by 1967 it had fallen to 748,000 [*see bottom illustration at left*]. There are reasons to believe, however, that the reports may understate the actual number of legal abortions by several hundred thousand. It is said that many Japanese physicians, in order to minimize their income taxes, do not report all the abortions they perform. Curiously, the number of reported abortions continued to decline instead of rising when the birthrate dropped precipitately in 1966, the "Year of the Fiery Horse." (According to tradition, girls born during such a year grow up to be bad-natured and hence difficult to marry off.) Since it is doubtful that the sharp drop in births can be attributed mainly to a surge in the effective use of contraception by the most tradition-bound segment of Japanese society, one can surmise that the number of abortions must actually have risen, rather than declining, in 1966.

A noteworthy fact about the legalized practice of abortion in the socialist countries of eastern Europe and in Japan is the remarkably low mortality. The mortality rates from abortion in these countries in recent years have been only one to four deaths per 100,000 legal abortions, far lower than the rates in the Scandinavian countries [*see top illustration on opposite page*]. One of the main reasons for the lower mortality is that Japan and the eastern European countries prohibit abortion after the third month of pregnancy except for medical reasons, whereas Sweden, for example, permits it up to the fifth month and Denmark up to the fourth month. Another reason is that the women applying for abortions in the Scandinavian countries are in a poorer state of health than the women undergoing abortion on request in the countries of eastern Europe and in Japan.

It is hard to determine the extent to which the legalization of abortion in these countries has reduced the incidence of criminal, or clandestine, abortions. Certainly illegal abortions have not been eliminated entirely, as is evidenced by the fact that a number of women with difficulties arising from such abortions turn up at hospitals for treatment. This stubborn survival of illegal abortion can probably be attributed to the fact that, for one reason or another, women sometimes want to conceal their pregnancy and therefore avoid the procedure of formal application and official approval. Be that as it may, there is little doubt that in the countries that permit abortion on request the practice of illegal abortion has declined greatly.

Little is known about the situation in the rest of the world. We have already noted that the number of illegal abortions in the U.S. is estimated to be in the hundreds of thousands a year. In most of the countries of western Europe illegal abortion is believed to be more common than in the U.S., and it has been said, although not verified, that in some of those countries the abortion rate exceeds the birthrate. Illegal abortion appears to be most prevalent in Latin America, particularly among the urban poor. In Chile, whose abortion laws are similar to those in the U.S., Rolando Armijo and Tegualda Monreal of the University of Chile School of Public Health estimated after a study of vital statistics that in a single year there were 20,000 illegal abortions, compared with 77,440 births, in Santiago, the capital city. A similar situation has been reported on the other side of the world. In Korea, which has a restrictive abortion law similar to those in most U.S. states, Sung-bong Hong estimated that Seoul, the capital city, had 55,000 illegal abortions in 1963—more than half as many abortions as births, 108,700.

The problem of illegal abortion is one that every nation must eventually deal with in some way. In a number of countries it is so serious that the medical profession and the government have joined forces to organize programs for family planning. It seems only reasonable that the U.S., particularly the state legislatures, should draw on the experiences now available from countries that are trying new approaches in the search for a rational solution to the problem.

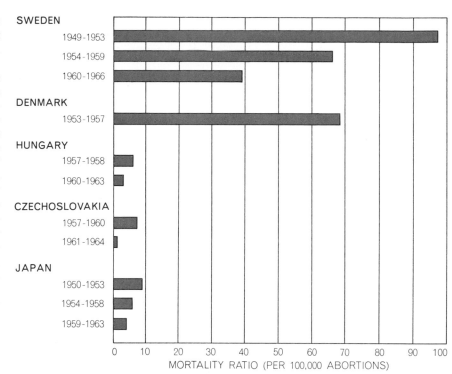

**MORTALITY RATIO** (deaths per 100,000 abortions) is known only where abortions are legal. It is much lower in permissive eastern Europe and Japan than in Sweden and Denmark.

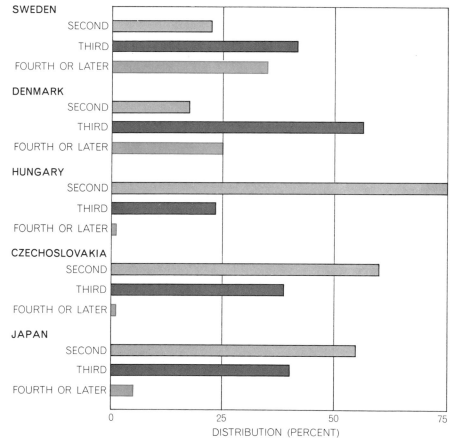

**LOW DEATH RATES** in eastern Europe and Japan are attributed to the discouragement there of abortions late in pregnancy and to the fact that many abortion patients in Scandinavia are already in poor health. In the chart abortions are distributed by month of gestation.

# 23

# Squatter Settlements

WILLIAM MANGIN

*October 1967*

Since the end of World War II squatter settlements around large cities have become a worldwide phenomenon. In the rapidly urbanizing but not yet industrialized countries millions of families from the impoverished countryside and from the city slums have invaded the outskirts of major cities and there set up enormous shantytowns. These illegal usurpations of living space have everywhere aroused great alarm, particularly among the more affluent city dwellers and government authorities. Police forces have made determined and violent efforts to repel the invasions, but the tide has been too much for them. The squatter settlements give every sign of becoming permanent.

The new shantytowns are without public services, unsanitary and in many respects almost intolerably insecure. Most middle-class and upper-class observers are inclined to regard them as a virulent social disease. Politicians and the police see them as dangerous defiance of law and order. Conservatives are certain that they are seedbeds of revolution and communism. City planners and architects view them as inefficient users of urban real estate and as sores on the landscape. Newspapers treat them as centers of crime and delinquency. Social workers are appalled by the poverty of many of the squatters, by the high incidence of underemployment and low pay, by the lack of medical treatment and sewage facilities and

by what they see as a lack of proper, decent, urban, middle-class training for the squatters' children.

The truth is that the shantytowns are not quite as they seem to outside observers. I first became acquainted with some of these settlements in Peru in 1952. Conducting studies in anthropology among villagers in the Peruvian mountains at that time, I occasionally visited some of their friends and relatives living in squatter settlements (they are called *barriadas* in Peru) on the fringes of the city of Lima. I was surprised to find that the squatter communities and the way the people lived differed rather widely from the outside impression of them. Since then I have spent 10 years in more or less continuous study of the *barriadas* of Peru, and it has become quite clear to me that many of the prevalent ideas about the squatter settlements are myths.

The common view is that the squatters populating the Peruvian shantytowns are Indians from the rural mountains who still speak only the Quechuan language, that they are uneducated, unambitious, disorganized, an economic drag on the nation—and also (consistency being no requirement in mythology) that they are a highly organized group of radicals who mean to take over and communize Peru's cities. I found that in reality the people of the *barriadas* around Lima do not fit this description at all.

Most of them had been city dwellers

for some time (on the average for nine years) before they moved out and organized the *barriadas*. They speak Spanish (although many are bilingual) and are far removed from the rural Indian culture; indeed, their educational level is higher than that of the general population in Peru. The *barriada* families are relatively stable compared with those in the city slums or the rural provinces. Delinquency and prostitution, which are common in the city slums, are rare in the *barriadas*. The family incomes are low, but most of them are substantially higher than the poorest slum level. My studies, based on direct observation, as well as questionnaires, psychological tests and other measurements, also indicate that the *barriada* dwellers are well organized, politically sophisticated, strongly patriotic and comparatively conservative in their sociopolitical views. Although poor, they do not live the life of squalor and hopelessness characteristic of the "culture of poverty" depicted by Oscar Lewis; although bold and defiant in their seizure of land, they are not a revolutionary "lumpenproletariat."

The squatters around the cities of Peru now number about 700,000, of whom 450,000 live in the *barriadas* of Lima itself. This is a substantial portion of the nation's entire population, which totals about 12 million. Like the squatter settlements in other countries, the *barriadas* of Peru represent the worldwide migration of people from the

country to the city and a revolt of the poor against the miserable, disorganized and expensive life in the city slums. In the shantytowns they find rent-free havens where they feel they can call their homes and the land their own.

The *barriadas* of Lima began some 20 years ago as clusters of families that had spontaneously fled from the city and set up communities of straw shacks on the rocky, barren land outside. The first, small settlements were short-lived, as the police forcibly drove the settlers off, sometimes with fatal beatings of men, women and children, and burned their shacks and household goods. Nevertheless, the squatters kept returning, as many as four times to the same place. They soon learned that there was greater safety in numbers, and the invasions of land and formation of *barriadas* became elaborately planned, secretly organized projects involving large groups.

The enterprise generally took the form of a quasi-military campaign. Its leaders were usually highly intelligent, articulate, courageous and tough, and often a woman was named the "secretary of defense" (a title borrowed from Peruvian labor organizations and provincial clubs). For the projected *barriada* community the leaders recruited married couples under 30 with children; single adults were usually excluded (and still are from most *barriadas*). Lawyers or near-lawyers among the recruited group searched land titles to find a site that was owned, or at least could be said to be owned, by some public agency, preferably the national government. The organizers then visited the place at night and marked out the lots assigned to the members for homes and locations for streets, schools, churches, clinics and other facilities.

After all the plans had been made in the utmost secrecy to avoid alerting the police, the organizers appealed confidentially to some prominent political or religious figure to support the invasion when it took place; they also alerted a friendly newspaper, so that any violent police reaction would be fully reported. On the appointed day the people recruited for the invasion, usually num-

**HILLSIDE SHANTYTOWN** in the Rimac district of Lima is seen in the photograph on the opposite page. Many squatter houses, originally straw shacks, are being rebuilt in brick and masonry whenever the earnings of the owners permit. Visible behind an unexcavated pre-Columbian mound (*top*) is one of the few public housing projects in Peru.

SQUATTERS BATTLE POLICE the morning after an "invasion" of unoccupied land near the Engineering School, north of Lima's city limits. The clash occurred in 1963; although police managed to clear the site temporarily, the squatters soon returned to build there.

bering in the hundreds and sometimes more than 1,000, rushed to the *barriada* site in taxis, trucks, buses and even on delivery cycles. On arriving, the families immediately began to put up shelters made of matting on their assigned lots.

More than 100 such invasions to set up *barriadas* have taken place in the Lima area in the past 20 years. The settlers have consistently behaved in a disciplined, courageous, yet nonprovocative manner, even in the face of armed attack by the police. In the end popular sympathy and the fear of the political consequences of too much police violence have compelled the government authorities to allow the squatters to stay. The present liberal regime of President Belaunde tries to prevent squatter invasions, but it does not attack them violently when they occur.

Once a *barriada* has established a foothold, it grows until it has used up its available land. The original settlers are joined by relatives and friends from the provinces and the city. From the relatively flat land where the first houses are built, new shacks gradually creep up the steep, rocky hillsides that overlook the city.

The surface appearance of the *barriadas* is deceptive. At first glance from a

distance they appear to be formless collections of primitive straw shacks. Actually the settlements are laid out according to plans, often in consultation with architectural or engineering students. As time goes on most of the shanties are replaced by more permanent structures. As soon as the residents can afford to, they convert their original straw shacks into houses of brick and cement. Indeed, the history of each *barriada* is plainly written in the mosaic of its structures. The new houses clinging to the high hillside are straw shacks; at the foot of the hill the older ones are built of masonry. One of the oldest *barriadas*, known as San Martin, has a paved main street, painted houses and elegant fronts on stores, banks and movie houses.

The squatters improve their houses as they accumulate a little extra money from employment and find spare time. At present the *barriada* communities are far too poor to afford the capital costs of utilities such as water systems and sewers. Water and fuel (mainly kerosene) are transported in bottles or drums by truck, bicycle or on foot. Some houses have electricity supplied by enterprising individuals who have invested in generators and run lines to their clients; a few of these entrepreneurs have gone so far as to acquire a television set (on time) and charge admission to the show. In some well-established *barriadas* the electric company of Lima has installed lines and service.

The major concern of the *barriada* people, and the greatest source of anxiety, is the problem of finding steady employment. The largest *barriadas* do provide considerable local employment, particularly in construction work. Many families obtain some income by operating stores, bars or shops in their homes; in the *barriada* I have studied most closely about a third of the households offer some kind of goods for sale. By and large, however, the people of the squatter settlements around Lima depend mainly on employment in the city. Most of the men and many of the women commute to jobs in Lima, working in personal services, factories, stores, offices and even in professional occupations. One *barriada* men's club includes among its members a physician, a bank branch manager, a police lieutenant, four lawyers, several businessmen and two Peace Corps volunteers.

The families that colonize a *barriada* are regarded as "owners" of their lots. As time goes on, many rent, trade or sell their lots and houses to others, using beautifully made titles with seals, lawyers' signatures and elaborate property descriptions—but in most cases with no legal standing. (Actually it appears that in Peru even private property is usually clouded by at least two titles, and much of the land is in litigation.) In the *barri-*

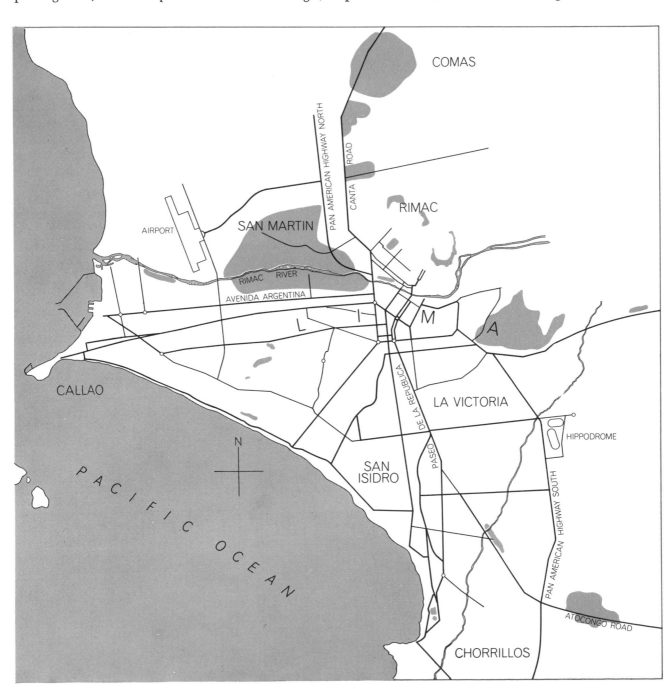

**BARRIADAS** of the city of Lima and its outskirts (*color*) shelter some 450,000 squatters who began to establish rent-free communities in 1945 on unoccupied hillsides north and south of the Rimac River. Now major *barriadas* also occupy both sides of the river downstream toward the port of Callao; a 20-kilometer stretch of the Pampa de Comas, including some agricultural land, along the road north to Canta, and hillsides bordering the road south to Atocongo, adjacent to the richest residential district in the Peruvian capital.

*adas,* as elsewhere in the nation, disputes over lot "ownership" arise; the claimants appeal variously to the association that runs the *barriada* or to the National Housing Authority, the Lima city government, the police or the courts. The decisions of these agencies generally have only a provisional character. A law adopted by the Peruvian national legislature in 1957 authorized the granting of land titles to *barriada* dwellers, but for several years it was ignored. In 1962 a group of engineers and architects in the National Housing Authority, taking advantage of the preoccupation of the military junta with other matters, passed out land titles to a few hundred families in two of the oldest *barriadas.* Even these titles, however, were marked "Provisional."

In most matters of public concern the *barriadas* are governed by their own membership associations. They hold elections about once a year—a rarity in Peru, where, except in the *barriadas,* no democratic elections of local officials had been held for more than 60 years before the present national government took office. The *barriada* associations levy taxes (in the form of "dues") on the residents, and they usually manage to collect them from most members. They also screen new applicants, resolve land disputes, try to prevent land speculation and organize cooperative projects. For official papers, such as voting registration and certificates of marriage, birth and death, the *barriada* people must resort to the city hall, and their reception by the town clerks is often so uncordial and whimsical that the quest for an essential document may be a heroic ordeal. (I have seen *barriada* birth certificates stamped "Provisional"!) Lacking authoritative police forces of their own, the *barriada* residents usually take their complaints of crimes and misdemeanors to the city police, but the latter seldom do anything more than register the complaint. For schooling of the children the *barriadas* depend mainly on the city's public and church schools. A few have elementary schools of their own, but generally students must commute to the city in the elementary grades as well as to high school and the university. The *barriada* people also have close connections with the city through their jobs, unions, social clubs, churches and services such as medical care, social security and unemployment insurance.

Many of the *barriada* associations have established working relations with city and national agencies and even

**SQUATTER ENTREPRENEURS,** residents of a Rimac *barriada,* run a sidewalk cobbler's shop complete with foot-powered stitching machines at the edge of the wholesale market.

**AGED BUS** is one of the many vehicles, some communally owned, that connect the outlying *barriadas* with downtown Lima. Many squatters commute to steady jobs in the city.

with international organizations such as the Peace Corps and the United Nations. Of the various agencies in a position to assist the *barriadas* perhaps the most important is Peru's National Housing Authority, known as the JNV. The JNV has been beset by power struggles between the national office and local city officials and by other confusions, so that its accomplishments are uneven. In some *barriadas* representatives of the JNV are cheered; in others they are stoned. (In one settlement the agency erected an impressive sign announcing that it was installing a water and sewage-disposal system; after six months had passed with no visible evidence of a start on the project, the residents began to pile fecal matter under the sign, whereupon JNV removed the sign.) Recently, however, the housing agency gave Lima officials authority to adopt and proceed with specific plans, and there is now considerable activity.

**WAITING FOR INVASION,** a squatter advance party at dawn inspects the previous night's work of blocking out the town plan for a new *barriada*. The rest of the invading squatters, as many as 1,000 in number, will soon arrive in trucks, buses and taxis.

**MAT-SHED SETTLEMENT** springs up within a few hours after an invasion and a new *barriada* is established. This squatter settlement on the Pampa de Comas is an unusual intrusion on cultivated land; the majority of invasions occupy idle or desert areas.

The *barriada* governments have not lacked the usual trouble of municipal administrations, including charges of corruption and factional splits. Moreover, their prestige and authority have declined as the need for community cohesion and defense against attack from outside has been reduced. There is a compensating trend, however, toward replacement of the original associations by full-fledged, official town governments. The two largest *barriadas* in the Lima area, San Martin and Pampa de Comas, now have elected mayors and town councils.

What, if anything, can be learned from the squatter settlements that will be of value in resolving the monumental problems of today's cities and their desperate people? I should like to present some conclusions from our own 10-year studies. They were carried out on a grant from the U.S. National Institute of Mental Health in cooperation with the Institute of Ethnology of the University of San Marcos and the Department of Mental Hygiene of the Ministry of Public Health in Peru, and with the assistance of a group of psychiatrists, anthropologists and social workers. We concentrated on an intensive study of a particular *barriada*, which I shall call Benavides. It consists of some 600 families. Over the 10-year period I have spent considerable time living in the community (in a rented room), interviewing a large sample of the population and examining their attitudes and feelings as indicated by various questionnaires and inventories, including the Rorschach and thematic apperception tests.

I am bound to say that I have been profoundly impressed by the constructive spirit and achievements of the *barriada* people. They have shown a really remarkable capacity for initiative, self-help and community organization. Visitors to the *barriadas*, many of them trained observers, remark on the accomplishments of the residents in home and community construction, on the small businesses they have created, on the degree of community organization, on how much the people have achieved without government help and on their friendliness. Most of the residents are neither resentful nor alienated; they are understandably cynical yet hopeful. They describe themselves as "humble people," abandoned by society but not without faith that "they" (the powers that be) will respond to people's needs for help to create a life of dignity for

DIGGING A SEWER is typical of squatters' communal ventures in self-improvement. The large brick structure beyond is another communal project, a partly finished church.

YEARLY ELECTIONS are a feature of *barriada* life scarcely known to other citizens of Peru. Until the Belaunde regime took office in 1963, democratic local elections were rare.

**SWIFTNESS** of a squatter invasion is exemplified by the settled appearance of this quiet lane in a new *barriada* outside Lima. None of these buildings had existed 24 hours earlier.

**STREET DOOR** of a mat-shed shelter consists of wooden frame and cloth drop that carries the house number. The resident wears conventional city dress. Many in the *barriadas* come to Lima from the country, but most are townfolk fleeing slum rents and slum conditions.

themselves. Recognizing fully that they are living in "infrahuman conditions," the *barriada* dwellers yearn for something better. Given any recognition or encouragement by the government, such as the paving of a street or even the collection of taxes from them, the people respond with a burst of activity in improvement of their homes.

This is not to say that either their spirit or their behavior is in any sense idyllic. There are tensions within the *barriada* and people take economic advantage of one another. They are victims of the same racial prejudice and class inequality that characterize Peruvian society in general. As in the world outside, the *barriada* people identify themselves as city people, country people, coastal people, mountaineers, Indians, Cholos, mestizos, Negroes—and cliques arise. With the passage of time and weakening of the initial *esprit de corps*, bickering within the community becomes more and more common. Charlatans and incompetents sometimes take over leadership of the *barriada*. Moreover, because of the poverty of their resources for financing major projects in community services, the people have a low estimate of their own capabilities and continually look to the government or other outside agencies for solutions to their problems.

Nevertheless, to an outside observer what is most striking is the remarkable progress the *barriada* people have made on their own. They have exhibited a degree of popular initiative that is seldom possible in the tightly controlled community-action programs in the U.S. The *barriadas* of Peru now represent a multimillion-dollar investment in house construction, local small businesses and public services, not to speak of the social and political investment in community organization. Such achievements hold lessons from which more advanced countries may well profit.

Particularly in house construction and land development the *barriada* people have done better than the government, and at much less cost. The failures of governments and private developers everywhere to provide low-cost housing for the poor are notorious. Administrative costs, bureaucratic restrictions and the high cost of materials and construction when government agencies do the contracting generally put the housing rentals beyond the reach of the lowest-income group. Equally disappointing are the failures in the design of this official public housing, which usually disregards the desires and style of life of the people for whom it is intended.

In the Peruvian *barriadas*, by avoiding government control and the requirements of lending institutions, the people have built houses to their own desires and on the basis of first things first. Because they needed shelter immediately, they built walls and a roof and left bathrooms and electricity to be added later. They want flat roofs and strong foundations so that they can add a second story. They want a yard for raising chickens and guinea pigs, and a front room that can serve as a store or a barroom. They have dispensed with the restrictive residential zoning and construction details that middle-class planners and architects consider essential for proper housing.

Like most rural people in Peru, the *barriada* settlers are suspicious of large-scale projects and wary of entering into loan or mortgage arrangements. Indeed, throughout South America there is a general dissatisfaction with large housing projects. Costly mistakes have been made in the construction of "satellite cities" and "superblocks." This has led the national governments and other interested agencies to give more attention to the possibilities in rehabilitating existing housing. In Peru the government is now initiating experiments in offering low-cost loans through credit cooperatives, providing optional technical assistance and other services and letting the prospective housebuilder do his own contracting. As John Turner, an architect with many years' experience in Peru, has pointed out, if people are sold land and allowed to do their own contracting and building with optional help, the costs go down for both the clients and the government.

Our studies of the *barriadas* of Peru show, in brief, that these settlements contain many constructive elements whose significance should not be ignored. The people believe that their present situation is far preferable to what they had in the provinces or the central city slums and that they have an investment in their future and that of their children. What we have learned in Peru is supported by investigations of squatter settlements around the world.

The squatters have produced their own answer to the difficult problems of housing and community organization that governments have been unable to solve. In Peru we may have a chance to study what can happen when a government works with popular initiative rather than fighting it.

TRANSFORMED *BARRIADA* was one of the first in Lima. Today most buildings are brick or stone and many have a second story. Although unsurfaced, its avenue is illuminated.

PROSPEROUS ENTERPRISE, the Restaurant Central, is located in the Pampa de Comas. In 1956 it was a one-story bar in a newly built *barriada* that had no electric power. Now there are streetlights and the restaurant has a second story, a coat of plaster and television.

# 24

# The Renewal of Cities

NATHAN GLAZER

*September 1965*

When we speak of the renewal of cities, we mean all the processes whereby cities are maintained or rebuilt: the replacement of old houses by new houses, of older streets by newer streets, the transformation of commercial areas, the relocation of industrial facilities, the rebuilding of public utilities; we refer to rehabilitation as well as demolition and rebuilding; we mean too the laws and administrative and financial mechanisms by which this rebuilding and rehabilitation are accomplished. The only way to discuss such an enormous subject is to consider all the elements of change in a city: its changing economic role, its changing population, decisions to buy or sell, stay or move, rehabilitate or demolish, and the larger market and political forces that affect all this.

Fortunately we can narrow our subject considerably. There exist, in this nation and others, specific public policies designed to plan and control at least some part of these vast processes.

PART OF A RENEWAL AREA in Philadelphia occupies the right side of the aerial photograph on the opposite page. On the left side of the photograph is an old area consisting largely of rundown three-story row houses. The cleared land is being used to put up new row houses, some of which appear at right. Construction has just begun in the block at upper right center. This project, known as the Southwest Temple urban renewal project, has involved the relocation of some 2,000 families and individuals, almost all low-income Negroes. The pattern of new occupancy suggests that the renewed area will house some 1,750 families and individuals, almost all middle-income Negroes. Wide streets bounding the renewal area are Broad Street (*left*), Columbia Avenue (*top*) and Girard Avenue (*bottom*).

In the U.S. such policies are expressed in the urban renewal program administered by the Urban Renewal Administration (a part of the Federal Housing and Home Finance Agency), which guides hundreds of local city agencies in the effort to transform urban renewal from a process dominated by the requirements and opportunities of the market to one guided by social intelligence—reflection on how the process might best create a better city.

The specific program that is the focus of this article began with the passage of the Housing Act of 1949 and has been expanded and modified continually since then. Before that time, of course, there were many mechanisms by which cities and states and the Federal Government attempted to affect the rebuilding of cities. The most significant Federal predecessor of urban renewal was public housing, that is, slum clearance and the building of subsidized Government housing for the poor. There remains a good deal of confusion between public housing and urban renewal. Indeed, the agency that is responsible for New York City's huge program of urban renewal, the largest in the nation, was until a few years ago called the Slum Clearance Commission. A similar agency in Chicago was called the Land Clearance Commission. And under the original Housing Act the effort to guide urban renewal was administered by a Division of Slums and Urban Redevelopment. All these agencies now have different names that foretell the sparkling new structures that will go on cleared land rather than the grimy ones that are to be cleared. Therein lies one of the great dilemmas of our approach to urban renewal: the fact that our program provides great powers and resources for clearing the way to get new

areas built but few resources for dealing with the people who live in the older areas that are to be cleared.

Federally supported public housing was only one of the ways in which government had tried to deal with urban problems before the development of a comprehensive renewal program. There were also Federally sponsored mortgage-insurance programs that helped to make possible the widespread construction of private, single-family houses in the suburbs of U.S. cities after World War II. In addition there were numerous efforts on the part of cities to control development and redevelopment with zoning regulations, health and building regulations concerning housing, and the establishment of local planning agencies. The urban renewal program made use of these local powers of planning and zoning, Federal credit mechanisms and the existing power to clear slums and build public housing; it added to these older approaches a powerful legal mechanism and a powerful financial mechanism, both designed to win the cooperation of private developers in the pursuit of public goals. The legal mechanism stipulated that a local renewal agency was empowered to condemn private property not only for public uses (which had long been permitted) and publicly owned housing but also for resale to private developers who agreed to fulfill the plan for the area that the local agency had drawn up. The financial mechanism, known as a write-down, committed the Federal Government to paying from two-thirds to three-quarters of the difference between, on the one hand, the cost of buying the land, clearing it and preparing it for the new development and, on the other, the price that private redevelopers would pay for it. The designers of the urban renewal legislation were proposing a compromise: public intelligence was to guide the rebuilding of cities, but the rebuilding would be carried out in such a way as to ensure significant private profits and ultimate private ownership of land the public had spent a great deal of money and effort to acquire.

The power of condemnation assured private developers that they could acquire large tracts. These were sought because they prevent the remaining slums from pressing too close on the renewed area, diminishing its desirability, value and profit for the owner. One social critic, Jane Jacobs, has dramatically questioned the need for such large tracts in her book *The Death and Life of Great American Cities*. Most modern planners, however, tend to endorse the developers' demand for large areas, citing the need for more parking and park space. As for the financial write-down, private developers sought it because the price of central-city slum areas was high, even if one took away from the property owners the right to raise their prices excessively. The slums were densely occupied and lucrative for the landlords, favorably located and well served by public

transportation and city facilities. In certain areas the financial power to write down the cost of land became far more important than the power to condemn. In Manhattan, for example, the redevelopment of urban renewal property has cost the public $1 million an acre—the difference between what was paid the owners of the land in order to clear it and what the developers paid to have the opportunity to redevelop it. In other areas developers were quite willing to pay the condemnation cost of the land, and it was the power to condemn and assemble that made redevelopment possible.

We have described the mechanisms of urban renewal; what were the objectives of the program? These can be ascertained if we examine the disparate elements in the alliance that forged it. There were first of all people committed to public planning and public housing. In 1949 these were the men and women who had participated in the great experiments of the New Deal, in which a modicum of European social imagination and concern in the area of housing had been introduced into the U.S. They saw urban renewal—even if they had qualms about the compromise

embodied in the legislation—as a means of extending the power of the people to affect through politics the growth of their cities and the quality of their housing and environment, thus reducing the power of the market to shape this for them. Tied to the original urban renewal legislation was provision for a good deal of public housing that would foreseeably accommodate those who had to be relocated from the demolished slums. It was unpleasant from the point of view of the reformers to have to pay the owners of slum property so much money for the privilege of replanning and rebuilding the areas, but the alternatives had been vetoed. One such alternative, put forward by Charles Abrams and Catherine Bauer Wurster, called for building more public housing on open and cheap land on the outskirts of a city and allowing the price of central slum properties to fall as they emptied. Such a solution was opposed by the big-city mayors and the commercial and financial interests dependent on maintaining business and property values in the centers of the big cities—in particular, department store owners and banks with mortgages on central-city property.

Urban renewal was created by an

CHICAGO SLUM was photographed in 1944 from a building on Federal Street. In the 1950's this neighborhood was demolished and rebuilt under the auspices of the U.S. Urban Renewal Authority, the Chicago Housing Authority and several other agencies.

alliance of those seeking reform and those seeking profit. The planners and advocates of public housing were trying to improve the environment of slum dwellers and the overall pattern of the city in terms of amenity and efficiency. The commercial and financial interests were trying to maintain the level of business and property values in downtown areas, jeopardized somewhat by an increasingly poor (and, incidentally, nonwhite) central-city populace. Both groups wanted to stem the rapid flow of the more prosperous citizens to the suburbs and hoped this could be done by remodeling the cities physically. The mayors, confronted with the increasing costs of urban government and threatened by the decline of property values and tax revenues, shared this hope. They saw in urban renewal the solution to the economic decline of central-city areas and an opportunity to build monuments and generally beautify the cities.

The alliance is no longer intact. The downtown commercial interests still support the program. The mayors still support it, seeing no alternative. The planners are split. Those who emphasize the social aims of planning, the problems of the poor and the slum dwellers, oppose the program on the grounds that it has done little for the poor and nothing to reverse the pattern of increased urban segregation. These planners are torn between their commitment to the ideal of the people shaping their own environment and their dismay at the actual environment that, under political and economic pressure, has been shaped. Most planners, however, support urban renewal; for one thing, the planners of today are not the planners of the 1940's who participated in the New Deal or whose ideas were molded by it. They are now in large part the professionals trained to fill needs created by the urban renewal program itself.

Let us review the present state of the program, taking our information from the report of the Housing and Home Finance Agency for 1964. By the end of that year local renewal agencies had acquired about 27,000 acres of urban land. "Redevelopers had been selected for 16,318 acres"; the rest was being cleared or was unsold. "Redevelopment had been completed or was actually under construction on more than 55 percent of that land," or about one-third of all the land that had been acquired. "By mid-1964, more than 72 percent of all land disposed of, exclusive of streets and alleys, had been purchased by private persons or organizations. More than half was intended for residential purposes. By mid-1964, 61,770 dwelling units of all kinds were completed and 18,300 more were under construction"—some 80,000 in all. The sum of Federal money involved in this effort—the capital grants that would eventually be required to complete this volume of urban renewal—was $4.3 billion. Midway through 1964 some 176,000 families and 74,000 individuals had been relocated from sites scheduled for urban renewal.

The scale of this undertaking seems different from various perspectives. Bernard Frieden, professor of city planning at the Massachusetts Institute of Technology and former editor of the *Journal of the American Institute of Planners*, estimates that deteriorated housing in New York City in 1960 covered 1,145 acres. The number of units of deteriorated housing recorded by the census of 1960 was 147,000. This suggests that the urban renewal program was of a sufficient order of magnitude to clear away all the slums of New York—if all of the program had been devoted to that city (and if it had been

**RENEWED NEIGHBORHOOD** was photographed from same perspective in 1965. Federal Street has been rerouted and is now adjacent to the railroad tracks. The development at center and right consists of eight units housing mostly middle-income families.

used to clear away slums, and if there had been policies to prevent new slums from forming). On the other hand, the 80,000 units of housing built or under construction since the beginning of urban renewal in 1949 is not an impressive total compared with the 7.3 million housing units built between 1960 and 1964, nor does the relocation of some 750,000 people seem highly significant in view of the fact that 40 million people move every year in the U.S.

Obviously one can say that renewal has just begun to scratch the surface of the need; there were, after all, 2.3 million substandard dwellings in our cities in 1960. It is also being said, however, that renewal has already gone too far, or at least too far in the wrong direction. Social critics allege that although the volume of building under the urban renewal program has been slight, its impact on certain parts of the population has been devastating. In some cities the designation for urban renewal of any area, no matter how decrepit the housing, arouses a desperate resistance among the people living there. Indeed, television dramas of daily life sometimes cast the local urban renewal agency in the role once played by the hardhearted banker. This ad-

verse reputation, a powerful comment on urban renewal, seems to arise from the real experience of the poor; it was not created by the social critics who now amplify it. The urban renewal agency does in fact represent a current threat to many: destroying small businessmen, evicting older people from their homes, forcing families from their tenements and then failing to relocate them in decent, safe, sanitary and reasonably priced housing as required by law, threatening buildings of historic or architectural value, and even attacking Bohemians and artists in their contemporary garrets. (These are the most dangerous opponents, because they know how to get publicity.) It is apparent that the urban renewal agency is a more vivid threat to security than the banker in these days of amortized mortgages.

Still, if the scale of urban renewal has been as small as I have indicated in terms of figures for voluntary movement of population, new dwellings built and people directly affected, how is it possible to argue that its effects on the city have been so damaging? Primarily because its impact has been on one segment of the urban population: the poor —those least able, materially or psychologically, to adapt to upheaval. The

people who live in old neighborhoods are, compared with the rest of the U.S. population, poor, old and more likely to be Negroes or members of other minority groups. They are often people with special ties to the neighborhood and special problems that keep them there. For many reasons, of which money is only one, they find it extremely difficult to find other housing in the city. Two-thirds of those relocated from urban renewal sites have been nonwhites (the program has sometimes been derisively termed "Negro removal"), whose problem of finding housing is compounded by the fact that few parts of the city will accept them. Many of the businesses on urban renewal sites were small and marginal; indeed, some provided for an aged couple a living no better than what they would get on welfare. Such people were nonetheless kept occupied, and they provided some of the social benefits of an old neighborhood that Jane Jacobs has described: places to leave messages, conversation to break the monotony and anonymity of city living, eyes to watch the street. Some 39,000 business properties had been acquired by urban renewal agencies as of September 30, 1963; studies have shown that a third do not survive relocation. Some of them would have

NEIGHBORHOOD DUE FOR RENEWAL on the upper West Side of Manhattan includes this block on 89th Street between Columbus Avenue and Central Park West. The brownstone houses were once one-family dwellings but have long since been converted to apartments. A slum by certain criteria, this block provides its residents with housing convenient to familiar institutions, stores

succumbed to the high death rate of small businesses in any case. Many that do relocate successfully move outside the city; thus ironically the city loses the taxes from business that urban renewal is meant to increase.

The urban renewal agency is required to demonstrate that enough housing is available for those whose homes are to be demolished, it is required to help them move and it has Federal resources to pay moving expenses for families and businesses. These requirements were much looser at the beginning of the urban renewal program than they are today, and the resources available were much scantier. Among the first large urban renewal projects in Manhattan were those undertaken by the energetic Robert Moses at a time when New York had a great shortage of housing, particularly low-cost housing. Relocation was unquestionably carried out in a businesslike and ruthless fashion (that is, rapidly on those sites where the developers were eager to move out the people and put up the new buildings, slowly on sites where they preferred to collect rents from the slum dwellers they were supposed to evict). Available aid, in the form of money or advice or social service, was slight. The image

and friends. Renewal plans call for moving the present tenants and selling the houses to people who can afford to renovate them.

of renewal, as of many things in this country, is largely set by what happens in New York, where most of the writers, publishers and television producers live; urban renewal began with a very poor image. It is uncertain whether enough has been done to correct the practices that created the nightmare one critic calls "the Federal bulldozer."

According to reports sent to Washington from local authorities, the dwellings of 87 percent of the families relocated from urban renewal sites are known and were inspected, and 92 percent of these are decent, safe and sanitary as required by law. These figures have been disputed by Chester Hartman, a city planner who worked on a major study of the impact of urban renewal conducted by the Center for Community Studies in Boston. Hartman argues that local authorities have loose standards in judging the quality of the housing into which people move from urban renewal sites. Thus the local agency reported that less than 2 percent of the families relocated from Boston's West End had moved into structurally substandard housing, whereas the Center for Community Studies placed the figure at 25 percent. Conversely, the local authorities tend to apply strict standards in judging the housing of an area they plan to demolish, because they have to satisfy Washington that the area is a genuine slum. Herbert Gans (in *The Urban Villagers*, a detailed description of the West End as an old, inner-city working-class district) has pointed out that what was a slum to the planners was good housing to those who lived there—housing they preferred to any other in the city, and in a neighborhood that contained the people and places they knew.

The West End study demonstrated that there was an improvement in the quality of the housing into which most families were relocated and an increase in the proportion of home owners. There was also an increase in rents: the median rent of the West Enders rose from $41 to $71 a month, and rent as a proportion of income rose from 14 to 19 percent. Similar studies have been completed in recent years, some of which indicate that before renewal the West End was a real bargain. Although the figures vary from survey to survey, the results of relocation form a pattern: housing is somewhat improved, rents go up, the proportion of rent to income goes up, home ownership increases.

How are we to evaluate such a pattern? There is currently great in-

terest among city planners and urban economists in developing a technique for quantitative comparison of costs and benefits, a technique that could in every case give an objective answer to the question: Is this urban renewal project worth it? Attempts at cost-benefit analysis have in the past been crude. For example, planners have compared the costs of police, welfare and other social services of an area to be leveled with the reduced costs after rebuilding, neglecting to take into account the fact that the costs are incurred not by neighborhoods or buildings but by people. The departure of the people does not, of course, reduce the costs; it merely changes the place where the costs are incurred. As Martin Anderson has shown in his critique of urban renewal, *The Federal Bulldozer,* even the simple analysis of tax returns from the property before and after redevelopment is often inadequate, since it may fail to take into account such elements as the loss of taxes during the long period of redevelopment and the possibility that the same new structures might have been built elsewhere in the city without redevelopment.

If the tangible aspects of renewal are difficult to evaluate in the balance sheet of a cost-benefit analysis, how can one assess such intangibles as the cost of relocating an old woman whose only remaining satisfactions in life are taking care of the apartment in which she has lived for many years, going to the church around the corner and exchanging a few words with the neighborhood merchants? Admittedly one can even work in these costs by reckoning the chance that she will require a nursing home when she moves, or some additional city service. Such tabulations may at times seem akin to dissecting a rainbow, but they are being made nonetheless. The major purpose of the West End study has been to determine the impact of relocation on the mental health of the participants. Reports by Marc Fried of the Harvard Medical School indicate that serious reactions of grief have exceeded, in depth and duration, most expectations.

Even if we can find a way of quantifying the intangible aspects of relocation, how are we to take them into account in making social policy? The decisions to renew or not to renew must be made by local governments responsive to the pressures of the different parts of the community. If the political costs of a certain course of action are great, they will certainly outweigh the results of any subtle analysis of psy-

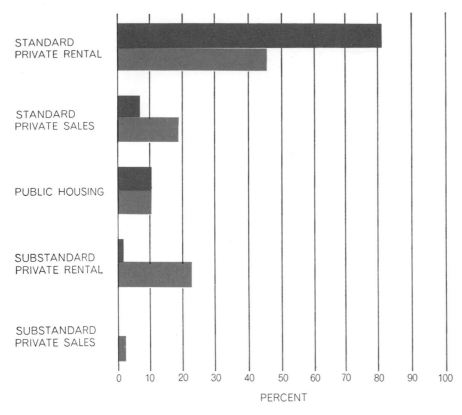

STANDARD
PRIVATE RENTAL

STANDARD
PRIVATE SALES

PUBLIC HOUSING

SUBSTANDARD
PRIVATE RENTAL

SUBSTANDARD
PRIVATE SALES

0  10  20  30  40  50  60  70  80  90  100

PERCENT

**SATISFACTORY RELOCATION was the subject of contradictory reports made by the Center for Community Studies in Boston (*colored bar in each grouping*) and the local urban renewal agency (*black bar*). Bars give percentage of families who found housing in each category. The discrepancies suggest different standards of the two agencies.**

chological, social and economic costs or benefits. Experience so far shows that almost invariably the despair in areas slated for demolition is not channeled into meaningful political opposition; it is outweighed by the arguments for renewal presented by planners to the city fathers and the prejudice among middle-class citizens against allowing what they consider slums to remain standing near them. The proponents of renewal have not, however, been oblivious to its reputation among the poor; with each subsequent housing act they have expanded the resources for relocating families and have heightened the obligation of local authorities to do the same. Let us review briefly the resources now available to the local urban renewal agency for dealing with this problem.

Families on sites scheduled for demolition have always had priority in moving into public housing. The amount of public housing built has approximated the amount demolished. In general, however, only half the families on a site are eligible for public housing, and all told only 20 percent of the relocated families have moved into it. Often there is not enough public

housing available at the precise time it is needed. The local public housing authority and the local renewal authority are two separate bodies; they deal with two separate agencies in Washington; they operate under separate laws, and although specific public housing projects theoretically could be built in anticipation of an old neighborhood being cleared, this has not often been done. In any case, many of those eligible for public housing will not accept it; this is particularly true of white families, who often refuse to move into projects in which they feel the proportion of Negroes is too high. Negroes and whites alike object to the institutional atmosphere of projects, with their regulations and requirements, and all share the apprehension that public housing attracts a concentration of problem families.

Since 1954 one of the major objectives of urban renewal has been the rehabilitation of old houses—a process that makes relocation unnecessary. Unfortunately rehabilitation, even with Federal loan programs to promote it, has rarely been successful. Renovating a house to meet the standards imposed by the program requires much more money than the occupants can raise; the property is then sold to a new own-

er. The general result is that poor people are moved out of houses that upper-income people can afford to renovate.

The sums available for relocating families and businesses were originally small, and they were provided only when they were needed to expedite development. These sums have been increased sharply and are now given more readily. The 1964 Housing Act for the first time recognizes and authorizes payment (of up to $500) to families, elderly individuals and small businessmen for the dislocations attendant on moving. It has taken 15 years for this principle, which is taken for granted in other countries, to be recognized by our government. Late but useful aid has also been extended by the Small Business Administration, which was authorized in 1961 to help businessmen reestablish themselves with loans, assistance and information.

Still other efforts have been made to ease the burdens of relocation. In the early 1950's special loans were designed to provide housing for those from urban renewal sites who were too poor to get regular housing but not poor enough to be eligible for public housing. The most successful type of loan was instituted in 1961; it permits nonprofit sponsors as well as limited-dividend corporations to get mortgages below the going rate to put up cooperative or rental housing for moderate-income families. There has also been a strengthening of Federal regulations requiring detailed reports from local agencies on the availability of housing (in different price ranges and for nonwhites as well as whites), on relocation plans and on current progress. Finally, the explosion of new social welfare programs for the poor provides additional resources. On the West Side of Manhattan, where extensive relocation is under way, a substantial number of social workers are engaged in various programs to help families find housing and settle in a new environment.

Gradually, after 15 years of putting so much energy into getting buildings down and so little into helping people up, we are beginning to develop the kind of program that should have existed from the beginning and that exists in the advanced European welfare state —a program whose emphasis is on providing housing. We are still faced by immense problems of segregation, institutionalism in public housing and human uprooting, but as of 1965 it should be possible for most local urban renewal authorities to carry out an effective relocation plan and even provide some of those benefits from relocation that the advocates of urban renewal maintain the process makes possible.

The question now becomes: What positive goals are we attempting to attain through renewal? How well does the renewal program make it possible to achieve them? It is not enough to say that we want new buildings instead of old buildings. Urban economists argue that in any event buildings will go up in response to market demands; urban renewal has merely shifted the location of new buildings rather than increased their actual number. Unquestionably renewal has done a good deal to bring investment into downtown areas, but what has the public gained by investing hundreds of millions of dollars for new street layouts, parking, open space and land write-downs for private developers—all for shoring up the center of the city? The answer is usually stated in terms of tradition or economics: The center must remain strong if a metropolitan area is to thrive. It must have good commercial and cultural facilities, and a significant proportion of middle- and upper-income residents. If private, unguided investment insists on going to the outlying suburbs (a tendency encouraged by the automobile, freeways and cheap suburban land), then public investment must redress the balance. Only in this way can the central city retain the middle- and upper-income people whose tax revenues enable it to provide services.

Both aspects of the defense of the central city have been challenged. Scott Greer of Northwestern University and Melvin Webber of the University of California at Berkeley observe that the form of the city is changing in such a way that Los Angeles will be the most likely model of the city of the future. They hold that behind the abandonment of the traditional city form is the fact that free citizens in an affluent society—particularly those with children—prefer to live in detached houses with some land. This seems to be true the world over; it is only where costs make such an arrangement impossible that people settle for apartment houses. To rebuild expensive inner-city land for residential purposes means building apartments, attractive only to such special elements as those without children—the young or the old. Certainly these groups represent an important market, but it does not follow that government should provide them with a subsidy. As for the economic argument—the need to attract the wealthy—it has been attacked as a form of discrimination against the poor. After all, the poor have come to the city's center because housing there is cheapest and most convenient for them. They are near their jobs, their friends and, in the case of immigrants, their families or countrymen. If the cities need subsidies to counter the increase in low-income residents, why must the subsidy take the form of urban renewal? Why not redistribute Federal taxes to cities on the basis of need and let the city choose how to spend it? If we do

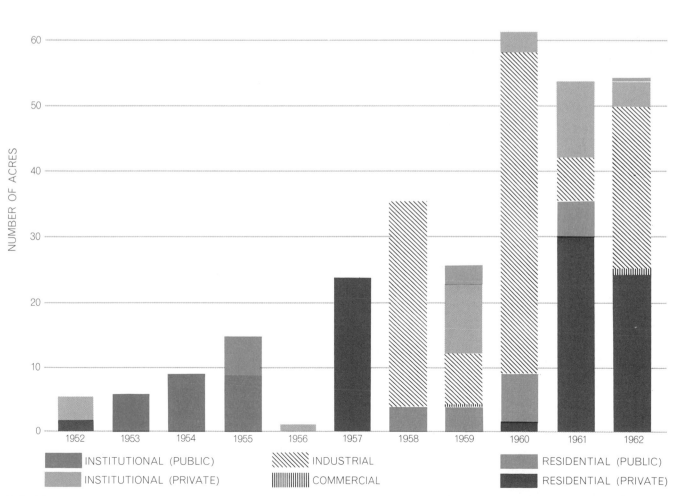

DISPOSITION OF LAND by the urban renewal agency in Philadelphia from 1952 to 1962 is charted. Bars give acreage devoted to types of reuse listed in key at bottom. A change in emphasis is implied by rise of developments for purposes other than housing.

this, a city made up largely of low-income people need not be a disaster.

A more basic challenge can be made to the argument that we need inner-city renewal to save the traditional centers. Why must we accept the present boundaries of cities as being permanent? These boundaries have been set by a variety of political accidents; as a result where one city (Boston, for example) may be a small part of a metropolitan area, another (Dallas) may embrace almost an entire metropolitan region. If the boundaries of each city could be redrawn to include most of the metropolitan area, the wealthy, who had abandoned the center for the outskirts, would again pay taxes to the city and the need for public investment in the center would be reduced. There are still other reasons why there should be some form of metropolitan government. Many problems in the provision of services could be solved more easily and effectively if they were examined from a metropolitan point of view rather than from the point of view of separate political entities within the metropolitan area . This is preeminently true of transportation, water supply, open space for recreation and air pollution. It seems

inordinately difficult to reorganize metropolitan governments in this country rationally; we can only envy the relative ease with which the government of London has been reorganized by an act of Parliament. The U.S. Government encourages metropolitan planning, but it can do little to create metropolitan governments to supplant the disparate governments within a metropolitan region.

One of the real virtues of urban renewal is that it has induced local communities to consider their needs and plan to meet them. In 1954 the Federal Government required that each city entering into an urban renewal program develop all the major operations of city government necessary to guide the rebuilding of the city and to submit a "workable program"—proper building codes and zoning ordinances, a comprehensive city plan, an administrative organization that could fulfill it, proof of interested citizens and the like. By 1959 the U.S. had instituted the Community Renewal Program, which provides substantial sums of money to cities to project their future development needs and policies. This program has supported much sophisticated work

involving simulation on computers of future urban development under alternative policies. Unfortunately too much of the current research and projection, no matter how imaginative, is oriented to the wrong scale: the city rather than the metropolitan area. Moreover, the major tool of the urban renewal program remains the specific project. It is still hoped that a better city can be achieved by supporting, by means of advantageous condemnation and land write-downs, specific projects based on the capacity to attract specific investment. This gives urban renewal an inherently spotty character.

Suppose it is—as I believe—essential that cities radically improve their function in inspecting buildings, requiring repairs and supporting them where necessary. Suppose a major way to improve a city is to root out substandard buildings wherever they are rather than demolish a huge area that is decrepit in spots. What Federal aid would be available for that? Much less than is available for the specific-project approach. Let me give an example.

A proposed project in San Francisco was going to cost $40 million. For this amount some 15,000 people would be relocated, their homes demolished and

**IMPACT OF RENEWAL ON NEGROES AND POOR is suggested by three maps of Philadelphia. Maps at left and center are based on 1960 census data. Map at left shows tracts where nonwhite population exceeded 80 percent (*color*) and tracts where it was 50 to 80 percent (*gray*). Map in center shows tracts where average family** **income was less than $4,720 annually (*color*) and where it fell between $4,720 and $6,000 (*gray*). Map at right shows zones scheduled for renewal by reconstruction (*color*) or renovation (*gray*). Dislocation of Negroes and the poor has been minimized in Philadelphia, but the problem is a continuing one in national program.**

the land turned over, somewhat improved by new streets, to builders. This is an enormous expense for a city the size of San Francisco, the total annual budget of which is only about $350 million. The money, however, was to come from the Federal Government and from the point of view of the city the undertaking was free. This would not be the case if San Francisco chose an alternative project, such as a major program of code enforcement, demolition of substandard housing or loans for rehabilitation. Urban renewal law and practice indicate that only a small fraction of $40 million would be extended for such efforts.

All the criticisms of urban renewal point to the fact that, whereas the program speaks of the whole city and all the ways in which it must be improved, provisions are made to influence only one aspect of the city—the physical nature of a given locale. The program as constituted and as practiced makes too little use of the traditional agencies of city government that must be depended on to improve cities. It also relates poorly to other large programs and expenditures in the city, such as the freeway program. When we consider the imaginative urban renewal that has been carried out in some European and Japanese cities by closely linking transportation arteries, housing, commercial and office facilities, we wonder why our projects are so often massive concentrations of a single function: all housing here, all concert halls there, all shopping there—and all poorly linked by transportation. This is the logical result, I would argue, of the fact that our urban renewal authority in Washington and the local agencies are oriented toward single missions—and the mission in every case is the individual project rather than the whole city.

After some 16 years of urban renewal we are still struggling with the problem of slums and still trying to formulate some alternative to the naïve image of the city beautiful in its middle-class version, an image that has increasingly lost its power to move people and solve problems. Under the pressure of a number of gifted critics, urban renewal has become an instrument that any city can use to develop policies well suited to its needs, and to carry out some of them. It is by no means a perfect instrument, but the source of its failings generally seems to be in the politics, the imagination and the structure of local government. It is there, I think, that we now need the chief efforts of our critics.

# 25

# The Social Power of the Negro

*April 1967*

The concept of "black power" is an inflammatory one. It was introduced in an atmosphere of militancy (during James Meredith's march through Mississippi last June) and in many quarters it has been equated with violence and riots. As a result the term distresses white friends of the Negro, frightens and angers others and causes many Negroes who are fearful of white disapproval to reject the concept without considering its rationale and its merits. The fact is that a form of black power may be absolutely essential. The experience of Negro Americans, supported by numerous historical and psychological studies, suggests that the profound needs of the poorest and most alienated Negroes cannot be met—and that there can therefore be no end to racial unrest—except through the influence of a unified, organized Negro community with genuine political and economic power.

Why are Negro efforts to achieve greater unity and power considered unnecessary and even dangerous by so many people, Negro as well as white, friends as well as enemies? I believe it is because the functions of group power —and hence the consequences of political and economic impotence—are not understood by most Americans. The "melting pot" myth has obscured the critical role of group power in the adjustment of white immigrant groups in this country. When immigrants were faced with discrimination, exploitation and

abuse, they turned in on themselves. Sustained psychologically by the bonds of their cultural heritage, they maintained family, religious and social institutions that had great stabilizing force. The institutions in turn fostered group unity. Family stability and group unity—plus access to political machinery, jobs in industry and opportunities on the frontier—led to group power: immigrants voted, gained political influence, held public office, owned land and operated businesses. Group power and influence expanded individual opportunities and facilitated individual achievement, and within one or two generations most immigrants enjoyed the benefits of first-class American citizenship.

The Negro experience has been very different, as I shall attempt to show in this article. The traumatic effects of separation from Africa, slavery and the denial of political and economic opportunities after the abolition of slavery created divisive psychological and social forces in the Negro community. Coordinated group action, which was certainly appropriate for a despised minority, has been too little evident; Negroes have seldom moved cohesively and effectively against discrimination and exploitation. These abuses led to the creation of an impoverished, undereducated and alienated group—a sizable minority among Negroes, disproportionately large compared with other ethnic groups. This troubled minority has a self-defeating "style" of life that leads to repeated fail-

ure, and its plight and its reaction to that plight are at the core of the continuing racial conflict in the U.S. Only a meaningful and powerful Negro community can help members of this group realize their potential, and thus alleviate racial unrest. The importance of "black power" becomes comprehensible in the light of the interrelation of disunity, impotence and alienation.

The roots of Negro division are of African origin. It is important to realize that the slave contingents brought out of Africa were not from a single ethnic group. They were from a number of groups and from many different tribes with different languages, customs, traditions and ways of life. Some were farmers, some hunters and gatherers, some traders. There were old animosities, and these were exacerbated by the dynamics of the slave trade itself. (Today these same tribal animosities are evident, as in Nigeria, where centuries-old conflict among the Ibo, Hausa and Yoruba tribes threatens to disrupt the nation. A significant number of slaves came from these very tribes.)

The cohesive potential of the captives was low to begin with, and the breakup of kinship groupings, which in Africa had defined people's roles and relations, decreased it further. Presumably if the Africans had been settled in a free land, they would in time have organized to build a new society meeting their own needs. Instead they were organized to meet the needs of their masters. The slaves were scattered in small groups (the average holding was only between two and five slaves) that were isolated from one another. The small number and mixed origins of each plantation's slaves made the maintenance of any oral tradition, and thus of any tribal or racial identity and pride, impossible. Moreover, any grouping that was potentially cohesive because of family, kinship or tribal connections was deliberately divided or tightly controlled to prevent rebellion. Having absolute power, the master could buy and sell, could decree cohabitation, punishment or death, could provide food, shelter and clothing as he saw fit. The system was engraved in law and maintained by the religious and political authorities and the armed forces; the high visibility of the slaves and the lack of places to hide made escape almost inconceivable.

The powerless position of the slave was traumatic, as Stanley M. Elkins showed in his study of Negro slavery. The male was not the respected provider, the protector and head of his household. The female was not rearing her child to take his place in a rewarding society, nor could she count on protection from her spouse or any responsible male. The re-ward for hard work was not material goods and the recognition of one's fellow men but only recognition from the master as a faithful but inferior being. The master—"the man"—became the necessary object of the slave's emotional investment, the person whose approval he needed. The slave could love or hate or have ambivalent feelings about the relationship, but it was the most important relationship of his life.

In this situation self-esteem depended on closeness or similarity to the master, not on personal or group power and achievement, and it was gained in ways that tended to divide the Negro population. House slaves looked down on field hands, "mixed-bloods" on "pure blacks," slaves with rich and important masters on slaves whose masters had less prestige. There was cleavage between the "troublemakers" who promoted revolt and sabotage and the "good slaves" who betrayed them, and between slave Negroes and free ones. The development of positive identity as a Negro was scarcely possible.

It is often assumed that with the end of the Civil War the situation of the free Negroes was about the same as that of immigrants landing in America. In re-ality it was quite different. Negroes emerging from slavery entered a society at a peak of racial antagonism. They had long since been stripped of their African heritage; in their years in America they had been unable to create much of a record of their own; they were deeply marked by the degrading experience of slavery. Most significant, they were denied the weapons they needed to become part of American life: economic and political opportunities. No longer of any value to their former masters, they were now direct competitors of the poor whites. The conditions of life imposed by the "Black codes" of the immediate postwar period were in many ways as harsh as slavery had been. In the first two years after the end of the war many Negroes suffered violence and death at the hands of unrestrained whites; there was starvation and extreme dislocation.

In 1867 the Reconstruction Acts put the South under military occupation and gave freedmen in the 11 Southern states the right to vote. (In the North, on the other hand, Negroes continued to be barred from the polls in all but nine states, either by specific racial qualifications or by prohibitive taxation. Until the Fifteenth Amendment was ratified in 1870, only some 5 percent of the Northern Negroes could vote.) The Reconstruction Acts also provided some military and legal protection, educational opportunities and health care. Reconstruction did not, however, make enough land available to Negroes to create an adequate power base. The plantation system meant that large numbers of Negroes remained under tight control and were vulnerable to economic reprisals. Although Negroes could outvote whites in some states and did in fact control the Louisiana and South Carolina legislatures, the franchise did not lead to real power.

This lack of power was largely due to the Negro's economic vulnerability, but the group divisions that had developed during slavery also played a part. It was the "mixed-bloods" and the house slaves of middle- and upper-class whites who had acquired some education and skills under slavery; now many of these people became Negro leaders. They often had emotional ties to whites and a need to please them, and they advanced the cause of the Negroes as a group most gingerly. Moreover, not understanding the causes of the apathy, lack of achievement and asocial behavior of some of their fellows, many of them found their Negro identity a source of shame rather than psychological support, and they were ready to subordinate the

SENEGAL
GAMBIA
SIERRA LEONE
LIBERIA

GUINEA
IVORY COAST
GHANA
NIGERIA
GABON
CONGO
REPUBLIC OF THE CONGO
ANGOLA

12,441
3,906
3,851
18,240
10,924
11,485

473 FROM MOZAMBIQUE AND EAST AFRICA

**VARIED ORIGIN of Negroes imported as slaves helps to explain divisions among Negro Americans. The map shows, as an example, the origin of slaves landed in South Carolina between 1733 and 1785. Even slaves from the same region were often from different tribes. Unlike white immigrants, Negroes had no common bonds of history, traditions and customs.**

SLAVERY made any organized Negro community impossible, stripped Negroes of racial pride and group traditions and created divisions among them. This engraving from *Harper's Weekly* shows house slaves lined up for auctioning by a trader in New Orleans.

needs of the group to personal gains that would give them as much social and psychological distance from their people as possible. The result was that Negro leaders, with some notable exceptions, often became the tools of white leaders. Throughout the Reconstruction period meaningful Negro power was being destroyed, and long before the last Negro disappeared from Southern legislatures Negroes were powerless.

Under such circumstances Negro economic and educational progress was severely inhibited. Negro-owned businesses were largely dependent on the impoverished Negro community and were operated by people who had little education or experience and who found it difficult to secure financing; they could not compete with white businesses. Negroes were largely untrained for anything but farm labor or domestic work, and a white social structure maintaining itself through physical force and economic exploitation was not likely to provide the necessary educational opportunities. Minimal facilities, personnel and funds were provided for the "Negro schools" that were established, and only the most talented Negroes were able—if they were lucky—to obtain an education comparable to that available to whites.

As John Hope Franklin describes it in *Reconstruction after the Civil War*, the Reconstruction was ineffective for the vast majority of Negroes, and it lasted only a short time: Federal troops had left most Southern states by 1870. While Negroes were still struggling for a first foothold, national political developments made it advisable to placate Southern

leaders, and the Federal troops were recalled from the last three Southern states in 1877. There was a brief period of restraint, but it soon gave way to violence and terror on a large scale. Threats and violence drove Negroes away from the polls. Racist sheriffs, legislators and judges came into office. Segregation laws were passed, buttressed by court decisions and law enforcement practices and erected into an institution that rivaled slavery in its effectiveness in excluding Negroes from public affairs—business, the labor movement, government and public education.

At the time—and in later years—white people often pointed to the most depressed and unstable Negro and in effect made his improvement in education and behavior a condition for the granting of equal opportunities to all Negroes. What kind of people made up this most disadvantaged segment of the Negro community? I believe it can be shown that these were the Negroes who had lived under the most traumatic and disorganized conditions as slaves. Family life had been prohibited, discouraged or allowed to exist only under precarious conditions, with no recourse from sale, separation or sexual violation. Some of these people had been treated as breeding stock or work animals; many had experienced brutal and sadistic physical and sexual assaults. In many cases the practice of religion was forbidden, so that even self-respect as "a child of God" was denied them.

Except for running away (and more tried to escape than has generally been

realized) there was nothing these slaves could do but adopt various defense mechanisms. They responded in various ways, as is poignantly recorded in a collection of firsthand accounts obtained by Benjamin A. Botkin. Many did as little work as they could without being punished, thus developing work habits that were not conducive to success after slavery. Many sabotaged the master's tools and other property, thus evolving a disrespect for property in general. Some resorted to a massive denial of the reality of their lives and took refuge in apathy, thus creating the slow-moving, slow-thinking stereotype of the Southern Negro. Others resorted instead to boisterous "acting out" behavior and limited their interests to the fulfillment of such basic needs as food and sex.

After slavery these patterns of behavior persisted. The members of this severely traumatized group did not value family life. Moreover, for economic reasons and by force of custom the family often lacked a male head, or at least a legal husband and father. Among these people irresponsibility, poor work habits, disregard for conventional standards and anger toward whites expressed in violence toward one another combined to form a way of life—a style—that caused them to be rejected and despised by whites and other Negroes alike. They were bound to fail in the larger world.

When they did fail, they turned in on their own subculture, which accordingly became self-reinforcing. Children born into it learned its way of life. Isolated and also insulated from outside influences, they had little opportunity to change. The values, behavior patterns and sense of alienation transmitted within this segment of the population from generation to generation account for the bulk of the illegitimacy, crime and other types of asocial behavior that are present in disproportionate amounts in the Negro community today. This troubled subgroup has always been a minority, but its behavior constitutes many white people's concept of "typical" Negro behavior and even tarnishes the image many other Negroes have of themselves. Over the years defensive Negro leaders have regularly blamed the depressed subgroup for creating a bad image; the members of the subgroup have blamed the leaders for "selling out." There has been just enough truth in both accusations to keep them alive, accentuating division and perpetuating conflicts, and impeding the development of group consciousness, cooperation, power and mutual gains.

It is surprising, considering the harsh conditions of slavery, that there were any Negroes who made a reasonable adjustment to freedom. Many had come from Africa with a set of values that included hard work and stability of family and tribal life. (I suspect, but I have not been able to demonstrate, that in Africa many of these had been farmers rather than hunters and gatherers.) As slaves many of them found the support and rewards required to maintain such values through their intense involvement in religion. From this group, after slavery, came the God-fearing, hardworking, law-abiding domestics and laborers who prepared their children for responsible living, in many cases making extreme personal sacrifices to send them to trade school or college. (The significance of this church-oriented background in motivating educational effort and success even today is indicated by some preliminary findings of a compensatory education program for which I am a consultant. Of 125 Negro students picked for the program from 10 southeastern states solely on the basis of academic promise, 95 percent have parents who are regular churchgoers, deeply involved as organizers and leaders in church affairs.)

For a less religious group of Negroes the discovery of meaning, fulfillment and a sense of worth lay in a different direction. Their creative talents brought recognition in the arts, created the blues and jazz and opened the entertainment industry to Negroes. Athletic excellence provided another kind of achievement. Slowly, from among the religious, the creative and the athletic, a new, educated and talented middle class began to emerge that had less need of white approval than the Negroes who had managed to get ahead in earlier days. Large numbers of Negroes should have risen into the middle class by way of these relatively stable groups, but because of the lack of Negro political and economic power and the barriers of racial prejudice many could not. Those whose aspirations were frustrated often reacted destructively by turning to the depressed Negro subgroup and its way of life; the subculture of failure shaped by slavery gained new recruits and was perpetuated by a white society's obstacles to acceptance and achievement.

In the past 10 years or so the "Negro revolt"—the intensified legal actions,

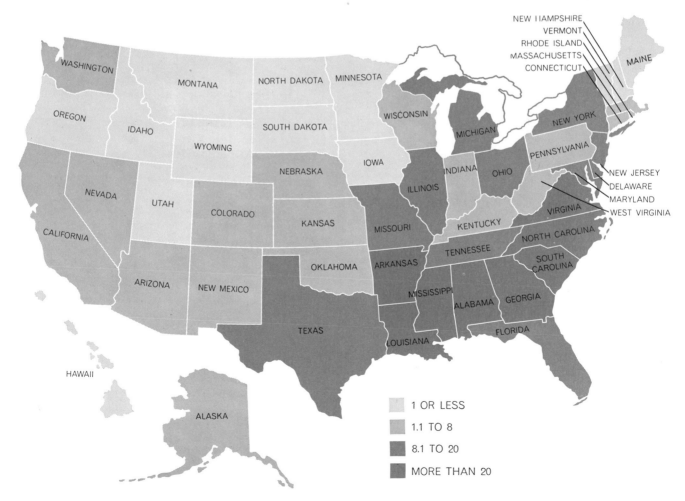

STATE-BY-STATE DISTRIBUTION of the Negro population is given as of 1960. The shading indicates each state's Negro population as a percent of the state's total population. In the North it is the big-city states that have the higher concentrations of Negroes.

nonviolent demonstrations, court decisions and legislation—and changing economic conditions have brought rapid
and significant gains for middle-class
Negroes. The mass of low-income Negroes have made little progress, however; many have been aroused by civil
rights talk but few have benefited. Of
all Negro families, 40 percent are classified as "poor" according to Social Security Administration criteria. (The figure for white families is 11 percent.)
Low-income Negroes have menial jobs
or are unemployed; they live in segregated neighborhoods and are exploited
by landlords and storekeepers; they are
often the victims of crime and of the
violent, displaced frustrations of their
friends and neighbors. The urban riots of
the past few years have been the reaction
of a small segment of this population to
the frustrations of its daily existence.

W hy is it that so many Negroes have
been unable to take advantage of
the Negro revolt as the immigrants did
of opportunities offered them? The major reason is that the requirements for

economic success have been raised. The
virtually free land on the frontier is gone.
The unskilled and semiskilled jobs that
were available to white immigrants are
scarce today, and many unions controlled
by lower-middle-class whites bar Negroes to keep the jobs for their present
members. The law does not help here
because Negroes are underrepresented
in municipal and state legislative bodies
as well as in Congress. Negroes hold
few policy-making positions in industry
and Negro small businesses are a negligible source of employment.

Employment opportunities exist, of
course—for highly skilled workers and
technicians. These jobs require education
and training that many Negroes, along
with many white workers, lack. The
training takes time and requires motivation, and it must be based on satisfactory education through high school.
Most poor Negroes lack that education,
and many young Negroes are not getting
it today. There are Negro children who
are performing adequately in elementary school but who will fail by the time
they reach high school, either because

their schools are inadequate or because
their homes and subculture will simply
not sustain their efforts in later years.

It is not enough to provide a "head
start"; studies have shown that gains
made as the result of the new preschool
enrichment programs are lost, in most
cases, by the third grade. Retraining
programs for workers and programs for
high school dropouts are palliative measures that have limited value. Some of
the jobs for which people are being
trained will not exist in a few years.
Many students drop out of the dropout
programs. Other students have such self-
defeating values and behavior that they
will not be employable even if they complete the programs.

A number of investigators (Daniel P.
Moynihan is one) have pointed to the
structure of the poorer Negro family as
the key to Negro problems. They point
to an important area but miss the crux
of the problem. Certainly the lack of a
stable family deprives many Negro children of psychological security and of the
values and behavior patterns they need
in order to achieve success. Certainly

REGIONAL DISTRIBUTION of the Negro population is shown.
The gray bars give each census region's Negro population as a
percent of the region's total population; the solid bars show
what percent of the total U.S. Negro population was in each region.

many low-income Negro families lack a father. Even if it were possible to legislate the father back into the home, however, the grim picture is unchanged if his own values and conduct are not compatible with achievement. A father frustrated by society often reacts by mistreating his children. Even adequate parents despair and are helpless in a subculture that leads their children astray. The point of intervention must be the subculture that impinges on the family and influences its values and style of behavior and even its structure.

How, then, does one break the circle? Many white children who found their immigrant family and subculture out of step with the dominant American culture and with their own desires were able to break away and establish a sense of belonging to a group outside their own—if the pull was strong enough. Some children in the depressed Negro group do this too. A specific pull is often needed: some individual or institution that sets a goal or acts as a model. The trouble is that racial prejudice and alienation from the white and Negro middle class often mean that there is little pull from the dominant culture on lower-class Negro children. In my work in schools in disadvantaged areas as a consultant from the Child Study Center of Yale University I have found that many Negro children perceive the outside culture as a separate white man's world. Once they are 12 or 14 years

old—the age at which a firm sense of racial identity is established—many Negroes have a need to shut out the white man's world and its values and institutions and also to reject "white Negroes," or the Negro middle class. Since these children see their problems as being racial ones, they are more likely to learn how to cope with these problems from a middle-class Negro who extends himself than from a white person, no matter how honest and free of hostility and guilt the white person may be.

Unfortunately the Negro community is not now set up to offer its disadvantaged members a set of standards and a psychological refuge in the way the white immigrant subcultures did. There is no Negro institution beyond the family that is enough in harmony with the total American culture to transmit its behavioral principles and is meaningful enough to Negroes to effect adherence to those principles and sufficiently accepted by divergent elements of the Negro community to act as a cohesive force. The church comes closest to performing this function, but Negroes belong to an exceptional number of different denominations, and in many cases the denominations are divided and antagonistic. The same degree of division is found in the major fraternal and civic organizations and even in civil rights groups.

There is a special reason for some of the sharp divisions in Negro organiza-

tions. With Negroes largely barred from business, politics and certain labor unions, the quest for power and leadership in Negro organizations has been and continues to be particularly intense, and there is a great deal of conflict. Only a few Negroes have a broad enough view of the total society to be able to identify the real sources of their difficulties. And the wide divergence of their interests often makes it difficult for them to agree on a course of action. All these factors make Negro groups vulnerable to divide-and-conquer tactics, either inadvertent or deliberate.

Viewing such disarray, altruistic white people and public and private agencies have moved into the apparent vacuum—often failing to recognize that, in spite of conflict, existing Negro institutions were meeting important psychological needs and were in close contact with their people. Using these meaningful institutions as vehicles for delivering new social services would have strengthened the only forces capable of supporting and organizing the Negro community. Instead the new agencies, public and private, have ignored the existing institutions and have tried to do the job themselves. The agencies often have storefront locations and hire some "indigenous" workers, but the class and racial gap is difficult to cross. The thong-sandaled, long-haired white girl doing employment counseling may be friendly and sympathetic to Negroes, but she cannot possibly tell a Negro youngster (indeed, she does not know that she should tell him): "You've got to look better than the white applicant to get the job." Moreover, a disadvantaged Negro —or any Negro—repeatedly helped by powerful white people while his own group appears powerless or unconcerned is unlikely to develop satisfactory feelings about his group or himself. The effects of an undesirable racial self-concept among many Negroes have been documented repeatedly, yet many current programs tend to perpetuate this basic problem rather than to relieve it.

A solution is suggested by the fact that many successful Negroes no longer feel the need to maintain psychological and social distance from their own people. Many of them want to help. Their presence and tangible involvement in the Negro community would tend to balance the pull—the comforts and the immediate pleasures—of the subculture. Because the functions of Negro organizations have been largely preempted by white agencies, however, no Negro institution is available through

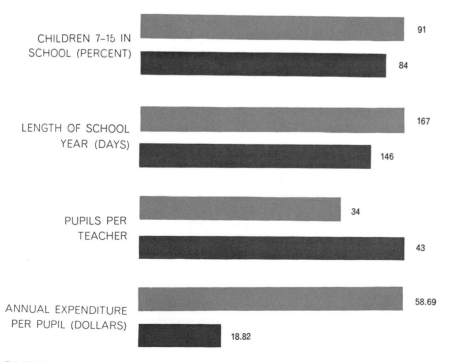

CHILDREN 7–15 IN SCHOOL (PERCENT)    91 / 84

LENGTH OF SCHOOL YEAR (DAYS)    167 / 146

PUPILS PER TEACHER    34 / 43

ANNUAL EXPENDITURE PER PUPIL (DOLLARS)    58.69 / 18.82

**LACK OF EDUCATION has handicapped Negroes. The charts compare segregated public school services for whites (*colored bars*) and Negroes (*gray bars*) in 17 Southern states and the District of Columbia in the years 1933–1934 (*three top charts*) and 1939–1940 (*bottom*).**

which such people can work to overcome a century of intra-Negro class alienation.

Recently a few Negroes have begun to consider a plan that could meet some of the practical needs, as well as the spiritual and psychological needs, of the Negro community. In Cleveland, New York, Los Angeles and some smaller cities new leaders are emerging who propose to increase Negro cohesiveness and self-respect through self-help enterprises: cooperatives that would reconstruct slums or operate apartment buildings and businesses providing goods and services at fair prices. Ideally these enterprises would be owned by people who mean something to the Negro community—Negro athletes, entertainers, artists, professionals and government workers—and by Negro churches, fraternal groups and civil rights organizations. The owners would share control of the enterprises with the people of the community.

Such undertakings would be far more than investment opportunities for well-to-do Negroes. With the proper structure they would become permanent and tangible institutions on which the Negro community could focus without requiring a "white enemy" and intolerable conditions to unify it. Through this mechanism Negroes who had achieved success could come in contact with the larger Negro group. Instead of the policy king, pimp and prostitute being the models of success in the subculture, the Negro athlete, businessman, professional and entertainer might become the models once they could be respected because they were obviously working for the Negro community. These leaders would then be in a position to encourage and promote high-level performance in school and on the job. At the same time broad measures to "institutionalize" the total Negro experience would increase racial pride, a powerful motivating force. The entire program would provide the foundation for unified political action to give the Negro community representatives who speak in its best interests.

That, after all, has been the pattern in white America. There was, and still is, Irish power, German, Polish, Italian and Jewish power—and indeed white Anglo-Saxon Protestant power—but color obviously makes these groups less clearly identifiable than Negroes. Churches and synagogues, cultural and fraternal societies, unions, business associations and networks of allied families and "clans" have served as centers of power that maintain group consciousness, provide jobs and develop new opportunities and join to form pressure and voting blocs.

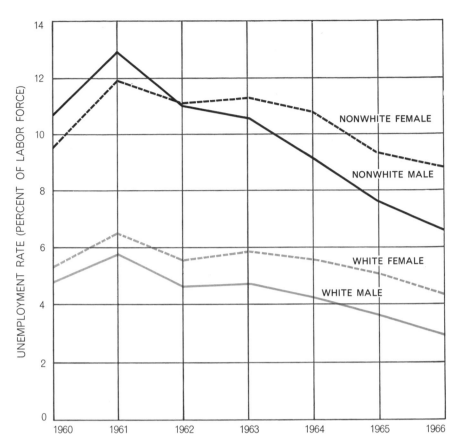

UNEMPLOYMENT RATE is higher among nonwhite workers than among white; the ratio has been about two to one in the past decade. Younger Negroes have been hardest hit.

The "nationality divisions" of the major parties and the balanced ticket are two reminders that immigrant loyalties are still not completely melted.

The idea of creating Negro enterprises and institutions is not intended as a rejection of genuinely concerned white people or as an indictment of all existing organizations. White people of good will with interest, skills and funds are needed and—contrary to the provocative assertions of a few Negroes—are still welcome in the Negro community. The kind of "black power" that is proposed would not promote riots; rather, by providing constructive channels for the energies released by the civil rights movement, it should diminish the violent outbursts directed against the two symbols of white power and oppression: the police and the white merchants.

To call for Negro institutions, moreover, is not to argue for segregation or discrimination. Whether we like it or not, a number of large cities are going to become predominantly Negro in a short time. The aim is to make these cities places where people can live decently and reach their highest potential with or without integration. An integrated society is the ultimate goal, but it may be a second stage in some areas.

Where immediate integration is possible it should be effected, but integration takes place most easily among educated and secure people. And in the case of immediate integration an organized and supportive Negro community would help its members to maintain a sense of adequacy in a situation in which repeated reminders of the white head start often make Negroes feel all the more inferior.

The power structure of white society—industry, banks, the press, government—can continue, either inadvertently or deliberately, to maintain the divisions in the Negro community and keep it powerless. Social and economic statistics and psychological studies indicate that this would be a mistake. For many reasons the ranks of the alienated are growing. No existing program seems able to meet the needs of the most troubled and troublesome group. It is generally agreed that massive, immediate action is required. The form of that action should be attuned, however, to the historically determined need for Negro political and economic power that will facilitate Negro progress and give Negroes a reasonable degree of control over their own destiny.

# 26

# Sickle Cells and Evolution

ANTHONY C. ALLISON

*August 1956*

Persevering study of small and seemingly insignificant phenomena sometimes yields surprising harvests of understanding. This article is an account of what has been learned from an oddly shaped red blood cell.

Forty-six years ago a Chicago physician named James B. Herrick, examining a Negro boy with a mysterious disease, found that many of his red blood cells were distorted into a crescent or sickle shape. After Herrick's report, doctors soon recognized many other cases of the same disease. They learned that it was hereditary and common in Negroes [see "Sickle-Cell Anemia," by George W. Gray; SCIENTIFIC AMERICAN, August, 1951]. The curious trait of the sickled blood cells gradually attracted the in-

terest of physiologists, biochemists, physical chemists, geneticists, anthropologists and others. And their varied investigations of this quirk of nature led to enlightenment on many unexpected subjects: the behavior of the blood's hemoglobin, inherited resistance to disease, the movements of populations over the world and the nature of some of the agencies that influence human evolution.

Let us review first what has been learned about the sickle cell phenomenon itself. As every student of biology knows, the principal active molecule in the red blood cells is hemoglobin, which serves as the carrier of oxygen. It appears that an unusual form of hemoglobin, pro-

duced under the influence of an abnormal gene, is responsible for the sickling of red cells. This hemoglobin molecule differs only slightly from the normal variety, and when there is an ample supply of oxygen it behaves normally: *i.e.*, it takes on oxygen and preserves its usual form in the red cells. But when the sickle cell hemoglobin (known as hemoglobin S)- loses oxygen, as in the capillaries where oxygen is delivered to the tissues, it becomes susceptible to a peculiar kind of reaction. It can attach itself to other hemoglobin S molecules, and they form long rods, which in turn attract one another and line up in parallel. These formations are rigid enough to distort the red cells from their normal disk shape into the shape

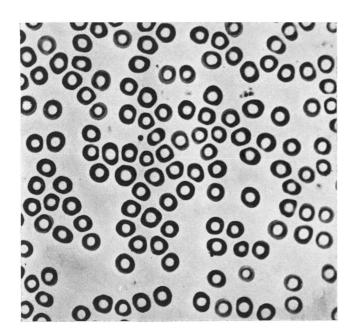

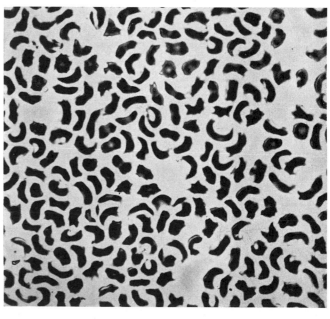

RED BLOOD CELLS of an individual with sickle cell trait, *i.e.*, a sickle cell gene from only one parent, are examined under the micro- scope. At the left are oxygenated red cells; they are disk-shaped. At the right are the same cells deoxygenated; they are sickle-shaped.

of a sickle [*see photomicrographs on page 179*]. Now the sickled cells may clog blood vessels; and they are soon destroyed by the body, so that the patient becomes anemic. The destruction of the hemoglobin converts it into bilirubin—the yellow pigment responsible for the jaundiced appearance often characteristic of anemic patients.

Most sufferers from sickle cell anemia die in childhood. Those who survive have a chronic disease punctuated by painful crises when blood supply is cut off from various body organs. There is no effective treatment for the disease.

From the first, a great deal of interest focused on the genetic aspects of this peculiarity. It was soon found that some Negroes carried a sickling tendency without showing symptoms of the disease. This was eventually discovered to mean that the carrier inherits the sickle cell gene from only one parent. A child who receives sickle cell genes from both parents produces only hemoglobin S and therefore is prone to sickling and anemia. On the other hand, in a person who has a normal hemoglobin gene from one parent and a hemoglobin S from the other sickling is much less likely; such persons, known as carriers of the "sickle cell trait," become ill only under exceptional conditions—for example, at high altitudes, when their blood does not receive enough oxygen.

The sickle cell trait is, of course, much more common than the disease. Among Negroes in the U. S. some 9 per cent carry the trait, but less than one fourth of 1 per cent show sickle cell anemia. In some Negro tribes in Africa the trait is present in as much as 40 per cent of the population, while 4 per cent have sickle cell genes from both parents and are subject to the disease.

The high incidence of the sickle cell gene in these tribes raised a most interesting question. Why does the harmful gene persist? A child who inherits two sickle cell genes (*i.e.*, is homozygous for this gene) has only about one fifth as much chance as other children of surviving to reproductive age. Because of this mortality, about 16 per cent of the sickle cell genes must be removed from the population in every generation. And yet the general level remains high without any sign of declining. What can be the explanation? Carriers of the sickle cell trait do not produce more children than those who lack it, and natural mutation could not possibly replace the lost sickle cell genes at any such rate.

The laws of evolution suggested a possible answer. Carriers of the sickle cell trait (a sickle cell gene from one parent and a normal one from the other) might have some advantage in survival over those who lacked the trait. If people with the trait had a lower mortality rate, counterbalancing the high mortality of sufferers from sickle cell anemia, then the frequency of sickle cell genes in the population would remain at a constant level.

What advantage could the sickle cell trait confer? Perhaps it protected its carriers against some other fatal disease—say malaria. The writer looked into the situation in malarious areas of Africa and found that children with the sickle cell trait were indeed relatively resistant to malarial infection. In some places they had as much as a 25 per cent better chance of survival than children without the trait. Children in most of Central Africa are exposed to malaria nearly all year round and have repeated infections during their early years. If they survive; they build up a considerable immunity to the disease. In some unknown way the sickle cell trait apparently protects young children against the malaria parasite during the dangerous years until they acquire an immunity to malaria.

On the African continent the sickle cell gene has a high frequency among people along the central belt, near the Equator, where malaria is common and is transmitted by mosquitoes through most

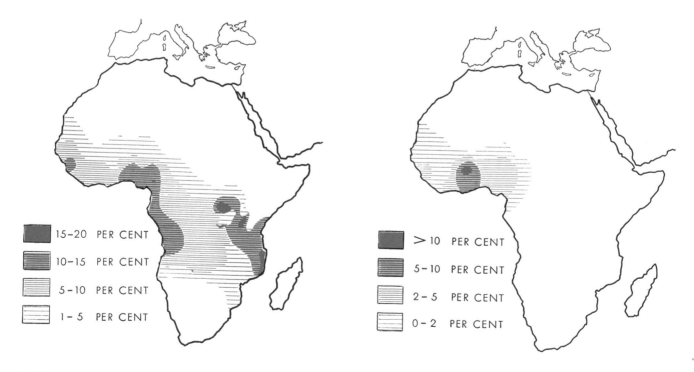

**FREQUENCY OF THE SICKLE CELL GENE** is plotted in per cent on the map of Africa. High frequencies are confined to a broad belt in which malignant tertian malaria is an important cause of death.

**FREQUENCY OF THE HEMOGLOBIN C GENE** is similarly plotted. Unlike the sickle cell gene, which has a widespread distribution, this gene is confined to a single focus in West Africa.

of the year. North and south of this belt, where malaria is less common and usually of the benign variety, the sickle cell gene is rare or absent. Moreover, even within the central belt, tribes in nonmalarious areas have few sickle cell genes.

Extension of the studies showed that similar situations exist in other areas of the world. In malarious parts of southern Italy and Sicily, Greece, Turkey and India, the sickle cell trait occurs in up to 30 per cent of the population. There is no reason to suppose that the peoples of all these areas have transmitted the gene to one another during recent times. The sickle cell gene may have originated independently in the several populations or may trace back to a few such genes passed along among them a thousand years ago. The high frequency of the gene in these populations today can be attributed mainly to the selective effect of malaria.

On the other hand, we should expect that when a population moves from a malarious region to one free of this disease, the frequency of the sickle cell gene will fall. The Negro population of the U. S. exemplifies such a development. When Negro slaves were first brought to North America from West Africa some 250 to 300 years ago, the frequency of the sickle cell trait among them was probably not less than 22 per cent. By mixed mating with Indian and white people this figure was probably reduced to about 15 per cent. In the absence of any appreciable mortality from malaria, the loss of sickle cell genes through deaths from anemia in 12 generations should have reduced the frequency of the sickle cell trait in the Negro population to about 9 per cent. This is, in fact, precisely the frequency found today.

Thus the Negroes of the U. S. show a clear case of evolutionary change. Within the space of a few hundred years this population, because of its transfer from Africa to North America, has undergone a definite alteration in genetic structure. This indicates how rapidly human evolution can take place under favorable circumstances.

Since the discovery of sickle cell hemoglobin (hemoglobin S), many other abnormal types of human hemoglobin have been found. (They are usually distinguished by electrophoresis, a separation method which depends on differences in the amount of the negative charge on the molecule.) One of the most

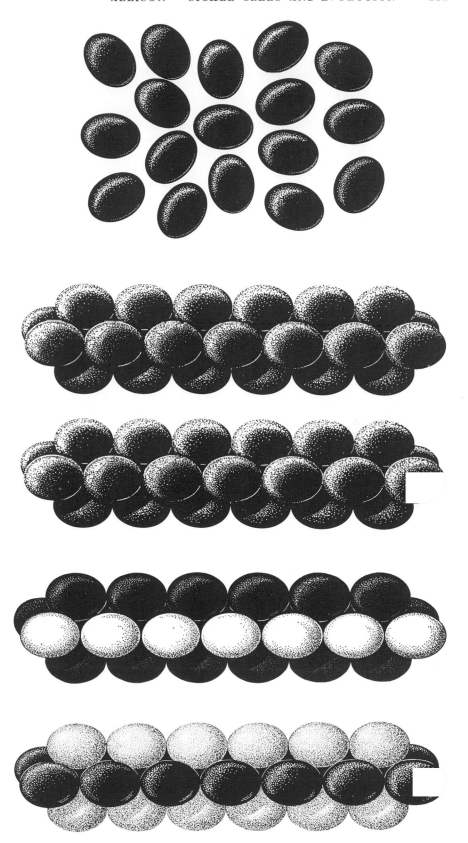

**HEMOGLOBIN MOLECULES** are represented as ellipsoids in these drawings. At the top are normal hemoglobin molecules, which are arranged almost at random in the red blood cell. Second and third from the top are sickle cell hemoglobin molecules, which form long helixes when they lose oxygen. Fourth is an aggregate of normal (*white*) and sickle cell molecules (*black*), in which every fourth molecule of the helix is normal. Fifth is an aggregate of hemoglobin C (*gray*) and sickle cell molecules; every other molecule is hemoglobin C.

common of these other varieties is called hemoglobin C. It, too, causes anemia in persons who have inherited the hemoglobin C gene from both parents. Moreover, the combination of hemoglobin S and hemoglobin C (one inherited from each parent) likewise leads to anemia. These two hemoglobins combine to form the rodlike structures that cause sickling of the red blood cells [see drawings on page 181].

The hemoglobin C gene is largely confined to West Africa, notably among people in the northern section of the Gold Coast, where the frequency of the trait runs as high as 27 per cent. Whether

hemoglobin C, like hemoglobin S, protects against malaria is not known. But the C gene must give some advantage, else it would not persist. Obviously inheritance of both C and S is a disadvantage, since it leads to anemia. As a consequence we should expect to find that where the C gene is present, the spread of the S gene is retarded. This does seem to be the case: in the northern Gold Coast the frequency of the S gene goes no higher than 5 per cent.

Another gene producing abnormal hemoglobin, known as the thalassemia gene, is common in Greece, Italy, Cyprus, Turkey and Thailand. The trait

is most prevalent in certain areas (e.g., lowlands of Sardinia) where malaria used to be serious, but there have not yet been any direct observations as to whether its carriers are resistant to malaria. The trait almost certainly has some compensating advantage, for it persists in spite of the fact that even persons who have inherited the gene from only one parent have a tendency to anemia. The same is probably true of another deviant gene, known as the hemoglobin E gene, which is common in Thailand, Burma and among some populations in Ceylon and Indonesia.

By now the identified hemoglobin types form a considerable alphabet: besides S, C, thalassemia and E there are D, G, H, I, J, K and M. But the latter are relatively rare, from which it can be inferred that they provide little or no advantage.

For anyone interested in population genetics and human evolution, the sickle cell story presents a remarkably clear demonstration of some of the principles at play. It affords, for one thing, a simple illustration of the principle of hybrid vigor. Hybrid vigor has been investigated by many breeding experiments with fruit flies and plants, but in most cases the crossbreeding involves so many genes that it is impossible to say what gene combinations are responsible for the advantages of the hybrid. Here we can see a human cross involving only a single gene, and we can give a convincing explanation of just how the hybridization provides an advantage. In a population exposed to malaria the heterozygote (hybrid) possessing one normal hemoglobin gene and one sickle cell hemoglobin gene has an advantage over either homozygote (two normal genes or two sickle cell genes). And this selective advantage, as we can observe, maintains a high frequency of a gene which is deleterious in double dose but advantageous in single dose.

Secondly, we see a simple example of inherited resistance to disease. Resistance to infection (to say nothing of disorders such as cancer or heart disease) is generally complex and unexplainable, but in this case it is possible to identify a single gene (the sickle cell gene) which controls resistance to a specific disease (malaria). It is an unusually di-

RATE OF CHANGE IN FREQUENCY of adults with sickle cells under different conditions is shown in this chart. The horizontal gray band represents the equilibrium frequency in a region where individuals with the sickle cell trait have an evolutionary advantage of about 25 per cent over individuals without the sickle cell trait. If a population of individuals with a low sickle cell frequency enters the region, the frequency will increase to an equilibrium value (long dashes). If hemoglobin C is already established in the same population, the frequency will increase to a lower value (short dashes). If a population of individuals with a high sickle cell frequency enters the region, the frequency will decrease (solid line). If this population enters a nonmalarious region, the frequency will fall to a low value (gray line).

rect manifestation of the fact, now universally recognized but difficult to demonstrate, that inheritance plays a large role in controlling susceptibility or resistance to disease.

Thirdly, the sickle cell situation shows that mutation is not an unmixed bane to the human species. Most mutations are certainly disadvantageous, for our genetic constitution is so carefully balanced that any change is likely to be for the worse. To adapt an aphorism, all is best in this best of all possible bodies. Nonetheless, the sickle cell mutation, which at first sight looks altogether harmful, turns out to be a definite advantage in a malarious environment. Similarly other mutant genes that are bad in one situation may prove beneficial in another. Variability and mutation permit the human species, like other organisms, to adapt rapidly to new situations.

Finally, the sickle cell findings offer a cheering thought on the genetic future of civilized man. Eugenists often express alarm about the fact that civilized societies, through medical protection of the ill and weak, are accumulating harmful genes: e.g., those responsible for diabetes and other hereditary diseases. The sickle cell history brings out the other side of the story: improving standards of hygiene may also *eliminate* harmful genes—not only the sickle cell but also others of which we are not yet aware.

| PHENOTYPE | GENOTYPE | ELECTROPHORETIC PATTERN | HEMOGLOBIN TYPES |
|---|---|---|---|
| NORMAL | $Hb^A$ — $Hb^A$ | | A |
| SICKLE CELL TRAIT | $Hb^S$ — $Hb^A$ | | SA |
| SICKLE CELL ANEMIA | $Hb^S$ — $Hb^S$ | | SS |
| HEMOGLOBIN C SICKLE CELL ANEMIA | $Hb^C$ — $Hb^S$ | | CS |
| HEMOGLOBIN C DISEASE | $Hb^C$ — $Hb^C$ | | CC |

HEMOGLOBIN SPECIMENS from various individuals are analyzed by electrophoresis. The phenotype is the outward expression of the genotype, which refers to the hereditary make-up of the individual. The H-shaped symbols in the genotype column are schematic representations of sections of human chromosomes, one from each parent. The horizontal line of the H represents a gene for hemoglobin type. $Hb^A$ is normal hemoglobin; $Hb^S$, sickle cell hemoglobin; $Hb^C$, hemoglobin C. This kind of electrophoretic pattern is made on a strip of wet paper between a positive and a negative electrode. The specimen of hemoglobin is placed on the line at the left side of each strip. In this experiment hemoglobin A migrates faster toward the positive electrode than sickle cell hemoglobin, which migrates faster than hemoglobin C. Thus the pattern for individuals with two types of hemoglobin is double.

# 27

# Teacher Expectations for the Disadvantaged

ROBERT ROSENTHAL and LENORE F. JACOBSON
*April 1968*

One of the central problems of American society lies in the fact that certain children suffer a handicap in their education which then persists throughout life. The "disadvantaged" child is a Negro American, a Mexican American, a Puerto Rican or any other child who lives in conditions of poverty. He is a lower-class child who performs poorly in an educational system that is staffed almost entirely by middle-class teachers.

The reason usually given for the poor performance of the disadvantaged child is simply that the child is a member of a disadvantaged group. There may well be another reason. It is that the child does poorly in school because that is what is expected of him. In other words, his shortcomings may originate not in his different ethnic, cultural and economic background but in his teachers' response to that background.

If there is any substance to this hypothesis, educators are confronted with some major questions. Have these children, who account for most of the academic failures in the U.S., shaped the expectations that their teachers have for them? Have the schools failed the children by anticipating their poor performance and thus in effect teaching them to fail? Are the massive public programs of educational assistance to such children reinforcing the assumption that they are likely to fail? Would the children do appreciably better if their teachers could be induced to expect more of them?

We have explored the effect of teacher expectations with experiments in which teachers were led to believe at the beginning of a school year that certain of their pupils could be expected to show considerable academic improvement during the year. The teachers thought the predictions were based on tests that had been administered to the student body toward the end of the preceding school year. In actuality the children designated as potential "spurters" had been chosen at random and not on the basis of testing. Nonetheless, intelligence tests given after the experiment had been in progress for several months indicated that on the whole the randomly chosen children had improved more than the rest.

The central concept behind our investigation was that of the "self-fulfilling prophecy." The essence of this concept is that one person's prediction of another person's behavior somehow comes to be realized. The prediction may, of course, be realized only in the perception of the predictor. It is also possible, however, that the predictor's expectation is communicated to the other person, perhaps in quite subtle and unintended ways, and so has an influence on his actual behavior.

An experimenter cannot be sure that he is dealing with a self-fulfilling prophecy until he has taken steps to make certain that a prediction is not based on behavior that has already been observed.

If schoolchildren who perform poorly are those expected by their teachers to perform poorly, one cannot say in the normal school situation whether the teacher's expectation was the cause of the performance or whether she simply made an accurate prognosis based on her knowledge of past performance by the particular children involved. To test for the existence of self-fulfilling prophecy the experimenter must establish conditions in which an expectation is uncontaminated by the past behavior of the subject whose performance is being predicted.

It is easy to establish such conditions in the psychological laboratory by presenting an experimenter with a group of laboratory animals and telling him what kind of behavior he can expect from them. One of us (Rosenthal) has carried out a number of experiments along this line using rats that were said to be either bright or dull. In one experiment 12 students in psychology were each given five laboratory rats of the same strain. Six of the students were told that their rats had been bred for brightness in running a maze; the other six students were told that their rats could be expected for genetic reasons to be poor at running a maze. The assignment given the students was to teach the rats to run the maze.

From the outset the rats believed to have the higher potential proved to be the better performers. The rats thought to be dull made poor progress and sometimes would not even budge from the starting position in the maze. A questionnaire given after the experiment showed that the students with the allegedly brighter rats ranked their subjects as brighter, more pleasant and more likable than did the students who had the allegedly duller rats. Asked about their methods of dealing with the rats, the students with the "bright" group turned out to have been friendlier, more enthusiastic and less talkative with the animals than the students with the "dull" group had been. The students with the "bright" rats also said they handled their animals more, as well as more gently,

than the students expecting poor performances did.

Our task was to establish similar conditions in a classroom situation. We wanted to create expectations that were based only on what teachers had been told, so that we could preclude the possibility of judgments based on previous observations of the children involved. It was with this objective that we set up our experiment in what we shall call Oak School, an elementary school in the South San Francisco Unified School District. To avoid the dangers of letting it be thought that some children could be expected to perform poorly we estab-

lished only the expectation that certain pupils might show superior performance. Our experiments had the financial support of the National Science Foundation and the cooperation of Paul Nielsen, the superintendent of the school district.

Oak School is in an established and somewhat run-down section of a middle-sized city. The school draws some students from middle-class families but more from lower-class families. Included in the latter category are children from families receiving welfare payments, from low-income families and from Mexican-American families. The

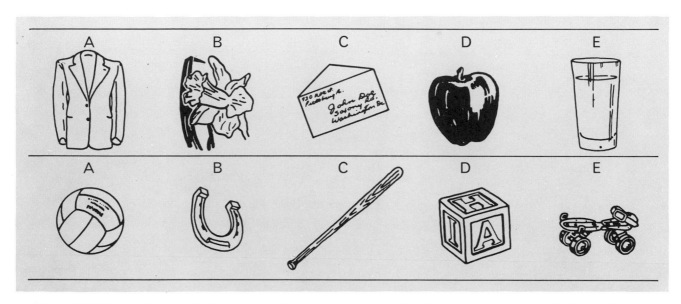

VERBAL ABILITY of children in kindergarten and first grade was tested with questions of this type in the Flanagan Tests of General Ability. In the drawings at top the children were asked to cross out the thing that can be eaten; in the bottom drawings the task was to mark "the thing that is used to hit a ball." The tests are published by Science Research Associates, Inc., of Chicago.

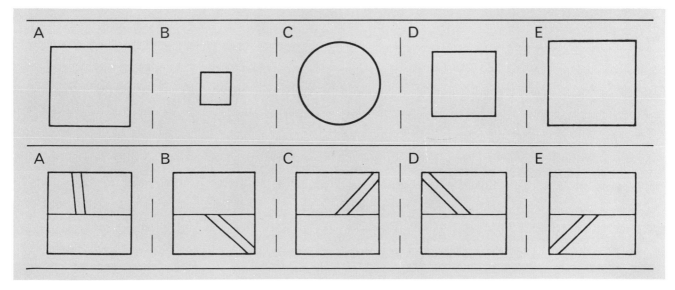

REASONING ABILITY of children in kindergarten and first grade was tested with abstract drawings. The children were told that four of the drawings in each example followed the same rule and one did not. The task was to mark the exception. In the drawings at top the exception is the circle; at bottom all the drawings except the first one have parallel lines that terminate at a corner.

school has six grades, each organized into three classes—one for children performing at above-average levels of scholastic achievement, one for average children and one for those who are below average. There is also a kindergarten.

At the beginning of the experiment in 1964 we told the teachers that further validation was needed for a new kind of test designed to predict academic blooming or intellectual gain in children. In actuality we used the Flanagan Tests of General Ability, a standard intelligence test that was fairly new and therefore unfamiliar to the teachers. It consists of two relatively independent subtests, one

focusing more on verbal ability and the other more on reasoning ability. An example of a verbal item in the version of the test designed for children in kindergarten and first grade presents drawings of an article of clothing, a flower, an envelope, an apple and a glass of water; the children are asked to mark with a crayon "the thing that you can eat." In the reasoning subtest a typical item consists of drawings of five abstractions, such as four squares and a circle; the pupils are asked to cross out the one that differs from the others.

We had special covers printed for the test; they bore the high-sounding ti-

tle "Test of Inflected Acquisition." The teachers were told that the testing was part of an undertaking being carried out by investigators from Harvard University and that the test would be given several times in the future. The tests were to be sent to Harvard for scoring and for addition to the data being compiled for validation. In May, 1964, the teachers administered the test to all the children then in kindergarten and grades one through five. The children in sixth grade were not tested because they would be in junior high school the next year.

Before Oak School opened the follow-

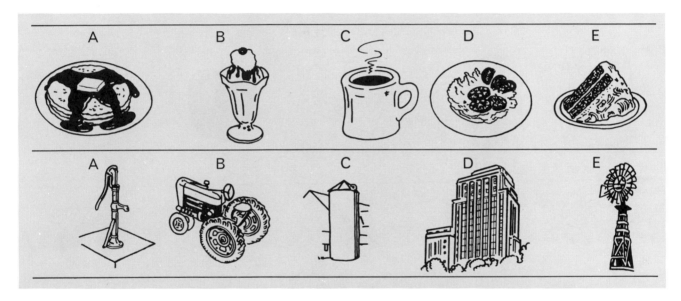

ADVANCED TESTS were given to children in second and third grades and grades four through six. Two examples from the test of verbal reasoning for grades four through six appear here. In

the example at top the children were asked to "find the beverage." In the bottom example the instruction that the pupils received from the teacher was "Find the one you are most likely to see in the city."

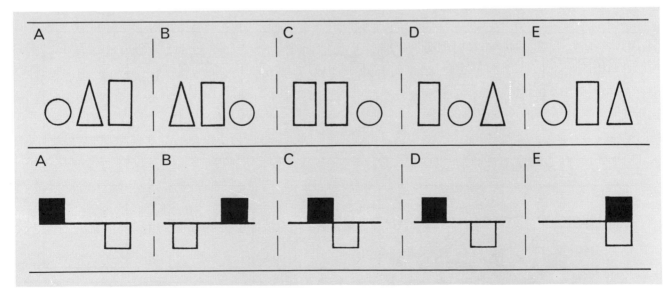

REASONING TEST for children in grades four through six was based on the same principles as the test for younger children but used more sophisticated examples. At top the exception is C, which

has no triangle. In the example at bottom the exception is E, because in all the other drawings the black and white squares are not aligned vertically. The tests were used to measure pupils' progress.

ing September about 20 percent of the children were designated as potential academic spurters. There were about five such children in each classroom. The manner of conveying their names to the teachers was deliberately made rather casual: the subject was brought up at the end of the first staff meeting with the remark, "By the way, in case you're interested in who did what in those tests we're doing for Harvard. . . ."

The names of the "spurters" had been chosen by means of a table of random numbers. The experimental treatment of the children involved nothing more than giving their names to their new teachers as children who could be expected to show unusual intellectual gains in the year ahead. The difference, then, between these children and the undesignated children who constituted a control group was entirely in the minds of the teachers.

All the children were given the same test again four months after school had started, at the end of that school year and finally in May of the following year. As the children progressed through the grades they were given tests of the appropriate level. The tests were designed

for three grade levels: kindergarten and first grade, second and third grades and fourth through sixth grades.

The results indicated strongly that children from whom teachers expected greater intellectual gains showed such gains [see illustration below]. The gains, however, were not uniform across the grades. The tests given at the end of the first year showed the largest gains among children in the first and second grades. In the second year the greatest gains were among the children who had been in the fifth grade when the "spurters" were designated and who by the time of the final test were completing sixth grade.

At the end of the academic year 1964–1965 the teachers were asked to describe the classroom behavior of their pupils. The children from whom intellectual growth was expected were described as having a better chance of being successful in later life and as being happier, more curious and more interesting than the other children. There was also a tendency for the designated children to be seen as more appealing, better adjusted and more affectionate, and as less

in need of social approval. In short, the children for whom intellectual growth was expected became more alive and autonomous intellectually, or at least were so perceived by their teachers. These findings were particularly striking among the children in the first grade.

An interesting contrast became apparent when teachers were asked to rate the undesignated children. Many of these children had also gained in I.Q. during the year. The more they gained, the less favorably they were rated.

From these results it seems evident that when children who are expected to gain intellectually do gain, they may be benefited in other ways. As "personalities" they go up in the estimation of their teachers. The opposite is true of children who gain intellectually when improvement is not expected of them. They are looked on as showing undesirable behavior. It would seem that there are hazards in unpredicted intellectual growth.

A closer examination revealed that the most unfavorable ratings were given to the children in low-ability classrooms who gained the most intellectually. When these "slow track" children were in the control group, where little intellectual gain was expected of them, they were rated more unfavorably by their teachers if they did show gains in I.Q. The more they gained, the more unfavorably they were rated. Even when the slow-track children were in the experimental group, where greater intellectual gains were expected of them, they were not rated as favorably with respect to their control-group peers as were the children of the high track and the medium track. Evidently it is likely to be difficult for a slow-track child, even if his I.Q. is rising, to be seen by his teacher as well adjusted and as a potentially successful student.

How is one to account for the fact that the children who were expected to gain did gain? The first answer that comes to mind is that the teachers must have spent more time with them than with the children of whom nothing was said. This hypothesis seems to be wrong, judging not only from some questions we asked the teachers about the time they spent with their pupils but also from the fact that in a given classroom the more the "spurters" gained in I.Q., the more the other children gained.

Another bit of evidence that the hypothesis is wrong appears in the pattern of the test results. If teachers had talked to the designated children more, which would be the most likely way of invest-

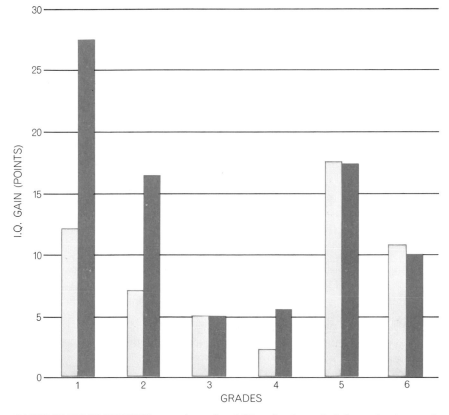

GAINS IN INTELLIGENCE were shown by children by the end of the academic year in which the experiment was conducted in an elementary school in the San Francisco area. Children in the experimental group (dark bars) are the ones the teachers had been told could be expected to show intellectual gains. In fact their names were chosen randomly. Control-group children (light bars), of whom nothing special was said, also showed gains.

ing more time in work with them, one might expect to see the largest gains in verbal intelligence. In actuality the largest gains were in reasoning intelligence.

It would seem that the explanation we are seeking lies in a subtler feature of the interaction of the teacher and her pupils. Her tone of voice, facial expression, touch and posture may be the means by which—probably quite unwittingly—she communicates her expectations to the pupils. Such communication might help the child by changing his conception of himself, his anticipation of his own behavior, his motivation or his cognitive skills. This is an area in which further research is clearly needed.

Why was the effect of teacher expectations most pronounced in the lower grades? It is difficult to be sure, but several hypotheses can be advanced. Younger children may be easier to change than older ones are. They are likely to have less well-established reputations in the school. It may be that they are more sensitive to the processes by which teachers communicate their expectations to pupils.

It is also difficult to be certain why the older children showed the best performance in the follow-up year. Perhaps the younger children, who by then had different teachers, needed continued contact with the teachers who had influenced them in order to maintain their improved performance. The older children, who were harder to influence at first, may have been better able to maintain an improved performance autonomously once they had achieved it.

In considering our results, particularly the substantial gains shown by the children in the control group, one must take into account the possibility that what is called the Hawthorne effect might have been involved. The name comes from the Western Electric Company's Hawthorne Works in Chicago. In the 1920's the plant was the scene of an intensive series of experiments designed to determine what effect various changes in working conditions would have on the performance of female workers. Some of the experiments, for example, involved changes in lighting. It soon became evident that the significant thing was not whether the worker had more or less light but merely that she was the subject of attention. Any changes that involved her, and even actions that she only thought were changes, were likely to improve her performance.

In the Oak School experiment the fact that university researchers, supported by

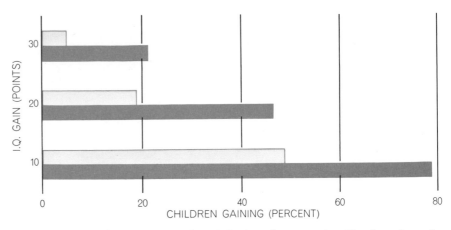

CHILDREN IN LOWER GRADES showed the most dramatic gains. The chart shows the percent of children in the first and second grades by amount of their gains in I.Q. points. Again dark bars represent experimental-group children, light bars control-group children. Two lower sets of bars include children from higher groups, so that lowest set sums results.

Federal funds, were interested in the school may have led to a general improvement of morale and effort on the part of the teachers. In any case, the possibility of a Hawthorne effect cannot be ruled out either in this experiment or in other studies of educational practices. Whenever a new educational practice is undertaken in a school, it cannot be demonstrated to have an intrinsic effect unless it shows some excess of gain over what Hawthorne effects alone would yield. In our case a Hawthorne effect might account for the gains shown by the children in the control group, but it would not account for the greater gains made by the children in the experimental group.

Our results suggest that yet another base line must be introduced when the intrinsic value of an educational innovation is being assessed. The question will be whether the venture is more effective (and cheaper) than the simple expedient of trying to change the expectations of the teacher. Most educational innovations will be found to cost more in both time and money than inducing teachers to expect more of "disadvantaged" children.

For almost three years the nation's schools have had access to substantial Federal funds under the Elementary and Secondary Education Act, which President Johnson signed in April, 1965. Title I of the act is particularly directed at disadvantaged children. Most of the programs devised for using Title I funds focus on overcoming educational handicaps by acting on the child—through remedial instruction, cultural enrichment and the like. The premise seems to be that the deficiencies are all in the child

and in the environment from which he comes.

Our experiment rested on the premise that at least some of the deficiencies—and therefore at least some of the remedies—might be in the schools, and particularly in the attitudes of teachers toward disadvantaged children. In our experiment nothing was done directly for the child. There was no crash program to improve his reading ability, no extra time for tutoring, no program of trips to museums and art galleries. The only people affected directly were the teachers; the effect on the children was indirect.

It is interesting to note that one "total push" program of the kind devised under Title I led in three years to a 10-point gain in I.Q. by 38 percent of the children and a 20-point gain by 12 percent. The gains were dramatic, but they did not even match the ones achieved by the control-group children in the first and second grades of Oak School. They were far smaller than the gains made by the children in our experimental group.

Perhaps, then, more attention in educational research should be focused on the teacher. If it could be learned how she is able to bring about dramatic improvement in the performance of her pupils without formal changes in her methods of teaching, other teachers could be taught to do the same. If further research showed that it is possible to find teachers whose untrained educational style does for their pupils what our teachers did for the special children, the prospect would arise that a combination of sophisticated selection of teachers and suitable training of teachers would give all children a boost toward getting as much as they possibly can out of their schooling.

# IV

## WHAT PRICE PROGRESS?

# IV

---

## What Price Progress?

*Introduction*

The articles in this section compel us to have second thoughts about numerous aspects of progress that have turned out to bear higher prices than were expected. H. J. Muller, in "Radiation and Human Mutation," discusses the human price of high-energy radiation, such as that from X-rays. At first, as is so often true of a new discovery, it seemed as though the X-ray was an asset with no attendant liabilities: It gave us the ability to see through the human body—an ability that has been extremely valuable in the practice of medicine. (Of course, even in the earliest days there were isolated examples of X-ray burns, but these were attributed to overdosage, and it was assumed that as long as we kept the dosage below a certain threshold nothing untoward would happen. It took almost a half of a century to relieve us of this pleasant illusion.)

The evaluation of the medical use of high-energy radiation can be carried out along utilitarian lines. X-rays have a certain probability of revealing a cancerous growth in time for it to be medically dealt with: this is a positive benefit. On the other hand, X-rays themselves have a certain probability of causing a cancer in the distant future: this is part of the cost of using them. Summarizing, we can write a simple equation:

$$\text{benefits} - \text{cost} = \text{utility (or value)},$$

or;

$$\text{benefits} - \text{disbenefits} = \text{utility},$$

or;

$$\Sigma\,(\text{benefits}) = \text{utility}.$$

That is, we can consider "benefits" as being the single variable which can take either negative or positive values, the negative values being what we ordinarily call "cost." In principle, the goal of utilitarian analysis is simple: to determine accurately *all* the benefits (both positive and negative), summarizing them into a single utility figure. If the utility is positive, we make use of the procedure; if it is negative, we avoid it.

In practice, of course, things are not quite that simple. When probability enters in—as, in principle, it always does—how do we reconcile the probability of a distant event with the probability of an event that is almost upon us? A cost that need not be paid until far in the future may never have to be paid by us, because we may die first. This is a lugubrious way to look on the bright side of things, but it should not be ignored. A fifty-five-year-old man with a cancerous growth may willingly subject himself to a large dosage of X-rays to get rid of it, even though he knows that those X-rays may cause another cancer to develop ten or fifteen years later. A twenty-year-old man suffering from cancer might evaluate the risks differently (because of his greater life expectancy) and seek some other means of getting rid of the growth.

Utilitarian analysis becomes even more difficult when we try to compare the benefits gained by the present generation with the probable costs to be paid by some future generation. To the humanitarian's demand that we take thought for posterity, a crudely practical man—according to legend, Uncle Joe Cannon, Speaker of the House of Representatives from 1903 to 1911—once retorted: "Why should I do anything for posterity? What has posterity done for me?"

To this there is no easy answer. As a society, however, we have evolved a sort of answer: we have invented institutions. "An institution," George Berg has said, "can be considered as an anticipating device designed to pay off its members now for behavior which will benefit and stabilize society later." A society functions poorly when it has no such institutions, and it is suffering from malfunction when its institutions produce disbenefits in the future. One can legitimately argue that the medical profession, as it practices at present, is being paid to produce disbenefits in the future in the form of a genetic burden that could be lifted sooner if doctors did not feel committed to saving every life, no matter how wretched.

A nagging problem in utilitarian analysis is that of comparing different qualities. Is an orange better than an apple? Better for what? And in whose mind? There is no abstract method of deciding whether oranges are better than apples. But if everyone has a limited amount of money and must make choices, in a free market the relative cost of two incommensurable goods will tend toward a stable ratio.

But this method is not always applicable: some goods are very difficult to compare. If smoking gives me pleasure, how do I compare the present pleasure of smoking with the probable shortening of my life span? Money cannot reconcile these incommensurables. E. Cuyler Hammond, in "The Effects of Smoking," gives evidence of the harmfulness of tobacco smoking that is quite convincing—to anyone who can look at it in an objective fashion. But if a person enjoys smoking—it happens that I don't, so I must use my imagination to put myself in the smoker's shoes—then his attempt to weigh the probabilities of the future is interfered with by the psychological process that Sigmund Freud called *denial.* Said Freud: "No one believes in his own death. In the unconscious everyone is convinced of his own immortality." Numerous other writers both before and after Freud have attested to the truth of this insight. To achieve objectivity a smoker must acknowledge the reality of denial and ask himself if he has made allowance for the bias it introduces into his thinking.

This is not to say that if all smokers were completely objective all would promptly quit smoking. After a completely objective analysis, a smoker might well elect to continue smoking. He might base this decision on observations of the effect of old age on others. Given a completely free choice and an indomitable "will" (whatever that word means) he might prefer the present pleasure of smoking to the dubious distant pleasure of senility. Right or wrong, such a decision is not necessarily irrational.

The article by Barron, Jarvik and Bunnell on "The Hallucinogenic Drugs" raises problems of frightening difficulty. The problems are grounded in philosophical puzzles that are unresolved, and perhaps unresolvable. Accepting the ordinary common-sense view of the reality of the external world, the connection between external events and subjective experience is this:

$$\text{external event} \longrightarrow \text{nervous phenomena} \longrightarrow \text{subjective experience.}$$

The existence of the nervous phenomena is of course something we have learned by studying other living beings, usually beings of another species. Ordinarily, we instinctively assume the reality of the external world, thus:

$$\text{external event} \longrightarrow \text{subjective experience.}$$

When we stop to think about it we may admit that all we *really* know is the subjective experience—that the external event is a "construct" that we make to account for the subjective experience. It would be unwise to get involved at this point in the many philosophical difficulties hidden in these simple statements. It is enough to say that these difficulties are connected with two problems that have names: *solipsism,* the belief that only *I* exist, and that all other people and objects are fictions that *I* choose, for some dubious and perhaps perverse reason, to invent and believe in; and the *egocentric predicament,* which each of us becomes aware of when he realizes that he never *really* knows what is going on in anybody else's head (even after rejecting the solipsistic position). Because I never really know what anybody else is thinking I have no way of knowing whether he is sincere. No matter how much I may want to cooperate with other people, and even when I do, there may always be a residual fear that others are not telling me the truth about their thoughts and beliefs.

If all I really know is what I subjectively experience, and if I want a pleasant subjective experience, why not bypass the external agent with which the experience is usually achieved, and produce it by some other means? Presumably, the most significant relationship is this:

$$\text{nervous phenomena} \longrightarrow \text{subjective experience.}$$

If I can produce the necessary nervous phenomena by the use of some external agent rather than by participating in the usual external event, why not?

In some special cases, the question "Why not?" is easily answered. If the subjective experience that I am speaking of is tasting food, and if this experience can be obtained by means other than eating energy-rich compounds, then it is easy to answer the question. Put more bluntly, we don't even have to answer the question: nature will do it for us. If some people produce the desired subjective experience by eating food, while others use a prosthetic shunt that bypasses food, the argument will quickly be settled. The bypassers will soon starve, and cease to keep the argument going.

In other cases the answer is not so easily reached. What of the euphoria that comes at the conclusion of hard work, or when a threatening challenge has been satisfactorily met? What of the mystic experiences traditionally produced by meditation coupled with long isolation from routine sensory inputs? What of the sexual (orgasmic) feelings that normally arise only from the sex act itself? And what of feelings of brotherly love that can be achieved by an earnest and empathetic inquiry into the thoughts and actions of others? Suppose that we can achieve each of these desirable feelings by some sort of a prosthetic shunt? Is the shunt as good as an authentic experience? How does one answer this question? By what criteria does one distinguish a good answer from a bad?

The problem has a weighty social component. The joint sharing of a biochemical shunt, e.g., peyote by the Indians of the Native American Church, no doubt gives a warm feeling of "togetherness," producing the virtues of sharing and love. Such is the report of those on the inside of experiences like this. But those who are outside, even if by their own choice, may have a shut-out feeling. The reactions of many nondrug takers to drug takers give clear evidence of a feeling of rejection.

On a less psychological and more objective level, the taking of mind-altering drugs raises a ubiquitous problem of side-effects—both physiological and characterological. To be at all widely used, a drug must not have drastic side-effects. This means that the awareness of any physiological dangers that are entailed will be slow in coming. It may take considerable time for damage to develop, and for statistical analysis to demonstrate that the damage occurs as a result of taking the drug.

Whether or not there are physiological dangers in the simple sense, there may be characterological dangers. If the individual, as the result of repeated use, finally comes to the conclusion that he prefers a prosthetic shunt to the genuine experience for which it is a substitute, should it be his right to exercise this preference? If he becomes incapable of supporting himself, should society support him and continue to supply him with the drug? Should society lock up such substances? And what effects must such a substance have before it becomes of concern to society? If LSD should be controlled, what about marijuana? If marijuana, what about alcohol? If alcohol, what about an ordinary tranquilizer? We dislike having to draw an arbitrary line across a continuum. But undoubtedly we must, and this question persists: "At what level?"

The public interest is entangled in the use of prosthetic shunts. Suppose the "ideal" drug were to be found, one which had no bad side-effects either at the time of taking or later, one that did not incapacitate but merely made life so pleasant for the taker that he settled for the sort of loving and competition-free existence that has been praised by many men of religion in the past, and by numerous leaders of the young today. Suppose such a paragon among drugs were found. Should society permit it to be used? And by whom?

The key word is "society." If society is indeed a singular, unitary thing (as implied in the preceding sentence), then it might be difficult to argue against a pure, side-effects-free, euphoric drug. But society is not unitary; there are many societies, the largest of which are called nations. There are many nations, of which ours is only one. What would happen if some nations permitted the use of a perfect drug while others forbade it? What would be the result of natural selection operating upon nations that varied in this regard? The answer should be obvious.

James Olds presents us with an interesting example of serendipity in research in his article "Pleasure Centers in the Brain." While investigating the sleep-control area of the brain, a group of physiologists discovered that pleasure—really only inferred pleasure in another animal, but an inference difficult to doubt—is indeed the result of activity in a certain region of the brain. No chemical prosthetic shunt was needed in Olds's experiment. Presumably direct electrical stimulation of this region brings with it no undesired side-effects: merely the full horror of illegitimate pleasure. Such at least is my reaction. The author of the experiments, however, views his results with equanimity: "enough of the brain-stimulating work has been repeated . . . to indicate that our general conclusions can very likely be generalized eventually to human beings—with modifications, of course."

LaMont C. Cole, in his review of Rachel Carson's *Silent Spring,* introduces us to a human problem of quite a different sort, the problem of readjusting our unconscious assumptions to take account of the insights provided by a sensitive worker in a comparatively new discipline—here ecology. Paul Sears has called ecology the "subversive science." Somewhat ruefully, somewhat gleefully, other ecologists have had to agree that he is right. Ecology subverts our traditional views of the simple relationship between cause and effect, and presents us with practical conclusions that threaten the established social and commercial order. An ecologist's lot is not a happy one.

*We can never do merely one thing.* This assertion has the air of a platitude about it, but it needs to be said to establish the usually undescribed framework within which the ecologist does his thinking, a framework that is notably and dangerously lacking in the minds of many men of action. The assertion points toward the idea of a "system," as understood in the fields of ecology, systems analysis, and operations analysis. Like most fundamental ideas, a "system" is hard to define in a few words. It is better defined *ostensively*, i.e., by an exhibition of instances of its operation. We can profitably examine an amusing example given in Darwin's *Origin of Species:*

> The number of humble-bees in any district depends in a great measure upon the number of field-mice which destroy their combs and nests; and Colonel Newman, who has long attended to the habits of humble-bees, believes that "more than two-thirds of them are thus destroyed all over England." Now the number of mice is largely dependent, as everyone knows, on the number of cats; and Colonel Newman says, "Near villages and small towns I have found the nests of humble-bees more numerous than elsewhere, which I attribute to the number of cats that destroy the mice." Hence it is quite credible that the presence of a feline animal in large numbers in a district might determine, through the intervention first of mice and then of bees, the frequency of certain flowers in that district!

The story was later embroidered upon by others who pointed out that (on the one hand) it is well known that old maids keep cats; and (on the other) that red clover, which requires bumble-bees as pollinators, is used to make the hay that feeds the horses of the British cavalry. From all of which "it logically follows" that the continuance of the British Empire is dependent on a bountiful supply of old maids! (Should a Ph.D. thesis be done on the relation between the marriage rate in England from 1920-1960 and the loss of India, Suez, and the African colonies? Worse topics have served!)

The $n$ elements of the biological world are united by approximately $n^2$ strands of interrelationships into what has been called the "web of life," a phrase derived from Darwin's works. The web of life is fantastically complex. Pluck where we will at this web, we produce multiple and widespread effects, most of which we have not the knowledge to foresee. *We can never do merely one thing,* and the damage caused by the unforeseen effects of our plucking at the web of life may far outweigh the gains from the foreseen and desired effects. This is the theme, as Norbert Wiener pointed out, of W. W. Jacobs' haunting story *The Monkey's Paw*. It is also the theme of *Silent Spring*.

Language often obstructs learning. An ecological system is multidimensional, but language, unless used precisely, can impose a one-dimensional train of thought upon us: *cause—effect*. Do I want to kill bacteria? Let me, then, use a bactericide. (I am surprised if a penicillin-sensitive animal that I am attempting to save from bacterial infection dies—it was outside the line of my thought.) Thalidomide is a tranquillizer, therefore it will tranquillize a nervous pregnant woman—until she finds that she has produced a child without arms. Insecticides kill insects; herbicides kill herbs: their other properties are forgotten. In coining goal-oriented words scientists are like magicians entrapped by their own magic. Rachel Carson disclosed the danger of such magic when she pointed out that the proper word for what is called an "insecticide" is "biocide."

Magicians do not give up easily, however. "Insecticides" is still the current word, to which is grudgingly appended a footnote admitting that there are "side-effects." A "side-effect" can realistically be defined as any effect we do not want, or do not wish to think about. The term denigrates reality. By coupling "side-effect" to a magic word we avoid acknowledging the omnipresence of the web of life. Like the epicycles that "saved the appearances" of Ptolemaic astronomy and so permitted an awkward theory to hang on a little longer, "side-effect" enables us to cling to the simple cause and effect paradigm when it is already clear that we should change to the paradigm of the *system*. ("Intervening variables" in psychology, and "modifier genes" in genetics serve a smiilar function.)

Seven years after the publication of *Silent Spring* a few people associated with the pesticide industry are still indignant about particular assertions made therein. They miss the point. The significance of *Silent Spring* lies not so much in the particular facts it presents as in its clarion call to return to Darwin's vision of the web of life, to the paradigm of the system. There is an important practical consequence of this intellectual shift: the burden of proving the safety of innovations has been shifted. When we adhere to the cause and effect paradigm we assume that a pesticide kills pests only, and the burden of proving that it does anything else lies on him who asserts that it has more than a single effect. But when our thoughts are governed by the paradigm of the system we recognize that we can never do merely one thing; whenever anyone offers us a new product that will kill petsts we immediately ask " . . . and what else?" The ecological assumption is that no introduction of new elements into the web of life should be tolerated until it has been proved that the harm caused is less than the benefits gained. This attempt to shift the burden of proof is always resented.

There can be no single solution to the problems posed by *Silent Spring*. The solutions will be many, and it will take much work to find them. A singularly beautiful solution is described by Carroll M. Williams in "Third-Generation Pesticides." This work had its origin in a singularly neat example of serendipity, and it is leading to developments that promise to satisfy both the biologist in his insistence on not disrupting the web of life, and the businessman in his equally understandable insistence on profits.

Many of the evils that man has introduced into his life are not the result of complete unawareness of side-effects, but are consequences of a limited awareness coupled with uncontrolled population growth. In the introduction to section III we discussed some of the problems that become more and more intolerable as the multiplication of human beings continues. Waste products produced by the imperfect combustion of the fuel used in most types of engines produce another such problem of scale. A. J. Haagen-Smit, in "The Control of Air Pollution," deals with what is proving to be a massive and persistent problem. In the United States, the automobile is the principal polluter of air, but outlawing it would disrupt our entire economic system. A national economy resembles the web of life: we dare not attempt to remove one of its major constituents for fear of seriously disrupting the whole system. Even slightly altering the performance (for example, the acceleration) of so important a constituent as the automobile would have widespread effects on the economy that are not easy to predict in detail. The slightest change would be resisted by millions of people whose short-term interests were adversely affected.

Although the automobile is a major producer of air pollution, it is tolerated because it brings so many benefits with it. That its physical disbenefits extend beyond smog production was first forcibly brought to the public's attention by Ralph Nader's *Unsafe At Any Speed*. The philosopher David Hawkins, using Nader's book as a springboard, analyzes the complex problem of automobile safety in the penultimate essay in this section. Nader's book, like Rachel Carson's, aroused a storm of protests and denial, and like *Silent Spring*, left a permanent imprint upon national affairs. The day of creative muck-raking is not over! (The term "muck-raker" was of course created by those who defend the status quo; the occupation remains a noble one in spite of the magicians who tamper with our language.)

To those who optimistically insist on assuming that corrective engineering will take care of any pollution introduced into the environment, John R. Clark's article on "Thermal Pollution and Aquatic Life" introduces a somber new note of warning. The usual response to the discovery of a new pollutant is to say, "Let's convert it to something harmless or useful." Noxious sulfur dioxide in flue-gas can be converted to sulfuric acid; smelly distillates from coke-making can be converted into a myriad of useful chemicals. But what if the pollutant is heat? What do you convert heat to? The answer is, *nothing*. The only thing to do with unwanted heat is to get rid of it—that is, move it somewhere else, perhaps foisting it off on someone else who is not well enough informed to know what you are doing to his environment.

This strategy may work for awhile. But as population increases, and as more of the world becomes industrialized, the burden imposed by heat becomes greater. Ultimately, we are unable to get rid of unwanted heat. Ultimately, unless population growth is controlled by other means, we shall all perish as did the people in the Black Hole of Calcutta. That is a long way off—though not so far in the future as the birth of Christ is in the past, given the present rate of growth.

# 28

# Radiation and Human Mutation

H. J. MULLER
*November 1955*

The revolutionary impact on men's minds brought about by the development of ways of manipulating nuclear energy, both for destructive and for constructive purposes, is causing a public awakening in many directions: physical, biological and social. Among the biological subjects attracting wide interest is the effect of radiation upon the hereditary constitution of mankind. This article will consider the part which may be played by radiation in altering man's biological nature, and also the no less important effects that may be produced on our descendants by certain other pertinent influences under modern civilization.

At the cost of being too elementary for readers who are already well informed on biological matters it must first be explained that each cell of the body contains a great collection—10,000 or more—of diverse hereditary units, called genes, which are strung together in a single-file arrangement to form the tiny threads, visible under the microscope, called chromosomes. It is by the interactions of the chemical products of these genes that the composition and structure of every living thing is determined. Before any cell divides, each of its genes reproduces itself exactly or, as we say, duplicates itself. Thus each chromosome thread becomes two, both structurally identical. Then when the cell divides, each of the two resulting cells has chromosomes exactly alike. In this way the descendant cells formed by successive divisions and, finally, the individuals of subsequent generations derived from such cells, tend to inherit genes like those originally present.

However, the genes are subject to rare chemical accidents, called gene mutations. Mutation usually strikes but one gene at a time. A gene changed by mutation thereafter produces daughter genes having the mutant composition. Thus descendants arise that have some abnormal characteristic. Since each gene is capable of mutating in numerous more or less different ways, the mutant characteristics are of many thousands of diverse kinds, chemically at least.

Very rarely a mutant gene happens to have an advantageous effect. This allows the descendants who inherit it to multiply more than other individuals in the population, until finally individuals with that mutant gene become so numerous as to establish the new type as the normal type, replacing the old. This process, continued step after step, constitutes evolution.

But in more than 99 per cent of cases the mutation of a gene produces some kind of harmful effect, some disturbance of function. This disturbance is sometimes enough to kill with certainty any individual who has inherited a mutant gene of the same kind from both his parents. Such a mutant gene is called a lethal. More often the effect is not fully lethal but only somewhat detrimental, giving rise to some risk of premature death or failure to reproduce.

Now in the great majority of cases an individual who receives a mutant gene from one of his parents receives from the other parent a corresponding gene that is "normal." He is said to be heterozygous, in contrast to the homozygous individual who receives like genes from both parents. In a heterozygous individual the normal gene is usually dominant, the mutant gene recessive. That is, the normal gene usually has much more influence than the mutant gene in determining the characteristics of the individual. However, exact studies show that the mutant gene is seldom completely recessive. It does usually have some slight detrimental effect on the heterozygous individual, subjecting him to some risk of premature death or failure to reproduce or, as we may term it, a risk of genetic extinction. This risk is commonly of the order of a few per cent, down to a fraction of 1 per cent.

If a mutant gene causes an average risk of extinction of, for instance, 5 per cent, that means there is one chance in 20 that an individual possessing it will die without passing on the same gene to offspring. Thus such a mutant gene will, on the average, pass down through about 20 generations before the line of descent containing it is extinguished. It is therefore said that the "persistence" of that particular gene is 20 generations. There is some reason to estimate that the average persistence of mutant genes in general may be something like 40 generations, although there are vast differences between genes in this respect.

## The Human Store of Mutations

Observations on the frequency of certain mutant characteristics in man, supported by recent more exact observations on mice by W. L. Russell at Oak Ridge, indicate that, on the average, the chance of any given human gene or chromosome region undergoing a mutation of a given type is one in 50,000 to

100,000 per generation. Moreover, studies on the fruit fly *Drosophila* show that for every mutation of a given type there are at least 10,000 times as many other mutations occurring. Now since it is very likely that man is at least as complicated genetically as Drosophila, we must multiply our figure of 1/100,000, representing our more conservative estimate of the frequency of a given type of mutation, by at least 10,000 to obtain a minimum estimate of the total number of mutations arising in each generation among human germ cells. Thus we find that at least every tenth egg or sperm has a newly arisen mutant gene. Taking the less conservative estimate of 1/50,000 for the frequency of a given type of mutation, our figure would become two in 10.

Every person, however, arises from both an egg and a sperm and therefore contains twice as many newly arisen mutant genes as the mature germ cells do, so the figure becomes two to four in 10. When we say that the per capita frequency of newly arisen mutations is .2 to .4, we mean that there are, among every 10 of us, some two to four mutant genes which arose among the germ cells

of our parents. This is the frequency of so-called spontaneous mutation, which occurs even without exposure to radiation or other special treatment.

Far more frequent than the mutant genes that have newly arisen are those that have been handed down from earlier generations and have not yet been eliminated from the population by causing death or failure to reproduce. The average per capita frequency of all the mutant genes present, new and old, is calculated by multiplying the frequency of newly arisen mutations by the persistence figure.

The greatly simplified diagram on page 8, in which we suppose the frequency of new mutations in each generation to be .2 and the persistence to be only four generations, shows why this relation holds. We start with 10 individuals. Let us suppose that in this first generation eight persons contain no mutant genes while each of the other two has one newly arisen mutant gene. In the second generation these two mutant genes are passed along and two new ones are added to the group, making the total frequency 4/10. By the fourth gen-

eration the frequency is 8/10. After that the frequency remains constant because each mutant gene lasts only four generations and is assumed to be replaced by a normal gene.

Of course in any actual case neither the multiplication nor the distribution of mutant genes among individuals is as regular as in this simplified illustration, but the general principle holds. However, as previously mentioned, the persistence of mutant genes is of the order of 40 generations, instead of only four. Thus the equilibrium frequency becomes not 8/10 but 8. In other words, each person would on the average contain, by this reckoning, an accumulation of about eight detrimental mutant genes.

It happens that this very rough, "conservative" estimate, made by the present writer six years ago, agrees well with the estimate arrived at a few months ago by Herman Slatis, in a study carried out in Montreal by a more direct method. His method was based on the frequency with which homozygous abnormalities appeared among the children of marriages between cousins.

The eight mutations estimated above,

HUMAN CHROMOSOMES, which are much more difficult to photograph than those of fruit flies, are clearly revealed as dark bodies in this photomicrograph by T. C. Hsu of the M. D. Anderson Hospital and Tumor Institute in Houston, Tex. The chromosomes, which are enlarged approximately 3,000 diameters, are in a human spleen cell. The cell was grown in a laboratory culture after spleen tissue had been removed from a four-month-old fetus. Human body cells normally contain 48 chromosomes; human germ cells, 24.

it should be understood, do not include most of the multitude of more or less superficial differences, sometimes conspicuous but very minor in the conduct of life, whereby, in the main, we recognize one another. The latter mutations probably arise relatively seldom yet become inordinately numerous because of their very high persistence. Thus the value that we arrive at for the frequency of mutant genes depends very much upon just where the line is drawn in excluding this mutational "froth." As yet little attention has been given to this point. The number eight, at any rate, includes only mutant genes which when homozygous give rather definite abnormalities. In the great majority of cases these genes are only heterozygous and usually are but slightly expressed. Yet they do become enough expressed to cause, in each individual, his distinctive pattern of functional weaknesses, depending upon which of these mutant genes he contains and what his environment has been. The influence of environment on gene expression is often important.

Even the genes that give only a trace of detrimental effect, or are detrimental only when homozygous, play an important role, because of their high persistence and consequent high frequency. When conditions change, certain combinations of these genes may occasionally happen to be more advantageous than the type previously prevailing, and so tend to become established.

### The Effects of Mutant Genes

In general each detrimental mutant gene gives rise to a succession of more or less slight impairments in the generations that carry it. Even if only slightly detrimental, it must finally result in extinction. Moreover, even though an individual suffers less from a slightly detrimental gene than from a markedly detrimental or lethal one, nevertheless the slightly detrimental gene, being passed down to a number of individuals which is inversely proportional to the amount of harm done to each individual, occasions a total amount of damage comparable to that produced by the very detrimental gene. Although each of us may be handicapped very little by any one of our detrimental genes, the sum of all of them causes a noticeable amount of disability, which is usually felt more as we grow older.

The frequency of mutant genes levels off at an equilibrium only when conditions for both mutation frequency and gene elimination have remained stable (or have at any rate fluctuated about a given average) for many generations. During such a period about as many mutations must be eliminated as are arising per generation. If, however, the mutation rate or the average persistence or both changed significantly because of increased radiation or a change in environmental conditions which made mutant genes more or less harmful than previously, then the frequency would move toward a new level. But it would be a long time before the new equilibrium was reached. If the average persistence of mutant genes was 40 generations, the new equilibrium would still be very incompletely attained after 1,000 years.

### The Effects of Radiation

We may next consider how a given dose of ionizing radiation would affect the population. Such radiation, when absorbed by the germ cells of animals, usually induces mutations which are similar to the spontaneous ones. They are in-

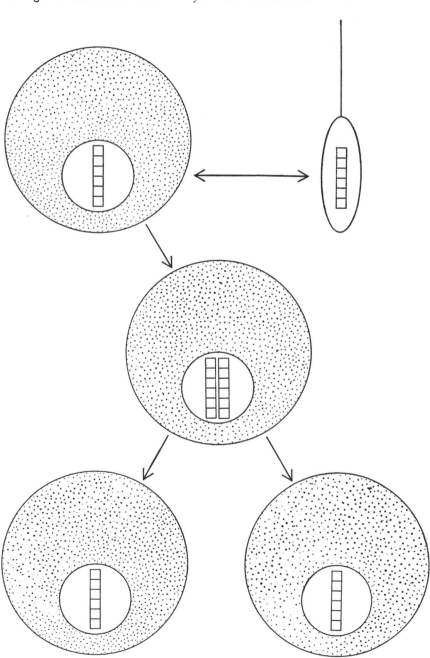

FUNDAMENTAL PROCESS of heredity is depicted in highly schematic form. At the upper left is an egg cell; at the upper right, a sperm cell. Each contains a single chromosome bearing only six genes (square segments of chromosome). The chromosomes are paired in the fertilized egg (center), resulting in an organism with a complete set of genes from each parent. When the organism produces its own germ cells (bottom), the chromosomes are separated, leaving one set of genes to combine with those of a mate in the new generation.

duced at a frequency which is proportional to the total amount of the radiation received, regardless of the duration or time-distribution of the exposure. Russell's data on mice—the nearest experimental object to man that has been used in such studies—show that it would take about 40 roentgen units of radiation to produce mutations in them at a frequency equal to their spontaneous frequency. If the frequency of spontaneous mutations is two new mutations per generation among each 10 individuals, a dose of 40 roentgens, by adding two induced mutations to the two spontaneous ones, would result in a total mutation frequency of four new mutations per generation among 10 individuals. Now assuming the total mutant gene content is eight per individual to begin with, the radiation dose would raise this figure from 8 to 8.2, an increase of only 2.5 per cent. This effect on the population would ordinarily be too small to produce noticeable changes in important characteristics. One must also bear in mind that an actual mean change in a population may be masked by the great genetic differences among individuals and by differences in environment between two groups that are to be compared. These considerations explain why even Hiroshima survivors who had been relatively close to the blast, and who may have absorbed several hundred roentgens, showed no statistically significant increase in genetic defects among their children. However, offspring of U. S. radiologists who (judging by the incidence of leukemia) probably were exposed during their work over a long period to about as much radiation as these Hiroshima survivors, do show a statistically significant increase in congenital abnormalities, as compared with the offspring of other medical specialists. This was recently established in a study by Stanley Macht and Philip Lawrence.

The toll taken by mutant genes upon the descendants of exposed individuals is spread out over more than a thousand years—40 generations. It is too small to be demonstrable in any one generation of descendants. In the first generation of offspring of a population exposed to 40 roentgens, where the induced mutation rate is .2 per individual and the average risk of extinction for any given mutant gene is 1/40, the frequency of extinctions occasioned by these mutant genes would be .2/40 or 1/200. This would mean, for example, that in a total population of 100 million some 500,000 persons would die prematurely or fail to reproduce as a result of having mutant genes that had been induced in their parents by the ex-

posure. Moreover, a much larger number would be damaged to a lesser extent. The total of induced extinctions in all generations subsequent to the exposure would be .2 times 100 million, or 20 million, and the disabilities short of extinction would be numbered in the hundreds of millions. And yet the amount of genetic deterioration in the population due to the exposure would be small in a relative sense, for the induced mutations would have added only 2.5 per cent to the load of mutant genes already accumulated by spontaneous mutation.

The situation would be very different if the doubling of the mutation frequency by irradiation in each generation were continued for many generations, say for 1,500 years. For after 1,500 years the mutant gene content would have been raised from eight to nearly 16 per individual. Along with this doubled frequency of detrimental genes there would of course be a corresponding increase in the amount of disability and in the frequency of genetically occasioned extinction of individuals.

It is possible that all this would be ruinous to a modern human population, even though in most kinds of animals it could probably be tolerated. For, in the first place, human beings multiply at a low rate which does not allow nearly as rapid replacement of mutant genes by normal ones as can occur in the great majority of species. Secondly, under modern conditions the rate of human multiplication is reduced much below its potential. Thirdly, the pressure of natural selection toward eliminating detrimental genes is greatly diminished, under present conditions at least, through the artificial saving of lives. Under these circumstances a long-continued doubling of the mutation frequency might eventually mean, if the situation persisted, total extinction of the population. However, we do not now have nearly enough knowledge of the strength of the various factors here involved to pass a quantitative judgment as to how high the critical mutation frequency would have to be, and how low the levels of multiplication and selection, to bring about this denouement. We can only see that danger lies in this direction, and call for further study of the whole matter.

## Bomb Effects

In the light of the facts reviewed, we are prepared to come to some conclusions concerning the problem of the genetic effects of nuclear explosions. Let us start with the test explosions. J. Rotblat of London has estimated that the

tests of the past year approximately doubled the background radiation for the year, in regions of the earth remote from the explosions. In the U. S. they raised the background radiation from about .1 to about .2 of a roentgen for the year. The natural background radiation of about .1 roentgen per year causes, we estimate, about 5 per cent of the spontaneous mutations in man. Hence a doubling of it would cause a rise of the same amount in the occurrence of new mutant genes. Although this influence, if continued over a generation, would induce an enormous number of mutations—of the order of 20 million in the world population of some two billion—nevertheless the effect, in relation to the already accumulated store of detrimental mutations, would be comparatively small. It would raise the per capita content of mutant genes at most by only a few tenths of 1 per cent.

Much more serious genetic consequences would follow from atomic warfare itself, in the regions subject to the fall-outs of the first few days. As for regions remote from the explosions (say the Southern Hemisphere), Rotblat and Ralph Lapp have reckoned that a hydrogen-uranium bomb like those tested in the Pacific would deliver an effective dose of about .04 roentgen throughout the whole period of radioactive disintegration. Thus 1,500 such bombs would deliver about 60 roentgens—an amount which might somewhat more than double the mutation frequency for one generation. Since there would be relatively little residual radioactivity in these remote regions after the passage of a generation, and since it is scarcely conceivable that such bombing would be repeated in many successive generations, it seems probable that most of the world's inhabitants below about the Tropic of Cancer would escape serious genetic damage. However, they would be likely in the course of centuries to become contaminated by extensive interbreeding with the survivors of the heavy irradiations in the North. For although an attempt might be made to establish a genetic quarantine, this would, for psychological reasons, be unlikely to be maintained with sufficient strictness for the hundreds of years required for the success of such a program.

In the regions subject to the more immediate fall-outs, pattern bombing could have resulted in practically all populous areas receiving several thousand roentgens of gamma radiation. Even persons well protected in shelters during the first week might subsequently be subjected to a protracted exposure

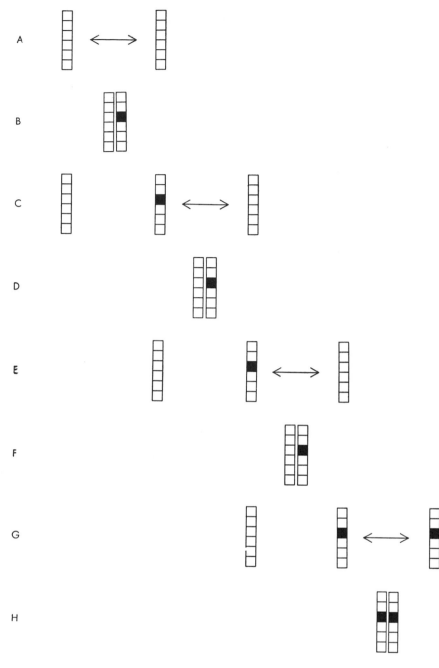

cause a 12-fold to 40-fold rise in the mutation frequency of that generation.

Such an increase, assuming that the population was already loaded with an accumulation of mutant genes amounting to 40 times the annual spontaneous mutation rate, would at one step cause a 30 to 100 per cent increase in the mutant gene content. In fact, the detrimental effect would be considerably greater than that indicated by these figures, because the newly added mutant genes, unlike those being "stored" at an equilibrium level, would not yet have been subjected to any selective elimination in favor of the less detrimental ones. It can be estimated that this circumstance might cause the total detrimental influence to be twice as strong for each new mutant gene, on the average, as for each old one. Therefore the increase in detrimental effect would be between 60 and 200 per cent.

Owing to these circumstances, an effect would be produced similar to that of a doubled accumulation of genes, such as we saw would ensue from a doubled mutation frequency after about a thousand years of repetition. Thus offspring of the fall-out survivors might have genetic ills twice or even three times as onerous as ours.

The worst of the matter is that the effects of this enormous sudden increase in the genetic load would by no means be confined to just one or two generations. Here is where the inertia of mutant-gene content, which in the case of a moderately increased mutation frequency works to spread out and thus to soften its impact, now shows the reverse side of its nature: its extreme prolongation of the effect. That is, the gene content is difficult to raise, but once raised, it is equally resistant to being reduced.

Supposing the average content of markedly detrimental genes per person to be only doubled, from 8 to 16, more than 50 per cent of the population would come to contain a number of these mutant genes (16 or more) that was as great or greater than that now present in the most afflicted 1 per cent, if the distribution followed the Poisson principle. When we consider the extent to which we are already troubled with ills of partly or wholly genetic origin, especially as we grow older, the prospect of so great an increase in them in the future is far from reassuring.

It is fortunate, in the long run, that sterility and death ensue when the accumulated dose has risen beyond about 1,000 to 3,000 roentgens. For the frequency of mutations received by the

**HARMFUL RECESSIVE MUTATION** may persist for generations before it is fully expressed. This diagram is based on the schematic chromosomes depicted on page 4. In row A the chromosomes of two parents are paired (*arrow*). In row B a gene of their offspring mutates (*black*). In row C the mutant gene has been transmitted to the next generation. If the mutant gene is recessive (*i.e.*, if the corresponding gene of the paired chromosome has a dominant effect), it is masked. Here a new set of genes is introduced (*second arrow*) from another line of descent. In rows D, E and F the mutant gene is passed along. In row G a mutant gene of the same character is introduced from still another line of descent. In offspring of this union (*row H*), the harmful effect of the paired recessive genes is expressed.

adding up to some 2,500 roentgens. Moreover, this estimate fails to take into account the soft radiation from inhaled and ingested materials which under some circumstances, as yet insufficiently dealt with in open publications, may become concentrated in the air, water or food and find fairly permanent lodgment in the body. Now although some 400

roentgens is the semilethal dose (that killing half its recipients) if received within a short time, a considerably higher dose can be tolerated if spread out over a long period. Thus it is quite possible that a large proportion of those who survive and reproduce will have received a dose of some 1,000 to 1,500 roentgens or even more. This would

descendants of an exposed population is thereby prevented from rising much beyond the amount which we have here considered. This being the case, it is probable that the offspring of the survivors, even though considerably weakened genetically, would nevertheless—some of them—be able to struggle through and reestablish a population which could continue to survive.

Yet, supposing the population were able to re-establish its stability of numbers within, say, a couple of centuries, what would be the toll among the later generations in terms of premature death and failure to reproduce? If 40 roentgens produce .2 new mutant genes per person, then 1,000 roentgens must on the average add five mutant genes to each person's composition. All of these five genes must ultimately lead to genetic extinction. But if, to be conservative, we suppose that two to three genes, on the average, work together in causing extinction, we reach the conclusion that, in a population whose numbers remain stable after the first generation following the exposure, there will ultimately be about two cases of premature death or failure to reproduce for each first-generation offspring of an exposed individual.

If, however, the descendants multiply and re-establish the original population size in a century or two, then the number of extinctions will be multiplied also. Over the long run the number of "genetic deaths" will be approximately twice as large, altogether, as the population total in any one generation. The future extinctions would in this situation be several times as numerous as the deaths that had occurred in the directly exposed generation.

Even though it is probable that mankind would revive ultimately after exposure to radiation, large or small, let us not make the all-too-common mistake of gauging whether or not such an exposure is genetically "permissible" merely by the criterion of whether or not humanity would be completely ruined by it. The instigation of nuclear war, or indeed of any other form of war, can hardly find a valid defense in the proposition, even though true, that it will probably not wipe out the whole of mankind.

### Radiation from Other Sources

It is by the standard of whether individuals are harmed, rather than whether the human race will be wiped out, that we should judge the propriety of everyday practices that may affect the human genetic constitution. We have to consider, for one thing, the amount of radiation which the population should be allowed to receive as a result of the peacetime uses of atomic energy.

How much effort, inconvenience and money are we willing to expend in the avoidance of one genetic extinction, one frustrated life and other partially frustrated lives, not to be beheld by us? Shall we accept the present official view that the "permissible" dose for industrially exposed personnel may be as high as .3 roentgen per week, that is, 300 roentgens in 20 years—a dose which would cause such a worker to transmit somewhere between .5 and 1.5 mutations per offspring conceived after that time?

Exactly the same questions apply in medical practice. A U. S. Public Health survey conducted three years ago showed that at that time Americans were receiving a skin dose of radiation averaging about two roentgens per year per person from diagnostic examinations alone. Of course only a small part of this could have reached the germ cells, but if the relative frequencies of the different types and amounts of exposure were similar to those enumerated in studies recently carried out in British hospitals, we may calculate that the total germ-cell dose was about a thirtieth of the total skin dose, namely, about .06 roentgen per person per year. This is about 12 times as much as the dose that had previously been estimated to reach the reproductive organs of the general population (not the hospital population) in England. However, the U. S. is notoriously riding "the wave of the future" in regard to the employment of X-rays; it is still expanding their use rapidly, while other countries are following as fast as they can.

Now this dose of .06 roentgen per year, the only estimate for the U. S. that we have, is of the same order of magnitude (perhaps twice as large) as the annual dose received in the U. S. over the past four years from all the nuclear test explosions. It seems rather disproportionate that so much furor should be raised about the genetic effects of the latter and so little about the former.

The writer's personal conviction is that, at the present stage of international relations or at least at the stage of the past several years, the tests have been fully justified as warnings and defensive measures against totalitarianism, despite the future sacrifices that they inexorably bring in their train, although it is to be hoped that this stage is now about to become obsolete. On the other hand, it seems impossible to find justification for the large doses to which the germ cells of patients are exposed in medical prac-

tice. It would involve comparatively little care or expense to shield the gonads or take other precautions to reduce the dose being received by the reproductive organs and other parts not being examined. And the deliberate irradiation of the ovaries to induce ovulation, and of the testes to provide an admittedly temporary means of avoiding pregnancies, should be regarded as malpractice.

We must remember that nuclear weapons tests and possibly nuclear warfare may be dangers of our own turbulent times only, whereas physicians will always be with us. It is easier and better to establish salutary policies with regard to any given practice early than late in its development. If we continue neglectful of the genetic damage from medical irradiations, the dose received by the germ cells will tend to creep higher and higher. It will also be joined by a rising dose from industrial uses of radioactivity. For the industrial and administrative powers-that-be will tend to take their cue in such matters from the physicians, not from the biologists, even as they do today. It should be our generation's concern to take note of this situation and to make further efforts to start off the expected age of radiation, if there is to be one, in a rational way as regards protection from this insidious agent, so

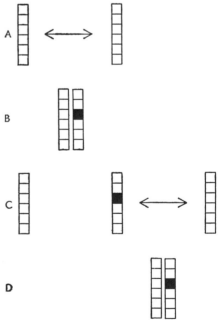

**HARMFUL DOMINANT MUTATION,** as opposed to a recessive one, is quickly eliminated. Here the mutation (*black*) occurs at the same stage as in the diagram on page 6. It occurs in a germ cell, so its effect is not expressed in that generation. When the chromosome bearing the mutant gene is paired with another, however, the mutation is expressed in the offspring of the union.

INDIVIDUALS

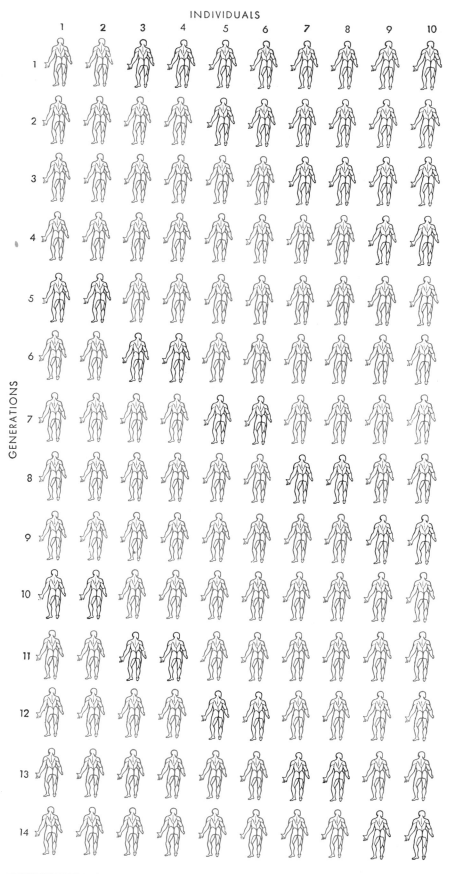

EQUILIBRIUM IS ATTAINED by recessive mutant genes under the conditions assumed here. The first assumption is that in each generation (*horizontal rows*) two new mutant genes (*colored figures*) arise among 10 individuals. The second assumption is that each line of descent bearing the mutant gene dies out four generations after the mutation has occurred. In the diagram this extinct line is then replaced by a new one. Thus after three generations the number of individuals bearing a mutant gene is stabilized at eight out of 10.

## Chemical Agents

as to avoid that permanent, significant raising of the mutation frequency which in the course of ages could do even more genetic damage than a nuclear war.

Radiation is by no means the only agent that is capable of drastically increasing the frequency of mutation. Diverse organic substances, such as the mustard gas group, some peroxides, epoxides, triazene, carbamates, ethyl sulfate, formaldehyde and so forth, can raise the mutation frequency as much as radiation.

The important practical question is: to what extent may man be unwittingly raising his mutation frequency by the ingestion or inhalation of such substances, or of substances which, after entering the body, may induce or result in the formation of mutagens that penetrate to the genes of the germ cells? As yet far too little is known of the extent to which our genes, under modern conditions of exposure to unusual chemicals, are being subjected to such mutagenic influences.

A surprising recent finding by Aaron Novick and Leo Szilard at the University of Chicago is that in coli bacteria the feeding of ordinary purines normal to the organism more than doubled the spontaneous mutation frequency, while methylated purines, and more especially caffeine (as had been found by other workers), had a much stronger mutagenic effect. Thus far, however, caffeine has not proved mutagenic in fruit flies, although it is possible that it is destroyed in their gut. In Novick and Szilard's work compounds of purines with ribose (*e.g.* adenosine) counteracted the mutagenic effect of the purines. Furthermore, adenosine and guanosine even acted as "antimutagens" when there had been no addition of purines to the nutrient medium, as though a considerable part, about a third, of the spontaneous mutations were being caused by the purines naturally present in the cells. This work, then, indicates both the imminence of the mutagenic risks to which we may be subject and also the fact that means of controlling these risks and, to some extent, even of controlling the processes of spontaneous mutation themselves, are already coming into view.

Other large differences in the frequency of so-called spontaneous mutations were found in my studies in 1946 on the mutation frequencies characterizing different stages in the germ-cell cycle of the fruit fly. Moreover, J. B. S. Haldane, dealing with data of others, adduced some evidence that the germ cells of older men have a much higher frequency of newly arisen mutant genes than those of young men. If this result for man, so different from what we have just noted for fruit flies, should be confirmed, it might prove to be more damaging, genetically, for a human population to have the habit of reproduction at a relatively advanced age than for its members to be exposed regularly to some 50 roentgens of ionizing radiation in each generation.

It is evident from these varied examples that the problem of maintaining the integrity of the genetic constitution is a much wider one than that of avoiding the irradiation of the germ cells, inasmuch as diverse other influences may play a mutagenic role equal to or greater in importance than that of radiation.

The view has been expressed that, since some chemical mutagenesis and even radiation mutagenesis occur naturally, the effects of such normal processes should cause us no great concern. Aside from the fact that not everything that is natural is desirable, we must always be conscious of the hazards added by civilization. Certain civilized practices, such as the use of X-rays and radioactivity (and possibly reproduction at an advanced age or the drinking of coffee and tea), are causing genetic damage to be done at a significantly more rapid rate than in olden times.

## Relaxed Selection

It is evident that the rate of elimination of mutant genes is just as important as the mutation frequency in the determination of the human genetic constitution. What we really mean here, of course, is "selective" elimination. The importance of this distinction is seen in the fact that in the ancestors of both men and mice much the same mutations must have occurred, but that the different conditions of their existence—the ever more mousy living of the mouse progenitors and the manlier living of the pre-men—caused a different group of genes to become selected from out of their common store.

A very distinctive feature of our modern industrial civilization is the tremendous saving of human lives which would have been sacrificed under primitive conditions. This is accomplished in part by medicine and sanitation but also by the abundant and diverse artificial aids to living supplied by industry and widely disseminated through the operation of modern social practices. The proportion of those who die prematurely is now so

small that it must be considerably below the proportion who would have to be eliminated in order to extinguish mutant genes as fast as new ones arise. In other words, many of the saved lives must represent persons who under more primitive conditions would have died as a result of genetic disabilities. Moreover, the genetically less capable survivors apparently do not have a much lower rate of multiplication than the more capable; in fact, there are certain oppositely working tendencies.

It is probably a considerable underestimate to say that half of the detrimental genes which under primitive conditions would have met genetic extinction, today survive and are passed on. On the basis of this conservative estimate we can calculate that in some 10 generations, or 250 to 300 years, the accumulated genetic effect would be much like that from exposure of a population to a sudden heavy dose of 200 to 400 roentgens, such as was received by the most heavily exposed survivors of Hiroshima. If the techniques of saving life in our civilization continue to advance, the accumulation of mutant genes will rise to ever higher levels. After 1,000 years the population in all likelihood would be as heavily loaded with mutant genes as though it were descended from the survivors of hydrogen-uranium bomb fallouts, and the passage of 2,000 years would continue the story until the system fell of its own weight or changed.

The process just depicted is a slow, invisible, secular one, like the damage resulting from many generations of exposure to overdoses of diagnostic X-rays. Therefore it is much less likely to gain credence or even attention than the sensational process of being overdosed by fall-outs from bombs. This situation, then, even more than the danger of fallouts, calls for basic education of the public and publicists, if they are to reshape their deep-rooted attitudes and practices as required.

It is necessary for mankind to realize that a species rises no higher, genetically, and stays no higher, than the pressure of selection forces it to, and that it responds to any relaxation of that pressure by sinking correspondingly. It will in fact take as much rope in sinking as we pay out to it. The policy of saving all possible genetic defectives for reproduction must, if continued, defeat its own purposes. The reason for this is evident as soon as we consider that when, by artificial devices, a moderately detrimental gene is made less detrimental, its frequency will gradually creep upward toward a new equilibrium level, at which

it is finally being eliminated anyway at the same rate as that at which it had been eliminated originally, namely, at the rate at which it arises by mutation. This rate of elimination, being once more just as high as before medicine began, will at the same time reflect the fact that as much suffering and frustration (except insofar as we may deaden them with opiates) will then be existing, in consequence of that detrimental gene, as existed under primitive conditions. Thus, with all our medicine and other techniques, we will be as badly off as when we started out.

Not all genetic disabilities, however, would simply be made less detrimental. Some of them would be rendered not detrimental at all under the circumstances of a highly artificial civilization, in the sense that they were enabled to persist indefinitely and thus to become established as the new norm of our descendants. The number of these disabilities would increase up to such a level that no more of them could be supported and compensated for by the technical means available and by the resources of the social system. The burden of the individual cases, up to that level, would have become largely shifted from the given individuals themselves to the whole community, through its social services (a form of insurance), yet the total cost would be divided among all individuals and that cost would keep on rising as far as it was allowed to rise.

Ultimately, in that Utopia of Inferiority in the direction of which we are at the moment headed, people would be spending all their leisure time in having their ailments nursed, and as much of their working time as possible in providing the means whereby the ailments of people in general were cared for. Thus we should have reached the acme of the benefits of modern medicine, modern industrialization and modern socialization. But, because of the secular time scale of evolutionary change and the inertia which retards changes in gene frequency, this condition would come upon the world with such insensible slowness that, except for a few long-haired cranks who took genetics seriously, and perhaps some archaeologists, no one would be conscious of the transformation. If it were called to their attention, they would be likely to rationalize it off as progress. It is hard to think of such a system not at length collapsing, as people lost the capabilities and the incentives needed to keep it going. Such a collapse could not be into barbarism, however, since the population would have become unable to survive primitive conditions; thus a collapse at this point would mean annihilation.

## Countermeasures

There is an alternative policy, and I am hopeful it will be adopted. The alternative does not by any means abandon modern social techniques or call for a return to the fabulous golden age of noble savages or even of rugged individualism. It makes use of all the science, skills and genuine arts we have, to ameliorate, improve and ennoble human life, and, so far as is consistent with its quality and well-being, to extend its quantity and range. Medicine, especially that of a far-seeing and a promoting kind, seeking actively to foster health, vigor and ability, becomes, on this policy, more developed than ever. Persons who nevertheless had defects would certainly have them treated and compensated for, so as to help them to lead useful, satisfying lives. But—and here is the crux of the matter—those who were relatively heavily loaded with genetic defects would consider it their obligation, even if these defects had been largely counteracted, to refrain from transmitting their genes, except when they also possessed genes of such unusual value that the gain for the descendants was likely to outweigh the loss. Only by the adoption of such an attitude towards genetics and reproduction, an attitude seldom encountered as yet, will it be possible for posterity indefinitely to sustain and extend the benefits of medicine, of technology, of science and of civilization in general.

With advance in realistic education should come a better realization of man's place in the great sweep of evolution, and of the risks and the opportunities,

genetic as well as nongenetic, which are increasingly opening up for him.

It is evident from these considerations that the same change in viewpoint that leads to the policy of voluntary elimination of detrimental genes would carry with it the recognition that there is no reason to stop short at the arrested norm of today. For all goods, genetic or otherwise, are relative, and, so far as the genetic side of things is concerned, our own highest fulfillment is attained by enabling the next generation to receive the best possible genetic equipment. What the implementation of this viewpoint involves, by way of techniques on the one hand, and of wisdom in regard to values on the other hand, is too large a matter for treatment here. Nevertheless, certain points regarding the genetic objectives to be more immediately sought do deserve our present notice.

For one thing, the trite assertion that one cannot recognize anything better than oneself, or in imagination rise above oneself, is merely a foolish vanity on the part of the self-complacent. Among the important objectives to be sought for mankind are all-around health and vigor, joy of life and longevity. Yet they are far from the supreme aims. For these aims we must search through the most rational and humane thought of those who have gone before us, and integrate with it thinking based on our present vantage point of knowledge and experience. In the light of such a survey it becomes clear that man's present paramount requirements are, on the one hand, a deeper and more integrated understanding and, on the other hand, a more heartfelt, keener sympathy, that is, a deeper fellow-feeling, leading to a stronger impulse to cooperation—more,

in a word, of love.

It is wishful thinking on the part of some psychologists to assert that these qualities result purely from conditioning or education. For although conditioning certainly plays a vital role, nevertheless *Homo sapiens* is both an intelligent and a cooperating animal. It is these two complex genetic characteristics, working in combination and serviced by the deftness of his hands, which above all others have brought man to his present estate. Moreover, there still exist great, diverse and numerous genetic differences in the biological bases of these traits within any human population. Although our means of recognition of these genetic differences are today very faulty and tend to confound differences of genetic origin with those derived from the environment, these means can be improved. Thus we can be enabled to recognize our betters. Yet even today our techniques are doubtless more accurate than the trials and errors whereby, after all, nature did manage to evolve us up to this point where we have become effective in counteracting nature. Certainly then it would be possible, if people once became aware of the genetic road that is open, to bring into existence a population most of whose members were as highly developed in regard to the genetic bases of both intelligence and social behavior as are those scattered individuals of today who stand highest in either separate respect.

If the dread of the misuse of nuclear energy awakens mankind not only to the genetic dangers confronting him but also to the genetic opportunities, then this will have been the greatest peacetime benefit that radioactivity could bestow upon us.

# 29

# The Effects of Smoking

E. CUYLER HAMMOND
*July 1962*

In 1560 Jean Nicot, the French ambassador to Portugal, wrote that an American Indian herb he had acquired had marvelous curative powers. For a time his view was widely accepted, and in his honor the herb was given the generic name *Nicotiana*. The species *Nicotiana rustica,* first introduced into Europe for smoking in pipes, was harsh and rather disagreeable. Later it was supplanted by *Nicotiana tabacum,* which produces a pleasanter smoke. *N. rustica* is still grown in the U.S.S.R. and other parts of Asia, but *N. tabacum* is now the chief source of smoking tobacco and is the only species cultivated in the U.S.

Skepticism about the medical value of tobacco developed near the end of the 16th century; not long thereafter smoking was condemned as a pernicious habit responsible for all manner of ills. This did not prevent smoking from becoming an almost universal habit among men in Europe and the American colonies. Actually there was no scientific evidence for any harmful effects of tobacco until the middle of the 19th century.

It appears that M. Bouisson, an obscure French physician, deserves credit for the first well-documented clinical study of the matter. In 1859, reporting on patients in the hospital at Montpellier, he observed that of 68 patients with cancer of the buccal cavity (45 of the lip, 11 of the mouth, seven of the tongue and five of the tonsil) 66 smoked pipes, one chewed tobacco and one apparently used tobacco in some form. He noted

that cancer of the lower lip ordinarily developed at the point where the pipe was held in the mouth. He further noted that lip cancer occurred more frequently among individuals who smoked short-stemmed pipes (then called "mouth burners") than among those who smoked long-stemmed clay pipes or pipes with stems made of a substance that does not conduct heat. He suggested that the cancer resulted from irritation of the tissue by tobacco products and heat.

Bouisson's observations were confirmed repeatedly over the next half-century, but since mouth cancer did not loom as a major medical problem the effect on smoking habits was insignificant. Another statistically unimportant problem early recognized as being associated with smoking was Buerger's disease, a rare affliction of the peripheral arteries. It was found to occur exclusively among smokers and to subside when the patient stopped smoking. In 1936, however, two New Orleans surgeons, Alton Ochsner and Michael E. De Bakey, observed that nearly all their lung cancer patients were cigarette smokers. Noting that lung cancer seemed to be on the increase and that it was paralleled by a general rise in cigarette smoking, they suggested a causal connection between the two phenomena. In 1938 Raymond Pearl, the noted Johns Hopkins University medical statistician, reported that smokers had a far shorter life expectancy than those who did not

use tobacco. The effect was so great as to indicate that smoking must be associated with diseases other than cancer. The first experimental evidence for an association between tobacco and cancer came in 1939, when A. H. Roffo of Argentina reported that he had produced cancer by painting tarlike tobacco extracts on the backs of rabbits. After World War II there was renewed interest in the subject of smoking and health, due partly to trends in tobacco consumption and partly to trends in death rates.

Before 1914 tobacco had been consumed mainly in pipes, cigars, chewing tobacco and snuff [*see illustration on page 209*]. Cigarettes began to be popular during World War I. In the period from the early 1920's to 1960 the consumption of manufactured cigarettes in the U.S. rose from about 750 per adult per year to 3,900 per adult per year. During the same period the consumption of tobacco in all other forms declined by about 70 per cent. The net result was that consumption of all tobacco products rose about 30 per cent.

The changes in smoking habits are more significant than the over-all rise in tobacco consumption. Smoke from cigars and pipes is heavy and as a rule slightly alkaline. Few people can inhale it without coughing or becoming dizzy or nauseated. Cigarette smoke, on the other hand, is relatively light, nearly neutral and can be inhaled readily. Most habit-

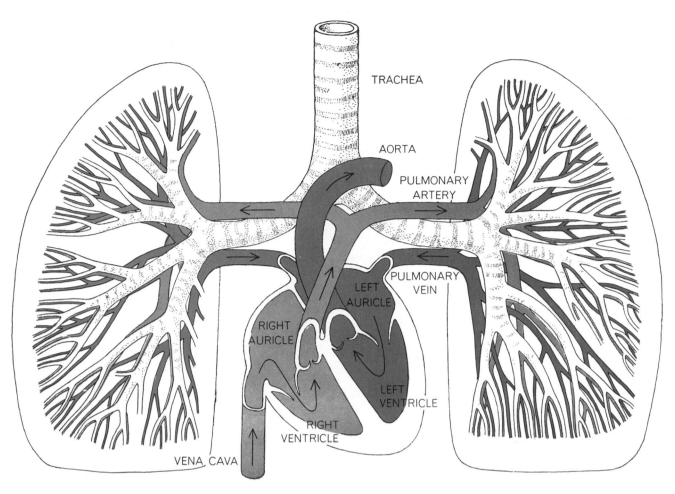

**HEART AND LUNGS** are both affected by inhaled tobacco smoke, which travels down the trachea, through the bronchial tubes to the alveoli. "Tars" deposit on the epithelium and lead to clogging of alveoli. These and the capillaries are often ruptured by coughing. The heart must then pump blood through a smaller number of capillaries, against increased pressure, on a reduced oxygen supply.

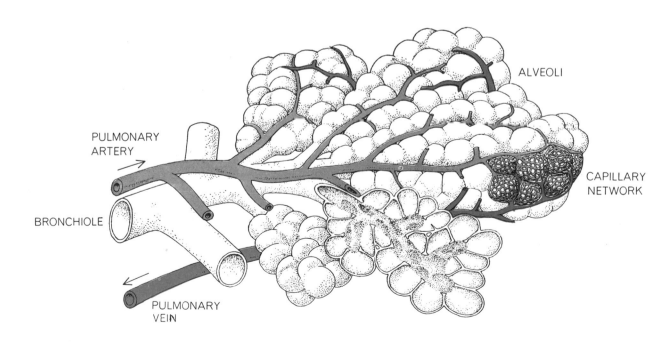

**ALVEOLI** of the lungs are air sacs formed by terminal expansion of the bronchioles. Oxygen is supplied to the blood through the capillaries embedded in the alveolar walls. Destruction of this tissue thus reduces the rate at which the lungs can take up oxygen.

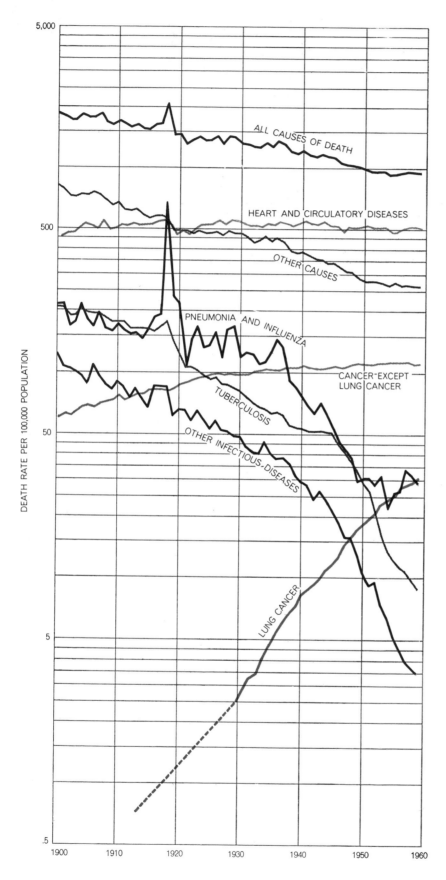

DEATH RATE PER 100,000 POPULATION

RISE IN LUNG CANCER DEATHS runs counter to the 60-year downtrend in total death rates among U.S. men. In 1959 lung cancer accounted for 29,335 deaths. Colon and rectal cancers, next in order of frequency, caused 19,129 male deaths. The nearly steady death rate for heart and circulatory diseases conceals a significant rise in coronary artery disease, which is offset by a long-term decline in other forms of heart disease. Curves are age-adjusted so that death rates are not spuriously shifted by changing age composition of the population.

ual cigarette smokers inhale to some degree, and heavy cigarette smokers tend to inhale deeply. In a recent study conducted by the American Cancer Society detailed information has been obtained on the smoking habits of 43,068 men and women. Only 7 per cent of the cigarette smokers among the men said that they did not inhale, whereas noninhalation was reported by 53 per cent of the pipe smokers and 71 per cent of the cigar smokers. Deep inhalation was reported by 24 per cent of the cigarette smokers compared with only 3 per cent of the pipe smokers and 1.5 per cent of the cigar smokers. Women who smoke inhale to a lesser degree than men smokers do. Furthermore, women over the age of 40 smoke far fewer cigarettes than men of the same age do, and few women over 55 smoke as much as a pack a day. Among current cigarette smokers now over 50, the majority of the men started the habit before they were 20, whereas the majority of the women did not begin until they were over 35.

During the past half-century total death rates—including death rates from almost all infectious diseases and some noninfectious ones—have declined rapidly. Lung cancer is a striking exception. Deaths from lung cancer in the U.S. have climbed from 4,000 in 1935 to 11,000 in 1945 and to 36,000 in 1960. The toll in 1960 was approximately equal to the number of deaths caused by traffic accidents. In 1960, 86 per cent of those who died from lung cancer were men. Between 1935 and 1960 the age-standardized death rate from lung cancer among U.S. men (the death rate adjusted for age differences in the composition of the population) increased 600 per cent; among women it increased 125 per cent. And for the past several years lung cancer has been the principal form of fatal cancer among men.

Painstaking studies have clearly demonstrated that the increase in lung cancer is real and not attributable merely to improvement in diagnosis. Lung cancer (that is, bronchogenic carcinoma) arises in the epithelium, or lining, of the bronchial tubes. The increase seems to be confined to two closely related forms of the disease: epidermoid carcinoma and undifferentiated carcinoma. There seems to be little, if any, increase in another form of the disease: adenocarcinoma. (In adenocarcinoma the diseased cells assume an arrangement resembling that of the cells in a gland.)

Lung cancer accounted for about 2 per cent of all U.S. deaths in 1960, and for about 6 per cent of deaths among men in their late 50's and 60's. The leading cause of death in the U.S. is coronary artery disease of the heart, which accounted in 1960 for nearly 29 per cent of all deaths, and for about 35 per cent of deaths among men still in their 40's and 50's. As in the case of lung cancer, coronary artery disease is less common among women, accounting for only about 16 per cent of the deaths occurring between the ages of 40 and 59.

In the late 1940's, when a number of investigators became concerned with lung cancer, cigarette smoking was only one of several factors suggested as possible causes for the increase in the disease. It was already well known that lung cancer could result from prolonged and heavy occupational exposure to certain industrial dusts and vapors. These include chromates, nickel carbonyl and dusts containing radioactive particles. Moreover, they result in epidermoid or undifferentiated carcinoma of the bronchial tubes and not in the less common adenocarcinoma.

This led to the hypothesis that the increase in lung cancer was due to increased exposure of the human population to air contamination of some sort. The factor involved had to be widespread and not confined to any particular occupational group. (In all countries with adequate mortality statistics lung cancer was found to have increased.) Three factors that met the requirements were: fumes from the combustion of solid and liquid fuels, dust from asphalt roads and the tires of motor vehicles, and cigarette smoking. The first two have not been ruled out as possibly contributing somewhat to the occurrence of lung cancer. It is the third that concerns us here.

As a first step a number of studies were made comparing the smoking habits of lung cancer patients with the smoking habits of individuals free of the disease. The results confirmed the 1936 observation of Ochsner and De Bakey. In every such study a far larger percentage of cigarette smokers was found in the lung cancer group than in the control group. Indeed, virtually all patients with epidermoid or undifferentiated carcinoma of the bronchial tubes admitted to smoking. There appeared to be less association, if there was any at all, between smoking habits and adenocarcinoma of the lung.

Cancer was not the only disease studied in relation to smoking habits. Knowing the acute effects of nicotine on the circulatory system, many physicians believed that smoking might be bad for patients with heart disease. In fact, a

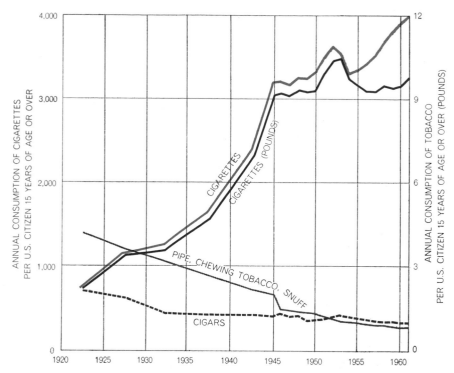

**CHANGES IN TOBACCO USE** produced a fivefold rise in cigarette consumption between the early 1920's and 1961, and a drop of nearly 70 per cent in consumption of all other tobacco products. Cigarettes are plotted both in units (*color*) and in pounds of unstemmed-tobacco equivalent. Other tobacco products are shown only in pounds. Filter cigarettes, which use less tobacco than nonfilter types, have been growing in popularity since 1954.

study made at the Mayo Clinic in 1940 by John P. English, Fredrick A. Willius and Joseph Berkson had indicated a considerable degree of association between smoking habits and coronary artery disease. Furthermore, many doctors were under the impression that smoking had a bad effect on patients with gastric and duodenal ulcers.

A number of investigators, myself among them, were uncertain as to the validity of these "clinical impressions" and "retrospective studies." A useful way to minimize bias and other difficulties in looking for an association between a disease and its possible causes is to employ the "prospective," or "follow up," method of investigation. The method consists of questioning a large number of presumably healthy individuals, keeping in touch with them for a number of years and finally ascertaining whether or not deaths in later years are associated with habits reported by the subjects before they became ill.

Two such prospective studies were undertaken in the fall of 1951, one in Britain by W. Richard Doll and A. Bradford Hill and the other in the U.S. by Daniel Horn and me. Under the auspices of the British Medical Research Council, Doll and Hill initiated their investigation by mailing questionnaires on smoking habits to all British physicians. They ob-

tained information on all deaths among British physicians by checking death certificates. Their study is still in progress. Several years later similar investigations were undertaken by Harold F. Dorn, who studied U.S. veterans holding life insurance; by E. W. R. Best, G. H. Josie and C. B. Walker, who are studying Canadian veterans and pensioners; and by John Edward Dunn, Jr., George Linden and Lester Breslow, who are studying men employed in certain occupations in California. In 1959 I started a new and more extensive prospective study in which smoking is included as only one of many factors under investigation.

The findings in all these investigations are remarkably similar; indeed, they are as close as could possibly be expected considering that the subjects were drawn from different populations and were of different ages. In the interest of brevity, therefore, I shall present data only from two studies with which I am personally concerned. The first of these was carried out as follows.

After designing and pretesting a questionnaire in the fall of 1951, we trained more than 22,000 American Cancer Society volunteers as researchers for the study. Between January 1 and May 31 of 1952 they enrolled subjects in 394 counties in nine states. The subjects, all men between the ages of 50 and 69, answered a simple confidential questionnaire on their smoking habits, both past and present. A total of 187,783 men were enrolled, filled out usable questionnaires and were successfully kept track of for the next 44 months. Death certificates were obtained for all who died, and additional medical information was gathered for those who were reported to have died of cancer. All together 11,870

deaths were reported, of which 2,249 were attributed to cancer.

The most important finding was that the total death rate (from all causes of death combined) is far higher among men with a history of regular cigarette smoking than among men who never smoked, but only slightly higher among pipe and cigar smokers than among men who never smoked. This is illustrated in the first of the series of charts on pages

8, 9 and 10. The death rates have been adjusted for age, and for ease of comparison the death rate of men who never smoked has been set at one.

Men who had smoked cigarettes regularly and exclusively were classified according to their cigarette consumption at the time they were enrolled in the study. It was found that death rates rose progressively with increasing number of cigarettes smoked per day, as shown in the second chart in the series. The death rate of those who smoked two or more packs of cigarettes a day was approximately two and a quarter times higher than the death rate of men who never smoked.

Being a heavy cigarette smoker myself at the time, I was curious to know the death rate of ex-cigarette smokers. This is shown in the third chart in the series. The death rate of men who had given up cigarette smoking a year or more prior to enrollment was considerably lower than that of men who were still smoking cigarettes when they were enrolled in the study.

Next we analyzed the data in relation to cause of death as reported on death certificates. Such information is subject to error, but on checking medical records we found that the diagnosis of cancer had been confirmed by microscopic examination of tissue in 79 per cent of the deaths ascribed to this disease. Even in some of these cases, however, the site of origin of the cancer was unknown or open to question. This is because cancer, unless successfully treated at an early stage, spreads through the body and its source is often difficult to determine. There is another difficulty that has to do with other causes of death. People in the older age groups not infrequently suffer from two or more diseases, one or another of which could be fatal. Since death can result from the combined effects of these diseases, it is difficult, and perhaps illogical, to ascribe death to one alone. These difficulties should be kept in mind in evaluating the following findings.

During the course of the study 7,316 deaths occurred among subjects with a history of regular cigarette smoking (some of whom smoked pipes and/or cigars as well as cigarettes). We divided these deaths according to primary cause as reported on death certificates. This is shown in the table on the opposite page under the heading "Observed deaths." Only 4,651 of these cigarette smokers would have died during the course of the study if their death rates had exactly matched those of men of the same age who had never smoked. This is shown in

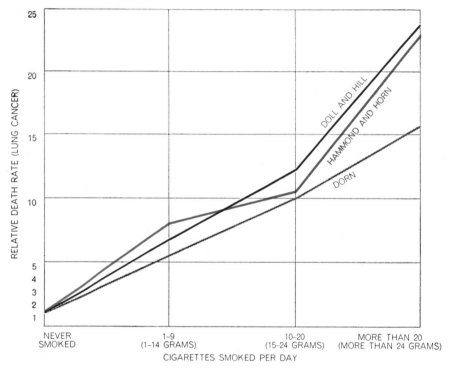

**LUNG CANCER** as a cause of death increases with the number of cigarettes (or gram equivalent) consumed per day, according to three major studies cited in the text. "Relative death rate" is the death rate among smokers divided by the death rate found among nonsmokers.

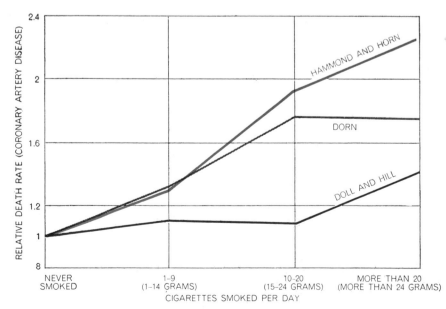

**CORONARY ARTERY DISEASE** as a cause of death also increases with the cigarettes smoked per day. The relative death rates are lower than for lung cancer because coronary artery disease is the leading cause of death among nonsmokers as well as among smokers.

the table under the heading "Expected deaths." The difference of 2,665 deaths (7,316 minus 4,651) can be considered the "Excess deaths" associated with a history of regular cigarette smoking. Of these excess deaths 52.1 per cent were attributed to coronary artery disease of the heart, 13.5 per cent to lung cancer and the remainder to other diseases. From this it is apparent that as a cause of death coronary artery disease is by far the most important disease associated with cigarette smoking.

From the standpoint of attempting to determine causal relations, it is best to study the figures in the table under the heading "Relative death rate." This is the observed number of deaths divided by the expected number of deaths, which in essence is the death rate of cigarette smokers divided by the death rate of subjects who never smoked.

Since coronary artery disease is the leading cause of death among men in the U.S. today, it is not surprising that we found it to be the leading cause of death among nonsmokers as well as among cigarette smokers. But the rate was 70 per cent higher among cigarette smokers. As shown in the fourth chart in the series on the next three pages, the death rate attributed to coronary artery disease increased progressively with the amount of cigarette smoking. We also found that ex-cigarette smokers had a lower death rate from this disease than did men who were still smoking cigarettes at the start of the study.

Lung cancer is an extremely rare cause of death among nonsmokers, except for those who have had prolonged and heavy occupational exposure to certain dusts and fumes. Taking death-certificate diagnosis at face value, the lung cancer death rate was more than 10 times higher among cigarette smokers than among nonsmokers. On obtaining medical records we found that, of 448 deaths attributed to this cause, the diagnosis of bronchogenic carcinoma was established by microscopic examination in addition to other evidence in 327 cases, of which 32 were adenocarcinoma. The fifth chart in the series shows age-standardized death rates by amount of cigarette smoking based on the 295 deaths from well-verified cases of bronchogenic carcinoma other than adenocarcinoma. The rate was very low for men who had never smoked, it increased with the amount of cigarette smoking, and it was very high for men who smoked two or more packs of cigarettes a day. When standardized both for age and for the amount of smoking, the rate for ex-cigarette smokers who had given up the habit for a year or more was considerably lower than the rate for men who were smoking cigarettes regularly at the start of the study. The lung cancer death rate of cigar and pipe smokers was very low compared with that of cigarette smokers, although higher than the rate for nonsmokers.

All together 127 deaths were attributed to cancer of other tissues (mouth, tongue, lip, larynx, pharynx and esophagus) that are directly exposed to tobacco smoke and material condensed from tobacco smoke. In 114 of these cases the diagnosis was confirmed by microscopic examination. Of these 114 men, 110 were smokers and only four had never smoked. The figures suggest that pipe and cigar smoking may be more important than cigarette smoking in relation to cancer of one or more sites included in this group, but the number of cases was not sufficient for a reliable evaluation of this point. Nevertheless, these cancers were the only causes of death for which the death rate of pipe and cigar smokers was found to be far higher than the death rate of nonsmokers.

Other reported causes of death showing a fairly high degree of association with cigarette smoking were gastric and duodenal ulcers, certain diseases of the arteries, pulmonary diseases (including pneumonia and influenza), cancer of the bladder and cirrhosis of the liver. Many other diseases appeared to be somewhat associated with cigarette smoking.

In 1959 I started a new study considerably larger than the first one. By securing the services of some 68,000 volunteer workers of the American Cancer Society in 1,121 counties in 25 states, we enrolled as subjects 1,079,000 men and women over the age of 30. Each of them filled out a lengthy confidential questionnaire including questions on family history, diseases and physical complaints, diet, smoking and other habits, residence history, occupational exposures and many other factors not included in previous studies. We plan to follow these subjects for six years. So far follow-up information is available only for the first 10½ months of observation.

The early findings on smoking are in close agreement with findings in all previous studies. In this study smokers were asked the degree to which they inhaled the smoke. It was found that, in relation to total death rates, the degree of inhalation is as important, and perhaps more important, than the amount of smoking [see illustration on page 215].

The new study has also revealed a high degree of association between cigarette smoking and a number of physical complaints, most particularly coughing, shortness of breath, loss of appetite and loss of weight [see illustration on page 216]. These complaints were related to the degree of inhalation as well as to the amount of smoking. They were reported less frequently by cigar and pipe smokers (most of whom do not inhale) than by cigarette smokers (most of whom

| CAUSE OF DEATH | OBSERVED DEATHS | EXPECTED DEATHS | EXCESS DEATHS | PERCENTAGE OF EXCESS | RELATIVE DEATH RATE |
|---|---|---|---|---|---|
| TOTAL DEATHS (ALL CAUSES) | 7,316 | 4,651 | 2,665 | 100.0 | 1.57 |
| CORONARY ARTERY DISEASE | 3,361 | 1,973 | 1,388 | 52.1 | 1.70 |
| OTHER HEART DISEASES | 503 | 425 | 78 | 2.9 | 1.18 |
| CEREBRAL VASCULAR LESIONS | 556 | 428 | 128 | 4.8 | 1.30 |
| ANEURYSM AND BUERGER'S DISEASE | 86 | 29 | 57 | 2.1 | 2.97 |
| OTHER CIRCULATORY DISEASES | 87 | 68 | 19 | 0.7 | 1.28 |
| LUNG CANCER | 397 | 37 | 360 | 13.5 | 10.73 |
| CANCER OF THE BUCCAL CAVITY, LARYNX OR ESOPHAGUS | 91 | 18 | 73 | 2.7 | 5.06 |
| CANCER OF THE BLADDER | 70 | 35 | 35 | 1.3 | 2.00 |
| OTHER CANCERS | 902 | 651 | 251 | 9.4 | 1.39 |
| GASTRIC AND DUODENAL ULCER | 100 | 25 | 75 | 2.8 | 4.00 |
| CIRRHOSIS OF THE LIVER | 83 | 43 | 40 | 1.5 | 1.93 |
| PULMONARY DISEASE (EXCEPT CANCER) | 231 | 81 | 150 | 5.6 | 2.85 |
| ALL OTHER DISEASES | 486 | 453 | 33 | 1.2 | 1.07 |
| ACCIDENT, VIOLENCE, SUICIDE | 363 | 385 | −22 | −0.8 | 0.94 |

**DEATHS AMONG REGULAR CIGARETTE SMOKERS**, labeled "Observed deaths," are compared with the number of deaths "expected" if the death rates for each age group among smokers had been the same as those found among nonsmokers. The table summarizes the results of the study conducted by the author and Daniel Horn. The column "Excess deaths" can be considered as the excess number of deaths associated with cigarette smoking. "Relative death rate" is the observed number of deaths divided by the expected number.

inhale either moderately or deeply).

Two prospective studies of smoking in relation to the occurrence of coronary artery disease have been carried out in Framingham, Mass., and Albany, N.Y. The combined findings from these studies were published on April 19 in *The New England Journal of Medicine* by Joseph T. Doyle, Thomas R. Dawber, William B. Kannel, A. Sandra Heslin and Harold A. Kahn. On enrollment in these studies each subject was given a medical examination. No symptoms of coronary artery disease were initially found in 4,120 men. These men were re-examined from time to time for a number of years. Symptoms of coronary artery disease (as well as death from this disease) were found far more frequently among those who smoked cigarettes regularly than among those who did not smoke. The total death rate was more than twice as high among men who smoked more than 20 cigarettes a day as among men who had never smoked. Ex-smokers and cigar and pipe smokers had morbidity and mortality records similar to the records of those who had never smoked. Thus the findings in this study based on medical examination of subjects were in close agreement with findings in the other U.S. studies.

Although all the studies have shown essentially the same results, there are some interesting differences between the results in Britain and the U.S. Lung cancer death rates are about twice as high in Britain as they are in the U.S.; chronic bronchitis is reported to be a common cause of death by British physicians but is seldom mentioned as a cause of death in the U.S.; death rates from coronary artery disease (as reported on death certificates) are far lower in Britain than they are in the U.S. No one really knows the reasons for these differences. Speculations on the subject may be briefly summarized as follows.

Climate, the method of heating houses, exposure to air pollutants and occupational exposure to dusts and fumes have all been suggested as possible reasons why both lung cancer and chronic bronchitis appear to occur more frequently in Britain than in this country. Differences in smoking habits have also been suggested as a possible factor. Doll and Hill have studied the length of discarded cigarette butts in England and Wales, and Ernest L. Wynder of the Sloan-Kettering Institute for Cancer Research and I have made similar studies on this side of the Atlantic. The average length of the butts was found to be 18.7 millimeters in England and Wales (where cigarettes are quite expensive), compared with 27.9 mm. in Canada and 30.9 mm. in the U.S. Therefore British smokers consume more of each cigarette and so receive a higher amount of nicotine and tobacco tar than Canadian and U.S. smokers do.

Diet has been suggested as a possible reason why death rates from coronary artery disease appear to be higher in the U.S. than they are in Britain. This apparent difference may be at least partly due to difference in diagnosis of the cause of death. Death can result from the combined effects of heart disease and lung ailments, particularly in older people. In the case of heart failure in a person suffering from a lung disease it is sometimes difficult to decide which to record as the principal cause of death. Thus the apparent high death rate reported as due to chronic bronchitis in Britain may be related to the comparatively low death rate reported as due to coronary artery disease in that country. Be that as it may, the Doll and Hill study showed less of a relation between smoking and coronary artery disease than did our U.S. study [*see lower illustration on page 210*]. On the other hand, Doll and Hill found a very high relation between smoking and death from chronic bronchitis.

In recent years considerable attention has been given to the chemical composition of tobacco smoke. A great many compounds have been identified, most of which are present in very small amounts. Some are distilled out of the tobacco and others are products of combustion. Included are numerous poisons (such as nicotine), various agents that are highly irritating to mammalian tissues, several carcinogenic (cancer-

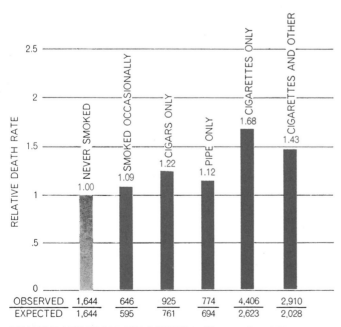

DEATH RATE FROM ALL CAUSES in Hammond and Horn study was far higher among cigarette smokers than among men who never smoked, but only slightly higher among pipe and cigar smokers.

| | NEVER SMOKED | SMOKED OCCASIONALLY | CIGARS ONLY | PIPE ONLY | CIGARETTES ONLY | CIGARETTES AND OTHER |
|---|---|---|---|---|---|---|
| | 1.00 | 1.09 | 1.22 | 1.12 | 1.68 | 1.43 |
| OBSERVED | 1,644 | 646 | 925 | 774 | 4,406 | 2,910 |
| EXPECTED | 1,644 | 595 | 761 | 694 | 2,623 | 2,028 |

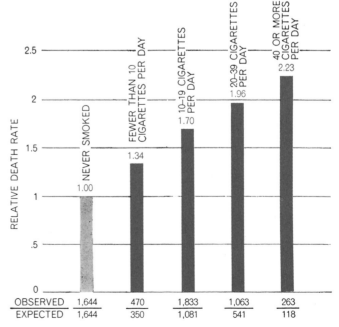

DAILY CIGARETTE CONSUMPTION showed a direct correlation with relative death rate from all causes. The study followed 187,783 men between the ages of 50 and 69 for 44 months.

| | NEVER SMOKED | FEWER THAN 10 CIGARETTES PER DAY | 10-19 CIGARETTES PER DAY | 20-39 CIGARETTES PER DAY | 40 OR MORE CIGARETTES PER DAY |
|---|---|---|---|---|---|
| | 1.00 | 1.34 | 1.70 | 1.96 | 2.23 |
| OBSERVED | 1,644 | 470 | 1,833 | 1,063 | 263 |
| EXPECTED | 1,644 | 350 | 1,081 | 541 | 118 |

producing) compounds and some co-carcinogenic compounds (materials that increase the potency of carcinogens). Most of this material is suspended in small particles, which together with carbon monoxide, air and other gases constitute tobacco smoke.

Ernest Wynder and his various collaborators have shown that tobacco-smoke condensate, or "tar," produces cancer in mice and rabbits if applied repeatedly to the skin over a long period of time. A number of investigators have confirmed these findings. The cancers so produced in rodents are of a type known as epidermoid carcinoma. (A synonym is squamous cell carcinoma, because the cells tend to be flattened, or squamous.) Different strains of animals vary in susceptibility, some being highly susceptible and others highly resistant.

Many investigators who have tried to produce lung cancer in rodents by exposing them to tobacco smoke have not succeeded in doing so. This may be because of two serious difficulties. Whereas a human smoker takes in smoke through his mouth, mice and other small rodents breathe through their noses, and in rodents this organ has developed into a remarkably efficient filter for preventing particulate matter from being drawn into the lung. Moreover, mice are sensitive to the acute toxic effects of tobacco smoke.

Several years ago I exposed mice to cigarette smoke under such conditions that they were forced to breathe smoke of approximately the same concentration as that of smoke taken in by human cigarette smokers. Unfortunately many of my animals went into convulsions and died within a few minutes. The remaining animals lived only a short time. By reducing the concentration of smoke the animals can be kept alive, but under such conditions it is doubtful whether or not their lungs are any more heavily exposed to the particulate matter of cigarette smoke than are the lungs of a nonsmoker sitting in a small room with several heavy smokers.

Nevertheless, by subjecting mice to tolerable concentrations of tobacco smoke Cecilie and Rudolph Leuchtenberger and Paul F. Doolon of the Children's Cancer Research Foundation in Boston have succeeded in producing various changes in the lining of the bronchial tubes of mice. These changes are similar to changes found in the bronchial tubes of human cigarette smokers. So far no cancers have been produced in mice thereby. This is consistent with the finding that lung cancer rarely occurs in human beings who are only slightly exposed to tobacco smoke.

During smoking the tissues first exposed to tobacco smoke are the lips, the tongue and the mucous membrane of the mouth. Some of the components of tobacco smoke (including known carcinogens) fluoresce under ultraviolet light. Robert C. Mellors of the Cornell University Medical College has shown that this material penetrates the cells of the lining of the mouth. The type of cancer that arises in this tissue is epidermoid carcinoma—the same type of cancer that is produced when tobacco tar is applied to the skin of experimental animals. Furthermore, the amount of tar required to produce epidermoid carcinoma of the skin in mice is roughly comparable to the exposure of a heavy smoker who develops epidermoid carcinoma of the lip or mouth.

In study after study a high degree of association has been found between smoking of all types (as well as the chewing of tobacco) and the occurrence of cancer of these tissues. It is hard to escape the conclusion that this association reflects a direct causal relation. This does not preclude the possibility that other factors (such as host susceptibility or exposure to other carcinogenic materials) are involved in at least some cases.

What has just been said of smoking in relation to cancer of the lips, mouth and tongue also applies to cancer of the pharynx and cancer of the larynx. The situation is slightly different in cancer of the esophagus; this passageway is exposed to ingested tobacco-smoke condensate but not directly to the smoke. The strong association between smoking and epidermoid carcinoma of the esophagus, however, would seem to point to the same conclusion.

When inhaled, tobacco smoke travels

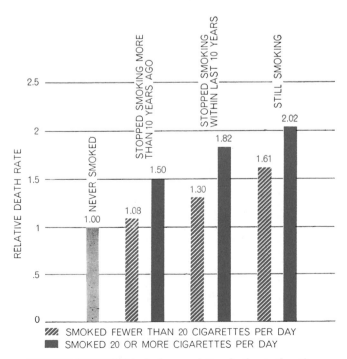

FORMER SMOKERS had a lower relative death rate than those still smoking, particularly if they had smoked fewer than 20 cigarettes a day and had stopped smoking for at least 10 years.

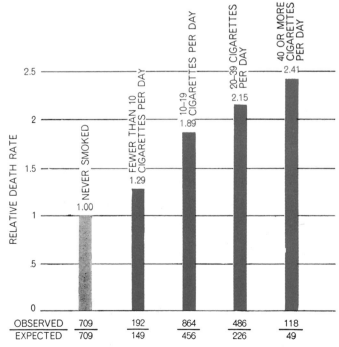

CORONARY ARTERY DISEASE, which accounted for 52.1 per cent of excess deaths among regular smokers of cigarettes, was also correlated very closely with the smoker's daily cigarette consumption.

down the trachea to the bronchial tubes of the lungs. All but a few cases of lung cancer originate in the lining, or epithelium, of these tubes. This is remarkable tissue, well worth describing here. Normally it consists of just two layers of cells that rest on a thin mat of tiny fibers called the basement membrane. This membrane separates the epithelium from the underlying tissue. Directly on top of the basement membrane is a layer of small, round cells with relatively small nuclei. They are called basal cells. On top of the basal cells is a single layer of cells known as columnar cells (because from the side they look like columns) interspersed with a few goblet cells (which look like little wine goblets). The goblet cells secrete a sticky fluid onto the surface. This is augmented by fluid secreted by glands located below the basement membrane. Protruding from the top of the columnar cells are short, hairlike cilia, which constantly move in a whiplike manner. This causes fluid on the epithelium to move up through the bronchial tubes and the trachea into the mouth, where it is either swallowed or expectorated.

The cilia and the fluid perform an extremely important function in cleansing the lungs. Small particles of dust or smoke that settle on the surface of the bronchial tubes are trapped in the fluid and, together with the fluid, are moved up and out of the lungs.

It has been shown by Anderson C. Hilding of St. Luke's Hospital in Duluth,

Minn., by Paul Kotin of the University of Southern California School of Medicine and by others that tobacco smoke inhibits the movement of the cilia to such a degree that the flow of fluid is slowed down, if not stopped altogether. This allows an accumulation of tobacco-smoke products and whatever other material happens to fall on the lining of the bronchial tubes. Smokers and nonsmokers alike—particularly those living in cities with polluted air and those engaged in certain occupations—inhale dust of various types, and some of the dusts contain carcinogenic substances.

For a number of years I have been cooperating in an extensive study of human lung tissue with Oscar Auerbach, a pathologist at the Veterans Administration Hospital in East Orange, N.J., and with Arthur Purdy Stout of the Columbia University College of Physicians and Surgeons. Some of our findings can be summarized as follows.

At the East Orange Veterans Hospital and at a number of hospitals in upstate New York the lungs are routinely removed at autopsy. The trachea and bronchial tubes are dissected out of the lungs and systematically divided into 208 portions, each of which is embedded in paraffin. A thin section of tissue is cut from each of these portions, mounted on a glass slide and stained with a suitable dye for microscopic examination. Independently, under the supervision of Lawrence Garfinkel of my staff, an interviewer is sent to the home of each patient

to obtain information on his or her occupational history, residence history and smoking habits. We do not include a case unless this information can be obtained. All told we have studied tissue from the bronchial tubes of more than 1,000 individuals.

In each of our studies microscope slides from a number of different patients have been put in completely random order by the use of a table of random numbers. They are then labeled with a serial number that gives no clue to their identity. All the slides are studied microscopically by Auerbach and samples of them are checked by Stout. After the slides are examined, the serial numbers are decoded so that the microscopic findings can be analyzed in relation to other information about the subjects.

Three major types of change occur in bronchial epithelium: hyperplasia (an increase in the number of layers of cells), loss of ciliated columnar cells and changes in the nuclei of cells [see illustration on page 217]. Hyperplasia is the usual reaction of surface tissues to almost any type of irritation, either chemical or mechanical. A familiar example is the formation of calluses on the hands. We found some degree of hyperplasia in 10 to 18 per cent of slides from nonsmokers, in more than 80 per cent of slides from light cigarette smokers and in more than 95 per cent of slides from heavy cigarette smokers. Extensive hyperplasia (defined as five or more layers of cells between the basement mem-

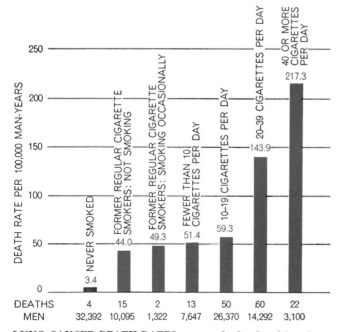

LUNG CANCER DEATH RATES, age-standardized and based on well-established cases exclusive of adenocarcinoma, rose sharply for men smoking more than a pack a day in the Hammond and Horn study. But the death rate among former smokers was much lower.

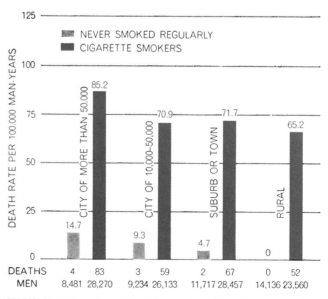

NONSMOKERS IN THE CITY sometimes die of lung cancer, but the death rate was only a fraction of that found among cigarette smokers who lived in the country. The death rates are based on well-established cases of lung cancer, exclusive of adenocarcinoma.

brane and the columnar cells) was frequently found in heavy cigarette smokers but rarely in other subjects.

Loss of ciliated columnar cells was observed in nonsmokers but far more frequently in cigarette smokers, and the frequency of this observation increased with the amount of cigarette smoking. The implication is that foreign material tends not to be removed, and thus can accumulate where the cilia have been destroyed.

An important finding was the occurrence of cells with atypical nuclei. The nuclei of cancer cells are usually large, irregular in shape and characteristically have an abnormal number of chromosomes. A few cells with nuclei that have such an appearance are occasionally found in the bronchial epithelium of men and women who have never smoked. Presumably they result from somatic mutation or some similar process. In nonsmokers the frequency of such cells does not increase with age.

Large numbers of cells with atypical nuclei of this kind were found in slides from cigarette smokers, and the number increased greatly with the amount of smoking. In heavy cigarette smokers we found many lesions composed entirely of cells with atypical nuclei and lacking cilia. Fewer such lesions were found in light cigarette smokers and none were found in nonsmokers. Among heavy cigarette smokers the number of cells with atypical nuclei increased markedly with advancing age.

In our latest study of bronchial epithelium we matched 72 ex-cigarette smokers, 72 men who had smoked cigarettes regularly up to the time of their terminal illness and 72 men who had never smoked. None of the men had died of lung cancer. Within each of the 72 triads, the three men were the same age, had similar employment histories and similar residence histories. Somewhat more changes were found in slides from ex-cigarette smokers than in slides from men who had never smoked. The important finding, however, was that the cellular changes, particularly the occurrence of cells with atypical nuclei, were fairly rare in ex-cigarette smokers compared with men who had smoked up to the time of their terminal illness. The study indicated that the number of cells with atypical nuclei declines when a cigarette smoker gives up the habit. This probably occurs slowly over a period of years.

The location of lesions is also significant and correlates with an observation one can make by passing cigarette smoke through glass tubing. Some years ago I found that when smoke was passed through a tube with a Y-shaped bifurcation, more tar precipitated where the tube branched than elsewhere. Acting on this lead, we have studied changes in bronchial epithelium in relation to bifurcations. There are numerous such points in the bronchial tree, because the tubes divide and redivide into smaller and smaller tubes. We found that lesions composed entirely of cells with atypical nuclei occur far more frequently at bifurcations than elsewhere.

In order to determine the significance of these changes we studied the bronchial epithelium of men who had died of bronchogenic carcinoma. Carcinoma is defined as a tumor, composed of cells with atypical nuclei, that originated in the epithelium and has penetrated the basement membrane and "invaded" the underlying tissue. Once such an invasion has occurred, the tumor grows—often to considerable size—and spreads to many parts of the body. In men who had died of lung cancer we found large numbers of cells with atypical nuclei, as well as many lesions composed entirely of such cells, scattered throughout the epithelium of the bronchial tubes of both lungs. In a few instances we found tiny independent carcinomas in which the tumor cells had broken through the basement membrane at just one small spot. These carcinomas looked exactly like many of the other lesions composed entirely of cells with atypical nuclei, except that in the other lesions we did not find any cells that had broken through the basement membrane. We are of the opinion that many, if not all, of the lesions composed entirely of atypical cells represent an early, preinvasive stage of carcinoma. This is a well-known occurrence in the cervix of the uteri of women and is called carcinoma *in situ*.

Judging from experimental evidence as well as from our findings in human beings, we are of the opinion that carcinoma of bronchial epithelium originates with a change in the nuclei of a few cells; that by cell division the number of such cells gradually increases; that finally lesions composed entirely of atypical cells are formed; and that occasionally cells in such a lesion penetrate the basement membrane, producing the disease known as carcinoma. Apparently the process is reversible up to the time the cells with atypical nuclei break through the basement membrane.

Where does the inhalation of tobacco smoke fit into this picture? There appear to be three possibilities:

1. It may be that exposure to tobacco smoke induces changes in the nuclei of

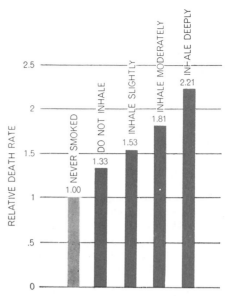

DEGREE OF INHALATION among cigarette smokers is charted against relative death rates from all causes. Rates are based on the author's new study of 1,079,000 men and women, which was begun in 1959.

cells. This would account for the increase of such cells both with the amount of smoking and with the number of years of smoking. It would not, however, in itself account for the finding of a decrease in the number of such cells when a cigarette smoker gives up the habit.

2. It may be that exposure to tobacco smoke simply increases the probability of changes taking place in the nuclei of cells as a result of exposure to inhaled carcinogenic agents other than those in tobacco smoke. The inhibition of ciliary movement by tobacco smoke may be the major factor involved in such a relation. Again this would not in itself account for the decrease in cells with atypical nuclei following cessation of cigarette smoking.

3. It may be that exposure to tobacco smoke produces a change in the local environment of bronchial epithelium so as to favor the survival and reproduction of certain mutant cells that have atypical nuclei of the type observed, as opposed to the survival and reproduction of normal cells. On this hypothesis the development of cancer results from natural selection under conditions of greatly altered environment. It is unnecessary to assume that tobacco smoke causes mutations, since a few cells with atypical nuclei are sometimes found in the bronchial epithelium of nonsmokers. This hypothesis suggests that normal cells are best adapted to an environment free of tobacco smoke, whereas cells with atypical nuclei are best adapted to an environment that includes smoke. The hypothesis thus accounts for the decline in the number of cells with atypical nu-

clei on the cessation of cigarette smoking.

I favor the last of these three hypotheses. It appears to account for all the findings, whereas the other two hypotheses account for only some of them. The three hypotheses are not, however, mutually exclusive.

To account for the association between cigarette smoking and certain other diseases, such as lung infections and coronary artery disease, other plausible mechanisms exist. On inhalation, air and any smoke it may contain passes through bronchial tubes of decreasing diameter, which finally deliver it to the tiny sacs called alveoli. The alveoli have thin walls supported by fibers of connective tissue. These walls contain capillary tubes through which blood flows from the pulmonary arteries to the pulmonary veins. During its passage through these capillaries the blood releases carbon dioxide and absorbs oxygen. At the same time carbon monoxide, nicotine and other impurities that may be present in the air or smoke are absorbed into the blood.

The small bronchial tubes are subject to being plugged with mucus. This frequently occurs in infectious diseases of the lung, with the result that secretions and bacteria are trapped in the alveolar spaces, thereby producing pneumonia. In cigarette smokers the interior diameter of the small bronchial tubes is considerably reduced by hyperplasia, so that the opening is very small indeed. In addition we find that smoking results in increased activity of the glands that secrete mucus into the bronchial tubes. This combination almost certainly increases the likelihood of the tubes being plugged by mucus. In my opinion this is enough to explain the finding that death rates from infectious diseases of the lung are considerably higher among cigarette smokers than among nonsmokers.

The occlusion of a bronchial tube by mucus (or by a spasm) often traps air in the alveoli to which that tube leads. If the person then happens to cough, the pressure of the trapped air can be increased to such a degree that the thin walls of the alveoli rupture. Coughing, excess mucus and reduction in the diameter of the small bronchial tubes increase the likelihood of such rupture.

Recently we have studied the alveoli in relation to cigarette smoking. We found extensive rupturing of the walls of a great many alveoli in the lungs of heavy cigarette smokers, a considerable amount in lighter cigarette smokers and very little in nonsmokers. The rupturing of the walls is usually accompanied by

a fibrous thickening of the remaining alveolar walls, together with a fibrous thickening of the walls of the small blood vessels in the vicinity. This probably results from the mechanism outlined above, since cigarette smoking produces coughing as well as hyperplasia of the bronchial tubes and increased secretion of mucus.

Ruptures in the walls of the alveoli destroy the capillary tubes located in the walls. If many are destroyed, far greater pressure is required to force the same quantity of blood through the remaining capillaries. All the blood must pass through them each time it circulates through the body, and the right ventricle of the heart has to supply the pressure. As a result the work load of the heart is increased in proportion to the degree of destruction of the alveoli.

Since oxygen is supplied to the blood through the capillaries in the alveoli, destruction of this tissue reduces the oxygen supply on which all the tissues of the body depend. In smokers this is compounded by the inhalation of carbon monoxide, which combines with hemoglobin more readily than oxygen does. This combination is enough to account for the shortness of breath often reported by cigarette smokers.

Because of its great activity heart muscle requires an abundant supply of oxygen. The inhalation of tobacco smoke increases the work load of this muscle and at the same time reduces the quantity of oxygen available to the muscle. In addition the action of nicotine on the

BRONCHIAL EPITHELIUM is the original site of almost all lung cancer, which often develops as shown on the opposite page. The photomicrographs (1 through 5), made by Oscar Auerbach of the East Orange, N.J., Veterans Administration Hospital, magnify human epithelial tissue 325, 250, 250, 75 and 110 diameters respectively. One of the first effects of smoking on normal epithelium (1) is hyperplasia (2), an increase in the number of basal cells. The cila disappear and the cells become squamous, or flattened (3). When the cells develop atypical nuclei and become disordered (4), the result is called carcinoma in situ. When these cells break through the basement membrane (5), the cancer may spread through lungs and to the rest of the body.

nervous system produces a temporary increase in the heart rate and a constriction of the peripheral blood vessels, which in turn produces a temporary increase in blood pressure. This also puts an added strain on the heart. Since a normal heart has extraordinary reserve powers, it can probably withstand these effects of smoking. A diseased heart may not be able to do so.

Autopsy studies (including a study of young men killed in the Korean war) have shown that the great majority of American men have at least some degree of atherosclerosis of the coronary arteries that supply blood to the muscle of the heart. Atherosclerosis consists of the progressive development of plaques (composed largely of cholesterol) with-

| COMPLAINT | CIGARETTE SMOKERS (PER CENT) | NONSMOKERS (PER CENT) | RATIO (SMOKERS TO NONSMOKERS) |
|---|---|---|---|
| COUGH | 33.2 | 5.6 | 5.9 |
| LOSS OF APPETITE | 3.3 | 0.9 | 3.7 |
| SHORTNESS OF BREATH | 16.3 | 4.7 | 3.5 |
| CHEST PAINS | 7.0 | 3.7 | 1.9 |
| DIARRHEA | 3.3 | 1.7 | 1.9 |
| EASILY FATIGUED | 26.1 | 14.9 | 1.8 |
| ABDOMINAL PAINS | 6.7 | 3.8 | 1.8 |
| HOARSENESS | 4.8 | 2.6 | 1.8 |
| LOSS OF WEIGHT | 7.3 | 4.5 | 1.6 |
| STOMACH PAINS | 6.0 | 3.8 | 1.6 |
| INSOMNIA | 10.2 | 6.8 | 1.5 |
| DIFFICULTY IN SWALLOWING | 1.4 | 1.0 | 1.4 |

PHYSICAL COMPLAINTS are more frequent among people who smoke a pack of cigarettes or more a day than among nonsmokers. The figures are from the author's large new study.

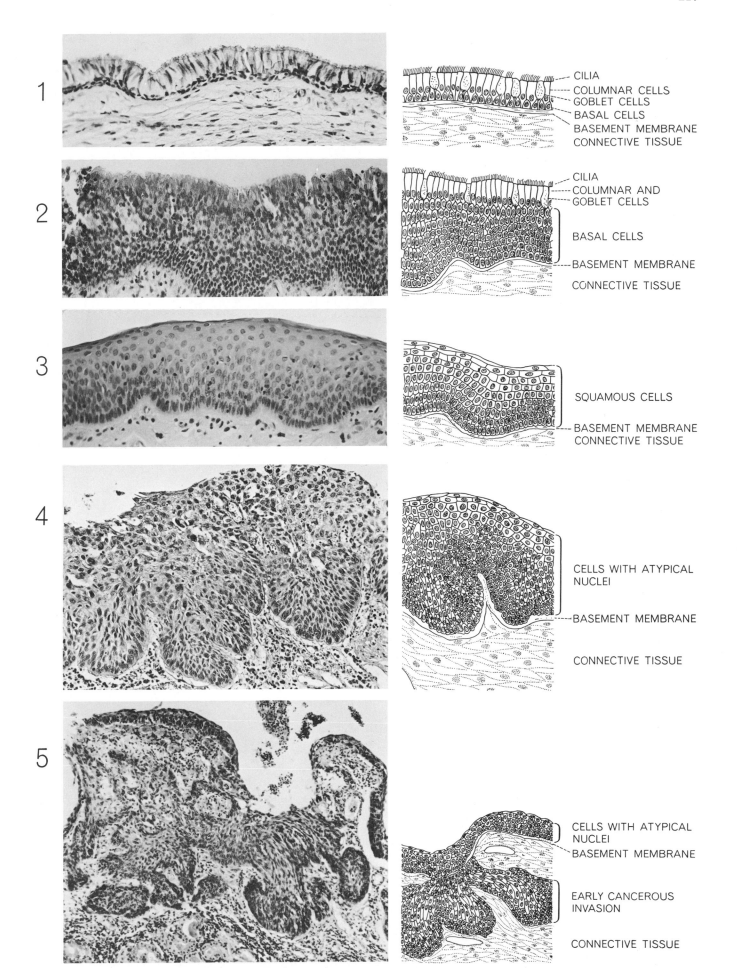

1. CILIA
COLUMNAR CELLS
GOBLET CELLS
BASAL CELLS
BASEMENT MEMBRANE
CONNECTIVE TISSUE

2. CILIA
COLUMNAR AND
GOBLET CELLS
BASAL CELLS
BASEMENT MEMBRANE
CONNECTIVE TISSUE

3. SQUAMOUS CELLS
BASEMENT MEMBRANE
CONNECTIVE TISSUE

4. CELLS WITH ATYPICAL
NUCLEI
BASEMENT MEMBRANE
CONNECTIVE TISSUE

5. CELLS WITH ATYPICAL
NUCLEI
BASEMENT MEMBRANE
EARLY CANCEROUS
INVASION
CONNECTIVE TISSUE

in the walls of these relatively small blood vessels, which thereby reduces their interior diameter. This in turn reduces the supply of blood to the heart muscle. Eventually it may completely cut off the supply of blood to a portion of the heart muscle, and this portion dies. Moreover, blood clots often form in diseased coronary arteries. This can also shut off the blood and cause the death of heart tissues. The common symptom of a stoppage in coronary blood flow is a heart attack.

As described above, cigarette smoking decreases the quantity of oxygen per unit volume of blood. Atherosclerosis of the coronary arteries tends to reduce the volume of blood delivered to the heart muscle per minute. Therefore if a person with atherosclerosis of the coronary arteries is also a cigarette smoker, his heart muscle receives far less than the normal supply of oxygen per minute. At the same time, because of the effects of smoking, a heavy work load is placed on his heart muscle. In my opinion this combination of conditions is sufficient to account for the finding that the death rate from coronary artery disease is higher in cigarette smokers than it is in men who never smoked, that the rate increases with the amount of cigarette smoking, and that it is lower in ex-cigarette smokers than it is in men who continue to smoke cigarettes.

Not only the heart but also all other organs of the body require oxygen obtained through the alveoli of the lungs and distributed by the blood. Thus a reduction in oxygen supply resulting from smoking may have a serious effect on any diseased organ, and in some instances it can make the difference between life and death. Perhaps this accounts for the finding that death rates from a multiplicity of chronic diseases are slightly higher among cigarette smokers than among nonsmokers.

I shall touch only briefly on two other diseases that appear to be significantly associated with cigarette smoking: gastric and duodenal ulcers and cancer of the bladder. In our first study cigarette smokers, compared with nonsmokers, had four times the relative death rate from the two kinds of ulcer and twice the death rate from cancer of the bladder. Doll and his associates in England recently performed a controlled clinical experiment demonstrating that smoking is indeed harmful to patients with gastric ulcer. Eighty patients who were regular smokers were divided at random into two groups, one allowed to continue smoking, the other advised to stop. Among the 40 patients who continued to smoke, the ulcers healed at a significantly slower rate than they did among the 40 patients who cut down on their smoking or stopped altogether. The mechanism by which smoking evidently retards recovery is unknown. It may be due to indirect effects, such as the effect of nicotine in the bloodstream, or to direct action of ingested tobacco smoke on the lining of the stomach.

As for cancer of the bladder, it is well known that exposure to carcinogenic agents can produce cancer in parts of the body remote from the tissue to which the agent is applied. For example, prolonged exposure to beta-naphthylamine often produced cancer of the bladder in workers in aniline dye plants. Conceivably some agent in tobacco smoke works in the same way, but until the problem is thoroughly investigated judgment should be deferred.

After reviewing the evidence, the mildest statement I can make is that, in my opinion, the inhalation of tobacco smoke produces a number of very harmful effects and shortens the life span of human beings. The simplest way to avoid these possible consequences is not to smoke at all. But one can avoid the most serious of them by smoking cigars or a pipe instead of cigarettes, provided that one does not inhale the smoke. An individual who chooses to smoke cigarettes can minimize the risks by restricting his consumption and by not inhaling.

The individual solution to the problem apparently requires more will power than many cigarette smokers have or are inclined to exert. I am confident, however, that more generally acceptable solutions can be found. There is good reason to suppose that the composition of tobacco smoke, both qualitative and quantitative, is a matter of considerable importance. Until several years ago the mainstream smoke of most U.S. cigarettes contained about 35 milligrams of "tar" per cigarette, of which about 2.5 milligrams was nicotine. The smoke from filter-tip cigarettes now on the market ranges in tar content from as low as 5.7 milligrams per cigarette to nearly 30 milligrams and the nicotine content from .4 to 2.5 milligrams. It is apparent that by selection of tobacco and by means of an effective filter, the nicotine and tar content of cigarette smoke can be markedly reduced. Some filters are selective in their action. For example, Wynder and Dietrich Hoffmann have recently found that a certain type of filter, which passes a reasonable amount of smoke, removes almost all the phenols. This may be important, since the same in-

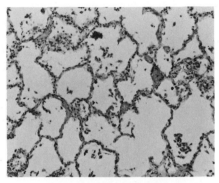

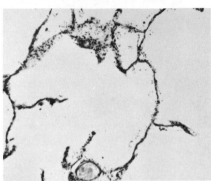

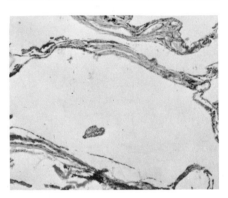

RUPTURE OF ALVEOLAR WALLS is a progressive process, from the normal state (*top*) to the rupture of some walls (*middle*) to the disappearance in certain areas of all the alveolar tissue (*bottom*). These photomicrographs, made by Auerbach, magnify the tissue approximately 120 diameters.

vestigators have reported that the phenols in cigarette smoke strongly inhibit the action of cilia in the bronchial tubes, and that some phenols increase the action of known carcinogenic agents. Furthermore, by various processes it is possible to alter the chemical composition of the smoke before it reaches the filter.

Considering this, I believe that extensive research should be undertaken to determine the effects of various constituents of cigarette smoke and to find means of removing those that are most harmful. Until this has been accomplished it seems advisable to reduce the total tar and nicotine content of cigarette smoke by the means now available.

# 30

# The Hallucinogenic Drugs

FRANK BARRON, MURRAY E. JARVIK,
and STERLING BUNNELL, JR.

*April 1964*

Human beings have two powerful needs that are at odds with each other: to keep things the same, and to have something new happen. We like to feel secure, yet at times we like to be surprised. Too much predictability leads to monotony, but too little may lead to anxiety. To establish a balance between continuity and change is a task facing all organisms, individual and social, human and non-human.

Keeping things predictable is generally considered one of the functions of the ego. When a person perceives accurately, thinks clearly, plans wisely and acts appropriately—and represses maladaptive thoughts and emotions—we say that his ego is strong. But the strong ego is also inventive, open to many perceptions that at first may be disorganizing. Research on the personality traits of highly creative individuals has shown that they are particularly alert to the challenge of the contradictory and the unpredictable, and that they may even court the irrational in their own make-up as a source of new and unexpected insight. Indeed, through all recorded history and everywhere in the world men have gone to considerable lengths to seek unpredictability by disrupting the functioning of the ego. A change of scene, a change of heart, a change of mind: these are the popular prescriptions for getting out of a rut.

Among the common ways of changing "mind" must be reckoned the use of intoxicating substances. Alcohol has quite won the day for this purpose in the U.S. and much of the rest of the world. Consumed at a moderate rate and in sensible quantities, it can serve simultaneously as a euphoriant and tranquilizing agent before it finally dulls the faculties and puts one to sleep. In properly disposed individuals it may dissolve sexual inhibitions, relieve fear and anxiety, or stimulate meditation on the meaning of life. In spite of its costliness to individual and social health when it is used immoderately, alcohol retains its rank as first among the substances used by mankind to change mental experience. Its closest rivals in popularity are opium and its derivatives and various preparations of cannabis, such as hashish and marijuana.

This article deals with another group of such consciousness-altering substances: the "hallucinogens." The most important of these are mescaline, which comes from the peyote cactus *Lophophora williamsii;* psilocybin and psilocin, from such mushrooms as *Psilocybe mexicana* and *Stropharia cubensis;* and d-lysergic acid diethylamide (LSD), which is derived from ergot (*Claviceps purpurea*), a fungus that grows on rye and wheat. All are alkaloids more or less related to one another in chemical structure.

Various names have been applied to this class of substances. They produce distinctive changes in perception that are sometimes referred to as hallucinations, although usually the person under the influence of the drug can distinguish his visions from reality, and even when they seem quite compelling he is able to attribute them to the action of the drug. If, therefore, the term "hallucination" is reserved for perceptions that the perceiver himself firmly believes indicate the existence of a corresponding object or event, but for which other observers can find no objective basis, then the "hallucinogens" only rarely produce hallucinations. There are several other names for this class of drugs. They have been called "psychotomimetic" because in some cases the effects seem to mimic psychosis [see "Experimental Psychoses," by six staff members of the Boston Psychopathic Hospital; SCIENTIFIC AMERICAN, June, 1955]. Some observers prefer to use the term "psychedelic" to suggest that unsuspected capacities of the imagination are sometimes revealed in the perceptual changes.

The hallucinogens are currently a subject of intense debate and concern in medical and psychological circles. At issue is the degree of danger they present to the psychological health of the person who uses them. This has become an important question because of a rapidly increasing interest in the drugs among laymen. The recent controversy at Harvard University, stemming at first from methodological disagreements

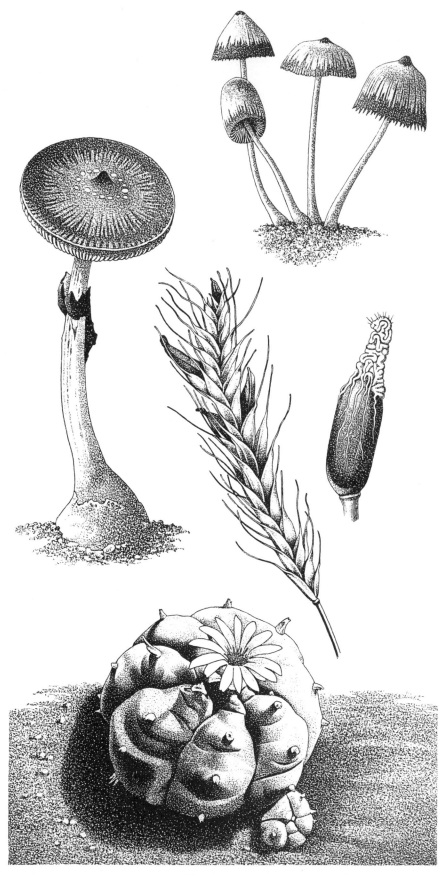

NATURAL SOURCES of the main hallucinogens are depicted. Psilocybin comes from the mushrooms *Stropharia cubensis* (*top left*) and *Psilocybe mexicana* (*top right*). LSD is synthesized from an alkaloid in ergot (*Claviceps purpurea*), a fungus that grows on cereal grains; an ergot-infested rye seed head is shown (*center*) together with a larger-scale drawing of the ergot fungus. Mescaline is from the peyote cactus *Lophophora williamsii* (*bottom*).

among investigators but subsequently involving the issue of protection of the mental health of the student body, indicated the scope of popular interest in taking the drugs and the consequent public concern over their possible misuse.

There are, on the other hand, constructive uses of the drugs. In spite of obvious differences between the "model psychoses" produced by these drugs and naturally occurring psychoses, there are enough similarities to warrant intensive investigation along these lines. The drugs also provide the only link, however tenuous, between human psychoses and aberrant behavior in animals, in which physiological mechanisms can be studied more readily than in man. Beyond this many therapists feel that there is a specialized role for the hallucinogens in the treatment of psychoneuroses. Other investigators are struck by the possibility of using the drugs to facilitate meditation and aesthetic discrimination and to stimulate the imagination. These possibilities, taken in conjunction with the known hazards, are the bases for the current professional concern and controversy.

In evaluating potential uses and misuses of the hallucinogens, one can draw on a considerable body of knowledge from such disciplines as anthropology, pharmacology, biochemistry, psychology and psychiatry.

In some primitive societies the plants from which the major hallucinogens are derived have been known for millenniums and have been utilized for divination, curing, communion with supernatural powers and meditation to improve self-understanding or social unity; they have also served such mundane purposes as allaying hunger and relieving discomfort or boredom. In the Western Hemisphere the ingestion of hallucinogenic plants in pre-Columbian times was limited to a zone extending from what is now the southwestern U.S. to the northwestern basin of the Amazon. Among the Aztecs there were professional diviners who achieved inspiration by eating either peyote, hallucinogenic mushrooms (which the Aztecs called *teo-nanacatyl*, or "god's flesh") or other hallucinogenic plants. *Teo-nanacatyl* was said to have been distributed at the coronation of Montezuma to make the ceremony seem more spectacular. In the years following the conquest of Mexico there were reports of communal mushroom rites among the Aztecs and other Indians of southern Mexico. The communal use has almost died out today, but in several

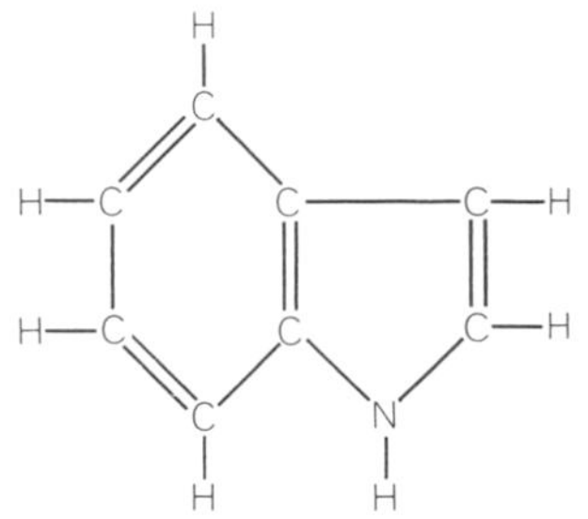

INDOLE RING

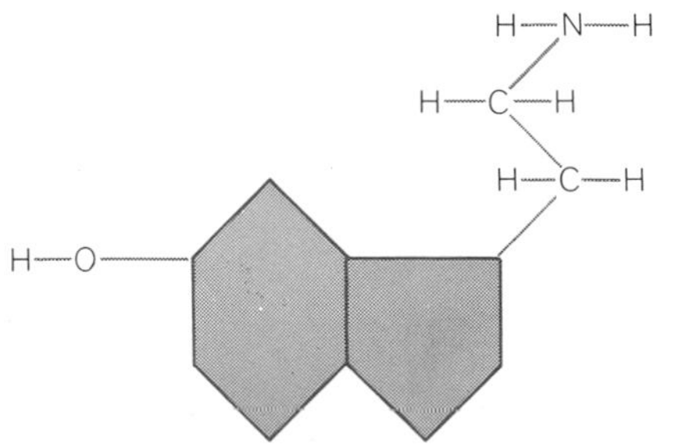

SEROTONIN

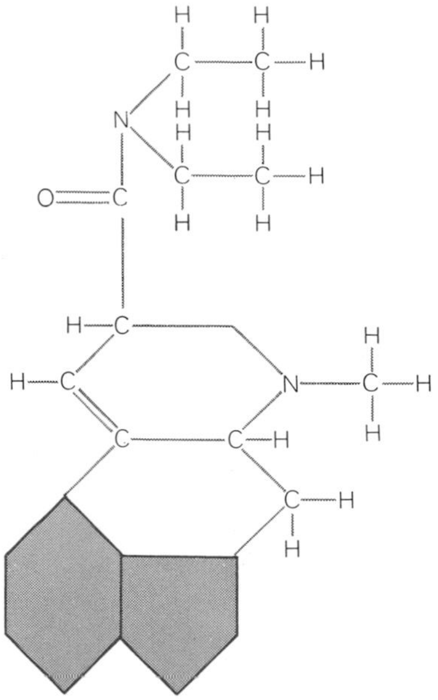

LSD

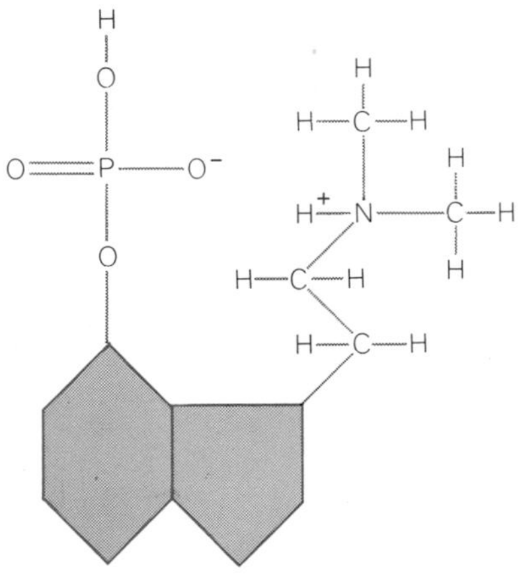

PSILOCYBIN

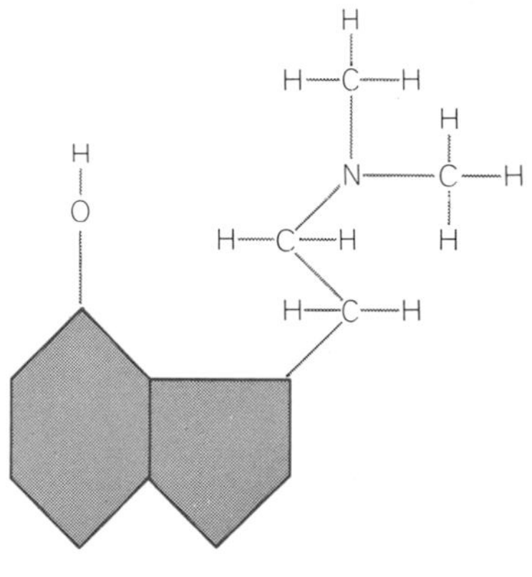

PSILOCIN

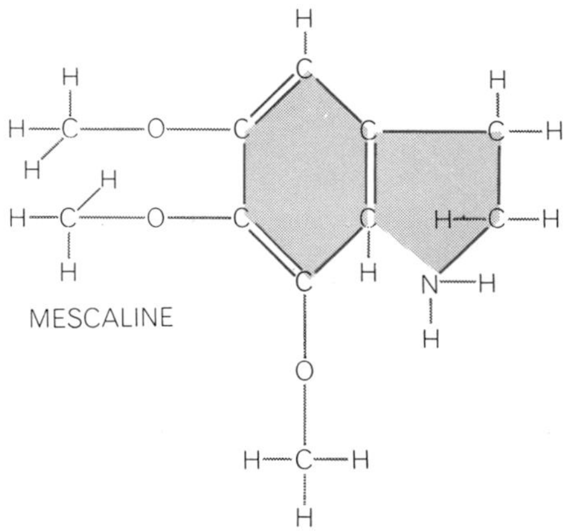

MESCALINE

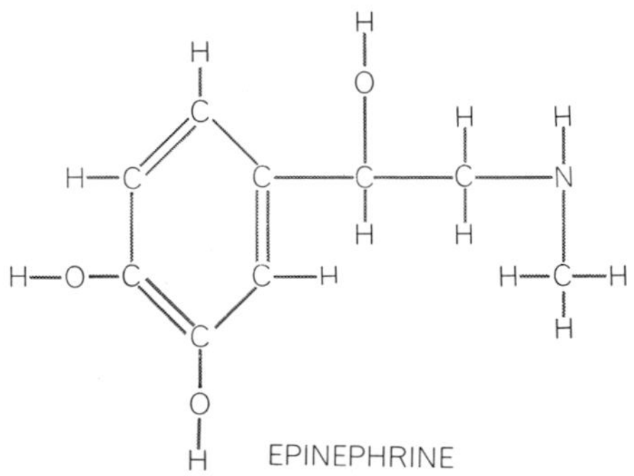

EPINEPHRINE

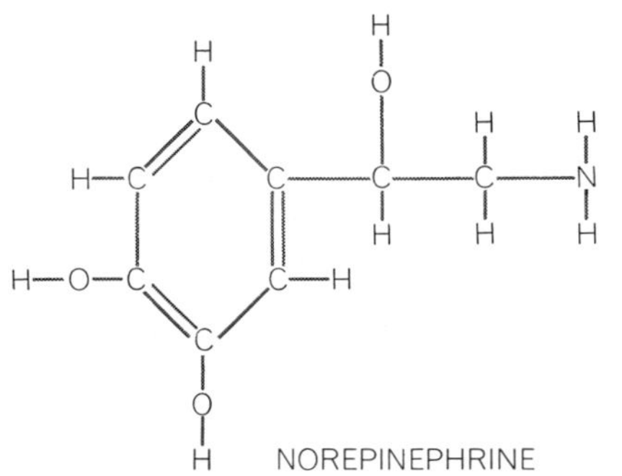

NOREPINEPHRINE

CHEMICAL RELATIONS among several of the hallucinogens and neurohumors are indicated by these structural diagrams. The indole ring (*in color at top*) is a basic structural unit; it appears, as indicated by the colored shapes, in serotonin, LSD, psilocybin and psilocin. Mescaline does not have an indole ring but, as shown by the light color, can be represented so as to suggest its relation to the ring. The close relation between mescaline and the two catechol amines epinephrine and norepinephrine is also apparent here.

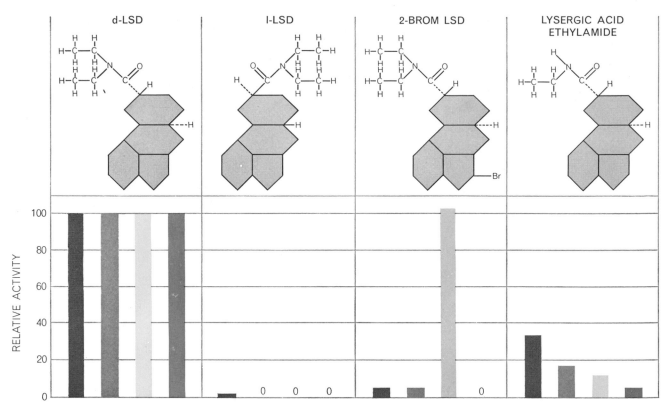

SLIGHT CHANGES in LSD molecule produce large changes in its properties. Here LSD (*left*) is used as a standard, with a "relative activity" of 100 in toxicity (*dark gray bar*), fever-producing effect (*light gray*), ability to antagonize serotonin (*light color*) and typical psychotomimetic effects (*dark color*). The stereoisomer of LSD (*second from left*) in which the positions of the side chains are reversed, shows almost no activity; the substitution of a bromine atom (*third from left*) reduces the psychotomimetic effect but not the serotonin antagonism; the removal of one of the two ethyl groups (*right*) sharply reduces activity in each of the areas.

tribes the medicine men or women (*curanderos*) still partake of *Psilocybe* and *Stropharia* in their rituals.

In the arid region between the Rio Grande and central Mexico, where the peyote cactus grows, the dried tops of the plants ("peyote buttons") were eaten by Indian shamans, or medicine men, and figured in tribal rituals. During the 19th century the Mescalero Apaches acquired the plant and developed a peyote rite. The peyotism of the Mescaleros (whence the name mescaline) spread to the Comanches and Kiowas, who transformed it into a religion with a doctrine and ethic as well as ritual. Peyotism, which spread rapidly through the Plains tribes, became fused with Christianity. Today its adherents worship God as the great spirit who controls the universe and put some of his power into peyote, and Jesus as the man who gave the plant to the Indians in a time of need. Saturday-night meetings, usually held in a traditional tepee, begin with the eating of the sacramental peyote; then the night is spent in prayer, ritual singing and introspective contemplation, and in the morning there is a communion breakfast of corn, game and fruit.

Recognizing the need for an effective organization to protect their form of worship, several peyote churches joined in 1918 to form the Native American Church, which now has about 225,000 members in tribes from Nevada to the East Coast and from the Mexican border to Saskatchewan. It preaches brotherly love, care of the family, self-reliance and abstinence from alcohol. The church has been able to defeat attempts, chiefly by the missionaries of other churches, to outlaw peyote by Federal legislation, and it has recently brought about the repeal of antipeyote legislation in several states.

The hallucinogens began to attract scholarly interest in the last decade of the 19th century, when the investigations and conceptions of such men as Francis Galton, J. M. Charcot, Sigmund Freud and William James introduced a new spirit of serious inquiry into such subjects as hallucination, mystical experience and other "paranormal" psychic phenomena. Havelock Ellis and the psychiatrist Silas Weir Mitchell wrote accounts of the subjective effects of peyote, or Anhalonium, as it was then called. Such essays in turn stimulated

the interest of pharmacologists. The active principle of peyote, the alkaloid called mescaline, was isolated in 1896; in 1919 it was recognized that the molecular structure of mescaline was related to the structure of the adrenal hormone epinephrine.

This was an important turning point, because the interest in the hallucinogens as a possible key to naturally occurring psychoses is based on the chemical relations between the drugs and the neurohumors: substances that chemically transmit impulses across synapses between two neurons, or nerve cells, or between a neuron and an effector such as a muscle cell. Acetylcholine and the catechol amines epinephrine and norepinephrine have been shown to act in this manner in the peripheral nervous system of vertebrates; serotonin has the same effect in some invertebrates. It is frequently assumed that these substances also act as neurohumors in the central nervous system; at least they are present there, and injecting them into various parts of the brain seems to affect nervous activity.

The structural resemblance of mescaline and epinephrine suggested a possible link between the drug and mental illness: Might the early, excited stage of schizophrenia be produced or at least triggered by an error in metabolism that produced a mescaline-like substance? Techniques for gathering evidence on this question were not available, however, and the speculation on an "M-substance" did not lead to serious experimental work.

When LSD was discovered in 1943, its extraordinary potency again aroused interest in the possibility of finding a natural chemical activator of the schizophrenic process. The M-substance hypothesis was revived on the basis of reports that hallucinogenic effects were produced by adrenochrome and other breakdown products of epinephrine, and the hypothesis appeared to be strengthened by the isolation from human urine of some close analogues of hallucinogens. Adrenochrome has not, however, been detected in significant amounts in the human body, and it seems unlikely that the analogues could be produced in sufficient quantity to effect mental changes.

The relation between LSD and serotonin has given rise to the hypothesis that schizophrenia is caused by an imbalance in the metabolism of serotonin, with excitement and hallucinations resulting from an excess of serotonin in certain regions of the brain, and depressive and catatonic states resulting from a deficiency of serotonin. The idea arose in part from the observation that in some laboratory physiological preparations LSD acts rather like serotonin but in other preparations it is a powerful antagonist of serotonin; thus LSD might facilitate or block some neurohumoral action of serotonin in the brain.

The broad objection to the serotonin theory of schizophrenia is that it requires an oversimplified view of the disease's pattern of symptoms. Moreover, many congeners, or close analogues, of LSD, such as 2-brom lysergic acid, are equally effective or more effective antagonists of serotonin without being significantly active psychologically in man. This does not disprove the hypothesis, however. In man 2-brom LSD blocks the mental effects of a subsequent dose of LSD, and in the heart of a clam it blocks the action of both LSD and serotonin. Perhaps there are "keyholes" at the sites where neurohumors act; in the case of those for serotonin it may be that LSD fits the hole and opens the lock, whereas the psychologically inactive analogues merely occupy the keyhole, blocking the action of serotonin or LSD without mimicking their effects. Certainly the re-

semblance of most of the hallucinogens to serotonin is marked, and the correlations between chemical structure and pharmacological action deserve intensive investigation. The serotonin theory of schizophrenia is far from proved, but there is strong evidence for an organic factor of some kind in the disease; it may yet turn out to involve either a specific neurohumor or an imbalance among several neurohumors.

The ingestion of LSD, mescaline or psilocybin can produce a wide range of subjective and objective effects. The subjective effects apparently depend on at least three kinds of variable: the properties and potency of the drug itself; the basic personality traits and current mood of the person ingesting it, and the social and psychological context, including the meaning to the individual of his act in taking the drug and his interpretation of the motives of those who made it available. The discussion of subjective effects that follows is compiled from many different accounts of the drug experience; it should be considered an inventory of possible effects rather than a description of a typical episode.

One subjective experience that is frequently reported is a change in visual perception. When the eyes are open, the perception of light and space is affected: colors become more vivid and seem to glow; the space between objects becomes more apparent, as though space itself had become "real," and surface details appear to be more sharply defined. Many people feel a new awareness of the physical beauty of the world, particularly of visual harmonies, colors, the play of light and the exquisiteness of detail.

The visual effects are even more striking when the eyes are closed. A constantly changing display appears, its content ranging from abstract forms to dramatic scenes involving imagined people or animals, sometimes in exotic lands or ancient times. Different individuals have recalled seeing wavy lines, cobweb or chessboard designs, gratings, mosaics, carpets, floral designs, gems, windmills, mausoleums, landscapes, "arabesques spiraling into eternity," statuesque men of the past, chariots, sequences of dramatic action, the face of Buddha, the face of Christ, the Crucifixion, "the mythical dwelling places of the gods," the immensity and blackness of space. After taking peyote Silas Weir Mitchell wrote: "To give the faintest idea of the perfectly satisfying intensity and purity of these gorgeous color fruits

WATER COLORS were done, while under the influence of a relatively large dose of a hallucinogenic drug, by a person with no art training. Originals are bright yellow, purple, green and red as well as black.

224

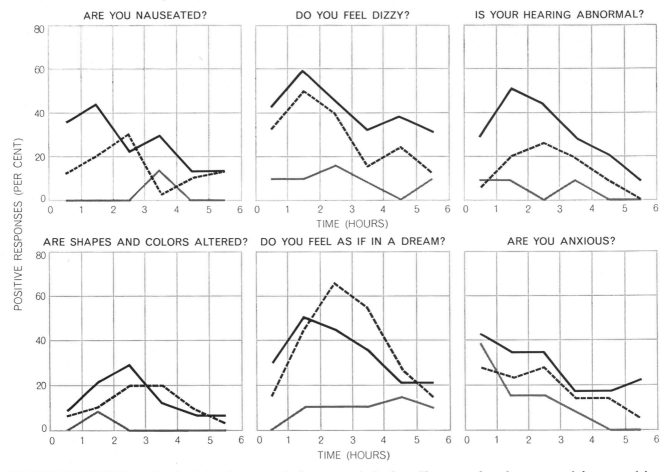

ARE YOU NAUSEATED?  DO YOU FEEL DIZZY?  IS YOUR HEARING ABNORMAL?

ARE SHAPES AND COLORS ALTERED?  DO YOU FEEL AS IF IN A DREAM?  ARE YOU ANXIOUS?

POSITIVE RESPONSES (PER CENT)

TIME (HOURS)

SUBJECTIVE REPORT on physiological and perceptual effects of LSD was obtained by means of a questionnaire containing 47 items, the results for six of which are presented. Volunteers were questioned at one-hour intervals beginning half an hour after they took the drug. The curves show the per cent of the group giving positive answers at each time. The gray curves are for those given an inactive substance, the broken black curves for between 25 and 75 micrograms and the solid black curves for between 100 and 225.

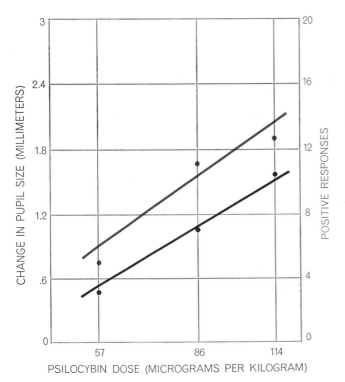

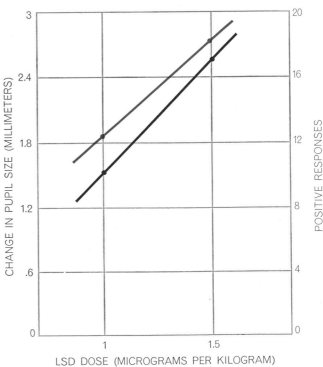

OBJECTIVE AND SUBJECTIVE effects vary with dosage as shown here. The data plotted in black are for the increase in size of pupil; the number of positive responses to questions like the ones at the top of the page are shown in color. The objective and subjective measures vary in a similar manner. The data are from an experiment done by Harris Isbell of the University of Kentucky.

is quite beyond my power." A painter described the waning hours of the effects of psilocybin as follows: "As the afternoon wore on I felt very content to simply sit and stare out of the window at the snow and the trees, and at that time I recall feeling that the snow, the fire in the fireplace, the darkened and book-lined room were so perfect as to seem almost unreal."

The changes in visual perception are not always pleasant. Aldous Huxley called one of his books about mescaline *Heaven and Hell* in recognition of the contradictory sensations induced by the drug. The "hellish" experiences include an impression of blackness accompanied by feelings of gloom and isolation, a garish modification of the glowing colors observed in the "heavenly" phase, a sense of sickly greens and ugly dark reds. The subject's perception of his own body may become unpleasant: his limbs may seem to be distorted or his flesh to be decaying; in a mirror his face may appear to be a mask, his smile a meaningless grimace. Sometimes all human movements appear to be mere puppetry, or everyone seems to be dead. These experiences can be so disturbing that a residue of fear and depression persists long after the effects of the drug have worn off.

Often there are complex auditory hallucinations as well as visual ones: lengthy conversations between imaginary people, perfectly orchestrated musical compositions the subject has never heard before, voices speaking foreign languages unknown to the subject. There have also been reports of hallucinatory odors and tastes and of visceral and other bodily sensations. Frequently patterns of association normally confined to a single sense will cross over to other senses: the sound of music evokes the visual impression of jets of colored light, a "cold" human voice makes the subject shiver, pricking the skin with a pin produces the visual impression of a circle, light glinting on a Christmas tree ornament seems to shatter and to evoke the sound of sleigh bells. The time sense is altered too. The passage of time may seem to be a slow and pleasant flow or to be intolerably tedious. A "sense of timelessness" is often reported; the subject feels outside of or beyond time, or time and space seem infinite.

In some individuals one of the most basic constancies in perception is affected: the distinction between subject and object. A firm sense of personal identity depends on knowing accurately the borders of the self and on being able to distinguish what is inside from what is outside. Paranoia is the most vivid pathological instance of the breakdown of this discrimination; the paranoiac attributes to personal and impersonal forces outside himself the impulses that actually are inside him. Mystical and transcendental experiences are marked by the loss of this same basic constancy. "All is one" is the prototype of a mystical utterance. In the mystical state the distinction between subject and object disappears; the subject is seen to be one with the object. The experience is usually one of rapture or ecstasy and in religious terms is described as "holy." When the subject thus achieves complete identification with the object, the experience seems beyond words.

Some people who have taken a large dose of a hallucinogenic drug report feelings of "emptiness" or "silence," pertaining either to the interior of the self or to an "interior" of the universe—or to both as one. Such individuals have a sense of being completely undifferentiated, as though it were their personal consciousness that had been "emptied," leaving none of the usual discriminations on which the functioning of the ego depends. One man who had this experience thought later that it had been an anticipation of death, and that the regaining of the basic discriminations was like a remembrance of the very first days of life after birth.

The effect of the hallucinogens on sexual experience is not well documented. One experiment that is often quoted seemed to provide evidence that mescaline is an anaphrodisiac, an inhibitor of sexual appetite; this conclusion seemed plausible because the drugs have so often been associated with rituals emphasizing asceticism and prayer. The fact is, however, that the drugs are probably neither anaphrodisiacs nor aphrodisiacs—if indeed any drug is. There is reason to believe that if the drug-taking situation is one in which sexual relations seem appropriate, the hallucinogens simply bring to the sexual experience the same kind of change in perception that occurs in other areas of experience.

The point is that in all the hallucinogen-produced experiences it is never the drug alone that is at work. As in the case of alcohol, the effects vary widely depending on when the drug is taken, where, in the presence of whom, in what dosage and—perhaps most important of all—by whom. What happens to the individual after he takes the drug, and his changing relations to the setting and the people in it during the episode, will further influence his experience.

Since the setting is so influential in these experiments, it sometimes happens that a person who is present when someone else is taking a hallucinogenic drug, but who does not take the drug himself, behaves as though he were under the influence of a hallucinogen. In view of this effect one might expect that a person given an inactive substance he thought was a drug would respond as though he had actually received the drug. Indeed, such responses have sometimes been noted. In controlled experiments, however, subjects given an inactive substance are readily distinguishable from those who take a drug; the difference is apparent in their appearance and behavior, their answers to questionnaires and their physiological responses. Such behavioral similarities as are observed can be explained largely by a certain apprehension felt by a person who receives an inactive substance he thinks is a drug, or by anticipation on the part of someone who has taken the drug before.

In addition to the various subjective effects of the hallucinogens there are a number of observable changes in physiological function and in performance that one can measure or at least describe objectively. The basic physiological effects are those typical of a mild excitement of the sympathetic nervous system. The hallucinogens usually dilate the pupils, constrict the peripheral arterioles and raise the systolic blood pressure; they may also increase the excitability of such spinal reflexes as the knee jerk. Electroencephalograms show that the effect on electrical brain waves is usually of a fairly nonspecific "arousal" nature: the pattern is similar to that of a normally alert, attentive and problem-oriented subject, and if rhythms characteristic of drowsiness or sleep have been present, they disappear when the drug is administered. (Insomnia is common the first night after one of the drugs has been taken.) Animal experiments suggest that LSD produces these effects by stimulating the reticular formation of the midbrain, not directly but by stepping up the sensory input.

Under the influence of one of the hallucinogens there is usually some reduction in performance on standard tests of reasoning, memory, arithmetic, spelling and drawing. These findings may not indicate an inability to perform well; after taking a drug many people simply refuse to co-operate with the tester. The very fact that someone should want to

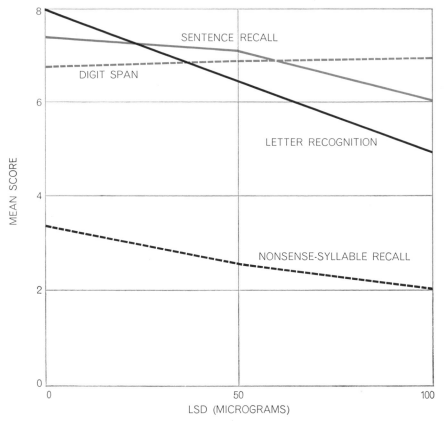

**EFFECT OF LSD** on memory was determined with standard tests. Curves show results of four tests for subjects given an inactive substance, 50 micrograms of the drug and 100 micrograms respectively. Effect of LSD was to decrease scores except in a test of digit-recall ability.

test them may seem absurd and may arouse either hostility or amusement. Studies by one of the authors in which tests of attention and concentration were administered to subjects who had been given different doses of LSD indicated that motivation was perhaps as important in determining scores as the subject's intellectual capacity.

The hallucinogenic drugs are not addictive—if one means by addiction that physiological dependence is established and the drug becomes necessary, usually in increasing amounts, for satisfactory physiological functioning. Some individuals become psychologically dependent on the drugs, however, and develop a "habit" in that sense; indeed, there is a tendency for those who ingest hallucinogens habitually to make the drug experience the center of all their activities. LSD, mescaline and psilocybin do produce physiological tolerance. If the same quantity of LSD is administered on three successive days, for example, it will not suffice by the third day to produce the same subjective or physiological effects; tolerance develops more slowly and less completely with mescaline and psilocybin. When an individual becomes tolerant to a given dos-

age of LSD, the ordinarily equivalent dose of psilocybin produces reduced effects. This phenomenon of cross-tolerance suggests that the two drugs have common pathways of action. Any tolerance established by daily administration of the drugs wears off rather rapidly, generally being dissipated within a few days if the drug is not taken.

The three major hallucinogens differ markedly in potency. The standard human doses—those that will cause the average adult male weighing about 150 pounds to show the full clinical effects—are 500 milligrams of mescaline, 20 milligrams of psilocybin and .1 milligram of LSD. It is assumed that in a large enough dose any of the hallucinogens would be lethal, but there are no documented cases of human deaths from the drugs alone. Death has been brought on in sensitive laboratory animals such as rabbits by LSD doses equivalent to 120 times the standard human dose. Some animals are much less susceptible; white rats have been given doses 1,000 times larger than the standard human dose without lasting harm. The maximum doses known by the authors to have been taken by human beings are 900 milligrams of mescaline, 70 milligrams

of psilocybin and two milligrams of LSD. No permanent effects were noted in these cases, but obviously no decisive studies of the upper limits of dosage have been undertaken.

There are also differences among the hallucinogens in the time of onset of effects and the duration of intoxication. When mescaline is given orally, the effects appear in two or three hours and last for 12 hours or more. LSD acts in less than an hour; some of its effects persist for eight or nine hours, and insomnia can last as long as 16 hours. Psilocybin usually acts within 20 or 30 minutes, and its full effect is felt for about five hours. All these estimates are for the standard dose administered orally; when any of the drugs is given intravenously, the first effects appear within minutes.

At the present time LSD and psilocybin are treated by the U.S. Food and Drug Administration like any other "experimental drug," which means that they can be legally distributed only to qualified investigators who will administer them in the course of an approved program of experimentation. In practice the drugs are legally available only to investigators working under a Government grant or for a state or Federal agency.

Nevertheless, there has probably been an increase during the past two or three years in the uncontrolled use of the drugs to satisfy personal curiosity or to experience novel sensations. This has led a number of responsible people in government, law, medicine and psychology to urge the imposition of stricter controls that would make the drugs more difficult to obtain even for basic research. These people emphasize the harmful possibilities of the drugs; citing the known cases of adverse reactions, they conclude that the prudent course is to curtail experimentation with hallucinogens.

Others—primarily those who have worked with the drugs—emphasize the constructive possibilities, insist that the hallucinogens have already opened up important leads in research and conclude that it would be shortsighted as well as contrary to the spirit of free scientific inquiry to restrict the activities of qualified investigators. Some go further, questioning whether citizens should be denied the opportunity of trying the drugs even without medical or psychological supervision and arguing that anyone who is mentally competent should have the right to explore the varieties of conscious experience if he can do so without harming himself or others.

The most systematic survey of the incidence of serious adverse reactions to hallucinogens covered nearly 5,000 cases, in which LSD was administered on more than 25,000 occasions. Psychotic reactions lasting more than 48 hours were observed in fewer than two-tenths of 1 per cent of the cases. The rate of attempted suicides was slightly over a tenth of 1 per cent, and these involved psychiatric patients with histories of instability. Among those who took the drug simply as subjects in experiments there were no attempted suicides and the psychotic reactions occurred in fewer than a tenth of 1 per cent of the cases.

Recent reports do indicate that the incidence of bad reactions has been increasing, perhaps because more individuals have been taking the hallucinogens in settings that emphasize sensation-seeking or even deliberate social delinquency. Since under such circumstances there is usually no one in attendance who knows how to avert dangerous developments, a person in this situation may find himself facing an extremely frightening hallucination with no one present who can help him to recognize where the hallucination ends and reality begins. Yet the question of what is a proper setting is not a simple one. One of the criticisms of the Harvard experiments was that some were conducted in private homes rather than in a laboratory or clinical setting. The experimenters defended this as an attempt to provide a feeling of naturalness and "psychological safety." Such a setting, they hypothesized, should reduce the likelihood of negative reactions such as fear and hostility and increase the positive experiences. Controlled studies of this hypothesis have not been carried out, however.

Many psychiatrists and psychologists who have administered hallucinogens in a therapeutic setting claim specific benefits in the treatment of psychoneuroses, alcoholism and social delinquency. The published studies are difficult to evaluate because almost none have employed control groups. One summary of the available statistics on the treatment of alcoholism does indicate that about 50 per cent of the patients treated with a combination of psychotherapy and LSD abstained from alcohol for at least a year, compared with 30 per cent of the patients treated by psychotherapy alone.

In another recent study the results of psychological testing before and after

LSD therapy were comparable in most respects to the results obtained when conventional brief psychotherapy was employed. Single-treatment LSD therapy was significantly more effective, however, in relieving neurotic depression. If replicated, these results may provide an important basis for more directed study of the treatment of specific psychopathological conditions.

If the hallucinogens do have psychotherapeutic merit, it seems possible that they work by producing a shift in personal values. William James long ago noted that "the best cure for dipsomania is religiomania." There appear to be religious aspects of the drug experience that may bring about a change in behavior by causing a "change of heart." If this is so, one might be able to apply the hallucinogens in the service of moral regeneration while relying on more conventional techniques to give the patient insight into his habitual behavior patterns and motives.

In the light of the information now available about the uses and possible abuses of the hallucinogens, common sense surely decrees some form of social

control. In considering such control it should always be emphasized that the reaction to these drugs depends not only on their chemical properties and biological activity but also on the context in which they are taken, the meaning of the act and the personality and mood of the individual who takes them. If taking the drug is defined by the group or individual, or by society, as immoral or criminal, one can expect guilt and aggression and further social delinquency to result; if the aim is to help or to be helped, the experience may be therapeutic and strengthening; if the subject fears psychosis, the drug could induce psychosis. The hallucinogens, like so many other discoveries of man, are analogous to fire, which can burn down the house or spread through the house life-sustaining warmth. Purpose, planning and constructive control make the difference. The immediate research challenge presented by the hallucinogens is a practical question: Can ways be found to minimize or eliminate the hazards, and to identify and develop further the constructive potentialities, of these powerful drugs?

NATIVE AMERICAN CHURCH members take part in a peyote ceremony in Saskatchewan, Canada. Under the influence of the drug, they gaze into the fire as they pray and meditate.

# 31

# Pleasure Centers in the Brain

*October 1956*

The brain has been mapped in various ways by modern physiologists. They have located the sensory and motor systems and the seats of many kinds of behavior—centers where messages of sight, sound, touch and action are received and interpreted. Where, then, dwell the "higher feelings," such as love, fear, pain and pleasure? Up to three years ago the notion that the emotions had specific seats in the brain might have been dismissed as naive—

akin perhaps to medieval anatomy or phrenology. But recent research has brought a surprising turn of affairs. The brain does seem to have definite loci of pleasure and pain, and we shall review here the experiments which have led to this conclusion.

The classical mapping exploration of the brain ranged mainly over its broad, fissured roof—the cortex—and there localized the sensory and motor systems and other areas which seemed to control

most overt behavior. Other areas of the brain remained mostly unexplored, and comparatively little was known about their functions. Particularly mysterious was the series of structures lying along the mid-line of the brain from the roof down to the spinal cord, structures which include the hypothalamus and parts of the thalamus [*see diagram on page 230*]. It was believed that general functions of the brain might reside in these structures. But they were difficult

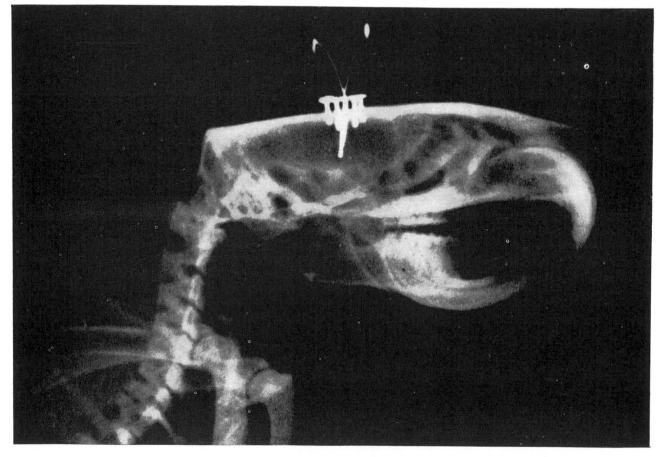

IMPLANTED ELECTRODES in the brain of a rat are shown in this X-ray photograph. The electrodes are held in a plastic carrier screwed to the skull. They can be used to give an electrical stimulus to the brain or to record electrical impulses generated by the brain.

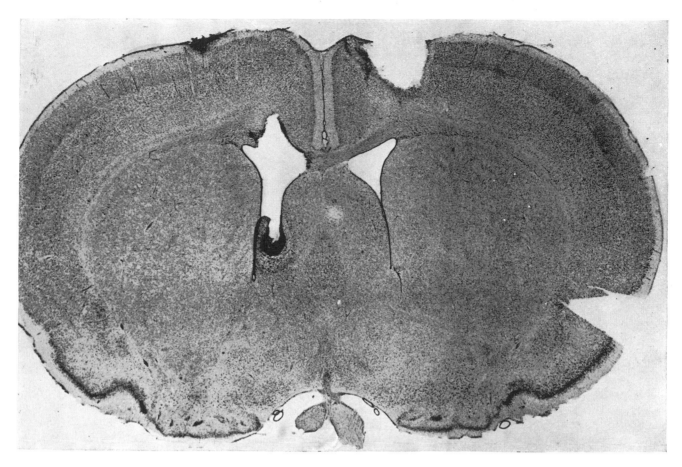

RAT'S BRAIN in a photomicrographic cross section shows a black spot to left of center, marking the point where electrical stimulus was applied. Such cross sections make it possible to tell exactly which center in the brain was involved in the animal's response.

to investigate, for two reasons. First, the structures were hard to get at. Most of them lie deep in the brain and could not be reached without damaging the brain, whereas the cortex could be explored by electrical stimulators and recording instruments touching the surface. Secondly, there was a lack of psychological tools for measuring the more general responses of an animal. It is easy to test an animal's reaction to stimulation of a motor center in the brain, for it takes the simple form of flexing a muscle, but how is one to measure an animal's feeling of pleasure?

The first difficulty was overcome by the development of an instrument for probing the brain. Basically the instrument is a very fine needle electrode which can be inserted to any point of the brain without damage. In the early experiments the brain of an animal could be probed only with some of its skull removed and while it was under anesthesia. But W. R. Hess in Zurich developed a method of studying the brain for longer periods and under more normal circumstances. The electrodes were inserted through the skull, fixed in position and left there; after the skin healed over the wound, the animal could be studied in its ordinary activities.

Using the earlier technique, H. W. Magoun and his collaborators at Northwestern University explored the region known as the "reticular system" in the lower part of the mid-brain [see page 230]. They showed that this system controls the sleep and wakefulness of animals. Stimulation of the system produced an "alert" electrical pattern, even from an anesthetized animal, and injury to nerve cells there produced more or less continuous sleep.

Hess, with his new technique, examined the hypothalamus and the region around the septum (the dividing membrane at the mid-line), which lie forward of the reticular system. He found that these parts of the brain play an important part in an animal's automatic protective behavior. In the rear section of the hypothalamus is a system which controls emergency responses that prepare the animal for fight or flight. Another system in the front part of the hypothalamus and in the septal area apparently controls rest, recovery, diges-

tion and elimination. In short, these studies seemed to localize the animal's brain responses in situations provoking fear, rage, escape or certain needs.

There remained an important part of the mid-line region of the brain which had not been explored and whose functions were still almost completely unknown. This area, comprising the upper portion of the middle system, seemed to be connected with smell, and to this day it is called the rhinencephalon, or "smell-brain." But the area appeared to receive messages from many organs of the body, and there were various other reasons to believe it was not concerned exclusively or even primarily with smell. As early as 1937 James W. Papez of Cornell University suggested that the rhinencephalon might control emotional experience and behavior. He based this speculation partly on the observation that rabies, which produces profound emotional upset, seems to attack parts of the rhinencephalon.

Such observations, then, constituted our knowledge of the areas of the brain until recently. Certain areas had

230

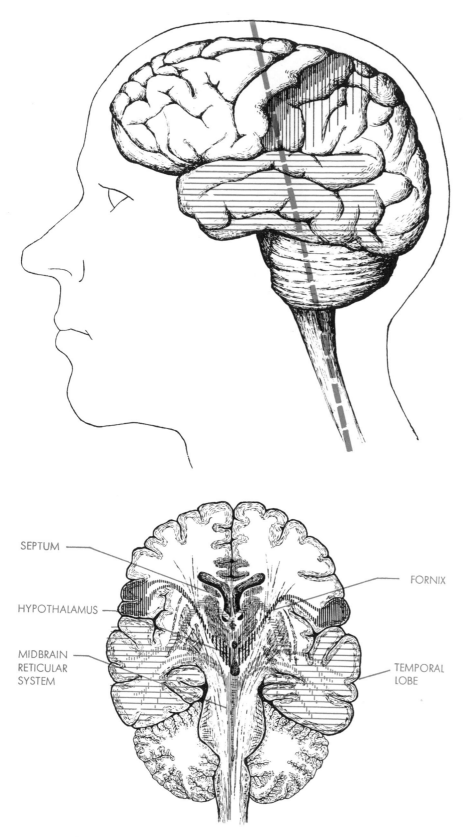

SEPTUM

HYPOTHALAMUS

MIDBRAIN
RETICULAR
SYSTEM

FORNIX

TEMPORAL
LOBE

**LOCATIONS OF FUNCTION** in the human brain are mapped in these two diagrams. The white areas in both diagrams comprise the motor system; the black crosshatched areas, the sensory system. Crosshatched in color are the "nonspecific" regions now found to be involved in motivation of behavior. The diagram at bottom shows the brain from behind, dissected along the heavy dashed line at top. The labels here identify the centers which correspond to those investigated in the rat. The fornix and parts of the temporal lobes, plus associated structures not labeled, together constitute the rhinencephalon or "smell-brain."

been found to be involved in various kinds of emotional behavior, but the evidence was only of a general nature. The prevailing view still held that the basic motivations—pain, pleasure and so on—probably involved excitation or activity of the whole brain.

Investigation of these matters in more detail became possible only after psychologists had developed methods for detecting and measuring positive emotional behavior—pleasure and the satisfaction of specific "wants." It was B. F. Skinner, the Harvard University experimental psychologist, who produced the needed refinement. He worked out a technique for measuring the rewarding effect of a stimulus (or the degree of satisfaction) in terms of the frequency with which an animal would perform an act which led to the reward. For example, the animal was placed in a bare box containing a lever it could manipulate. If it received no reward when it pressed the lever, the animal might perform this act perhaps five to 10 times an hour. But if it was rewarded with a pellet of food every time it worked the lever, then its rate of performing the act would rise to 100 or more times per hour. This increase in response frequency from five or 10 to 100 per hour provided a measure of the rewarding effect of the food. Other stimuli produce different response rates, and in each case the rise in rate seems to be a quite accurate measure of the reward value of the given stimulus.

With the help of Hess's technique for probing the brain and Skinner's for measuring motivation, we have been engaged in a series of experiments which began three years ago under the guidance of the psychologist D. O. Hebb at McGill University. At the beginning we planned to explore particularly the midbrain reticular system—the sleep-control area that had been investigated by Magoun.

Just before we began our own work, H. R. Delgado, W. W. Roberts and N. E. Miller at Yale University had undertaken a similar study. They had located an area in the lower part of the mid-line system where stimulation caused the animal to avoid the behavior that provoked the electrical stimulus. We wished to investigate positive as well as negative effects—that is, to learn whether stimulation of some areas might be sought rather than avoided by the animal.

We were not at first concerned to hit very specific points in the brain, and in fact in our early tests the electrodes did not always go to the particular areas in the mid-line system at which they were aimed. Our lack of aim turned out to be a fortunate happening for us. In one animal the electrode missed its target and landed not in the mid-brain reticular system but in a nerve pathway from the rhinencephalon. This led to an unexpected discovery.

In the test experiment we were using, the animal was placed in a large box with corners labeled A, B, C and D. Whenever the animal went to corner A, its brain was given a mild electric shock by the experimenter. When the test was performed on the animal with the electrode in the rhinencephalic nerve, it kept returning to corner A. After several such returns on the first day, it finally went to a different place and fell asleep. The next day, however, it seemed even more interested in corner A.

At this point we assumed that the stimulus must provoke curiosity; we did not yet think of it as a reward. Further experimentation on the same animal soon indicated, to our surprise, that its response to the stimulus was more than curiosity. On the second day, after the animal had acquired the habit of returning to corner A to be stimulated, we began trying to draw it away to corner B, giving it an electric shock whenever it took a step in that direction. Within a matter of five minutes the animal was in corner B. After this, the animal could be directed to almost any spot in the box at the will of the experimenter. Every step in the right direction was paid with a small shock; on arrival at the appointed place the animal received a longer series of shocks.

Next the animal was put on a T-shaped platform and stimulated if it turned right at the crossing of the T but not if it turned left. It soon learned to turn right every time. At this point we reversed the procedure, and the animal had to turn left in order to get a shock. With some guidance from the experimenter it eventually switched from the right to the left. We followed up with a test of the animal's response when it was hungry. Food was withheld for 24 hours. Then the animal was placed in a T both arms of which were baited with mash. The animal would receive the electric stimulus at a point halfway down the right arm. It learned to go there, and it always stopped at this point, never going on to the food at all!

After confirming this powerful effect of stimulation of brain areas by experiments with a series of animals, we set out to map the places in the brain where

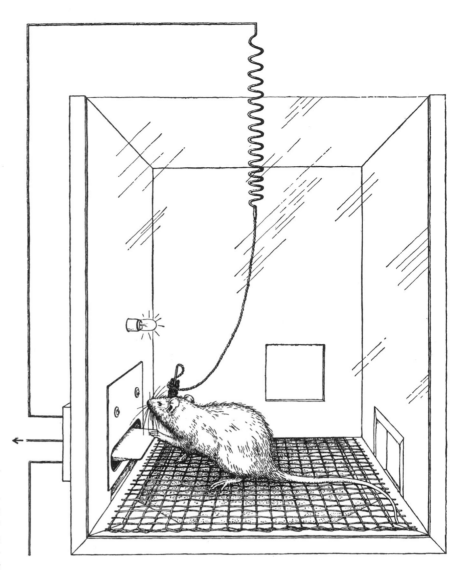

SELF-STIMULATION CIRCUIT is diagrammed here. When the rat presses on treadle it triggers an electric stimulus to its brain and simultaneously records action via wire at left.

such an effect could be obtained. We wanted to measure the strength of the effect in each place. Here Skinner's technique provided the means. By putting the animal in the "do-it-yourself" situation (*i.e.*, pressing a lever to stimulate its own brain) we could translate the animal's strength of "desire" into response frequency, which can be seen and measured.

The first animal in the Skinner box ended all doubts in our minds that electric stimulation applied to some parts of the brain could indeed provide reward for behavior. The test displayed the phenomenon in bold relief where anyone who wanted to look could see it. Left to itself in the apparatus, the animal (after about two to five minutes of learning) stimulated its own brain regularly about once every five seconds, taking a stimulus of a second or so every time. After 30 minutes the experimenter turned off the current, so that the animal's pressing of the lever no longer stimulated the brain. Under these conditions the animal pressed it about seven times and then went to sleep. We found that the test was repeatable as often as we cared to apply it. When the current was turned on and the animal was given one shock as an *hors d'oeuvre*, it would begin stimulating its brain again. When the electricity was turned off, it would try a few times and then go to sleep.

The current used to stimulate was ordinary house current reduced by a small transformer and then regulated between one and five volts by means of a potentiometer (a radio volume control). As the resistance in the brain was approximately 12,000 ohms, the current ranged from about .000083 to .000420 of an ampere. The shock lasted up to about a second, and the animal had to release the lever and press again to get more.

We now started to localize and quantify the rewarding effect in the brain by planting electrodes in all parts of the brain in large numbers of rats. Each rat had a pair of electrodes consisting of insulated silver wires a hundredth of an inch in diameter. The two stimulating tips were only about one 500th of an inch apart. During a test the animal was placed in a Skinner box designed to produce a chance response rate of about 10 to 25 bar-presses per hour. Each animal was given about six hours of testing with the electric current turned on and one hour with the current off. All responses were recorded automatically, and the animal was given a score on the basis of the amount of time it spent stimulating its brain.

When electrodes were implanted in the classical sensory and motor systems, response rates stayed at the chance level of 10 to 25 an hour. In most parts of the mid-line system, the response rates rose to levels of from 200 to 5,000 an hour, definitely indicative of a rewarding effect of the electric stimulus. But in some of the lower parts of the mid-line system there was an opposite effect: the animal would press the lever once and never go back. This indicated a punishing effect in those areas. They appeared to be the same areas where Delgado, Roberts and Miller at Yale also had discovered the avoidance effect—and where Hess and others had found responses of rage and escape.

The animals seemed to experience the strongest reward, or pleasure, from stimulation of areas of the hypothalamus and certain mid-brain nuclei—regions which Hess and others had found to be centers for control of digestive, sexual, excretory and similar processes. Animals with electrodes in these areas would stimulate themselves from 500 to 5,000 times per hour. In the rhinencephalon the effects were milder, producing self-stimulation at rates around 200 times per hour.

Electric stimulation in some of these regions actually appeared to be far more rewarding to the animals than an ordinary satisfier such as food. For example, hungry rats ran faster to reach an electric stimulator than they did to reach food. Indeed, a hungry animal often ignored available food in favor of the pleasure of stimulating itself electrically. Some rats with electrodes in these places stimulated their brains more than 2,000 times per hour for 24 consecutive hours!

Why is the electric stimulation so rewarding? We are currently exploring this question, working on the hypothesis that brain stimulation in these regions must excite some of the nerve cells that would be excited by satisfaction of the basic drives—hunger, sex, thirst and so forth. We have looked to see whether some parts of the "reward system" of the brain are specialized; that is, there may be one part for the hunger drive, another for the sex drive, etc.

In experiments on hunger, we have found that an animal's appetite for electric stimulation in some brain regions increases as hunger increases: the animal will respond much faster when hungry than when full. We are performing similar tests in other places in the brain with variations of thirst and sex hormones. We have already found that there are areas where the rewarding effects of a brain stimulus can be abolished by castration and restored by injections of testosterone.

Our present tentative conclusion is that emotional and motivational mechanisms can indeed be localized in the brain; that certain portions of the brain are sensitive to each of the basic drives. Strong electrical stimulation of these areas seems to be even more satisfying than the usual rewards of food, etc. This finding contradicts the long-held theory that strong excitation in the brain means punishment. In some areas of the brain it means reward.

The main question for future research is to determine how the excited "reward" cells act upon the specific sensory-motor systems to intensify the rewarded

RAT IS CONNECTED to electrical circuit by a plug which can be disconnected to free the animal during rest periods. Presence of electrodes does not pain or discommode the rat.

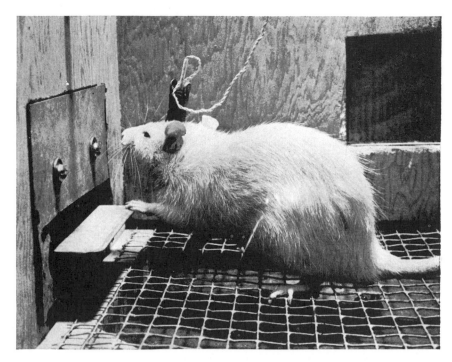

RAT SEEKS STIMULUS as it places its paw on the treadle. Some of the animals have been seen to stimulate themselves for 24 hours without rest and as often as 5,000 times an hour.

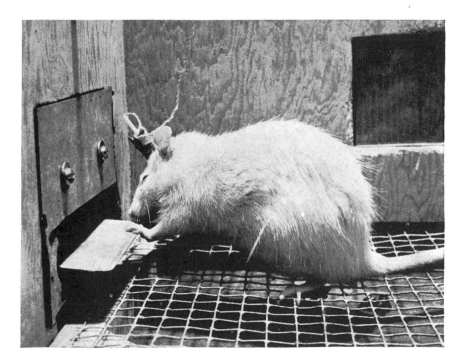

RAT FEELS STIMULUS as it presses on treadle. Pulse lasts less than a second; the current is less than .0005 ampere. The animal must release lever and press again to renew the stimulus.

behavior.

At the moment we are using the self-stimulating technique to learn whether drugs will selectively affect the various motivational centers of the brain. We hope, for example, that we may eventually find one drug that will raise or lower thresholds in the hunger system, another for the sex-drive system, and so forth. Such drugs would allow control of psy-chological disorders caused by surfeits or deficits in motivational conditions.

Enough of the brain-stimulating work has been repeated on monkeys by J. V. Brady and J. C. Lilly (who work in different laboratories in Washington, D. C.) to indicate that our general con-clusions can very likely be generalized eventually to human beings—with modi-fications, of course.

# 32

# An Indictment of
# the Wide Use of Pesticides

LAMONT C. COLE

*December 1962*

A review of *Silent Spring*, by Rachael Carson. Boston: Houghton Mifflin, 1962.

As an ecologist I am glad this provocative book has been written. That is not to say I consider it a fair and impartial appraisal of all the evidence. On the contrary, it is a highly partisan selection of examples and interpretations that support the author's thesis. The fact remains that the extreme opposite has been impressed on the public by skilled professional molders of public opinion. It is surely time for laymen to take an objective interest in what man is doing to alter his environment, and *Silent Spring* provides many dreadful examples of how the environment has been damaged by the indiscriminate application of chemicals. Miss Carson gives little attention to poisoning from such chemicals as radioactive fallout, detergents, industrial poisons and food additives; she concentrates her attack on the pesticides, primarily insecticides and secondarily herbicides.

Miss Carson and her publisher are to be praised for devoting 55 pages of a nontechnical book to a list of sources so that the reader can look up the original accounts and judge the evidence on which the author bases her statements. Errors of fact are so infrequent, trivial and irrelevant to the main theme that it would be ungallant to dwell on them. I shall merely express the hope that chemists will not be shocked into abandoning the book when, on page 16, they find rotenone and pyrethrum listed among the "simpler inorganic insecticides of prewar days." Actually it might be interesting to reverse the argument and consider the proposition that the complex structure of pyrethrum, which has frustrated the analysis of the ablest organic chemists, is also responsible for frustrating the abilities of insects to develop resistance.

As Miss Carson describes it, all over the world, but particularly in the U.S., agricultural land, forests, gardens, lawns, roadsides and even urban centers are being dusted and sprayed with an assortment of violent poisons for the principal purpose of increasing agricultural production in a land already plagued by overproduction. The instigators of this abuse are chemical manufacturers seeking increased markets and a Department of Agriculture seeking increased appropriations.

The rain of poison seldom accomplishes its purpose of eradicating the target organism; indeed, the organism often becomes more destructive than before. If the undesirable species is temporarily controlled, it is likely to be replaced by an even more destructive form. People and game and domestic animals may suffer acute poisoning from contact with the drifting pesticide or objects coated with it. Those who escape this fate accumulate the toxin in their body fat, from which it may be released by dieting at a later date. Even without detectable poisoning our livers may be suffering impairment of the ability to detoxify other substances, and other tissues may be suffering damage that will subsequently appear as cancer or as defects in succeeding generations.

Meanwhile the poisoned insects are poisoning insectivorous birds or rendering them sterile. When the sprayed vegetation becomes litter on the soil, the poison is absorbed by earthworms. The worms then become poisonous to robins and other birds, or they may pass the toxins on to moles and shrews, which in turn become poisonous to predators such as hawks and owls. Residues get into water and kill stock drinking it; they also kill fish via the organisms on which they feed. Residues of different toxins react in the water to produce mixtures "that no responsible chemist would think of combining in his laboratory." This

witches' brew enters the ground water, rendering the soil toxic for years and having little-understood effects on the organisms that are essential to soil fertility. It percolates into underground reservoirs, from which it may emerge years later. Eventually it reaches rivers and estuaries, where additional damage is possible. Young shrimp are very susceptible, but snails, the hosts of some dangerous parasites, thrive on their relative immunity and the food provided by the remains of their more vulnerable neighbors.

Anyone will recognize this as an argument that is bound to arouse the ire of powerful elements in our society. Indeed, the counterattack has already started, prompted by the prepublication of about a third of *Silent Spring* in *The New Yorker*. A reviewer for *Time* refers to Miss Carson's "emotional and inaccurate outburst" and, without citing evidence, proclaims the merits of pesticides in a statement with which, in my opinion, no responsible scientist would want to associate himself. Similarly, a magazine published by a major chemical manufacturer describes the "Desolate Year" that would overtake the U.S. if we should have to do without chemical pesticides. As one reads this skillful fantasy it is easy to become persuaded that years like those just before World War II could not possibly have occurred: no chlorinated hydrocarbons, no organic phosphates, payments to farmers to reduce production and still crop surpluses!

Where does the truth lie? It lies in part with Miss Carson, who presents enough solidly established facts to justify some alarm. It has been a long time since any of us has eaten a meal free of pesticides or has deposited any body fat uncontaminated by them. People have been fatally poisoned as well as wildlife and domestic animals. The long-term effects of these chemicals are quite unknown; some biologists, including myself, have shied away from 2,4-D and its relatives for 15 years on the grounds that it is only discreet to regard a known mutagen as a potential carcinogen. The campaigns led by the Department of Agriculture against the gypsy moth in the East and the fire ant in the South are sadly reminiscent of the old Western conflict over the control of coyotes by poisoning.

On the other hand, insecticides that have a residual action have led man to some notable triumphs. Typhus is such a scourge that Hans Zinsser, in his fascinating book *Rats, Lice and History,* was able to develop a strong case for its governing role in human history. When, in 1944, DDT stopped an incipient typhus epidemic in Italy in its tracks, many of us then in Government service regarded it as the first authentic triumph of man over epidemic disease since smallpox vaccination. The road ahead looked bright, because we realized that

the almost superhuman effort and exorbitant expense borne by the Rockefeller Foundation and the people of Brazil to exterminate the dangerous malarial mosquito *Anopheles gambiae* with pyrethrum could not often be duplicated. But soon insects began to develop resistance to the new insecticides. So began a race in which chemists sought to develop new insecticides faster than the insects could develop resistance.

In such circumstances it was out of

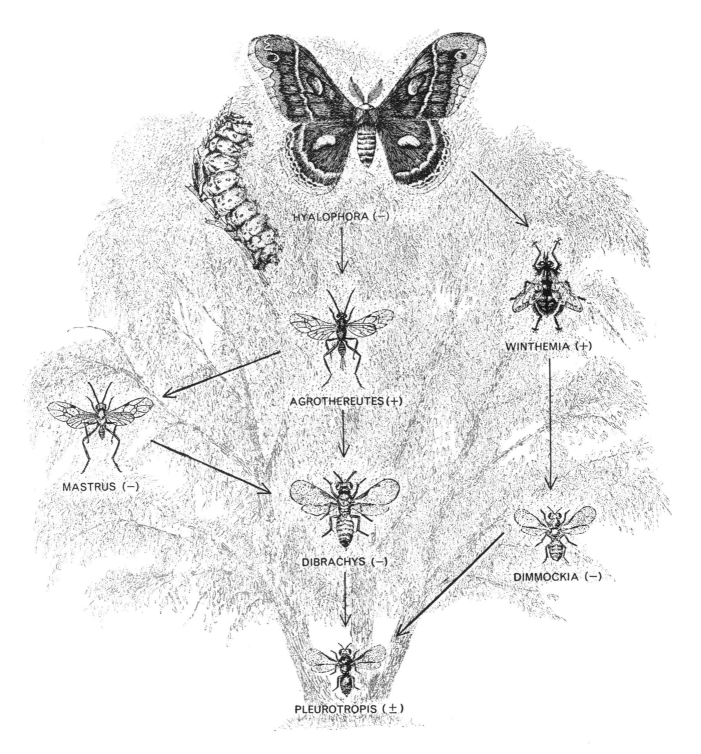

*A simple ecological system in which it is difficult to distinguish friend* ( + ) *from enemy* ( — )

the question to test a chemical for long-term chronic effects before putting it into use. Many sincere and competent individuals used their best judgment in recommending precautions and, as Miss Carson admits, a large proportion of the unfortunate cases of poisoning have stemmed from improper use of and disregard for precautions. I suspect that the inevitable way to progress for man, as for nature, is to try new things in an almost haphazard manner, discarding the failures and building on the successes. In man's case there will always be some who deplore certain aspects of progress; to choose a chemical example at random, I myself deplore the curing of hams and bacon by a process more akin to embalming than to anything Grandfather did in his smokehouse. But in the long run, unless a new suggestion can be discarded on the basis of prior knowledge, the only way to evaluate it is to try it out. I am just old enough to recall campaigns against poisoning by iodized salt, ethyl gasoline and aluminum cooking utensils. On the other hand, I also recall the "testimonials" of persons cured by medicine-show nostrums.

Although Miss Carson is fully aware that the basis for control measures lies in the field of ecology, both she and her opponents have failed to place the controversy in a genuine ecological context. The "balance of nature," to which she refers several times, is an obsolete concept among ecologists. It is nonetheless true that the consequences of removing even a single species from a biotic community may be drastic. Moreover, as the development of resistance to insecticides has shown, alteration of the physical and chemical environment can alter the course of evolution. In the following paragraphs I shall try to suggest the complexity of the problem as it appears to an ecologist, hoping that this will convince the reader that nobody knows enough to adopt an extreme position in the matter. I shall use insects in my argument because I know most about them, but among ecologists there are weed fanciers who could make the same points with botanical examples.

Miss Carson's book appeals explicitly to cat lovers, bird watchers and sportsmen but contains little to comfort insect fanciers. She fails to mention that bees are probably less threatened by the modern insecticides than they were by the older arsenicals. If it were possible to exterminate insects without direct damage to man and other animals, I am not sure Miss Carson would not grieve chiefly for the starving birds and fishes that depend on insects for food. I too like woodpeckers, nuthatches and trout, but insects are more important to man. It is true that our modern economy could survive without silk, honey, beeswax and shellac, and I do not expect the layman to be impressed by the fact that biological science could not have achieved its present state so rapidly if our ancestors had succeeded in exterminating the fruit fly *Drosophila*.

But what would happen to the world of plant life without insects? The mere recognition that flowers on flowering plants evolved as adaptations for pollination, usually by insects, suggests the magnitude of the adjustment that would have to take place in an insect-free world. Agriculturists could still grow potatoes and corn, and it is possible, although expensive, to pollinate apples by hand. Cherries might conceivably become an expensive delicacy, as truffles are today, so that it would be economically feasible to pollinate the trees by hand. But it is unthinkable that man deprived of bees would attempt to pollinate alfalfa and clover. Substitutes for these crops would have to be found, and we would probably give up growing cotton and many fruits and vegetables.

When trees shed their leaves, and whenever vegetation dies, organic matter is added to the surface and below the surface of the soil. Much of it consists of decay-resistant materials such as cellulose and lignin. This plant debris must be broken down into simpler compounds that can be reused. Continued accumulation, such as occurs in peat bogs, would be highly detrimental to plant life. Therefore the rate of decomposition of organic matter must on the average equal the rate of accumulation, which in a dense forest may amount to several tons per acre per year. Insects, and their relatives the mites, play a large and perhaps indispensable role in this process. The despised termite is an essential part of tropical biotic communities. We are continuing to learn about these processes, but we do not yet know enough to predict with confidence the long-term effects of tampering with the insect populations of the soil. Here I am with Miss Carson. The chemical residues disturb me, and blind interference with such a complex system impresses me as being irresponsible.

I think, however, that Miss Carson misinterprets the evolutionary significance of the development of hereditary resistance to insecticides. Resistance comes about because members of the population differ in susceptibility to the toxin and, unless treatment is intensive enough to cause extinction, it is the individuals with the highest intrinsic resistance that survive to become the progenitors of the next generation. It will normally be impossible to exterminate a population with insecticides except in a confined situation such as one would encounter on a boat or a small island. The edge of the treated area will usually provide a region of low dosage where conditions are right for selecting resistant strains.

As Miss Carson interprets the situation, "spraying kills off the weaklings . . . it is the 'tough' insects that survive chemical attack." On this premise she reasons that we must forever seek more potent poisons. This line of reasoning is almost certainly incorrect. When members of a population are selected for one particular attribute, the resultant genetic constitution automatically changes other characteristics. The new form can be expected to be in some respects "tougher" and in others weaker, perhaps much weaker. Genes that cause damage can spread as long as incidental benefits outweigh the harm.

A striking example of this phenomenon is furnished by the very harmful gene causing "sickle cell" hemoglobin in man. Individuals receiving this gene from both parents are doomed to sterility and premature death. Those receiving it from only one parent are subnormal in their ability to tolerate reduced oxygen tension; they would not make good mountaineers, airplane pilots or divers, and it is possible that they are handicapped in their normal activities. Nevertheless, in West Africa native populations have been found in which 40 per cent or more of the people carry the sickle-cell gene. The reason for this high incidence is that the gene conveys resistance to the often fatal falciparum malaria. The resistance benefits the population enough to offset the damaging effect of premature death for a small proportion of the children.

Our domestic animals possess traits that have been selected to make them superior to their wild ancestors. But nobody can doubt that these animals are in other ways inferior; most of them probably could not survive in the wild. Similarly, houseflies that are DDT-resistant seem to grow more slowly and to be less prolific than susceptible flies. The resistant flies are adapted to an environment containing DDT (and often related compounds), but they appear to be inferior to susceptible flies in their adaptations to a DDT-free environment. Thus a new chemical that is unrelated to DDT either structurally or in mode of action is likely to be more effective against DDT-resistant insects than against susceptible strains. I do not for a moment believe that the chemicals are producing superinsects.

Ecologists have more fundamental reservations about pest control than anything I have mentioned so far. These involve, first, the basic question of what regulates the size of any population in nature. The details are a matter of current controversy, but I think that if I choose my words carefully I can convey the essence of the problem without upsetting colleagues in any of the opposing camps.

The basic problem lies in this theorem: All sorts of environmental factors can affect fertility and mortality, but only certain types of factors can *regulate* population size in the sense of holding it within definite bounds. To do this a factor must exhibit what in electronics is called negative feedback. The average population member must be more liable to suffer death or sterility from the factor when he is a member of a large population than when he belongs to a small population.

An analogy may help to clarify this point. Imagine that you fill a bucket with water halfway to the top and set it outdoors with the expectation of having the water level become stabilized by a balance between precipitation and evaporation. For the population of molecules in the bucket these processes represent births and deaths respectively. Any student of probability theory can tell you that if these are the only processes acting on the population, it is 100 per cent certain that, no matter what the climate, at some later date the container will either overflow or be completely dry; no "regulation" is possible. If, however, the container leaks, and leaks faster the fuller it gets, a more or less stable water level may be attainable. The leaks represent a negative feedback mechanism that makes a molecule, on the average, more liable to loss by leakage as the number of molecules in the container increases.

Now, ecologists are universally agreed that shortages of essential environmental resources such as food, space and suitable nesting sites have this negative feedback property and set upper bounds to population size. Social interactions have this quality for some species in which the individuals cannot tolerate crowding beyond a certain point. It is also quite generally agreed that communicable diseases and parasites spread more readily when populations are dense and so have negative feedback qualities.

The situation for predators is much less clear. Often the predator is merely living on surplus individuals that have been forced by population pressures out of the most secure parts of the habitat. By and large the muskrats that are eaten by minks would die from exposure to the weather or something else if minks were not present. That is why ecologists hesitate to advocate spiders or "insect friends" such as the praying mantis or "our feathered friends" for pest control. No matter how friendly the control organisms are, it would take a great deal of careful study in any particular case to tell if their presence means much one way or the other to the populations on which they feed.

Predators are seldom able to increase in numbers as rapidly as their prey. If the cats in a farmer's barn had to depend on mice for food, they could never control the mouse population. They are effective only because of what has been called subsidized predation; the farmer subsidizes the cats to maintain a cat population large enough to turn to mice and exert control when the mouse population increases. In natural situations it is impossible to tell without careful quantitative study the extent to which something such as subsidized predation may be operating. When we subject an area to a nonselective insecticide, we run the risk of knocking out some predator currently subsisting on other foods but maintaining numbers adequate to suppress an incipient outbreak of a destructive form. No farmer would think of trying to control mice in his barn by indiscriminately slaughtering mice and cats, but we risk doing something analogous when the application of insecticide is routine and not based on a demonstrated immediate need.

Some ecologists contend that poisons lack the feedback qualities necessary to regulate population size, except when man creates the feedback by poisoning more intensively when populations increase. In this view poisoning should be carefully integrated with continuous study of population trends. I prefer not to commit myself as being for or against this generalization, but once again it illustrates the complexity of the problem of pest control. Most users and distributors of insecticides have probably never even heard this argument.

The situation becomes even more complex when we look into the network of interactions of a given species and parasites that really do have the ability to regulate population size. Numerous cases have been observed of pest species made more abundant by spraying because their parasites were destroyed. Miss Carson gives examples, and the papers by Paul DeBach and W. E. Ripper that she cites explore the matter more thoroughly.

A simplified description of an actual multispecies system may help to clarify the problem. The example I present here was worked out by F. L. Marsh and published in *Ecology* in 1937. It is surely typical of relatively simple natural situations and is unusual only in having received careful study.

The larvae of the moth *Hyalophora cecropia* feed on the leaves of trees. It is not necessarily harmful to a tree to have insects feeding on it, any more than it is to have a gardener pruning it, but to simplify the argument we will assume that *Hyalophora* is an "enemy" of the trees.

It is also not necessarily harmful to a population to have parasites and predators thinning its numbers; this may sometimes be highly beneficial. But again let us regard any species that reduces the numbers of *Hyalophora* as a "friend" of the trees. Marsh found two species of woodpeckers, two species of mice and two insects—the tachinid fly *Winthemia* and the ichneumon wasp *Agrothereutes*—to be important in this respect. I shall say no more about the vertebrates but will describe the relations of the two "beneficial" insects and four additional species of insects. All six are "parasitoids" in which the females skillfully seek out immature stages of a host species and arrange to have their own young devour those of the host.

In the accompanying illustration [*page 235*] the arrows indicate the direction of flow of matter and energy: from tree to moth to consumers of the moth, and so on. Since we are putting ourselves in the position of partisans of the tree, we use plus signs to label influences that benefit the tree and minus signs to represent detrimental influences. Thus *Hyalophora* is a "harmful" insect and *Winthemia* and *Agrothereutes* are "beneficial."

*Winthemia* serves as a host for the eulophid wasp *Dimmockia*, which is here classed as harmful because it destroys a beneficial insect. *Agrothereutes* is host to no less than five parasitoids, but only two of these need concern us here: the pteromalid wasp *Dibrachys* and the ichneumon wasp *Mastrus*. These are by definition harmful because they destroy the beneficial *Agrothereutes*. But *Dibrachys* also attacks *Mastrus* and in this respect is beneficial. Evidently

*Dibrachys* can be either good or bad for the trees depending on which host is most heavily attacked, and this may depend on which is most abundant.

Even that is not the end. Another eulophid wasp, *Pleurotropis,* attacks both *Dimmockia* and *Dibrachys.* Its influence on *Dimmockia* must benefit the tree, but what can we say of its influence on *Dibrachys?* This is beneficial or not depending on what *Dibrachys* is.

The diagram illustrates on a small scale the sort of thing Miss Carson has in mind when she refers to the balance of nature. Most ecologists today object to using a static term like "balance" to describe such a dynamic system of whirling wheels within wheels. Monkey wrenches thrown at random into such an intricate machine can be expected to necessitate repairs far beyond any mere shifting of weight to restore a balance.

What should one expect from spraying a system such as this? To speculate a bit, flies tend to be more susceptible than wasps to some of the modern insecticides; as a result *Winthemia* might be the first to feel the effects. But we have decided that this is a beneficial form, so the spray damages the tree in this case. Marsh found that *Dibrachys,* unlike the others, has some individuals present throughout the year in the pupal stage, which is likely to be resistant to poisons. Therefore if the spray is applied at the wrong time, *Dibrachys,* which we have been unable to classify as good or bad, is likely to be the least affected. In the unlikely event that all the insects should be killed, the tree would be deprived of insect protectors and would be wide open to attack by the moths flying the next year.

This example, although simplified to the point of being unnatural, illustrates the difficulties of foreseeing the consequences of control measures. The suburbanite who goes to the store and buys something "to stop the bugs from eating up my trees" is likely to be in for a rude shock. Even if he does not poison his lawn, his cat, the songbirds or himself, he may unwittingly set in motion a train of events with no predictable destination. Perhaps it will be just as well if the insects are resistant.

To summarize, man is doing a great deal to alter the face of the earth and much of what he is doing is alarming to biologists. In recent years a number of able writers have brought some aspects of this threatening situation to public attention. Miss Carson is continuing the trend with an attack on chemical pesticides that is so vigorous it is easy to overlook a rare hedging sentence. (On page 12, for example, she disavows any contention that the chemicals should never be used.) She does not, however, convey an appreciation of the really great difficulty of the general problem. She underplays the importance of insects to man and probably overstates the importance of birds and other forms with many human admirers. But what I interpret as bias and oversimplification may be just what it takes to write a best seller, and Miss Carson has already proved that she knows how to do that. If the message of *Silent Spring* is widely enough read and discussed, it may help us toward a much needed reappraisal of current policies and practices.

# 33

# Third-Generation Pesticides

CARROLL M. WILLIAMS

*July 1967*

Man's efforts to control harmful insects with pesticides have encountered two intractable difficulties. The first is that the pesticides developed up to now have been too broad in their effect. They have been toxic not only to the pests at which they were aimed but also to other insects. Moreover, by persisting in the environment—and sometimes even increasing in concentration as they are passed along the food chain—they have presented a hazard to other organisms, including man. The second difficulty is that insects have shown a remarkable ability to develop resistance to pesticides.

Plainly the ideal approach would be to find agents that are highly specific in their effect, attacking only insects that are regarded as pests, and that remain effective because the insects cannot acquire resistance to them. Recent findings indicate that the possibility of achieving success along these lines is much more likely than it seemed a few years ago. The central idea embodied in these findings is that a harmful species of insect can be attacked with its own hormones.

Insects, according to the latest estimates, comprise about three million species—far more than all other animal and plant species combined. The number of individual insects alive at any one time is thought to be about a billion billion ($10^{18}$). Of this vast multitude 99.9 percent are from the human point of view either innocuous or downright helpful. A few are indispensable; one need think only of the role of bees in pollination.

The troublemakers are the other .1 percent, amounting to about 3,000 species. They are the agricultural pests and the vectors of human and animal disease. Those that transmit human disease are the most troublesome; they have joined with the bacteria, viruses and protozoa in what has sometimes seemed like a grand conspiracy to exterminate man, or at least to keep him in a state of perpetual ill health.

The fact that the human species is still here is an abiding mystery. Presumably the answer lies in changes in the genetic makeup of man. The example of sickle-cell anemia is instructive. The presence of sickle-shaped red blood cells in a person's blood can give rise to a serious form of anemia, but it also confers resistance to malaria. The sickle-cell trait (which does not necessarily lead to sickle-cell anemia) is appreciably more common in Negroes than in members of other populations. Investigations have suggested that the sickle cell is a genetic mutation that occurred long ago in malarial regions of Africa. Apparently attrition by malaria-carrying mosquitoes provoked countermeasures deep within the genes of primitive men.

The evolution of a genetic defense, however, takes many generations and entails many deaths. It was only in comparatively recent times that man found an alternative answer by learning to combat the insects with chemistry. He did so by inventing what can be called the first-generation pesticides: kerosene to coat the ponds, arsenate of lead to poison the pests that chew, nicotine and rotenone for the pests that suck.

Only 25 years ago did man devise the far more potent weapon that was the first of the second-generation pesticides. The weapon was dichlorodiphenyltrichloroethane, or DDT. It descended on the noxious insects like an avenging angel. On contact with it mosquitoes, flies, beetles—almost all the insects—were stricken with what might be called the "DDT's." They went into a tailspin, buzzed around upside down for an hour or so and then dropped dead.

The age-old battle with the insects appeared to have been won. We had the stuff to do them in—or so we thought. A few wise men warned that we were living in a fool's paradise and that the insects would soon become resistant to DDT, just as the bacteria had managed to develop a resistance to the challenge of sulfanilamide. That is just what happened. Within a few years the mosquitoes, lice, houseflies and other noxious insects were taking DDT in their stride. Soon they were metabolizing it, then they became addicted to it and were therefore in a position to try harder.

Fortunately the breach was plugged by the chemical industry, which had come to realize that killing insects was —in more ways than one—a formula for

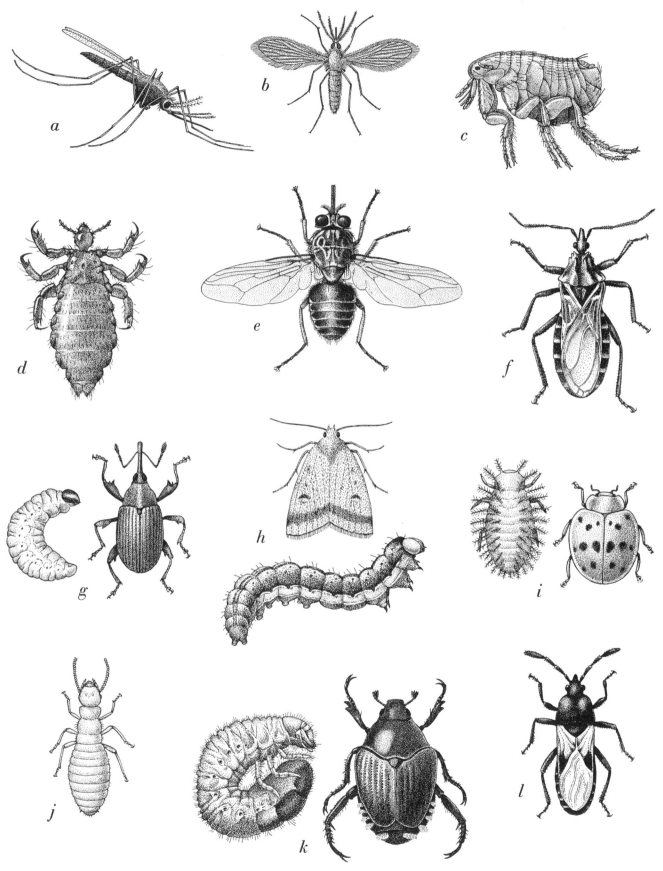

INSECT PESTS that might be controlled by third-generation pesticides include some 3,000 species, of which 12 important examples are shown here. Six (*a–f*) transmit diseases to human beings; the other six are agricultural pests. The disease-carriers, together with the major disease each transmits, are (*a*) the *Anopheles* mosquito, malaria; (*b*) the sand fly, leishmaniasis; (*c*) the rat flea, plague; (*d*) the body louse, typhus; (*e*) the tsetse fly, sleeping sickness, and (*f*) the kissing bug, Chagas' disease. The agricultural pests, four of which are depicted in both larval and adult form, are (*g*) the boll weevil; (*h*) the corn earworm; (*i*) the Mexican bean beetle; (*j*) the termite; (*k*) the Japanese beetle, and (*l*) the chinch bug. The species in the illustration are not drawn to the same scale.

getting along in the world. Organic chemists began a race with the insects. In most cases it was not a very long race, because the insects soon evolved an insensitivity to whatever the chemists had produced. The chemists, redoubling their efforts, synthesized a steady stream of second-generation pesticides. By 1966 the sales of such pesticides had risen to a level of $500 million a year in the U.S. alone.

Coincident with the steady rise in the output of pesticides has come a growing realization that their blunderbuss toxicity can be dangerous. The problem has attracted widespread public attention since the late Rachel Carson fervently described in *The Silent Spring* some actual and potential consequences of this toxicity. Although the attention thus aroused has resulted in a few attempts to exercise care in the application of pesticides, the problem cannot really be solved with the substances now in use.

The rapid evolution of resistance to pesticides is perhaps more critical. For example, the world's most serious disease in terms of the number of people afflicted continues to be malaria, which is transmitted by the *Anopheles* mosquito—an insect that has become completely resistant to DDT. (Meanwhile the protozoon that actually causes the disease is itself evolving strains resistant to antimalaria drugs.)

A second instance has been presented recently in Vietnam by an outbreak of plague, the dreaded disease that is conveyed from rat to man by fleas. In this case the fleas have become resistant to pesticides. Other resistant insects that are agricultural pests continue to take a heavy toll of the world's dwindling food supply from the moment the seed is planted until long after the crop is harvested. Here again we are confronted by an emergency situation that the old technology can scarcely handle.

The new approach that promises a way out of these difficulties has emerged during the past decade from basic studies of insect physiology. The prime candidate for developing third-generation pesticides is the juvenile hormone that all insects secrete at certain stages in their lives. It is one of the three internal secretions used by insects to regulate growth and metamorphosis from larva to pupa to adult. In the living insect the juvenile hormone is synthesized by the corpora allata, two tiny glands in the head. The corpora allata are also responsible for regulating the flow of the hormone into the blood.

At certain stages the hormone must be

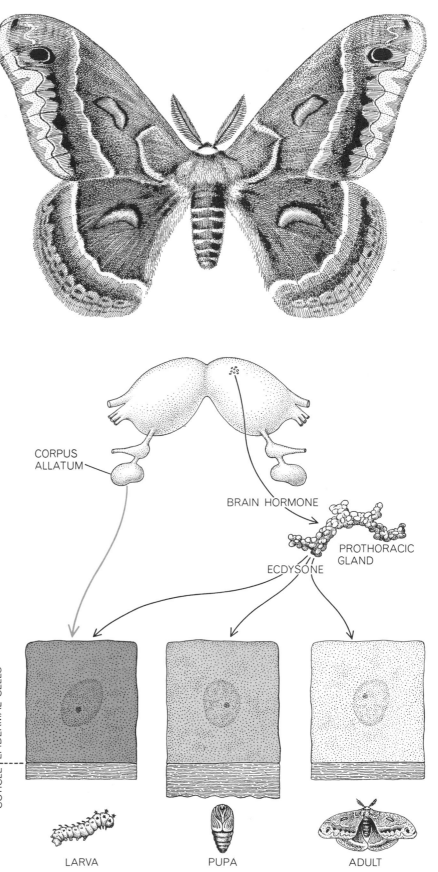

**HORMONAL ACTIVITY** in a Cecropia moth is outlined. Juvenile hormone (*color*) comes from the corpora allata, two small glands in the head; a second substance, brain hormone, stimulates the prothoracic glands to secrete ecdysone, which initiates the molts through which a larva passes. Juvenile hormone controls the larval forms and at later stages must be in low concentration or absent; if applied then, it deranges insect's normal development. The illustration is partly based on one by Howard A. Schneiderman and Lawrence I. Gilbert.

CHEMICAL STRUCTURES of the Cecropia juvenile hormone (*left*), isolated this year by Herbert Röller and his colleagues at the University of Wisconsin, and of a synthetic analogue (*right*) made in 1965 by W. S. Bowers and others in the U.S. Department of Agriculture show close similarity. Carbon atoms, joined to one or two hydrogen atoms, occupy each angle in the backbone of the molecules; letters show the structure at terminals and branches.

JUVENILE HORMONE ACTIVITY has been found in various substances not secreted by insects. One (*left*) is a material synthesized by M. Romanuk and his associates in Czechoslovakia. The other (*right*), isolated and identified by Bowers and his colleagues, is the "paper factor" found in the balsam fir. The paper factor has a strong juvenile hormone effect on only one family of insects, exemplified by the European bug *Pyrrhocoris apterus*.

secreted; at certain other stages it must be absent or the insect will develop abnormally [*see illustration on preceding page*]. For example, an immature larva has an absolute requirement for juvenile hormone if it is to progress through the usual larval stages. Then, in order for a mature larva to metamorphose into a sexually mature adult, the flow of hormone must stop. Still later, after the adult is fully formed, juvenile hormone must again be secreted.

The role of juvenile hormone in larval development has been established for several years. Recent studies at Harvard University by Lynn M. Riddiford and the Czechoslovakian biologist Karel Sláma have resulted in a surprising additional finding. It is that juvenile hormone must be absent from insect eggs for the eggs to undergo normal embryonic development.

The periods when the hormone must be absent are the Achilles' heel of insects. If the eggs or the insects come into contact with the hormone at these times, the hormone readily enters them and provokes a lethal derangement of further development. The result is that the eggs fail to hatch or the immature insects die without reproducing.

Juvenile hormone is an insect invention that, according to present knowledge, has no effect on other forms of life. Therefore the promise is that third-generation pesticides can zero in on insects to the exclusion of other plants and animals. (Even for the insects juvenile hormone is not a toxic material in the usual sense of the word. Instead of killing, it derails the normal mechanisms of development and causes the insects to kill themselves.) A further advantage is self-evident: insects will not find it easy to evolve a resistance or an insensitivity to their own hormone without automatically committing suicide.

The potentialities of juvenile hormone as an insecticide were recognized 12 years ago in experiments performed on the first active preparation of the hormone: a golden oil extracted with ether from male Cecropia moths. Strange to say, the male Cecropia and the male of its close relative the Cynthia moth remain to this day the only insects from which one can extract the hormone. Therefore tens of thousands of the moths have been required for the experimental work with juvenile hormone; the need has been met by a small but thriving industry that rears the silkworms.

No one expected Cecropia moths to supply the tons of hormone that would be required for use as an insecticide. Obviously the hormone would have to be synthesized. That could not be done, however, until the hormone had been isolated from the golden oil and identified.

Within the past few months the difficult goals of isolating and identifying the hormone have at last been attained by a team of workers headed by Herbert Röller of the University of Wisconsin. The juvenile hormone has the empirical formula $C_{18}H_{36}O_2$, corresponding to a molecular weight of 284. It proves to be the methyl ester of the epoxide of a previously unknown fatty-acid derivative [*see upper illustration on this page*]. The apparent simplicity of the molecule is deceptive. It has two double bonds and an oxirane ring (the small triangle at lower left in the molecular diagram), and it can exist in 16 different molecular configurations. Only one of these can be the authentic hormone. With two ethyl groups ($CH_2 \cdot CH_3$) attached to carbons No. 7 and 11, the synthesis of the hormone from any known terpenoid is impossible.

The pure hormone is extraordinarily active. Tests the Wisconsin investigators have carried out with mealworms suggest that one gram of the hormone would result in the death of about a billion of these insects.

A few years before Röller and his colleagues worked out the structure of the authentic hormone, investigators at several laboratories had synthesized a number of substances with impressive juvenile hormone activity. The most potent of the materials appears to be a crude mixture that John H. Law, now at the University of Chicago, prepared by a simple one-step process in which hydrogen chloride gas was bubbled through an alcoholic solution of farnesenic acid. Without any purification this mixture was 1,000 times more active than crude Cecropia oil and fully effective in killing all kinds of insects.

One of the six active components of Law's mixture has recently been identified and synthesized by a group of workers headed by M. Romaňuk of the Czechoslovak Academy of Sciences. Romaňuk and his associates estimate that from 10 to 100 grams of the material would clear all the insects from 2½ acres. Law's original mixture is of course even more potent, and so there is much interest in its other five components.

Another interesting development that preceded the isolation and identification of true juvenile hormone involved a team of investigators under W. S. Bowers of the U.S. Department of Agriculture's laboratory at Beltsville, Md. Bowers and his colleagues prepared an analogue of juvenile hormone that, as can be seen in the accompanying illustration [*top of opposite page*], differed by only two carbon atoms from the authentic Cecropia hormone (whose structure was then, of course, unknown). In terms of the dosage required it appears that the Beltsville compound is about 2 percent as active as Law's mixture and about .02 percent as active as the pure Cecropia hormone.

All the materials I have mentioned are selective in the sense of killing only insects. They leave unsolved, however, the problem of discriminating between the .1 percent of insects that qualify as pests and the 99.9 percent that are helpful or innocuous. Therefore any reckless use of the materials on a large scale could constitute an ecological disaster of the first rank.

The real need is for third-generation pesticides that are tailor-made to attack only certain predetermined pests. Can such pesticides be devised? Recent work that Sláma and I have carried out at Harvard suggests that this objective is by no means unattainable. The possibility arose rather fortuitously after Sláma arrived from Czechoslovakia, bringing with him some specimens of the European bug *Pyrrhocoris apterus*—a species that had been reared in his laboratory in Prague for 10 years.

To our considerable mystification the bugs invariably died without reaching sexual maturity when we attempted to rear them at Harvard. Instead of metamorphosing into normal adults they continued to grow as larvae or molted into adult-like forms retaining many larval characteristics. It was evident that the bugs had access to some unknown source of juvenile hormone.

Eventually we traced the source to the paper toweling that had been placed in the rearing jars. Then we discovered that almost any paper of American origin—including the paper on which *Scientific American* is printed—had the same effect. Paper of European or Japanese manufacture had no effect on the bugs. On further investigation we found that the juvenile hormone activity originated in the balsam fir, which is the principal source of pulp for paper in Canada and the northern U.S. The tree synthesizes what we named the "paper factor," and this substance accompanies the pulp all the way to the printed page.

Thanks again to Bowers and his associates at Beltsville, the active material of the paper factor has been isolated and characterized [*see lower illustration on opposite page*]. It proves to be the methyl ester of a certain unsaturated fatty-acid derivative. The factor's kinship with the other juvenile hormone analogues is evident from the illustrations.

Here, then, is an extractable juvenile hormone analogue with selective action against only one kind of insect. As it happens, the family Pyrrhocoridae includes some of the most destructive pests of the cotton plant. Why the balsam fir should have evolved a substance against only one family of insects is unexplained. The most intriguing possibility is that the paper factor is a biochemical memento of the juvenile hormone of a former natural enemy of the tree—a pyrrhocorid predator that, for obvious reasons, is either extinct or has learned to avoid the balsam fir.

In any event, the fact that the tree synthesizes the substance argues strongly that the juvenile hormone of other species of insects can be mimicked, and perhaps has been by trees or plants on which the insects preyed. Evidently during the 250 million years of insect evolution the detailed chemistry of juvenile hormone has evolved and diversified. The process would of necessity have gone hand in hand with a retuning of the hormonal receptor mechanisms in the cells and tissues of the insect, so that the use as pesticides of any analogues that are discovered seems certain to be effective.

The evergreen trees are an ancient lot. They were here before the insects; they are pollinated by the wind and thus, unlike many other plants, do not depend on the insects for anything. The paper factor is only one of thousands of terpenoid materials these trees synthesize for no apparent reason. What about the rest?

It seems altogether likely that many of these materials will also turn out to be analogues of the juvenile hormones of specific insect pests. Obviously this is the place to look for a whole battery of third-generation pesticides. Then man may be able to emulate the evergreen trees in their incredibly sophisticated self-defense against the insects.

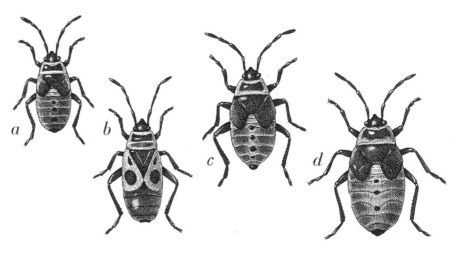

**EFFECT OF PAPER FACTOR** on *Pyrrhocoris apterus* is depicted. A larva of the fifth and normally final stage (*a*) turns into a winged adult (*b*). Contact with the paper factor causes the insect to turn into a sixth-stage larva (*c*) and sometimes into a giant seventh-stage larva (*d*). The abnormal larvae usually cannot shed their skin and die before reaching maturity.

# 34

# The Control of Air Pollution

A. J. HAAGEN-SMIT

*January 1964*

The past decade has seen a change in the public's attitude toward air pollution. Formerly the tendency was to deplore smog but to regard it as one of the inescapable adjuncts of urban life. Now there is a growing realization that smog, beyond being a vexatious nuisance, may indeed present hazards to health, and that in any case the pollution of the air will inevitably grow worse unless something is done about it. As a result many communities have created agencies to deal with air pollution and have, with varying degrees of effectiveness, backed the agencies with laws.

Going considerably beyond these efforts is the program in Los Angeles, a city rather widely regarded as the smog capital of the U.S. There the authorities have adopted the attitude that it is not enough to know smog exists; they have undertaken extensive studies to ascertain its components and to understand something of the complex processes by which it is created. Moreover, with help from the state they have taken pioneering steps toward curbing the emissions of the automobile, which is both a major cause of air pollution and a far more difficult source to control than such stationary installations as petroleum refineries

LOS ANGELES SMOG, shown in photograph on opposite page, casts thick pall over city. Persistence and severity of smogs led the city to undertake pioneering and extensive programs to curb air pollution.

and electric power plants. As a result of California's activities a device to control the emissions from the crankcases of automobiles is now standard equipment on all new cars in the U.S. The state is also working toward a program that will result in a measure of control over emissions from the automobile exhaust.

Complaints about polluted air go far back in time. As long ago as 1661 the English diarist John Evelyn declared in a tract entitled *Fumifugium, or the inconvenience of the Aer and Smoak of London* that the city "resembles the face Rather of Mount Aetna, the Court of Vulcan, Stromboli, or the Suburbs of Hell than an Assembly of Rational Creatures and the Imperial seat of our Incomparable Monarch." Air pollution has drawn similar complaints in many cities over the centuries.

For a long time, however, these complaints were like voices in the wilderness. Among the few exceptions in the U.S. were St. Louis and Pittsburgh, where the residents decided at last that they had inhaled enough soot and chemicals and took steps several years ago to reduce air pollution, primarily by regulating the use of coal. These, however, were isolated cases that did not deeply penetrate the consciousness of people in other parts of the country.

It was probably the recurrence of crises over smog in Los Angeles that awakened more of the nation to the possibility that the same thing could happen elsewhere and to the realization that air,

like water, should be considered a precious resource that cannot be used indiscriminately as a dump for waste materials. By the time residents of Washington, D.C., complained of eye irritation and neighboring tobacco growers suffered extensive crop damage, it was clear that Los Angeles smog was not just a subject for jokes but a serious problem requiring diligent efforts at control. As a result the pace of antipollution activity has quickened at all levels of government. In addition to the community efforts already mentioned, a national air-sampling network now exists to assemble data on the extent of air pollution, and extensive studies of the effects of smog on health and the economy are under way.

Still, these efforts seem modest when viewed against the size of the problem. Surgeon General Luther L. Terry spoke at the second National Conference on Air Pollution late in 1962 of "how far we have to go." He said: "Approximately 90 per cent of the urban population live in localities with air-pollution problems—a total of about 6,000 communities. But only half of this population is served by local control programs with full-time staffs. There are now about 100 such programs, serving 342 local political jurisdictions. The median annual expenditure is about 10 cents per capita, an amount clearly inadequate to do the job that is necessary."

Enough has been done, however, to demonstrate that a concerted attack on the smog problem can produce a clearing of the air. Los Angeles, which Terry has called "the area in the United States that's devoting more money and more effort toward combating the problem than any other city," provides an example of the possibilities, the difficulties and the potential of such an attack.

Los Angeles certainly qualifies as a community where air pollution has created an annoying and at times dangerous situation. Two-thirds of the year

smog is evident through eye irritation, peculiar bleachlike odors and a decrease in visibility that coincides with the appearance of a brownish haze. According to the California Department of Public Health, 80 per cent of the population in Los Angeles County is affected to some extent.

The city's decision to attack the smog problem dates from a report made in 1947 by Raymond R. Tucker, who as an investigator of air-pollution problems played a major role in the St. Louis smog

SMOG CURTAIN falls over the view from the campus of the California Institute of Technology. At top is the scene on a weekday morning; at bottom, the same scene that afternoon.

battle and is now the mayor of that city. His report on Los Angeles enumerated the sources of pollution attributable to industry and to individuals through the use of automobiles and the burning of trash. The report recommended immediate control of known sources of pollution and a research program to determine if there were any other things in the air that should be controlled.

Largely on the basis of the Tucker report, *The Los Angeles Times* started with the aid of civic groups a campaign to inform and arouse the public about smog. As a result the state legislature in 1948 passed a law permitting the formation of air-pollution control districts empowered to formulate rules for curbing smog and endowed with the necessary police power for enforcement of the rules. Los Angeles County created such a district the same year.

The district began by limiting the dust and fumes emitted by steel factories, refineries and hundreds of smaller industries. It terminated the use of a million home incinerators and forbade the widespread practice of burning in public dumps. These moves reduced dustfall, which in some areas had been as much as 100 tons per square mile per month, by two-thirds, bringing it back to about the level that existed in 1940 before smog became a serious problem in the community. That achievement should be measured against the fact that since 1940 the population of Los Angeles and the number of industries in the city have doubled.

Although the attack on dustfall produced a considerable improvement in visibility, the typical smog symptoms of eye irritation and plant damage remained. The district therefore undertook a research program to ascertain the origin and nature of the substances that caused the symptoms. One significant finding was that the Los Angeles atmosphere differs radically from that of most other heavily polluted communities. Ordinarily polluted air is made strongly reducing by sulfur dioxide, a product of the combustion of coal and heavy oil. Los Angeles air, on the other hand, is often strongly oxidizing. The oxidant is mostly ozone, with smaller contributions from oxides of nitrogen and organic peroxides.

During smog attacks the ozone content of the Los Angeles air reaches a level 10 to 20 times higher than that elsewhere. Concentrations of half a part of ozone per million of air have repeatedly been measured during heavy smogs. To establish such a concentration directly would require the dispersal of about

1,000 tons of ozone in the Los Angeles basin. No industry releases significant amounts of ozone; discharges from electric power lines are also negligible, amounting to less than a ton a day. A considerable amount of ozone is formed in the upper atmosphere by the action of short ultraviolet rays, but that ozone does not descend to earth during smog conditions because of the very temperature inversion that intensifies smog. In such an inversion warm air lies atop the cold air near the ground; this stable system forms a barrier not only to the rise of pollutants but also to the descent of ozone.

Exclusion of these possibilities leaves sunlight as the only suspect in the creation of the Los Angeles ozone. The cause cannot be direct formation of ozone by sunlight at the earth's surface because that requires radiation of wavelengths shorter than 2,000 angstrom units, which does not penetrate the atmosphere to ground level. There was a compelling reason, however, to look for an indirect connection between smog and the action of sunlight: high oxidant or ozone values are found only during daylight hours. Apparently a photochemical reaction was taking place when one or more ingredients of smog were exposed to sunlight—which is of course abundant in the Los Angeles area.

In order for a substance to be affected by light it has to absorb the light, and the energy of the light quanta has to be sufficiently high to rupture the chemical bonds of the substance. A likely candidate for such a photochemical reaction in smog is nitrogen dioxide. This dioxide is formed from nitrogen oxide, which originates in all high-temperature combustion through a combining of the nitrogen and oxygen of the air. Nitrogen dioxide has a brownish color and absorbs light in the region of the spectrum from the blue to the near ultraviolet. Radiation from the sun can readily dissociate nitrogen dioxide into nitric oxide and atomic oxygen. This reactive oxygen attacks organic material, of which there is much in the unburned hydrocarbons remaining in automobile exhaust. The result is the formation of ozone and various other oxidation products. Some of these products, notably peracylnitrates and formaldehyde, are eye irritants. Peracylnitrates and ozone also cause plant damage. Moreover, the oxidation reactions are usually accompanied by the formation of aerosols, or hazes, and this combination aggravates the effects of the individual components in the smog complex.

The answer to the puzzle of the oxi-

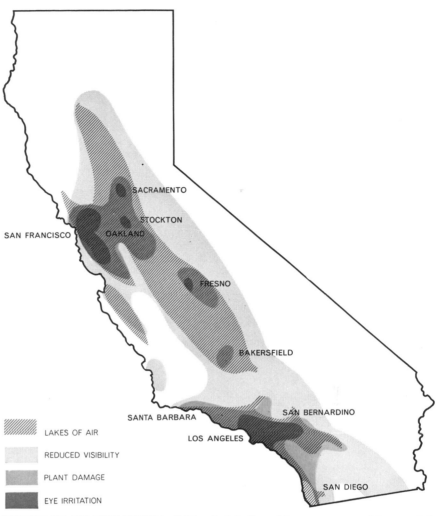

EXTENT OF AIR POLLUTION in California is indicated by gray areas on this map. Colored areas show the main natural airsheds, or lakes of air, into which pollutants flow. Sunlight acting on pollutants produces substances that irritate eyes and damage plants.

dizing smog of the Los Angeles area thus lay in the combination of heavy automobile traffic and copious sunlight. Similar photochemical reactions can of course occur in other cities, and the large-scale phenomenon appears to be spreading.

The more or less temporary effects of smog alone would make a good case for air-pollution control; there is in addition the strong likelihood that smog has adverse long-range effects on human health [see "Air Pollution and Public Health," by Walsh McDermott; SCIENTIFIC AMERICAN Offprint 612]. Workers of the U.S. Public Health Service and Vanderbilt University reported to the American Public Health Association in November that a study they have been conducting in Nashville, Tenn., has established clear evidence that deaths from respiratory diseases rise in proportion to the degree of air pollution.

For the control of air pollution it is of central importance to know that

organic substances—olefins, unsaturated hydrocarbons, aromatic hydrocarbons and the derivatives of these various kinds of molecules—can give rise to ozone and one or more of the other typical manifestations of smog. Control measures must be directed against the release of these volatile substances and of the other component of the smog reaction: the oxides of nitrogen. The organic substances originate with the evaporation or incomplete combustion of gasoline in motor vehicles, with the evaporative losses of the petroleum industry and with the use of solvents. A survey by the Los Angeles Air Pollution Control District in 1951 showed that losses at the refineries were more than 400 tons a day; these have since been reduced to an estimated 85 tons.

This reduction of one source was offset, however, by an increase in the emissions from motor vehicles. In 1940 there were about 1.2 million vehicles in the Los Angeles area; in 1950 there were

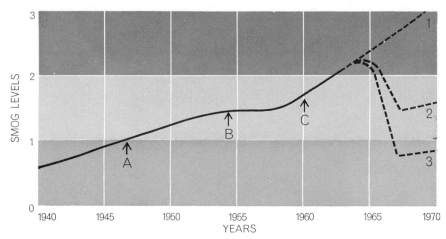

POLLUTION LEVELS in Los Angeles are plotted on scale (*left*) where *1* is 1947 level, *2* double and *3* triple that. *A* represents state pollution control law; *B*, control over refineries; *C*, motor vehicle controls. Broken lines indicate smog potential without new controls (*1*), with hydrocarbon controls (*2*) and with both hydrocarbon and nitrogen oxide controls (*3*).

two million; today there are 3.5 million. These vehicles burn about seven million gallons, or 21,500 tons, of gasoline a day. They emit 1,800 tons of unburned hydrocarbons, 500 tons of oxides of nitrogen and 9,000 tons of carbon monoxide daily. These emissions outweigh those from all other sources.

When motor vehicles emerged as a major source of air pollution, it was evident that state rather than local government could best cope with these moving sources. As a first step, and a pioneering one for the U.S., the California Department of Public Health adopted community standards for the quality of the air [*see top illustration on page 251*].

The adoption of these standards provided a sound basis for a program of controlling automobile emissions. Of special importance for that program was the establishment of the figure of .15 part per million by volume as the harmful level of oxidant. Years of observation have demonstrated that when the oxidant goes above .15 part per million, a significant segment of the population complains of eye irritation, and plant damage is readily noticeable. The standards also set the harmful level for carbon monoxide at 30 parts per million by volume for eight hours, on the basis of observations that under those conditions 5 per cent of the human body's hemoglobin is inactivated. A further stipulation of the standards was that these oxidant and carbon monoxide levels should not be reached on more than

four days a year. To attain such a goal in Los Angeles by 1970 would require the reduction of hydrocarbons and carbon monoxide by 80 and 60 per cent.

On the basis of these standards the California legislature in 1960 adopted the nation's first law designed to require control devices on motor vehicles. The law created a Motor Vehicle Pollution Control Board to set specifications and test the resulting devices. In its work the board has been concerned with two kinds of vehicular emission: that from the engine and that from the exhaust.

About 30 per cent of the total emission of the car, or 2 per cent of the supplied fuel, escapes from the engine. This "blowby" loss results from seepage of gasoline past piston rings into the crankcase; it occurs even in new cars. Evaporation from the carburetor and even from the fuel tank is substantial, particularly on hot days. Until recently crankcase emissions were vented to the outside through a tube. California's Motor Vehicle Polution Control Board began in 1960 a process leading to a requirement that all new cars sold in the state have by 1963 a device that carries the emissions back into the engine for recombustion. The automobile industry thereupon installed the blowby devices in all 1963 models, so that gradually crankcase emissions will come under control throughout the U.S. California is going a step further: blowby devices will have to be installed soon on certain used cars and commercial vehicles.

Two-thirds of the total automobile emission, or 5.4 per cent of the supplied fuel, leaves through the tail pipe as a result of incomplete combustion. For complete combustion, which would produce harmless gases, the air-fuel ratio should be about 15 to 1. Most cars are built to operate on a richer mixture, containing more gasoline, for smoother operation and maximum power; consequently not all the gasoline can be burned in the various driving cycles.

The exhaust gases consist mainly of nitrogen, oxygen, carbon dioxide and water vapor. In addition there are lesser quantities of carbon monoxide, partially oxidized hydrocarbons and their oxidation products, and oxides of nitrogen and sulfur. Most proposals for control of these gases rely on the addition of an afterburner to the muffler. Two approaches appear most promising. The direct-flame approach uses a spark plug or pilot light to ignite the unburned gases. The catalytic type passes them through a catalyst bed that burns them at lower temperatures than are possible

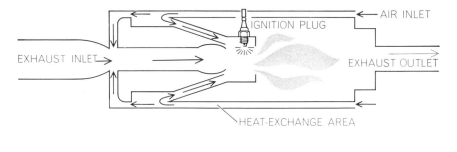

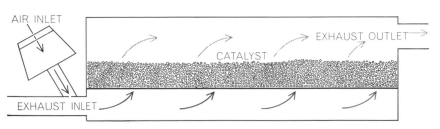

SECOND TYPE of afterburner involves leading exhaust gases through a catalyst bed; they can then be burned at lower temperatures than are possible in a direct-flame afterburner.

with direct-flame burners [*see bottom illustrations on opposite page*].

Building a successful afterburner presents several problems. The high temperatures require more costly materials, thereby increasing initial and replacement costs. Complications in operation arise from the burning of a mixture of gases and air of highly variable concentration. During deceleration the mixture may be so rich that without a bypass ceramics and catalysts will melt. In other cycles of operation there may not be enough fuel to keep the flame going. Moreover, the California law on exhaust-control devices stipulates that they must not be a fire hazard, make excessive noise or adversely affect the operation of the engine by back pressure.

Nine makes of afterburner—six catalytic and three direct-flame—are now under test by the California Motor Vehicle Pollution Control Board. Much testing and modification will be necessary before they are ready for the rough treatment to which they will be subjected when they are attached to all cars. Even after they have been installed a rigorous inspection program will be necessary to make certain that they are properly maintained and periodically replaced.

A preferable method of controlling hydrocarbon emissions from automobile tail pipes would be better combustion in the engine. Automobile engineers have indicated that engines of greater combustion efficiency will appear in the next few years. How efficient these engines will be remains to be seen; so does the effect of the prospective changes on emissions of oxides of nitrogen.

From all the emissions of an automobile the total loss in fuel energy is about 15 per cent; in the U.S. that represents a loss of about $3 billion annually. It is remarkable that the automobile industry, which has a reputation for efficiency, allows such fuel waste. Perhaps pressure for greater efficiency and for control of air pollution will eventually produce a relatively smogless car.

In any case it appears that the proposed 80 per cent control over motor vehicle emissions is a long way off. An alternative is to accept temporary controls at lower levels of effectiveness. It is possible to reduce unburned hydrocarbons and carbon monoxide by modification of the carburetor in order to limit the flow of fuel during deceleration, and by changing the timing of the ignition spark. Proper maintenance can reduce emissions by 25 to 50 per cent, depending on the condition of the car.

Accepting more practical but less ef-ficient means of curbing vehicular emissions requires making up the deficiency in the smog control program some other way. This can be done by control of the other smog ingredient: oxides of nitrogen. At one time it was thought that control of these oxides would be very difficult, and that was why the California law concentrated on curbing emissions of hydrocarbons. It has now been shown, however, that control of oxides of nitrogen, from stationary sources as well as gen, from stationary sources as well as from motor vehicles, is feasible. Oil-burning electric power plants have reduced their contribution by about 50 per cent through the use of a special two-phase combustion system. Research on automobiles has shown that a substantial reduction of oxides of nitrogen is feasible with a relatively simple method of recirculating some of the exhaust gases through the engine.

To arrive at an acceptable quality of air through the limitation of hydrocar-

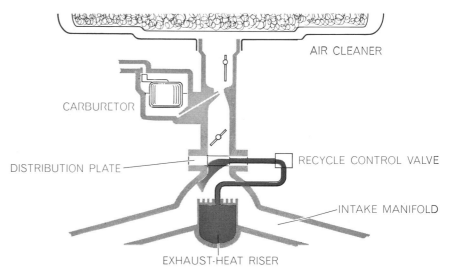

AIR CLEANER

CARBURETOR

DISTRIBUTION PLATE

RECYCLE CONTROL VALVE

INTAKE MANIFOLD

EXHAUST-HEAT RISER

**OXIDES OF NITROGEN emitted by automobiles may be curbed by this system, which takes exhaust gases before they leave the engine and recycles them through the combustion process.**

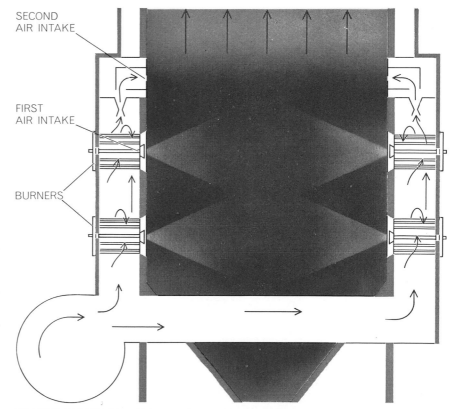

SECOND AIR INTAKE

FIRST AIR INTAKE

BURNERS

**INDUSTRIAL FURNACES have curbed emissions of oxides of nitrogen by two-phase combustion. It lowers temperatures by introducing air at two stages of the burning process.**

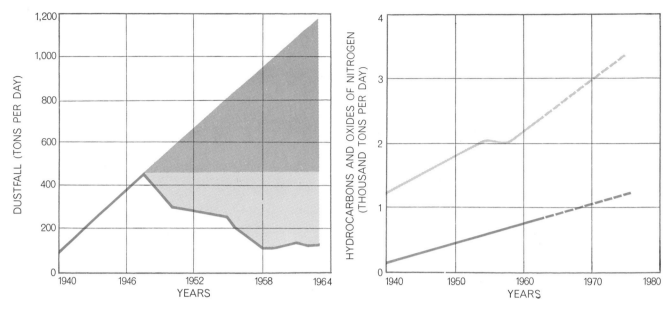

MAJOR POLLUTANTS in Los Angeles County are charted. Dustfall (*left*) has been visibly reduced (*light color*) by control measures; potential without controls is indicated by darker color. At right, light line shows actual and potential levels of hydrocarbons; dark line similarly represents oxides of nitrogen. Rises in spite of controls reflect growth of population and number of vehicles.

bons alone would require a reduction in the hydrocarbons of about 80 per cent, which could be achieved only with rigorous and efficient controls. The plateau of clean air can also be reached, however, by dealing with both hydrocarbons and oxides of nitrogen. The advantage of such an approach is that each one of the reductions would have to be less complete. An over-all reduction of the two major smog components by half would achieve the desired air quality [*see bottom illustration on opposite page*].

This combined approach offers the only practically feasible way to return to a reasonably smog-free atmosphere in Los Angeles as well as in other metropolitan areas plagued by photochemical smog. The California Department of Public Health is now considering the ex-

pansion of the smog control program to include curbs on emission of oxides of nitrogen. For such a program to succeed, however, there would have to be regular inspection of motor vehicles, control of carburetor and fuel tank losses, stringent additional controls over industry and the co-operation of citizens. Moreover, these efforts must be organized in such a way that they take into account the area's rapid population growth, which will mean proportionate rises in motor vehicle and industrial emissions.

Beyond the efforts to control industries and vehicles lie some other possibilities, all of which would have the broad objective of reducing the amount of gasoline burned in the area. They include electric propulsion, economy cars, increased use of public transportation

and improvement of traffic flow. A strong argument for resorting to some of or all these possibilities can be found in an examination of the carbon monoxide readings at a monitoring station in downtown Los Angeles. The readings show clear peaks resulting from commuter traffic. The carbon monoxide increase during a rush period is about 200 tons, representing the emission of about 100,-000 cars. That figure agrees well with vehicular counts made during the hours of heavy commuting.

Greater use of public transportation would produce a considerable reduction of peak pollution levels. So would improved traffic flow, both on the main commuter arteries and on the roads that connect with them. Reduction of the frequent idling, acceleration and deceleration characteristic of stop-and-go driving

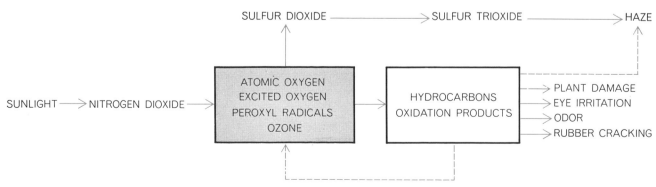

PHOTOCHEMICAL REACTION playing a major role in smog formation begins with sunlight acting on nitrogen dioxide, a product of combustion, to yield oxidants (*gray box*). They attack hydrocarbons, which come mainly from automobile exhausts, to produce irritating materials. Oxidants also attack sulfur dioxide, a product of coal and oil burning. Broken lines indicate interactions.

—the very cycles that produce the most hydrocarbons and oxides of nitrogen— could curb vehicular emissions by 50 per cent or more over a given distance. Detroit has a system of computing the optimum speed on certain freeways according to the density and flow of traffic; the speed is then indicated on large lighted signs. The result is a smoother flow. More techniques of this kind, more imaginative thinking about transportation in general, are necessary for a successful attack on smog.

There can be no doubt that the smogs of Los Angeles represent an extreme manifestation of a problem that is growing in every heavily populated area. Similarly, the control steps taken by Los Angeles will have to be duplicated to some degree in other cities. In those cities, as in Los Angeles, there will be difficulties. One is the cost of air-pollution control for communities that already find their budgets stretched; the Detroit City Council annually votes down an ordinance to ban the burning of leaves because it believes the city cannot afford the estimated cost of $500,-000 for carting the leaves off to dumps. Industry also may balk at smog controls out of concern for maintaining a competitive position. There is a related problem of co-ordination: industries are reluctant to install devices for curbing smoke while the city burns trash in open dumps.

Another problem involves mobilizing the public behind air-pollution control programs. Even though smog looks unpleasant, is occasionally offensive to the smell and irritating to the eye, and sometimes precipitates a public health disaster (as in Donora, Pa., in 1948 and in London in 1952), it nonetheless tends to be regarded as a fact of urban life and something that communities can live with if they must. Moreover, so many political jurisdictions must be involved in an effective attack on air pollution that any one community attempting a clean-up may find its efforts vitiated by another community's smog.

Nevertheless, a growing segment of the public is alert to the dangers of air pollution and determined to do something about it. If anything effective is to be done, however, it will require intelligent planning, aggressive public-education programs and resoluteness on the part of public officials. Then leadership by government and civic groups at all levels, united behind well-designed plans, could generate progress toward the goal of cleaner air.

| POLLUTANT | PARTS PER MILLION FOR ONE HOUR | | |
|---|---|---|---|
| | "ADVERSE" LEVEL | "SERIOUS" LEVEL | "EMERGENCY" LEVEL |
| CARBON MONOXIDE | | 120 | 240 |
| ETHYLENE | .5 | | |
| HYDROGEN SULFIDE | .1 | 5 | |
| SULFUR DIOXIDE | 1 | 5 | 10 |
| HYDROCARBONS | | | |
| NITROGEN DIOXIDE | | | |
| OXIDANT | 15 ON "OXIDANT INDEX" | NOT ESTABLISHED | NOT ESTABLISHED |
| OZONE | | | |
| AEROSOLS | | | |

AIR-QUALITY STANDARDS adopted by California set three levels of pollution: "adverse," at which sensory irritation and damage to vegetation occur; "serious," where there is danger of altered bodily function or chronic disease; "emergency," where acute sickness or death may occur in groups of sensitive persons. Blanks mean "not applicable." Pollutants listed in colored type are involved in or are the products of photochemical reaction. These standards, the first adopted by any state, provided a basis for pollution control measures.

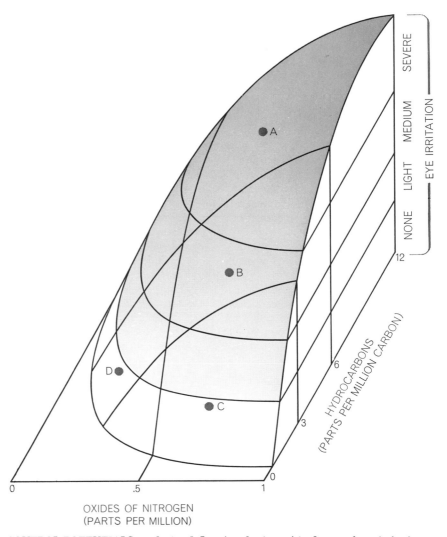

CONTROL POTENTIALS are depicted. Los Angeles is at *A* in degree of eye irritation on a day of heavy smog. Controls reducing hydrocarbons by 50 per cent would bring city down the slope to *B*, still not in clear zone shown in white. Hydrocarbon controls to *C* are impractical; control of both hydrocarbons and oxides of nitrogen would attain clear zone at *D*.

# 35

# The Safety of
# the American Automobile

DAVID HAWKINS

*May 1966*

A review of *Unsafe at Any Speed: The Designed-in Dangers of the American Automobile*, by Ralph Nader. New York: Grossman, 1965.

Automobile accidents accounted for about 50,000 deaths, or 4 percent of all deaths, in the U.S. last year. They caused 16 percent of the nonfatal injuries and 36 percent of the injuries involving permanent impairment. The average age at death by automobile is 38, in contrast with that in other categories, which is 62. Thus if we count man-years destroyed rather than men, the 4 percent figure goes up to 7 percent.

Such is the ranking of the automobile accident among the causes of death and disablement, yet we neglect to use known means of reducing injury rates, or to study new ones. For such research there are no national fund drives and few research institutes, public or private. Still, in spite of its destructiveness the automobile is of the essence of American life, and the automobile industry is a pillar of the affluent society. It seems proper that we confront ourselves with a new definition of the problems and responsibilities of accidental death on the highways. Affluence brings with it many resources for an attack on the problems, but it does not automatically generate the will to attack them.

Such an attack will not come easily or quickly. Competition among advocates is certainly going to be the order of the day. The President promises support for major traffic-safety legislation, including support for basic research on accidents. The manufacturers and the insurance companies are bestirring themselves. Everyone is in favor of greater safety, but to infer a corre-

sponding political unanimity would be excessively naïve. Deeper issues are involved.

For decades we have followed the policy that greater automobile safety was to be achieved primarily by campaigns of driving legislation, law enforcement, technical education and moral exhortation. This view has had, and for many it still has, the force of an ideological commitment. Nader's book can be described as an analysis and a critique of this ideology. It is an adversary work that points an accusing finger at the automobile manufacturers, charging them with indifference, callousness and arrogance in the face of genuine possibilities of safer automobile design. In his conclusion Nader advocates publicly defined and Federally enforced standards of safety in design and manufacture, supported by continuous research. Because of the adversary character of the book I wish to put it in the somewhat wider context necessary, I think, to any sane resolution of the issues Nader is attempting to define and dramatize.

Certainly I cannot adjudicate all the issues as he defines them. Reactions I have sampled, from the industry and its supporters, range from scorn and outrage to the charge of half-truth or the suggestion that a far more qualified statement would be more effective. In defense of the industry one may offer this much: Detroit is part of America, and its engineers and policy-makers drive cars. Corporations are highly adaptive entities, and we should not be surprised if the car manufacturers' view of safety were essentially that of the general public, one of indifference or unfocused anxiety. If there is blame, Detroit can share it with the rest of us. For the prosecution, however, one can

make a charge appropriate for any who hold a public trust: that they should be ahead of all others in their concern for the public welfare. Where safety is involved car manufacture should be judged as a profession, not as a trade. By such a standard, and without regard to Nader's book, the Detroit score is not high. The score surely did not rise with the news that private detectives, employed by General Motors, had harassed Nader by extensive inquiries into his private life, his views and associations. James M. Roche, the president of General Motors, apologized for this act of his company before the Ribicoff subcommittee of the Senate on March 22. This investigation, presumably initiated by Roche's subordinates in the General Motors legal division, suggests the prevalence of attitudes toward external critics that are scarcely those of self-conscious searchers after the truth about automobile safety. I am sure that such attitudes are not shared in the General Motors research division. Could Roche not find a way of letting us hear of the views and activities of *that* division? It would be, I think, a redemptive sequel.

The dominant view of automobile safety—"It's the driver's fault"—has obvious political implications. One implication is that a greater concentration of effort on the design of vehicles and traffic systems from the standpoint of safety will produce only minor results. Another is that the car, and behind the car its manufacturer, should not be regarded as a significant factor in accidents, except perhaps for a special and rare class of them. Accordingly Government-imposed safety standards are sure to be ineffective. Moreover, such standards are morally wrong; they imply responsibility for accidents on the part of innocent manufacturers and exculpate guilty drivers. This, of course, is a classic stance of free enterprise: *Caveat emptor!*

Having such obvious political relevance, this dominant view must be taken very seriously. When we take it so, we discover that it is (1) irrefutable, (2) irreproachable and (3) irrelevant to any question involving the safety design of vehicles. The dominant view is irrefutable because in all but a special class of cases it is proper to say—in retrospect—that a given accident could have been avoided had one or more drivers involved behaved differently and more nearly in accordance with fine prudential maxims such as "Don't drive too fast for the conditions." This view is irreproachable because in one important sense we are all morally bound to subscribe to it: *given* the state of the cars, the highways and other drivers, we must always accept the responsibility for driving as alertly, intelligently and cautiously as we know how. The view is irrelevant because it implies that a reduction of accidents can be achieved by the improved performance of drivers but in no way implies that improvements in design could not bring about a much greater reduction—as in fact appears to be the case. Let us take at face value the statement that "90 percent of accidents are due to faulty driving." Let us now hypothetically consider the reduction of fatalities to a third of their present level through the radical redesign of vehicles, assuming no change in the skill of drivers or their dedication to safety. It might then remain true that "90 percent of accidents are due to faulty driving." Meanwhile there would be 30,000 fewer Americans killed each year, including some careless drivers.

The 90 percent statement is an interesting one. What does it mean? I suppose it means that police investigators, looking for driver fault defined in some commonsense way, find it 90 percent of the time. It is quite clear, however, that another investigator, looking for machine-design fault, highway-design fault and so on might also find each of these to be present in 90 percent of the cases. "Faults" talked of in this way are not mutually exclusive sufficient conditions for accidents but are mutually consistent necessary conditions. The glib transition from one interpretation to the other marks the whole line of argument as invalid.

But is the dominant view really only an inglorious *non sequitur* serving the negligence of manufacturers and the servility of what Nader calls "the safety establishment"? I feel its roots are deeper than that. We as a people (and others even more so) are not yet equipped to think well about the problem. It may sound odd to say it in this age of aeronautical and astronautical achievements, but the automobile has outstripped us in the intellectual, moral and aesthetic capacities we need for coping with it, and the result is chronic foolishness, bad taste and venality. It is essentially a 19th-century invention and represents a jump in vehicle speeds by only a factor of 10 or so over the speeds available before. A tenfold increase in speed, however, is a hundredfold increase in kinetic energy and peak forces. Our predominantly pre-Newtonian commonsense intuitions do not stand up against so great a change. Thus far we have done little to replace our primitive near-zero-velocity intuitions with properly Newtonian ones.

A part of the machine, considered as an operative system, is the driver, whose cognitive and control processes must operate in this Newtonian milieu. High-velocity operation places demands of a novel kind on the organism, requiring discriminations of size, distance, velocity, direction and acceleration for which we are poorly equipped. It also implies a need for feedback mechanisms that has hardly been explored. I have a fantasy of electronic sense organs at the interface of the tire and the road, for example, coupled to the driver's tactile sense and pain responses. This would surely have a marvelous effect on cornering and braking. The evolution of the car-driver interaction is still quite primitive, and the sense in which the driver is the car's master is correspondingly limited. In this connection let me abstract from Nader's book a quotation from H. E. Humphreys, Jr., of the U.S. Rubber Company, that restates the dominant view in a way metaphysically appealing to pre-Newtonian—even Aristotelian—intuitions: "Being inanimate, no car, truck or bus can by itself cause an accident any more than a street or highway can do so. A driver is needed to put it into motion—after which it becomes an extension of his will."

The confrontation we need is a matter of the most obvious humanity and utility, but it goes deeper. We should find imaginative ways of cultivating our Newtonian intuitions, of refining our discrimination and taste in things automotive. We ought not to require statistics, as defenders of the dominant view are likely to urge, to prove that poorly anchored seats or glare from chromium-plated decorations is dangerous. We

ought to dislike such things without need of grisly evidence. We should cultivate the easy awareness of forces, masses, accelerations and our own limitations as high-speed feedback controls. And we should appreciate some subtleties of higher order—some feeling for the phenomenon of gyration, of static and dynamic friction, for the behavior of rubber under stress, for the elastic limits of steel frames or flesh and bone. It implies an improvement of taste to accept the seat belt and reject the knobby instrument panel. After two decades of soft tops called "hardtops" might we not come to see roll bars as rather elegant? It is certainly an improvement of taste to reject the jagged external decor recently so characteristic of our vehicles and see it as a collection of potential meat choppers. As a symbol of the worst I offer those charming hub decorations of flashing knives, presumably inspired by Ben-hur or James Bond.

In his examination of the dominant ideology, Nader has dug extensively into a variety of public and private sources: court records, patent applications, automobile magazines, the publications of standards groups and safety groups, engineering journals, the records of hearings and the transcripts of meetings. It is unfortunate, in this connection, that the book has a "popular" format, without adequate references. A fair number of essential references can be found in the text or inferred, but it is not easy. Some circumstantial accounts in the book are evidently based on private communications. The book goes principally in three directions: toward the record of past performance, to study failures in design or quality control; toward the research frontier, to define potential levels of vehicle safety, and finally, toward the record of the attitudes and policies of the automobile manufacturers, the tire manufacturers, the insurance companies and the many automobile associations and groups concerned in one way or another with safety. Therefore although the book contains a great variety of information about specific design problems and possible solutions, its focus throughout is on the entire system of beliefs and attitudes that penetrate and surround this remarkable industry. It is thus primarily a case study in economic and political sociology. It is not, however, detached and academic in tone. Nader is a lawyer, and his tone is relentless. His case is carefully constructed, and it brings together for the first time an impressive collection of factual information. The line of argument is clear but not lacking in subtlety, as in the interplay of contrasts in the treatment of Detroit's engineer-spokesmen and its "stylists." The relentlessness is qualified by occasional counterexamples. In his treatment of research organizations that are responsible for valuable safety information but that he sees as improperly influenced in their publication policies by company support, Nader manages both praise and blame; he also manages to make clear how little support there is as yet for first-class research. Having described this work as adversary and praised it, I must add that the proceeding will not be complete until we have had some careful and equally analytical replies that do their best with denials, counterexamples and reinterpretations. In the old days this book would have been called muckraking, but it formulates real issues that ought to be joined.

A good deal of the force of Nader's presentation depends on a case-study technique. His first chapter is devoted to design weaknesses in the Corvair of 1960 to 1963, specifically the combination of its rear engine and rear swing-axle suspension, which under certain circumstances produce a sudden "tuck-under" of the rear wheels and a loss of control. The story is that of a lawsuit brought against General Motors as an aftermath of such an accident, a prototype of many other such suits. The account produces evidence that the dangers involved in this car design are well known to accident investigators, to dealers and independent accessory manufacturers—and to General Motors. There follows a history of the Corvair, to show that both the hazards and the means of avoiding them (finally adopted in 1964 and 1965) were fully known ahead of time.

In the succeeding chapters Nader makes similar use of a variety of materials: brake failures in a 1953 Buick series, a defective suspension arm in a 1965 Ford series, a defective steering-system bracket in 1965 Chryslers; the "PNDLR" automatic gearshift pattern common to the Cadillac, Buick, Oldsmobile, Pontiac, Studebaker and Rambler for many years. This pattern makes possible a dangerous confusion between L (low gear) and R (reverse), as many records of "driver negligence" will show. I myself am fortunate in not being included in the records, having once "shifted down" into reverse at the beginning of a steep hill! The negligence was real, as is the fact that it would not have happened with the now approved "PRNDL" pattern.

Next there is a series of case studies involving "the driver's right to see": optical distortion, poor wiper and defroster design, inadequate mirror design, glare from reflecting surfaces, obstructing corner posts, tinted windshields. At a time when car stylists lean heavily on glamorous lines borrowed from aircraft, it would be pleasant to see cars embody one universal feature of aircraft: the elimination of shiny surfaces in front of the pilot. A good many specific items in the list have been corrected, more or less, presumably to the credit of the manufacturers. Nader makes his case here by a running stream of references and quotations to show that changes have often been delayed, grudging and (recently) impelled by the appearance of legislated safety standards on Federally purchased cars, by Congressional hearings and the threat of legislative controls.

A most interesting discussion in the book has directly to do with the consequences of high collision-accelerations, in the range of tens of $g$'s. Put in terms of forces, an unrestrained passenger strikes the windshield, the instrument panel or the steering wheel with what Leonardo da Vinci called "a weight in the direction of motion" of several tons. Such forces can be transiently endured (as Colonel John Paul Stapp showed in special deceleration apparatus) when forces are properly distributed over the body area. Energy-absorbing bumpers—not the prevailing sheet-metal décor—can considerably decrease peak forces by lengthening the collision time. Many simple interior-design changes can reduce peak pressures on the body by eliminating projections with which the body can collide. The steering wheel, the most lethal of all these projections, can be designed against being displaced to the rear and impaling the driver, and also to absorb the energy of his forward motion. Seat belts and even more effective restraints are an obvious necessity.

A completely different aspect of the design problem is the reduction of hydrocarbons and carbon monoxide in automobile exhaust gases. Hydrocarbons have been known for 15 years to be major contributors to photochemical smog, and carbon monoxide is highly toxic, possibly contributing to driver failure on crowded highways. In a separate chapter Nader tells the story of the struggles of the Los Angeles Air Pollution Control District, and of legislative bodies, to elicit from Detroit the acknowledgment of the reality of the problem, contributions to research on it or innovative action without the necessity of legislative compulsion.

Nader's final chapters deal with the expressed outlook of Detroit's safety engineers, whom he sees as public relations spokesmen assigned to the task of keeping the public unalarmed about safety problems; with the standardization policies of the Society of Automotive Engineers and the American Standards Association, and with the publicly stated position of the Automobile Manufacturers Association. In contrast to the enforced servility Nader ascribes to safety engineers who work for automobile companies he sets the increasing status of the design stylists, who are chiefly responsible for maintaining a high annual volume of sales.

To round out Nader's picture there is a survey of the system of organizations that he collectively labels "the traffic safety establishment." This consists of numerous national groups, united behind the dominant ideology of driver error. There is a good deal of space devoted to the President's Committee on Traffic Safety, largely financed by Federal funds and devoted, according to Nader, to the maintenance of "safe" Federal policies approved by its industry-sponsored policy makers. Finally, he describes the work of the National Safety Council, which he sees as a hotbed of indifference to any matters of safety other than its own extensive driver-safety program. Perhaps this program, coordinating the efforts of regional and local organizations, has had some effect over the years. Our record is better, as far as I can ascertain, than that of other countries. The main reason, of course, is that we have been at the business longer than other countries. Conscious safety propaganda, and thus the National Safety Council, may still count for something. The council is in fact a relatively small organization, best known to the public for its holiday-accident forecasts. These urge drivers to ever greater caution at just those times that are, as I calculate, statistically the safest: on national holidays the accident rate goes up by considerably less than the volume of traffic, thus decreasing in terms of vehicle miles.

In his last chapter Nader reviews the beginnings of Federal action on the issues of safety, starting with the Roberts subcommittee hearings begun in 1956 and continuing to the Ribicoff hearings of last year. He devotes considerable attention to the General Services Administration standards on Government cars, brought into existence as a result of work by the Roberts subcommittee. What Nader advocates is publicly defined and Federally enforced manufacturing safety standards, maintained and extended by continued short- and long-range research. In order to be successful such a program must have a "supportive constituency" of independent engineers, physicians, lawyers and psychologists, all well informed and devoted to the interests of safety.

Reluctance on the part of Detroit to acknowledge the need for or the possibility of far higher standards than those that now prevail can only intensify the persuasiveness of Nader's position. Is such reluctance inevitable? Could not the manufacturers decide that the potentialities of automobile safety are a new source of sales-stimulating annual innovation? I will not be outdone in skepticism that this decision is likely, but in fact there is no historical neces-

sity in the manufacturers' traditional view. That view includes the belief that a strong emphasis on safety design brings no competitive advantage. In actuality the belief has never been tested. Cannot the community of automobile manufacturers find ways to commit its members to such an emphasis? Legislation can force some such commitment, but it is better to lead than to be dragged.

# 36

# Thermal Pollution and Aquatic Life

JOHN R. CLARK

*March 1969*

Ecologists consider temperature the primary control of life on earth, and fish, which as cold-blooded animals are unable to regulate their body temperature, are particularly sensitive to changes in the thermal environment. Each aquatic species becomes adapted to the seasonal variations in temperature of the water in which it lives, but it cannot adjust to the shock of abnormally abrupt change. For this reason there is growing concern among ecologists about the heating of aquatic habitats by man's activities. In the U.S. it appears that the use of river, lake and estuarine waters for industrial cooling purposes may become so extensive in future decades as to pose a considerable threat to fish and to aquatic life in general. Because of the potential hazard to life and to the balance of nature, the discharge of waste heat into the natural waters is coming to be called thermal pollution.

The principal contributor of this heat is the electric-power industry. In 1968 the cooling of steam condensers in generating plants accounted for about three-quarters of the total of 60,000 billion gallons of water used in the U.S. for industrial cooling. The present rate of heat discharge is not yet of great consequence except in some local situations; what has aroused ecologists is the ninefold expansion of electric-power production that is in prospect for the coming years with the increasing construction of large generating plants fueled by nuclear energy. They waste 60 percent more energy than fossil-fuel plants, and this energy is released as heat in condenser-cooling water. It is estimated that within 30 years the electric-power industry will be producing nearly two million megawatts of electricity, which will require the disposal of about 20 million billion B.T.U.'s of waste heat per day. To carry off that heat by way of natural waters would call for a flow through power plants amounting to about a third of the average daily freshwater runoff in the U.S.

The Federal Water Pollution Control

HEATED EFFLUENT from a power plant on the Connecticut River is shown in color thermograms (*opposite page*), in which different temperatures are represented by different hues. At the site (*above*) three large pipes discharge heated water that spreads across the river at slack tide (*top thermogram*) and tends to flow downriver at ebb tide (*bottom*). An infrared camera made by the Barnes Engineering Company scans the scene and measures the infrared radiation associated with the temperature at each point in the scene (350 points on each of 180 horizontal lines). Output from an infrared radiometer drives a color modulator, thus changing the color of a beam of light that is scanned across a color film in synchrony with the scanning of the scene. Here the coolest water (*black*) is at 59 degrees Fahrenheit; increasingly warm areas are shown, in three-degree steps, in blue, light blue, green, light green, yellow, orange, red and magenta. The effluent temperature was 87 degrees. A tree (*dark object*) appears in lower thermogram because the camera was moved.

**NUCLEAR POWER PLANT** at Haddam on the Connecticut River empties up to 370,000 gallons of coolant water a minute through a discharge canal (*bottom*) into the river. In this aerial thermogram, made by HRB-Singer, Inc., for the U.S. Geological Survey's

Administration has declared that waters above 93 degrees Fahrenheit are essentially uninhabitable for all fishes in the U.S. except certain southern species. Many U.S. rivers already reach a temperature of 90 degrees F. or more in summer through natural heating alone. Since the waste heat from a single power plant of the size planned for the future (some 1,000 megawatts) is expected to raise the temperature of a river carrying a flow of 3,000 cubic feet per second by 10 degrees, and since a number of industrial and power plants are likely to be constructed on the banks of a single river, it is obvious that many U.S. waters would become uninhabitable.

A great deal of detailed information is available on how temperature affects the life processes of animals that live in the water. Most of the effects stem from the impact of temperature on the rate of metabolism, which is speeded up by heat in accordance with the van't Hoff principle that the rate of chemical reaction increases with rising temperature. The acceleration varies considerably for particular biochemical reactions and in different temperature ranges, but gener-

ally speaking the metabolic rate doubles with each increase of 10 degrees Celsius (18 degrees F.).

Since a speedup of metabolism increases the animal's need for oxygen, the rate of respiration must rise. F. E. J. Fry of the University of Toronto, experimenting with fishes of the salmon family, found that active fish increased their oxygen consumption as much as fourfold as the temperature of the water was raised to the maximum at which they could survive. In the brown trout the rate of oxygen consumption rose steadily until the lethal temperature of 79 degrees F. was reached; in a species of lake trout, on the other hand, the rate rose to a maximum at about 60 degrees and then fell off as the lethal temperature of 77 degrees was approached. In both cases the fishes showed a marked rise in the basal rate of metabolism up to the lethal point.

The heart rate often serves as an index of metabolic or respiratory stress on the organism. Experiments with the crayfish (*Astacus*) showed that its heart rate increased from 30 beats per minute at a water temperature of 39 degrees F. to 125 beats per minute at 72 degrees and then slowed to a final 65 beats per min-

ute as the water approached 95 degrees, the lethal temperature for this crustacean. The final decrease in heartbeat is evidence of the animal's weakening under the thermal stress.

At elevated temperatures a fish's respiratory difficulties are compounded by the fact that the hemoglobin of its blood has a reduced affinity for oxygen and therefore becomes less efficient in delivering oxygen to the tissues. The combination of increased need for oxygen and reduced efficiency in obtaining it at rising temperatures can put severe stress even on fishes that ordinarily are capable of living on a meager supply of oxygen. For example, the hardy carp, which at a water temperature of 33 degrees F. can survive on an oxygen concentration as low as half a milligram per liter of water, needs a minimum of 1.5 milligrams per liter when the temperature is raised to 95 degrees. Other fishes can exist on one to two milligrams at 39 degrees but need three to four milligrams merely to survive at 65 degrees and five milligrams for normal activity.

The temperature of the water has pronounced effects on appetite, digestion and growth in fish. Tracer experiments

Water Resources Division, temperature is represented by shades of gray. The hot effluent (*white*) is at about 93 degrees F.; ambient river temperature (*dark gray*) is 77 degrees. The line across the thermogram is a time marker for a series of absolute measurements.

with young carp, in which food was labeled with color, established that they digest food four times as rapidly at 79 degrees F. as they do at 50 degrees; whereas at 50 degrees the food took 18 hours to pass through the alimentary canal, at 79 degrees it took only four and a half hours.

The effects of temperature in regulating appetite and the conversion of the food into body weight can be used by hatcheries to maximize fish production in terms of weight. The food consumption of the brown trout, for example, is highest in the temperature range between 50 and 66 degrees. Within that range, however, the fish is so active that a comparatively large proportion of its food intake goes into merely maintaining its body functions. Maximal conversion of the food into a gain in weight occurs just below and just above that temperature range. A hatchery can therefore produce the greatest poundage of trout per pound of food by keeping the water temperature at just under 50 degrees or just over 66 degrees.

It is not surprising to find that the activity, or movement, of fish depends considerably on the water temperature. By and large aquatic animals tend to raise their swimming speed and to show more spontaneous movement as the temperature rises. In many fishes the temperature-dependent pattern of activity is rather complex. For instance, the sockeye salmon cruises twice as fast in water at 60 degrees as it does in water at 35 degrees, but above 60 degrees its speed declines. The brook trout shows somewhat more complicated behavior: it increases its spontaneous activity as the temperature rises from 40 to 48 degrees, becomes less active between 49 and 66 degrees and above 66 degrees again goes into a rising tempo of spontaneous movements up to the lethal temperature of 77 degrees. Laboratory tests show that a decrease in the trout's swimming speed potential at high temperatures affects its ability to feed. By 63 degrees trout have slowed down in pursuing minnows, and at 70 degrees they are almost incapable of catching the minnows.

That temperature plays a critical role in the reproduction of aquatic animals is well known. Some species of fish spawn during the fall, as temperatures drop; many more species, however, spawn in the spring. The rising temperature induces a seasonal development of their gonads and then, at a critical point, triggers the female's deposit of her eggs in the water. The triggering is particularly dramatic in estuarine shellfish (oysters and clams), which spawn within a few hours after the water temperature reaches the critical level. Temperature also exerts a precise control over the time it takes a fish's eggs to hatch. For example, fertilized eggs of the Atlantic salmon will hatch in 114 days in water at 36 degrees F. but the period is shortened to 90 days at 45 degrees; herring eggs hatch in 47 days at 32 degrees and in eight days at 58 degrees; trout eggs hatch in 165 days at 37 degrees and in 32 days at 54 degrees. Excessive temperatures, however, can prevent normal development of eggs. The Oregon Fish Commission has declared that a rise of 5.4 degrees in the Columbia River could be disastrous for the eggs of the Chinook salmon.

Grace E. Pickford of Yale University has observed that "there are critical temperatures above or below which fish will not reproduce." For instance, at a temperature of 72 degrees or higher the banded sunfish fails to develop eggs. In the case of the carp, temperatures in the range of 68 to 75 degrees prevent cell division in the eggs. The possum shrimp *Neomysis,* an inhabitant of estuaries, is blocked from laying eggs if the temperature rises above 45 degrees. There is also the curious case of the tiny crustacean *Gammarus,* which at temperatures above 46 degrees produces only female offspring.

Temperature affects the longevity of fish as well as their reproduction. D'Arcy Wentworth Thompson succinctly stated this general life principle in his classic *On Growth and Form:* "As the several stages of development are accelerated by warmth, so is the duration of each and all, and of life itself, proportionately curtailed. High temperature may lead to a short but exhausting spell of rapid growth, while the slower rate manifested at a lower temperature may be the best in the end." Thompson's principle has been verified in rather precise detail by experiments with aquatic crustaceans. These have shown, for example, that *Daphnia* can live for 108 days at 46 degrees F. but its lifetime at 82 degrees is 29 days; the water flea *Moina* has a lifetime of 14 days at 55 degrees, its optimal temperature for longevity, but only five days at 91 degrees.

Other effects of temperature on life processes are known. For example, a century ago the German zoologist Karl Möbius noted that mollusks living in cold waters grew more slowly but attained larger size than their cousins living in warmer waters. This has since been found to be true of many fishes and other water animals in their natural habitats.

Fortunately fish are not entirely at the mercy of variations in the water temperature. By some process not yet understood they are able to acclimate themselves to a temperature shift if it is moderate and not too sudden. It has been found, for instance, that when the eggs of largemouth bass are suddenly transferred from water at 65 or 70 degrees to water at 85 degrees, 95 percent of the eggs perish, but if the eggs are acclimated by gradual raising of the temperature to 85 degrees over a period of 30 to 40 hours, 80 percent of the eggs will survive. Experiments with the possum shrimp have shown that the lethal temperature for this crustacean can be raised by as much as 24 degrees (to a high of 93 degrees) by acclimating it through a series of successively higher temperatures. As a general rule aquatic animals can acclimate to elevated temperatures faster than they can to lowered temperatures.

Allowing for maximum acclimation (which usually requires spreading the gradual rise of temperature over 20 days), the highest temperatures that most fishes of North America can tolerate range from about 77 to 97 degrees F. The direct cause of thermal death is not known in detail; various investigators have suggested that the final blow may be some effect of heat on the nervous system or the respiratory system, the coagulation of the cell protoplasm or the inactivation of enzymes.

Be that as it may, we need to be concerned not so much about the lethal temperature as about the temperatures that may be *unfavorable* to the fish. In the long run temperature levels that adversely affect the animals' metabolism, feeding, growth, reproduction and other vital functions may be as harmful to a fish population as outright heat death.

Studies of the preferred, or optimal, temperature ranges for various fishes have been made in natural waters and in the laboratory [*see illustration on page 262*]. For adult fish observed in nature the preferred level is about 13 degrees F. below the lethal temperature on the average; in the laboratory, where the experimental subjects used (for convenience) were very young fish, the preferred level was 9.5 degrees below the lethal temperature. Evidently young fish need warmer waters than those that have reached maturity do.

The optimal temperature for any water habitat depends not only on the preferences of individual species but also on the well-being of the system as a whole. An ecological system in dynamic balance is like a finely tuned automobile engine, and damage to any component can disable or impair the efficiency of the entire mechanism. This means that if we are to expect a good harvest of fish, the temperature conditions in the water medium must strike a favorable balance for all the components (algae and other plants, small crustaceans, bait fishes and so on) that constitute the food chain producing the harvested fish. For example, above 68 degrees estuarine eelgrass does not reproduce. Above 90 degrees there is extensive loss of bottom life in rivers.

So far there have been few recorded instances of direct kills of fish by thermal pollution in U.S. waters. One recorded kill occurred in the summer of 1968 when a large number of menhaden acclimated to temperatures in the 80's became trapped in effluent water at 93 to 95 degrees during the testing of a new power plant on the Cape Cod Canal. A very large kill of striped bass occurred

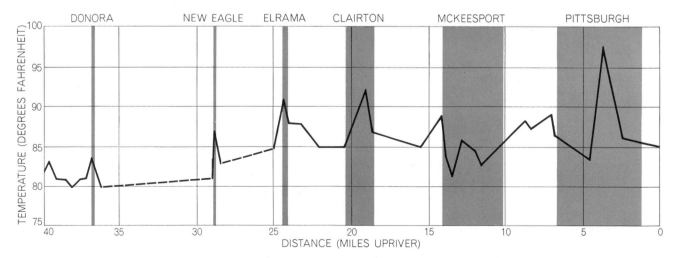

WATER TEMPERATURES can become very high, particularly in summer, along rivers with concentrated industry. The chart shows the temperature of the Monongahela River, measured in August, along a 40-mile stretch upriver from its confluence with the Ohio.

in the winter and early spring of 1963 at the nuclear power plant at Indian Point on the Hudson River. In that instance the heat discharge from the plant was only a contributing factor. The wintering, dormant fish, attracted to the warm water issuing from the plant, became trapped in its structure for water intake, and they died by the thousands from fatigue and other stresses. (Under the right conditions, of course, thermal discharges benefit fishermen by attracting fish to discharge points, where they can be caught with a hook and line.)

Although direct kills attributable to thermal pollution apparently have been rare, there are many known instances of deleterious effects on fish arising from natural summer heating in various U.S. waters. Pollution by sewage is often a contributing factor. At peak summer temperatures such waters frequently generate a great bloom of plankton that depletes the water of oxygen (by respiration while it lives and by decay after it dies). In estuaries algae proliferating in the warm water can clog the filtering apparatus of shellfish and cause their death. Jellyfish exploding into abundant growth make some estuarine waters unusable for bathing or other water sports, and the growth of bottom plants in warm waters commonly chokes shallow bays and lakes. The formation of hydrogen sulfide and other odorous substances is enhanced by summer temperatures. Along some of our coasts in summer "red tides" of dinoflagellates occasionally bloom in such profusion that they not only bother bathers but also may poison fish. And where both temperature and sewage concentrations are high a heavy toll of fish may be taken by proliferating microbes.

This wealth of evidence on the sensitivity of fish and the susceptibility of aquatic ecosystems to disruption by high temperatures explains the present concern of biologists about the impending large increase in thermal pollution. Already last fall 14 nuclear power plants, with a total capacity of 2,782 megawatts, were in operation in the U.S.; 39 more plants were under construction, and 47 others were in advanced planning stages. By the year 2000 nuclear plants are expected to be producing about 1.2 million megawatts, and the nation's total electricity output will be in the neighborhood of 1.8 million megawatts. As I have noted, the use of natural waters to cool the condensers would entail the heating of an amount of water equivalent to a third of the yearly freshwater runoff in the U.S.; during low-flow periods in summer the requirement would be 100 percent of the runoff. Obviously thermal pollution of the waters on such a scale is neither reasonable nor feasible. We must therefore look for more efficient and safer methods of dissipating the heat from power plants.

One might hope to use the heated water for some commercial purpose. Unfortunately, although dozens of schemes have been advanced, no practicable use has yet been found. Discharge water is not hot enough to heat buildings. The cost of transmission rules out piping it to farms for irrigation even if the remaining heat were enough to improve crop production. More promising is the idea of using waste heat in desalination plants to aid in the evaporation process, but this is still only an idea. There has also been talk of improving the efficiency of sewage treatment with waste heat from power plants. Sea farm-ing may offer the best hope of someday providing a needed outlet for discharges from coastal power plants; pilot studies now in progress in Britain and the U.S. are showing better growth of fish and shellfish in heated waters than in normal waters, but no economically feasible scheme has yet emerged. It appears, therefore, that for many years ahead we shall have to dispose of waste heat to the environment.

The dissipation of heat can be facilitated in various ways by controlling the passage of the cooling water through the condensers. The prevailing practice is to pump the water (from a river, a lake or an estuary) once through the steam-condensing unit, which in a 500-megawatt plant may consist of 400 miles of one-inch copper tubing. The water emerging from the unit has been raised in temperature by an amount that varies from 10 to 30 degrees F., depending on the choice of manageable factors such as the rate of flow. This heated effluent is then discharged through a channel into the body of water from which it was taken. There the effluent, since it is warmer and consequently lighter than the receiving water, spreads in a plume over the surface and is carried off in the direction of the prevailing surface currents.

The ensuing dispersal of heat through the receiving water and into the atmosphere depends on a number of natural factors: the speed of the currents, the turbulence of the receiving water (which affects the rate of mixing of the effluent with it), the temperature difference between the water and the air, the humidity of the air and the speed and direction of the wind. The most variable and most important factor is wind: other things being equal, heat will be dissipated from

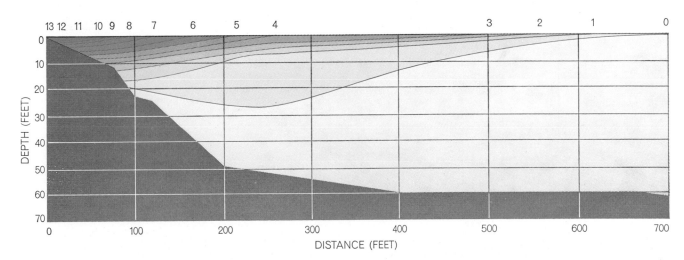

HOT-WATER "PLUME" that would result from an Indian Point nuclear power plant mixes with cooler Hudson River as shown by the one-degree contour lines. This section across the river shows temperature structure that was predicted by engineering studies.

the water to the air by convection three times faster at a wind speed of 20 miles per hour than at a wind speed of five miles.

In regulating the rate of water flow through the condenser one has a choice between opposite strategies. By using a rapid rate of flow one can spread the heat through a comparatively large volume of cooling water and thus keep down the temperature of the effluent; conversely, with a slow rate of flow one can concentrate the heat in a smaller volume of coolant. If it is advantageous to obtain good mixing of the effluent with the receiving water, the effluent can be discharged at some depth in the water rather than at the surface. The physical and ecological nature of the body of water will determine which of these strategies is best in a given situation. Where the receiving body is a swift-flowing river, rapid flow through the condenser and dispersal of the low-temperature effluent in a narrow plume over the water surface may be the most effective way to dissipate the heat into the atmosphere. In the case of a still lake it may be best to use a slow flow through the condenser so that the comparatively small volume of effluent at a high tem-

perature will be confined to a small area in the lake and still transfer its heat to the atmosphere rapidly because of the high temperature differential. And at a coastal site the best strategy may be to discharge an effluent of moderate temperature well offshore, below the ocean-water surface.

There are many waters, however, where no strategy of discharge will avail to make the water safe for aquatic life (and where manipulation of the discharge will also be insufficient to avoid dangerous thermal pollution), particularly where a number of industrial and power plants use the same body of water for cooling purposes. It therefore appears that we shall have to turn to extensive development of devices such as artificial lakes and cooling towers.

Designs for such lakes have already been drawn up and implemented for plants of moderate size. For the 1,000-megawatt power plant of the future a lake with a surface area of 1,000 to 2,000 acres would be required. (A 2,000-acre lake would be a mile wide and three miles long.) The recommended design calls for a lake only a few feet deep at one end and sloping to a depth of 50

feet at the other end. The water for cooling is drawn from about 30 feet below the surface at the deep end and is pumped through the plant at the rate of 500,000 gallons per minute, and the effluent, 20 degrees higher in temperature than the inflow, is discharged at the lake's shallow end. The size of the lake is based on a pumping rate through the plant of 2,000 acre-feet a day, so that all the water of the lake (averaging 15 feet in depth) is turned over every 15 days. Such a lake would dissipate heat to the air at a sufficient rate even in prolonged spells of unfavorable weather, such as high temperature and humidity with little wind.

Artificial cooling lakes need a steady inflow of water to replace evaporation and to prevent an excessive accumulation of dissolved material. This replenishment can be supplied by a small stream flowing into the lake. The lake itself can be built by damming a natural land basin. A lake complex constructed to serve not only for cooling but also for fishing might consist of two sections: the smaller one, in which the effluent is discharged, would be stocked with fishes tolerant to heat, and the water from this basin, having been cooled by exposure to

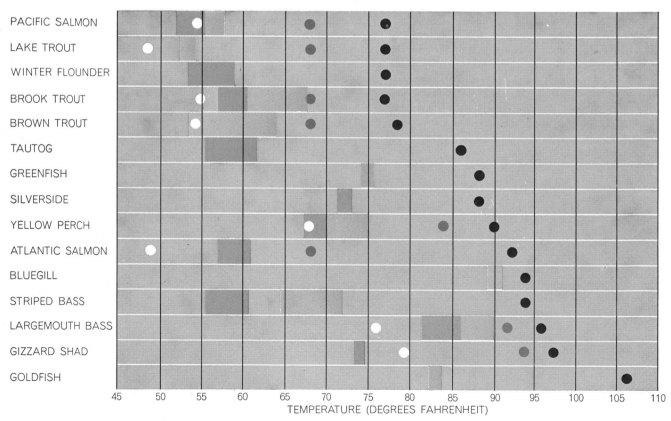

**FISHES VARY WIDELY** in temperature preference. Preferred temperature ranges are shown here for some species for which they have been determined in the field (*dark colored bands*) and, generally for younger fish, in the laboratory (*light color*). The chart also indicates the upper lethal limit (*black dot*) and upper limits recommended by the Federal Water Pollution Control Administration for satisfactory growth (*colored dot*) and spawning (*white dot*). Temperatures well below lethal limit can be in stress range.

the air, would then flow into a second and larger lake where other fishes could thrive.

In Britain, where streams are small, water is scarce and appreciation of aquatic life is high, the favored artificial device for getting rid of waste heat from power plants has been the use of cooling towers. One class of these towers employs the principle of removing heat by evaporation. The heated effluent is discharged into the lower part of a high tower (300 to 450 feet) with sloping sides; as the water falls in a thin film over a series of baffles it is exposed to the air rising through the tower. Or the water can be sprayed into the tower as a mist that evaporates easily and cools quickly. In either case some of the water is lost to the atmosphere; most of it collects in a basin and is pumped into the waterway or recirculated to the condensers. The removal of heat through evaporation can cool the water by about 20 degrees F.

The main drawback of the evaporation scheme is the large amount of water vapor discharged into the atmosphere. The towers for a 1,000-megawatt power plant would eject some 20,000 to 25,000 gallons of evaporated water per minute— an amount that would be the equivalent of a daily rainfall of an inch on an area of two square miles. On cold days such a discharge could condense into a thick fog and ice over the area in the vicinity of the plant. The "wet" type of cooling tower may therefore be inappropriate in cold climates. It is also ruled out where salt water is used as the coolant: the salt spray ejected from a single large power plant could destroy vegetation and otherwise foul the environment over an area of 160 square miles.

A variation of the cooling-tower system avoids these problems. In this refinement, called the "dry tower" method, the heat is transferred from the cooling water, through a heat exchanger something like an enormous automobile radiator, directly to the air without evaporation. The "dry" system, however, is two and a half to three times as expensive to build as a "wet" system. In a proposed nuclear power plant to be built in Vernon, Vt., it is estimated that the costs of operation and amortization would be $2.1 million per year for a dry system and $800,000 per year for a wet system. For the consumer the relative costs would amount respectively to 2.6 and 1 percent of the bill for electricity.

The public-utility industry, like other industries, is understandably reluctant to incur large extra expenses that add sub-

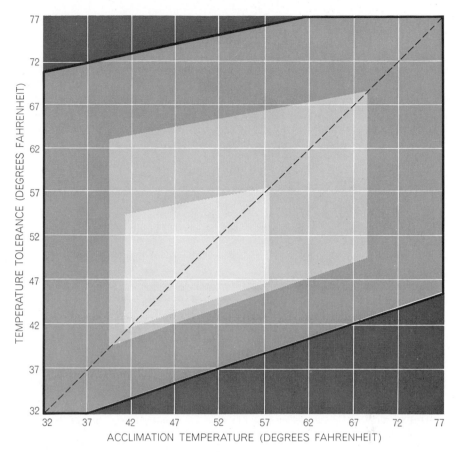

ACCLIMATION extends the temperature range in which a fish can thrive, but not indefinitely. For a young sockeye salmon acclimated as shown by the horizontal scale, spawning is inhibited outside the central (*light color*) zone and growth is poor outside the second zone (*medium color*); beyond the outer zone (*dark color*) lie the lethal temperatures.

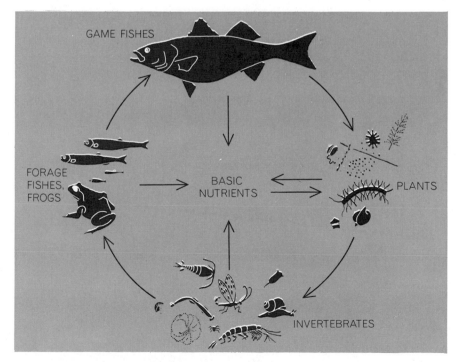

AQUATIC ECOSYSTEM is even more sensitive to temperature variation than an individual fish. A single game-fish species, for example, depends on a food chain involving smaller fishes, invertebrates, plants and dissolved nutrients. Any change in the environment that seriously affects the proliferation of any link in the chain can affect the harvest of game fish.

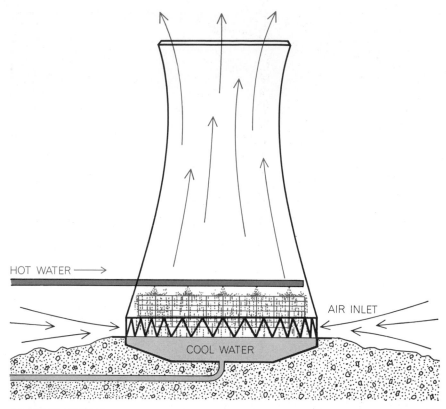

COOLING TOWER is one device that can dissipate industrial heat without dumping it directly into rivers or lakes. This is a "wet," natural-draft, counterflow tower. Hot water from the plant is exposed to air moving up through the chimney-like tower. Heat is removed by evaporation. The cooled water is emptied into a waterway or recirculated through the plant. In cold areas water vapor discharged into the atmosphere can create a heavy fog.

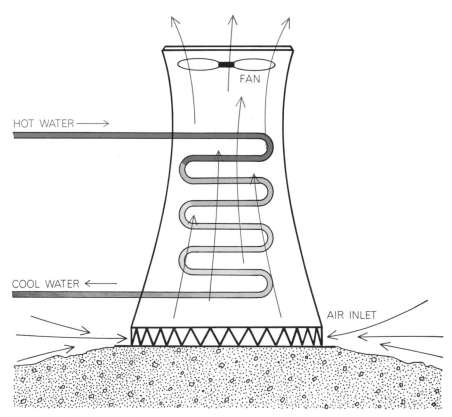

"DRY" COOLING TOWER avoids evaporation. The hot water is channeled through tubing that is exposed to an air flow, and gives up its heat to the air without evaporating. In this mechanical-draft version air is moved through the tower by a fan. Dry towers are costly.

stantially to the cost of its product and services. There is a growing recognition, however, by industry, the public and the Government of the need for protecting the environment from pollution. The Federal Water Pollution Control Administration of the Department of the Interior, with help from a national advisory committee, last year established a provisional set of guidelines for water quality that includes control of thermal pollution. These guidelines specify maximum permissible water temperatures for individual species of fish and recommend limits for the heating of natural waters for cooling purposes. For example, they suggest that discharges that would raise the water temperature should be avoided entirely in the spawning grounds of cold-water fishes and that the limit for heating any stream, in the most favorable season, should be set at five degrees F. outside a permitted mixing zone. The size of such a zone is likely to be a major point of controversy. Biologists would limit it to a few hundred feet, but in one case a power group has advocated 10 miles.

It appears that the problem of thermal pollution will receive considerable attention in the 91st Congress. Senator Edward M. Kennedy has proposed that further licensing of nuclear power-plant construction be suspended until a thorough study of potential hazards, including pollution of the environment, has been made. Senator Edmund S. Muskie's Subcommittee on Air and Water Pollution of the Senate Committee on Public Works last year held hearings on thermal pollution in many parts of the country.

Thermal pollution of course needs to be considered in the context of the many other works of man that threaten the life and richness of our natural waters: the discharges of sewage and chemical wastes, dredging, diking, filling of wetlands and other interventions that are altering the nature, form and extent of the waters. The effects of any one of these factors might be tolerable, but the cumulative and synergistic action of all of them together seems likely to impoverish our environment drastically.

Temperature, Gordon Gunter of the Gulf Coast Research Laboratory has remarked, is "the most important single factor governing the occurrence and behavior of life." Fortunately thermal pollution has not yet reached the level of producing serious general damage; moreover, unlike many other forms of pollution, any excessive heating of the waters could be stopped in short order by appropriate corrective action.

POWER PLANT being completed by the Tennessee Valley Authority on the Green River in Kentucky will be the world's largest coal-fueled electric plant. Its three wet, natural-draft cooling towers, each 437 feet in height and 320 feet in diameter at ground level, will each have a capacity of 282,000 gallons of water a minute, which they will be able to cool through a range of 27.5 degrees.

INDUSTRIAL PLANTS of various kinds can use towers to cool process water. These two five-cell cooling towers were built by the Marley Company for a chemical plant. They are wet, mechanical-draft towers of the cross-flow type: a fan in each stack draws air in through the louvers, across films of falling water and then up. The towers cool 120,000 gallons a minute through a 20-degree range.

# V

---

# WAR: THE ANGUISH
# OF RENUNCIATION

# V

---

# War: The Anguish of Renunciation

## Introduction

One of the most difficult things in the world is to recognize that a problem we are faced with is genuinely new. Innumerable thoughtful men have said that the quantitative change brought about by the atomic bomb and the intercontinental ballistic missile has created a situation that is qualitatively different from any we have faced before, a situation in which war is no longer a defensible method of settling international conflicts. We have difficulty accepting this view because we cannot help but recall comparable statements made previously, and previously disproved. For example, Stanley Leathes, one of the editors of the *Cambridge Modern History*, wrote in 1910, more than thirty years before the atomic bomb but only four years before the outbreak of the First World War: "On the whole, the existence of this tremendous military equipment makes for peace. The consequences of war would be felt in every household; and statesmen, as well as nations, shrink from the thought of a conflict so immense." Statesmen may still shrink from the thought—we hope they do—but we have no assurance that they shrink enough.

Looking to the animal kingdom for illumination we find that "The Fighting Behavior of Animals," as described by Irenäus Eibl-Eibesfeldt, is significantly different from war as we now know it. Among nonhuman animals, intraspecific conflict is limited and conventional. From time to time in the past human conflicts have enjoyed these characteristics; but not always. Looking to the future we see no way of assuring that future conflicts will be limited ones fought for conventional rewards.

What is it about the threat of future wars that poses a genuinely new problem? "Total war" has occurred before; in fact, the convention of sparing women and children has been observed only occasionally throughout man's history. The complete destruction of an entire national civilization is not novel either: Rome obliterated Carthage, and the Spaniards destroyed the Aztec and Incan civilizations.

The novelty we face is a product of the speed at which things can happen in the modern era. It is quite literally true that the only objective warning that we may have of the complete destruction of our national civilization may be given us no more than fifteen minutes in advance. So short a warning period (perhaps it should be called "a warning moment") is completely without precedent in the history of warfare. In the 1950's, when the RAND Corporation was carrying out strategic studies of modern warfare for the United States government, it considered that a "long war" would be one that lasted two days or more. What a contrast with the Hundred Years' War of the fourteenth century, or even the Thirty Years' War three centuries later! Earlier wars were part of the world's culture; a thermonuclear war will be the end of culture.

Connected with the suddenness of missile-borne warfare is the lack of time for feedback processes. A study made by Rosemary Coffey showed only one nation that continued a war beyond the point at which five per cent of its population had been lost. (The exception was Paraguay in its war against Brazil, Argentina, and Uruguay in 1865-1870. In this period an incredible 84 per cent of the male population was killed, and the end did not come until Francisco Lopez, the Paraguayan dictator, lost his life in battle.) If people are going to surrender, they must have time to reach the decision to do so. They must have time to assess their losses, and to conclude that they no longer want to live by that noble motto "It is better to die on your feet than to live on your knees." We can argue plausibly that most of the wars of the last hundred years have been won by the side with the greater industrial productivity. We cannot assume that this will be true in the future; as the economist Kenneth Galbraith has said: "We have probably had the last of the conflicts that, even remotely, could be called a Gross National Product war."

The feedback that the slow wars of the past made possible worked in another way as well. Seeing what he had done to the enemy, the victor may sometimes (but certainly not always) have grown sick of what he was doing. With the enemy thoroughly beaten, some of the "fun" may have passed out of the game. In a sense, when battle is joined face-to-face, some slight responsibility for what they are doing may be felt by the combatants. But it takes an act of great and improbable imagination for the man who pushes the button that launches ICBM's toward an unseen target to feel anything like a sense of responsibility. Even if he does, it won't make any difference.

We seem to be locked into a system from which there is no escape. It is a system with positive feedback; with runaway feedback; with too much "gain"— to use the radio engineer's term. The system has these characteristics:

*We* always tend to overestimate the extent and the efficacy of the enemy's weapons. In the 1950's, we Americans believed in the "bomber gap"—that is, that Russia had more bombers than we did. In the 1960's, we believed in an equally mythical "missile gap." What next?

Whatever weapon we are talking about, *our* supply of that weapon and our defenses against it must be "second to none."

We reject mirror-thinking: that is, we refuse to accept any suggestion to "put yourself in his place." In the moral sphere this is probably an inevitable concomitant of modern warfare. During the Second World War Himmler justified the Germans' use of human guinea pigs in testing the effects of sub-zero temperatures with these words: "I regard as traitors to their country those people who even today reject these human experiments and would rather see brave German soldiers die of the effect of chilling." (*Our* soldiers are human beings whose lives must be conserved; enemy soldiers seem less than human). From this it follows that any conceivable treatment of the enemy is morally right if it saves the lives of our soldiers. This, as a moral stance, is bad enough, but on a purely practical level the failure to resort to mirror-thinking is truly dangerous. Were we to use mirror-thinking systematically we would soon recognize that chronic overestimation of the enemy's strength plus a determination to be "second to none" produces a dynamic system that suffers from destructive feedback. The only stability in such a system is the stability of the grave.

To what extent is the danger of war still with us because of irrational feelings that are religious in character? Adolf Hitler, recalling the outbreak of the First World War, said: "I am not ashamed to acknowledge today that I was carried away by the enthusiasm of the moment and that I sank down upon my knees and thanked Heaven out of the fullness of my heart for the favor of having been permitted to live in such a time." Better men than Hitler have confessed to similar feelings. However ignoble its aim—and about that opinions differ—warfare has in the past brought out the best in many of its participants: whether we like it or not, we must not forget this fact. It was this paradox that led William James to pen his classic statement in 1901:

> Yet the fact remains that war is a school of strenuous life and heroism; and, being in the line of aboriginal instinct, is the only school that as yet is universally available. But when we gravely ask ourselves whether this wholesale organization of irrationality and crime be our only bulwark against effeminacy, we stand aghast at the thought, and think more kindly of ascetic religion. One hears of the mechanical equivalent of heat. What we now need to discover in the social realm is the moral equivalent of war: something heroic that will speak to men as universally as war does, and yet will be as compatible with their spiritual selves as war has proved itself to be incompatible.

James's "moral equivalent of war," though appealing, can hardly be the whole answer to the problem now facing us. James's approach might be called cosmical; the details of any scheme that enables us to avoid another world war must be pursued at what one might call a molecular level. Some of the more important of these schemes are discussed in Newman's review of four significant books on various aspects of the threat of war. Among other things, these books raise the question of the applicability of "game theory" to the settling of international disputes.

In the game of threat and counterthreat the nationalist wants to know only one thing: "How can I win?" The internationalist wants to know "How can we prevent the onset of destructive positive feedback?" Anatol Rapoport in "The Use and Misuse of Game Theory" discusses these matters. Scientists, by temperament theory-loving, enjoy the intellectual challenge of game theory. Whether game theory, or indeed any other rational approach, is capable of solving the practical problem of maintaining the peace is a moot point. Ralph E. Lapp, in his book *The Weapons Culture*, made an interesting observation:

> Many books have been written about deterrent doctrine, and the literature abounds with analyses made by operations analysts, war-gamers, and defense intellectuals. However, when the chips are down and hard decisions have to be made, the men at the center of the stage are the political leaders. These hardheaded men are apt to find war-gaming [merely] a heady intellectual exercise. We should recall that Kennedy's Cuban crisis, as chronicled by at least three men close to the scene, fails to illuminate a single war-gamer on the stage or in the wings.

Probably the most famous book review ever carried by *Scientific American* is James R. Newman's "Two Discussions of Thermonuclear War." The beginning of this review is unforgettable: "Is there really a Herman Kahn? It is hard to believe." Newman's revulsion from Kahn's book *On Thermonuclear War* is apparent in every line. In another journal (for instance, *The New York Review of Books*) Newman's impassioned prose might not have evoked much comment. But the writing in *Scientific American* is for the most part rather aseptic in its moral tone. Newman's book review stood out like a sore thumb; but it was much enjoyed as a literary production even by those who were not sure that he was right in his moral assessment.

It can be argued that Herman Kahn was (and is) continuing the great scientific tradition of being rational about everything, including irrational and devastating events. We do not approve of death by cancer, but we expect medical men to approach the problem rationally to see what can be done about it. Closer to the point, we do not approve of war, but when it comes to dealing with the casualties of war we see no way to dispute the rationality of the system called *triage*. Following any great battle there are always more wounded than can be dealt with by the limited number of medical personnel available. How should the doctors deal with them? On a first-come-first-served basis? This seems hardly defensible. The military medical man's answer to the problem is to divide the casualties into three groups: the slightly wounded, who will get along well enough by themselves without any attention; the seriously wounded, whose prospects of survival will be greatly improved by medical attention; and finally the desperately wounded, for whom a doctor can do little or nothing or whose wounds will require a disproportionately large amount of medical attention. It is the tradition of military medicine that when a choice must be made the first and the third groups are abandoned, and all attention is given to the second group. Many people find it morally repugnant that anyone should make such a choice—should "play God," as the saying goes. But rationally, is any other course of action preferable?

Is the moral repugnance implicit in Newman's question, "Is there really a Herman Kahn?" really defensible? Don't we need theorists who try to deal rationally with catastrophes? Or—to take the other side—do we, by encouraging such theorists to "think about the unthinkable," (to borrow a phrase from another book of Kahn's) thereby actuate the mechanism of the "self-fulfilling prophecy" and increase the probability that the unthinkable, having become thinkable, will in fact come into being? These are hard questions.

The threat posed by missile-delivered atomic bombs is discussed by P.M.S. Blackett in "Steps Toward Disaramament." To control this threat there clearly must be cooperation among sovereign nations. But how can sovereign nations be persuaded to cooperate, particularly when (as is almost invariably the rule) they distrust each other? The problem is clearly tied up closely with the reliability of the information that each has about the other.

We are naturally reluctant to sign a mutual restraint pact with a potential enemy if we think him capable of violating the terms of the pact without our being able to discover it. Sir Edward Bullard discusses part of the problem in "The Detection of Underground Explosions." This is a highly technical subject, and one in which much of the essential information is "classified," i.e., secret. There is, however, a clear trend in this field: year by year our technical ability to detect clandestine high-energy explosions in foreign territory has become more effective.

When the reliability of the detection methods had became sufficiently great, "National Security and the Nuclear-Test Ban" became reconcilable with each other, leading to the international agreement of 1963, here discussed by Jerome B. Wiesner and Herbert F. York. Because these two men have been unusually close to the center of scientific knowledge and power in this country, it is most significant that they should have reached the conclusion given in the next to last paragraph of their paper.

> Both sides in the arms race are thus confronted by the dilemma of steadily increasing military power and steadily decreasing national security. *It is our considered professional judgment that this dilemma has no technical solution.*

Thermonuclear war, then, is like the game of tick-tack-toe: there is no technical strategy that will permit one to win the game against an experienced opponent. The game of tick-tack-toe can never be won. The game of nuclear warfare must always be lost. If we are to survive, our strategy must transcend the level of technical devices. Moral and political innovations are indispensable.

A quasi-solution to the problem of living in a nuclear world is described by Arthur I. Waskow in "The Shelter-Centered Society." We see this as a quasi-solution because it suffers from two mammoth defects: one material, the other psychological. The material defect is the sheer cost of a really effective shelter program; estimates vary tremendously, but all are stated in billions of dollars. It is doubtful if any adequate shelter program could be put into effect in the United States without the expenditure of at least 25 per cent of one year's Gross National Product. One might argue that, given the necessity, this would be an easy sacrifice to make—that we all would merely have to cut down our scale of living by 2.5 per cent per year for ten successive years. However, the matter is not quite so simple. The Gross National Product includes not only such productive outputs as steel and food, but *all* transactions in which money changes hands, including the most trivial and wasteful, e.g., all the betting that is done in Las Vegas, the commissions of all real estate salesmen, and the fees of all advertising agencies. None of these "products" can build atomic shelters.

The psychological problems of converting to a shelter-centered society can be divided into two categories: external and internal. The external problem is strategic: to build a good system would probably take five or ten years of massive sacrifice. How could we make such sacrifices without alerting our putative enemies? Were we to learn that Russia had embarked on such a comprehensive program we would naturally conclude that she was doing so preparatory to unleashing an all-out missile-and-atom attack on us. If that is what we would conclude, we must also conclude that Russia would react in the same way to a similar action by us. Let us not forget to look in that strategic mirror.

The internal psychological problem is this: what would living in such a world do to us? Man is a truly adaptable animal, but even adaptation has its limits. For literally hundreds of millions of years, natural selection has been selecting for those genes that make for success in an animal that can move freely on the earth's surface—ultimately the animal we call *Homo sapiens*. The genes that produce any species constitute an intricately organized system, and no simple change in that system can produce successful adaptation to a markedly different existence. To adapt to a shelter existence would, even with the most intelligent genetic intervention by man himself, probably take thousands of years. This is a silly approach to a crisis.

A distressing aspect of the pathological response we call modern war is that it is rooted in the normal physiological processes of society. (There is nothing novel about this statement: all pathology is rooted in physiology. We sometimes forget this and emotionally revert to diatribes against "evil.") Normal processes may push us toward suicide—individually or collectively. Many of the people who inveigh most strongly against these processes are marginal members of society—but by no means all. Now and then a member of the Establishment calls attention to our errors. A most surprising (and gratifying) clarion call of this sort came from President Eisenhower when he gave his farewell address on 17 January 1961. The relevant section of his address follows:

A vital element in keeping the peace is our military establishment. Our arms must be mighty, ready for instant action, so that no potential aggressor may be tempted to risk his own destruction.

Our military organization today bears little relation to that known by any of my predecessors in peacetime, or indeed by the fighting men of World War II or Korea.

Until the latest of our world conflicts, the United States had no armaments industry. American makers of plowshares could, with time and as required, make swords as well. But now we can no longer risk emergency improvisation of national defense; we have been compelled to create a permanent armaments industry of vast proportions. Added to this, three and a half million men and women are directly engaged in the defense establishment. We annually spend on military security more than the net income of all United States corporations.

This conjunction of an immense military establishment and a large arms industry is new in the American experience. The total influence—economic, political, even spiritual—is felt in every city, every state house, every office of the federal government. We recognize the imperative need for this development. Yet we must not fail to comprehend its grave implications. Our toil, resources, and livelihood are all involved; so is the very structure of our society.

In the councils of government we must guard against the acquisition of unwarranted influence, whether sought or unsought, by the military-industrial complex. The potential for the disastrous rise of misplaced power exists and will persist.

We must never let the weight of this combination endanger our liberties or democratic processes. We should take nothing for granted. Only an alert and knowledgeable citizenry can compel the proper meshing of the huge industrial and military machinery of defense with our peaceful methods and goals, so that security and liberty may prosper together.

Nothing that has happened since 1961 has given the slightest cause for doubting the reality of the military-industrial complex and the power it exerts. The only addition that one might make now—and this is saddening to admit—is to change the phrase to "the military-industrial-scientific complex." Year by year more and more scientists and engineers have been caught up in the making of arms, and the salaries of such men have risen steadily—even faster than the Gross National Product. There more than a few indications that it is as difficult for scientists as for other men to be objective when their rice bowls are endangered. Their knowledge and their prestige make them unusually powerful, for good or for evil.

The gravest military threat to the continued survival of the United States is its own anti-ballistic missiles system. If the missiles ever "work" at all, they will do so only by releasing fantastic quantities of energy and fallout products, not only from the enemy's missiles but also from our own. To paraphrase an old joke: with such a defense, who needs enemies?

To survive our own "defense," let alone any attacking missiles that might penetrate it, we would have to build a massive fallout shelter system. But, as we saw earlier, building this system would alert and alarm our enemies. Thus we are brought right back again to Wiesner and York's conclusion that there is no "technical solution" to the problem of the arms race within the framework of the game we call "war."

An anti-ballistic missile system, like a shelter system, would be ruinously expensive. But the word "expensive" is a slanted word: that's the way the average citizen sees it. That which seems merely an expensive expedient to him is a source of profit to manufacturers, a new toy to the military whose lives are committed to the game of war, and an economic necessity to the limited cadre of scientists and technologists who fear that society would have no place for them did it not embark on some such nonproductive project. In 1966 and 1967 the pressure to build an ABM system grew month by month, reaching a peak in the summer of 1967, as one can easily see by reading the pages of *The Wall Street Journal* of that period. From the time of his appointment in 1961 Secretary of Defense Robert McNamara had resolutely opposed an ABM system, holding it to be a fantastically wasteful method of committing national suicide. As the summer of 1967 wore on, rumors that McNamara was on his way out of office appeared with increasing frequency, coupled with rumors that the military-industrial-scientific complex was going to have its way despite the conclusions of the best scientific analysis of the problem.

Finally, on 18 September 1967, McNamara capitulated, just six years, seven months, and one day after President Eisenhower's warning. Secretary McNamara's statement, which was given as an address at the United Press International Editors and Publishers Convention in San Francisco, probably deserves recognition as one of the great tragic statements of history. Unfortunately, the events that would prove to all that it deserves this title will—if they materialize—put an end to history.

According to legend, when Galileo was found guilty of teaching that the earth moved around the sun, he outwardly accepted the sentence of recantation imposed upon him by the Inquisition, but at the same moment muttered in his beard, "*Eppur si muove*"—"Nevertheless it moves." The story is apocryphal; but the fiction of the seventeenth century had its factual counterpart in the twentieth.

McNamara's speech in San Francisco is that counterpart. Secretary McNamara, probably the most competent chief of the defense establishment we have ever had, is an "organization man" to the very core of his being. He is also a man who understands the scientist's high regard for the truth. The unresolved stresses produced by the combination of these two traits are clearly evident in his San Francisco speech. Purely as a study in rhetoric the speech is unique and fascinating. Unintentionally it reveals the torment of a modern Galileo, the penitent heretic who recants, but still maintains: *Eppur si muove.*

Shortly after delivering the speech Robert McNamara resigned.

A technical reply to the euphoric dream of protecting the nation by "Anti-Ballistic-Missile Systems" is given in the article by Richard L. Garwin and Hans A. Bethe. All discussions of the contemporary technology of war suffer from the disadvantage of dealing with "classified" information. We can never be sure that a published view is justified by the unpublishable facts. The correspondence appended to Garwin and Bethe's article shows the sort of "stand-off" between opponents that results.

When Richard M. Nixon became President in 1969 the quiescent proposals for construction of an ABM system were revitalized, but with several significant changes. It was now admitted that it would probably cost more than the previously estimated five billion dollars—perhaps eight billion. It was also hinted that eight billion might not be the final bill. Whereas in McNamara's day it had always been insisted that the system was designed only to defend us against a possible Chinese attack, it was now admitted that it would be designed to ward off the Russians as well. (Rationalizations, no less than war, can escalate.) Finally, the system was rechristened. Its earlier name was the "Sentinel System," which had a rather military ring to it. As a concession to the peace-loving elements of the nation, President Nixon and his advisers judiciously called their burgeoning proposal the "Safeguard System." What could sound more pacific than that? Madison Avenue had evidently joined with the military-industrial-scientific complex to protect us from all evil, including anxiety. Very considerate of them.

Where does that leave us? In a quandary—a quandary whose boundaries are described by George W. Rathjens in "The Dynamics of the Arms Race." Our ability to plan for the future seems to suffer from asymmetry: we know a great deal about how to make our prospects of survival worse, but very little about how to make them better. Strategies of warfare, which were capable of producing a winner in the days before the missile-delivered atomic bomb, can now produce nothing but losers. We must find new strategies—strategies that are categorically different from the unimaginative ones we now compulsively pursue. Never before has mankind faced so serious a threat to its existence; never before have imaginative men been presented with so great a challenge to their ingenuity and wisdom.

# 37

# The Fighting Behavior of Animals

IRENAUS EIBL-EIBESFELDT

*December 1961*

Fighting between members of the same species is almost universal among vertebrates, from fish to man. Casual observation suggests the reason: Animals of the same kind, occupying the same niche in nature, must compete for the same food, the same nesting sites and the same building materials. Fighting among animals of the same species therefore serves the important function of "spacing out" the individuals or groups in the area they occupy. It thereby secures for each the minimum territory required to support its existence, prevents overcrowding and promotes the distribution of the species. Fighting also arises from competition for mates, and thus serves to select the stronger and fitter individuals for propagation of the species. It is no wonder, then, that herbivores seem to fight each other as readily as do carnivores, and that nearly all groups of vertebrates, except perhaps some amphibians, display aggressive behavior.

A complete investigation of fighting behavior must take account, however, of another general observation: Fights between individuals of the same species almost never end in death and rarely result in serious injury to either combatant. Such fights, in fact, are often highly ritualized and more nearly resemble a tournament than a mortal struggle. If this were not the case—if the loser were killed or seriously injured—fighting would have grave disadvantages for the species. The animal that loses a fight is not necessarily less healthy or less viable; it may simply be an immature animal that cannot withstand the attack of a mature one.

In view of the disadvantages of serious injury to a member of the species, evolution might be expected to have exerted a strong selective pressure against aggres-

sive behavior. But spacing out through combat was apparently too important to permit a weakening of aggressive tendencies; in fact, aggressiveness seems to have been favored by natural selection. It is in order to allow spacing out—rather than death or injury—to result from fighting that the ceremonial combat routines have evolved.

Investigators of aggressive behavior, often strongly motivated by concern about aggressive impulses in man, have usually been satisfied to find its origin in the life experience of the individual animal or of the social group. Aggressiveness is said to be learned and so to be preventable by teaching or conditioning. A growing body of evidence from observations in the field and experiments in the laboratory, however, points to the conclusion that this vital mode of behavior is not learned by the individual but is innate in the species, like the organs specially evolved for such combat in many animals. The ceremonial fighting routines that have developed in the course of evolution are highly characteristic for each species; they are faithfully followed in fights between members of the species and are almost never violated.

All-out fights between animals of the same species do occur, but usually in species having no weapons that can inflict mortal injury. Biting animals that can kill or seriously injure one another are usually also capable of quick flight. They may engage in damaging fights, but these end when the loser makes a fast getaway. They may also "surrender," by assuming a submissive posture that the winner respects. Konrad Z. Lorenz of the Max Planck Institute for the Physiology of Behavior in Germany has described such behavior in wolves and dogs. The fight begins with an ex-

change of bites; as soon as one contestant begins to lose, however, it exposes its vulnerable throat to its opponent by turning its head away. This act of submission immediately inhibits further attack by its rival. A young dog often submits by throwing itself on its back, exposing its belly: a pet dog may assume this posture if its master so much as raises his voice. Analogous behavior is common in birds: a young rail attacked by an adult turns the back of its head —the most sensitive part of its body— toward the aggressor, which immediately stops pecking. Lorenz has pointed out that acts of submission play a similar role in fights between men. When a victim throws himself defenseless at his enemy's feet, the normal human being is strongly inhibited from further aggression. This mechanism may now have lost its adaptive value in human affairs, because modern weapons can kill so quickly and from such long distances that the attacked individual has little opportunity to appeal to his opponent's feelings.

Most animals depend neither on flight nor on surrender to avoid damaging fights. Instead they engage in a ceremonial struggle, in the course of which the contestants measure their strength in bodily contact without harming each other seriously. Often these contests begin with a duel of threats—posturings, movements and noises—designed to cow the opponent without any physical contact at all. Sometimes this competition in bravado brings about a decision; usually it is preliminary to the remainder of the tournament.

On the lava cliffs of the Galápagos Islands a few years ago I observed such contests between marine iguanas (*Amblyrhynchus cristatus*), large algae-eating lizards that swarm by the hun-

*a*

*b*

*c*

*d*

CICHLID FISH (*Aequidens portalegrensis*) perform a ritual fight that begins with a threat and proceeds to bodily contact without damage to either. After a formal display (*a*) the fish fan their tails to propel currents of water at each other (*b*). Then the rivals grasp each other with their thick-lipped mouths and push and pull (*c*) until one gives up and swims away (*d*).

dreds over the rocks close to shore. During the breeding season each male establishes a small territory by defending a few square yards of rock on which he lives with several females. If another male approaches the territorial border, the local iguana responds with a "display." He opens his mouth and nods his head, presents his flank to his opponent and parades, stiff-legged, back and forth, his apparent size enlarged by the erection of his dorsal crest. If this performance does not drive the rival off, the resident of the territory attacks, rushing at the intruder with his head lowered. The interloper lowers his head in turn and the two clash, the tops of their heads striking together. Each tries to push the other backward. If neither gives way, they pause, back off, nod at each other and try again. (In an apparent adaptation to this mode of combat the head of the marine iguana is covered with horn-like scales.) The struggle ends when one of the iguanas assumes the posture of submission, crouching on its belly. The winner thereupon stops·charging and waits in the threatening, stiff-legged stance until the loser retreats [*see illustration on following page*]. A damaging fight is triggered only when an invader does not perform the ceremonies that signal a tournament; when, for example, the animal is suddenly placed in occupied territory by a man, or crosses another animal's territory in precipitous flight from an earlier contest. On these occasions the territory owner attacks by biting the intruder in the nape of the neck. Female iguanas, on the other hand, regularly engage in damaging fights for the scarce egg-laying sites, biting and shaking each other vigorously.

The lava lizard (*Tropidurus albemarlensis*) of the larger Galápagos Islands engages in a similar ceremonial fight that begins with the rivals facing each other, nodding their heads. Suddenly one of them rushes forward, stands alongside his opponent and lashes him with his tail once or several times, so hard that the blows can be heard several yards away. The opponent may reply with a

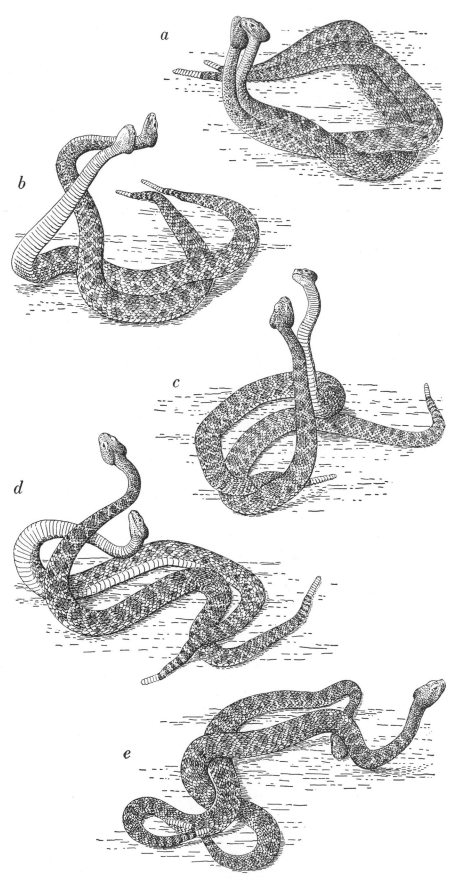

RATTLESNAKES (*Crotalus ruber*) perform the combat dance shown in these drawings based on a study by Charles E. Shaw of the San Diego Zoo. The rivals move together (*a*) and then "Indian wrestle" head to head (*b*). Sometimes they face each other, weaving and rubbing their ventral scales (*c*). Finally one lashes out and throws (*d*) and pins (*e*) the other.

MALE MARINE IGUANA (*Amblyrhynchus cristatus*) of the Galápagos Islands defends his territory against intruding males. As the rival approaches (*a*), the territory owner struts and nods his head. Then the defender lunges at the intruder and they clash head on (*b*), each seeking to push the other back. When one iguana (*left at "c"*) realizes he cannot win, he drops to his belly in submission.

tail-beating of his own. Then the attacker turns and retreats to his original position. The entire procedure is repeated until one of the lizards gives up and flees.

According to Gertraud Kitzler of the University of Vienna, fights between lizards of the central European species *Lacerta agilis* may terminate in a curious manner. After an introductory display one lizard grasps the other's neck in his jaws. The attacked lizard waits quietly for the grip to loosen, then takes his turn at biting. The exchange continues until one lizard runs away. Often, however, it is the biter, not the bitten, that does the fleeing. The loser apparently recognizes the superiority of the

winner not only by the strength of the latter's bite but also by his unyielding resistance to being bitten.

Beatrice Oehlert-Lorenz of the Max Planck Institute for the Physiology of Behavior has described a highly ritualized contest between male cichlid fish (*Aequidens portalegrensis*). The rivals first perform a display, presenting themselves head on and side on, with dorsal fins erected. Then they beat at each other with their tails, making gusty currents of water strike the other's side. If this does not bring about a decision, the rivals grasp each other jaw to jaw and pull and push with great force until the loser folds his fins and

swims away [see *illustration on page 276*]. John Burchard of the same institution raised members of another cichlid species (*Cichlasoma biocellatum*) in isolation from the egg stage and found that they fought each other in the manner peculiar to their species.

The ritualization of fighting behavior assumes critical importance in contests between animals that are endowed with deadly weapons. Rattlesnakes, for example, can kill each other with a single bite. When male rattlesnakes fight, however, they never bite. Charles E. Shaw of the San Diego Zoo has described the mode of combat in one species (*Crotalus ruber*) in detail. The two snakes glide along, side by side, each with the forward third of its length raised in the air. In this posture they push head against head, each trying to force the other sideways and to the ground, in accordance with strict rules reminiscent of those that govern "Indian wrestling." The successful snake pins the loser for a moment with the weight of his body and then lets the loser escape [see the illustration on page 277]. Many other poisonous snakes fight in a similar fashion.

Among mammals, the fallow deer (*Dama dama*) engages in a particularly impressive ceremonial fight. The rival stags march along side by side, heads raised, watching each other out of the corners of their eyes. Suddenly they halt, turn face to face, lower their heads and charge. Their antlers clash and they wrestle for a while. If this does not lead to a decision, they resume their march. Fighting and marching thus alternate until one wins. What is notable about this struggle is that the stags attack only when they are facing each other. A motion picture made by Horst Siewert of the Research Station for German Wildlife records an occasion on which one deer turned by chance and momentarily exposed his posterior to his opponent. The latter did not take advantage of this opportunity but waited for the other to turn around before he attacked. Because of such careful observance of the rules, accidents are comparatively rare.

Mountain sheep, wild goats and antelopes fight similar duels with their horns and foreheads, the various species using their horns in highly specific ways. From observation of clashes between rapier-horned oryx antelope (*Oryx gazella beisa*) and other African antelope, Fritz Walther of the Opel Open-Air Zoo for Animal Research concludes that the function of the horns is to lock the heads of the animals together as they engage in a pushing match. In one instance a

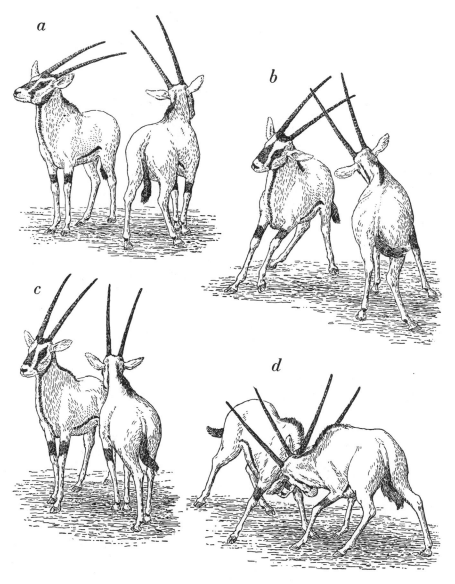

ORYX ANTELOPE (*Oryx gazella beisa*) has rapier-shaped horns but does not gore his rival. Two bulls begin combat with a display (*a*), then fence with the upper portion of their horns (*b*). After a pause (*c*) the rivals clash forehead to forehead (*d*) and push each other, using their horns to maintain contact. Drawings are based on observations by Fritz Walther.

duel between two oryx antelope in the wilderness was observed to begin with a display in which the two bulls stood flank to flank, heads held high. Then they came together in a first clash, only the upper third of their horns making contact. After a pause the animals charged again, this time forehead to forehead. They maintained contact by touching and beating their horns together [see illustration on preceding page]. Oryx antelope never use their horns as daggers in intraspecies fights. One hornless bull observed by Walther carried out the full ritual of combat as if he still had horns. He struck at his opponents' horns and missed by the precise distance at which his nonexistent horns would have made contact. Equally remarkable, his opponents acted as though his horns were in place and responded to his im-

NORWAY RATS (*Rattus norvegicus*) fight in a species-specific manner whether they are raised in isolation or with a group of other rats. The aggressor approaches, displaying his flank and arching his back (*a*). Then, standing on their hind legs, the two rats wrestle. They push with their forelegs (*b*) and sometimes kick with their hind legs (*c*). If one rat is forced to his back as they tussle (*d*), he sometimes gives up; otherwise the tournament phase ends and the real fight begins with a serious exchange of bites (*e*).

aginary blows.

Until field observations of this kind had accumulated in support of the innateness of fighting behavior, laboratory experiments had made a strong case for the notion that such behavior is learned. Experiments by J. P. Scott of the Roscoe B. Jackson Memorial Laboratory in Bar Harbor, Me., had indicated, for example, that a rat or a mouse reacts aggressively toward another rat or mouse primarily because of pain inflicted by a nestmate early in life. Scott suggests that aggressiveness should therefore be controllable by a change in environment; in other words, rats that have had no early experience of pain inflicted by another rat should be completely unaggressive.

To test this conclusion I raised male Norway rats in isolation from their 17th day of life, an age at which they do not show any aggressive behavior. When each was between five and six months old, I put another male rat in the cage with him. At first the hitherto isolated rat approached the stranger, sniffed at him and sometimes made social overtures. But this never lasted long. The completely inexperienced rat soon performed the species-specific combat display—arching his back, gnashing his teeth, presenting his flank and uttering ultrasonic cries. Then the two rats pushed, kicked and wrestled, standing on their hind legs or falling together to the ground. Sometimes the fights ended at this point, the rat that landed on his back giving up and moving away. But usually the rats went on to exchange damaging bites. The patterns of display, tussling and biting were essentially the same in the case of the inexperienced rats as in the case of those who had

been brought up with other rats and were faced by an outsider [*see illustration on page 280*]. The steps in the ritual are apparently innate and fixed behavior patterns; many of the movements seem to be available to each rat like tools in a toolkit.

Raising rats in groups, where there was an opportunity for young rats to undergo early painful experiences, provided another check on the Scott theory. The members of a group displayed almost no aggressive behavior toward each other. The few fights that took place rarely included biting; for the most part the animals merely pushed each other with their paws. But when a stranger was introduced into the group, he was attacked viciously and was hurt. This agrees with observations of wild Norway rats; they live peacefully together in large packs but attack any rat not a member of their group. Because the attacked animal is able to escape, the species has not developed a tournament substitute for biting. In the laboratory a strange rat introduced into a colony from which it cannot escape is likely to be killed. In sum, the experiments demonstrate that aggressiveness is aroused in adult male rats whenever a stranger enters the territory, even when the defender has had no painful experience with members of its species. Similar experiments on polecats (*Putorius putorius*) have shown the same results.

The view that aggressiveness is a basic biological phenomenon is supported by physiological studies of the underlying neural and hormonal processes. Some investigators have actually elicited fighting behavior in birds and mammals by stimulating specific areas of the brain

with electrical currents. The mind of a newborn animal is not a blank slate to be written on by experience. Aggressive behavior is an adaptive mechanism by which species members are spaced out and the fittest selected for propagation. Learning is no prerequisite for such behavior, although it probably has an influence on the intensity and detailed expression of aggressiveness.

In the human species, it seems likely, aggressive behavior evolved in the service of the same functions as it did in the case of lower animals. Undoubtedly it was useful and adaptive thousands of years ago, when men lived in small groups. With the growth of supersocieties, however, such behavior has become maladaptive. It will have to be controlled—and the first step in the direction of control is the realization that aggressiveness is deeply rooted in the history of the species and in the physiology and behavioral organization of each individual.

In this connection, it should be emphasized that aggressiveness is not the only motive governing the interaction of members of the same species. In gregarious animals there are equally innate patterns of behavior leading to mutual help and support, and one may assert that altruism is no less deeply rooted than aggressiveness. Man can be as basically good as he can be bad, but he is good primarily toward his family and friends. He has had to learn in the course of history that his family has grown, coming to encompass first his clan, then his tribe and his nation. Perhaps man will eventually be wise enough to learn that his family now includes all mankind.

# 38

# Preventing a Third World War

JAMES R. NEWMAN

*February 1959*

A review of *The Causes of World War Three*, by C. Wright Mills. New York: Simon and Schuster, 1958; *No More War!*, by Linus Pauling. New York: Dodd, Mead, 1958; *Inspection for Disarmament*, edited by Seymour Melman. New York: Columbia University Press, 1958; and *Peace or Atomic War?*, by Albert Schweitzer. New York: Holt, Reinhart and Winston, 1958.

Where there is an ear for reasoned argument these books will do a service. They are quite different books by quite different men, but they share a common concern: the continued existence of the human species. A sociologist examines the forces thrusting the world toward war; a chemist assesses the effects of radioactive fallout; a group of scientists and engineers surveys the feasibility of an inspection system for disarmament; a physician pleads for an end of the nuclear-arms race. The tone of the books is also quite different. One sounds like a trumpet, one is cautious and analytical, one is restrained but remorseless, one is eloquent but gentle. All, however, reflect a deep awareness of man's peril: the probable annihilation of the human race in the event of another war.

One may differ with Mills's analysis of the causes of the next war, but it cannot be denied that he is a man of courage and of conscience. He has written an angry essay. It is none the worse for that; a torch must burn to give light. He confronts three main questions. Do men make history, and so make wars? Are we drifting blindly into the final catastrophe, or are certain decisions and policies thrusting us toward it? Assuming that it lies within our power to avert war, what ought we to do?

Take the first question. It is often said

that certain great events such as wars are inevitable. Men are trapped by circumstances; fate makes the decisions. This ancient idea, gradually transformed into the belief that the Deity has fixed a great plan which determines every sparrow's fall is, among thinking men at least, obsolete. Nature is hard but not malicious; the stars are indifferent. Irresistible forces, it may be admitted, shape the universe, but man can see and judge and know and create and decide for himself. His freedom is bounded, but within its small sphere it is sovereign. The "thinking reed" is both the weakest and the strongest thing in nature. There is, however, another conception of fate that is not obsolete. It is in fact, Mills says, indispensable for adequate reflection on human affairs. According to this conception history is "the summary and unintended result of innumerable decisions of innumerable men." The men do not form an identifiable class; the decisions are not in themselves consequential enough for the results to have been foreseen. Like crowds of particles the decisions collide, coalesce and add up to the blind result—the historical event, which, as it were, is autonomous. This is the fate Karl Marx had in mind when he wrote in *The Eighteenth Brumaire of Louis Bonaparte,* "Men make their own history, but they do not make it just as they please."

Thus understood, fate is not, in Mills's words, a universal constant. It depends upon social structure and especially upon the concentration of power. Fate is a name we give to power so diffused and fragmented that we cannot discern the mechanics of its use; it is a name like probability, which baptizes our ignorance. But in a society where power is visibly concentrated, history is not drift or fate. It is the sum of the vital decisions made by the groups of men which hold

power. What they do—or fail to do—makes history; what others do is of little account: the others are the "utensils" of history-makers and the "objects" of history-making.

Given this diagnosis, we shall with Mills take it for granted that a handful of "high and mighty" of the U.S.S.R. make history; but what about the U. S.? Mills's thesis, already set forth in *The Power Elite*, is here reaffirmed. In our country history is made, he says, by a small group of men: "the high military, the corporation executives, the political directorate." This is the top level. The middle level is composed of "a drifting set of stalemated forces," and at the bottom is a passive, increasingly powerless "mass society." The power elite makes the decisions even though the formal democratic machinery has been relegated to the middle groups. The "great public" votes and even elects, but to what purpose? Of the men it elects only a few help make decisions, and then only when they are admitted to the elite.

Mills now warms to his argument. The power elite (in the U.S.S.R. as well as in the U. S.) is possessed by the "military metaphysic" that the constant threat of violence and the maintenance of a "balance of fright" are the essentials to a condition of peace. Many of the elite also believe that a war economy is the indispensable underpinning of economic prosperity. They are committed therefore to the arms race, which may in itself be the immediate cause of World War III. Political apathy of "publics" and moral insensibility of "masses" in both communist and capitalist worlds allow the economic and military causes of war to operate. Leading intellectual, scientific and religious circles are either confused and permissive, or join in the cold war. Few intellectuals put pressure on the elite to change its policies; fewer still set forth alternatives. In framing this indictment Mills is careful to say that the thrust to war is not an elite "plot," either here or in the U.S.S.R. The elite of each country has its "war parties" and its "peace parties," and both elites have what Mills calls "crackpot realists." They are men so rigidly focused on the next step "that they become creatures of whatever the next step brings"; they are also men who cling rigidly to general principles and who "join a high-flying moral rhetoric with an opportunist crawling among a great scatter of unfocused fears and demands." Practical next steps and round, hortatory principles, but no program: this is the main content, in Mills's view, of today's struggles of "politics."

Steadily we move towards the abyss. What is to be done? Mills's appeal is addressed mainly to intellectuals. They must stop fighting the cold war. They must make contact with their opposite numbers "among those now officially defined as our enemy." ("With them, we ought to make our own separate peace.") They must help educate one another. They must also remedy the default of religion and help awaken the public conscience; for religion itself is "morally dead" in the U. S., and ministers of God, who are responsible for the moral cultivation of conscience, "with moral nimbleness blunt conscience, covering it up with peace of mind." Scientists should honor publicly those, like the 18 German physicists, who have made their declarations for peace and against working on the new weaponry. Scientists should attempt "to deepen the split among themselves and to debate it." They should denounce secrecy. They should refuse to become members of a "Science Machine" under military authority. They should refuse to make weapons and boycott all research projects directly or indirectly relevant to the military. These are among the steps that Mills says would begin the practice of a professional code. The scientist, by adopting such a code, would reject "fate," for he would thereby declare his resolve to take at least his own fate into his own hands.

Mills's book is primarily a polemic and a sermon. It is, as intended, provoking as well as provocative. Patriotic groups will not clasp his thesis to their bosoms, and many disinterested students of society and power will differ strongly with his analysis of the causes of war. No open-minded person, however, will mistake this for the work of a hack or pitchman for the official lines of either side. Mills is his own man.

Linus Pauling, though gentler, is also his own man. His book is a powerful statement of the case for nuclear disarmament.

That war is today an insane method of solving disputes is a truth so obvious that it is hard to prove. Men are apt to acknowledge it, as they acknowledge their mortality, and then go about their business. But the proposition that we all have to die some day is not the same as that we all have to die the same day. Until now it had always been assumed that, though men were mortal, man would endure. This assumption, as Pauling shows, has become questionable.

Five chapters of his book discuss fallout. They give one of the clearest summaries of the problem I have seen. While there is considerable disagreement as to what constitutes "safe" limits of radiation with respect both to somatic and genetic injury, no responsible scientist doubts that wide "margins of uncertainty" (these words are from the United Nations report on radiation) surround all estimates. Before the subject became a hot military and political potato, it was accepted that even small amounts of radiation produce mutations, and that almost all mutant genes are bad. But because quantitative estimates of population exposure and of effects are so wide-ranging, it is easy for those who for official reasons wish to minimize the danger to becloud and confuse the issues. It cannot be foretold whose genes will be altered, or in what way, so let's all stop worrying and keep our fissionables dry.

The physicist Edward Teller, who has sought and gained wide publicity for his views, is Pauling's prime example of a public misinformer. Teller regards radiation from nuclear tests as a negligible threat to world health. World-wide fallout is, in his phrase, "as dangerous as being an ounce overweight." So far as possible genetic damage is concerned, Teller has stated that "radiation in small doses need not necessarily be harmful—indeed may conceivably be helpful." This puts fallout in roughly the same category as Lydia Pinkham's remedy. His views on radiation are of course essential to his general political position. He is convinced that the bigger the bombs the better the chance for peace. If the next war kills everyone, the world will again be peaceful; but this is not, I think, what Teller has in mind. He pins his hopes to "clean" bombs, which would leave survivors. Continued tests are needed, he says, to develop such bombs, and the tests themselves are harmless. Another reason he gives for not suspending tests is that atomic explosions can be concealed; we could never

be sure therefore that an international agreement prohibiting tests would not be secretly violated by the Soviets. Because of his eminence as "father of the H-bomb," Teller's opinions have carried considerable weight. Lately their prestige has waned, Pauling's arguments having contributed to this result.

The statement that fallout is as dangerous as being an ounce overweight Pauling characterizes as "ludicrous." It is based, he says, on a gross statistical blunder and represents a 1,500-fold underestimate of the hazard. One of Teller's points is that the people of Tibet, though exposed to much more intense cosmic radiation than people who live at lower altitudes, are as healthy as we are (or, at any rate, used to be). Pauling replies that the Tibetan exposure to increased amounts of cosmic radiation should produce an increase in the incidence of seriously defective children from 2 per cent—the average U. S. figure—to 2.3 per cent. But since there are no medical statistics for Tibet, Teller's statement is simply out of the blue. (He advanced it in a magazine article in order, as he said in a television debate with Pauling, merely to "quiet excessive fears.") Another of Teller's assertions is that the radiation danger to the average person is 10 times as great from luminous-dial wrist watches as from fallout. Pauling demolishes this statistic. Teller, it may be felt, is better at bombs than at arithmetic.

Pauling gives examples of official misstatements. Merril Eisenbud, an Atomic Energy Commission official, wrote an article in 1955 which stated that "the total fallout to date from all tests would have to be multiplied by a million to produce visible, deleterious effects except in areas close to the explosion itself." He was later asked by a chairman of a Congressional subcommittee if he remembered how much radioactivity had fallen on the city of Troy, N. Y., following a test held in Nevada a few days before. Eisenbud estimated the amount as "something under .1 roentgen" and finally settled on .01 roentgen. Ralph Lapp, who was present at the hearing, pointed out that a million times .01 roentgen is 10,000 roentgens. Whereupon one of the Senators observed that this amount would have the "visible and deleterious" effect of killing everybody in the region.

Commissioner Willard F. Libby of the A.E.C. said in 1955 that "the fallout dosage rate as of January 1 of this year could be increased 15,000 times without hazard." This is not as cheering as Eisenbud's figure, but cheering enough.

Pauling is less reassuring. He notes that 10 hours after the detonation of a small fission bomb at the Nevada test site, the level of gamma radiation at St. George, Utah, was .004 roentgen per hour, which is .1 roentgen a day. Multiplied by 15,000, this gives an exposure of 1,500 roentgens—a dose that will produce death within a few days from radiation sickness.

Pauling's own estimates of the effects of fallout are somber. He predicts an annual death rate of 8,000 if tests continue at the present rate. Moreover, 15,000 seriously defective children will be born each year whose defects must be attributed to the tests (this number does not include embryonic and neonatal deaths and stillbirths). It is possible, he admits, that his estimates are 10 times too large or 10 times too small. Perhaps then only 1,500 children will have to be sacrificed annually to the maintenance of peace; on the other hand, if the figures are too small, 150,000 children may be required. Even if no more tests are carried out, a lethal legacy of radioactive carbon 14 already released into the atmosphere will be handed on for thousands of years. According to Pauling's calculations, based to a large extent on data provided by Libby and others, the quantities of this long-lived isotope discharged in the period 1952 to 1958 will ultimately produce about one million seriously defective children and about two million embryonic and neonatal deaths, "and will cause many millions of people to suffer from minor hereditary defects." Our generation will be remembered.

When in April of last year Pauling first called attention to the menace of bomb-produced carbon 14, he was sharply criticized by three geneticists in a letter to *The New York Times* for making "erroneous" and "exaggerated" statements that "only add to the public's confusion and do not contribute to the solution of the problem." Libby dismissed Pauling's warning by saying that the effect of carbon 14 from nuclear tests "is equivalent to an increase in altitude of a few inches." But the ridicule backfired when the A.E.C.'s Division of Biology and Medicine issued a document, "The Biological Hazard to Man of Carbon 14 from Nuclear Weapons," whose conclusions agree quite closely with Pauling's. This document was unaccompanied by a press release and no enterprising reporter nosed it out. Lapp brought it to public attention in a letter to *The New York Times*.

Like Mills, Pauling makes a general appeal for peace and for international agreements on the cessation of bomb tests and on disarmament. His words are a *cri de cœur*, sustained by moral authority and reason. But there is little evidence that such pleas, uttered by those who know most about the dangers involved in a nuclear war, have persuaded leaders of states of the necessity of making radical changes in their policies. The argument is often heard that those responsible for the safety of the state are in a hopeless dilemma, for even if they are sincere in wanting peace and favor disarmament it is impossible to frame enforceable disarmament agreements. If no inspection system is workable, as Teller and others have claimed, what is the practical value of test bans and treaties to outlaw weapons?

There are at least two replies to this argument. One is that of the noted German physicist Max von Laue. "Suppose," he said, "I live in a big apartment house and burglars attack me; I am allowed to defend myself and, if need be, I may even shoot, but under no circumstances may I blow up the house. It is true that to do so would be an effective defense against burglars, but the resulting evil would be much greater than any I could suffer. But what if the burglars have explosives to destroy the whole house? Then I would leave them with the responsibility for the evil, and would not contribute anything to it."

The other reply is less noble, but is no less to the point in countering the untruthful and misleading statements that have made men skeptical about the possibility of enforcing disarmament.

It is given in the Melman book, a co-operative study made at Columbia University, which provides a searching examination of inspection techniques. One of the contributors, the Columbia physicist Jay Orear, effectively refutes the claim that atomic tests can be "bootlegged." He demonstrates that a comparatively small number of monitoring stations uniformly distributed throughout the U.S.S.R. and the U. S., using the combined techniques for picking up acoustic waves, seismic waves, electromagnetic radiation and radioactivity, would constitute an adequate inspection system for test suspension. This, together with a provision that U.N. inspectors "be invited to all large chemical explosions, should make it possible," Orear says, "to detect all nuclear tests unless they are of such ultra-low yield as to be in the class of World War II blockbusters." This view is substantially the same as the one adopted by the U. S. and the U.S.S.R. at the Geneva Conference of Experts held last summer.

Some 20 papers, together with an excellent general summary by Melman, make up the Columbia report. They deal with inspection of several major classes of activities: the production of heavy weapons of a conventional type, such as tanks, artillery and trucks; ship and submarine building; aircraft, missile and fissionable-material manufacture; the preparation of biological-warfare weapons. Anyone who has thought about the problem quickly realizes that inspection of the manufacture of heavy weapons and naval vessels is relatively easy. The sheer mass of metal that must be moved and processed, the size of the plants and yards, the large labor force required— all facilitate controls. One cannot hide a whale in the backyard. Inspectors with access to major factories can ensure against large-scale evasions: the Columbia group estimates that 5,000 to 10,000 inspectors are needed for 493 ordnance and accessories plants in the U. S. This number seems high, especially in view of the multiple approach recommended for every inspection assignment; but at least it represents a feasible operation that would not interfere with normal industrial activity. Monitoring the development of biological weapons presents greater difficulties. Because their dependability as mass killers is uncertain, biological weapons are unlikely to be the first choice of countries with large stores of atom bombs; but small countries with limited laboratory and manufacturing facilities may, it is suggested, avail themselves of what is called the "poor man's atom bomb"— dispensers of virulent material. "Continuous and close attention" should therefore be given these weapons.

These, however, are essentially secondary matters. The main worry is inspection of the production of fissionables and missiles. In addition to Orear's analysis of clandestine bomb-testing, the book contains reports on the possibility of using established radiation-protection services in an inspection scheme embracing the manufacture of nuclear explosives, on the possible "theft" of fissionable materials, on inspection of missile components, propellants and guidance systems, on the "amenability of the air-borne propulsion systems industry to production inspection."

The contributors weave a pretty tight net. It is hard to imagine how any militarily significant evasion of a disarmament agreement could slip unnoticed through the inspection system here envisaged. The analysis is based entirely on publicly available information; no attempt was made to gain access to "secrets." In Melman's opinion this self-imposed limitation has given the report a "conservative bias"; that is, "more access and more knowledge might have revealed more strategic control points for inspection" and thus made possible the elimination of many points now regarded as essential.

The basic assumption is that between 200 and 400 large missiles could be used "to devastate effectively" any one of the larger land areas of the earth. A useful inspection system must cope with possible efforts to produce these weapons in clandestine ways and also with the problem of hidden inventories of arms produced before the inspection system was instituted. In Congressional hearings some years ago J. R. Oppenheimer and other witnesses made the point that enough fissionable material to devastate a city the size of New York could be toted around in a violin case. The purpose of this testimony was to point up both the danger of atomic weapons and the desperate necessity for nations getting together to prohibit their manufacture and use. But this somewhat overheated illustration had unfortunate results. Those who accepted it at face value became convinced that no inspection system would work, that disarmament was altogether unfeasible and that the only prudent course for the U. S. was to amass an enormous stockpile of weapons. The Columbia report presents a more rational perspective. The contributors do not minimize the dangers, nor do they claim that any inspection system, however searching, is without loopholes; but the report offers convincing evidence that a comprehensive scheme can probably be framed, which would make secret rearmament so extraordinarily difficult "as to be virtually impossible." An inspection system is, after all, an alarm system. It is intended to give timely warning not of minor infractions but of illegal activities on a substantial scale, for these alone carry the threat of a decisive surprise attack.

The multiple-inspection approach recommended by the Columbia group includes, among other things, aerial reconnaissance (useful to detect the production rather than the existence of missiles); analysis and auditing of government budgets; monitoring stations; checks on the whereabouts and activities of engineers and scientists; surveillance of plants producing airframes, chemicals and fissionables; inventory validation. The contributors have tried hard to fore-

see how ingenious men might contrive to fool the inspectors. This carries the analysis pretty far. When serious consideration is given, for example, to testing—in connection with inventory validation—the alteration of records and the age of papers and inks, the reader begins to feel as if Dick Tracy had taken over. But these are minor aberrations, more than counterbalanced by a knowledgeable and sensible treatment of the various aspects of a tricky business.

It is no small thing to question, as this book does, the validity of the ruling notion of deterrence. It is widely held, and not only by Mills's "elite," that genocidal weapons offer a reasonable guarantee of peace because no nation would deliberately commit suicide. But neither history nor social psychology unequivocally supports this opinion. People do not vote on going to war, and children are never asked. Deterrents may not deter because the deliberate judgment that is essential to the "if-we-kill-them-they'll-kill-us-so-let's-not-kill-them" sequence rarely comes into play. Small causes may have large effects; moreover the dropping of even a single nuclear weapon is manifestly more provocative than slicing off Jenkins's ear or assassinating an archduke. An accident can set a catastrophic nuclear war in motion, and as nuclear weapons are increasingly available and dispersed in more hands, the probabilities of such an accident must necessarily increase. "One aberrant, psychotic person or person gone momentarily out of control," Melman writes, "could explode nuclear weapons at a random place, or over any populated area. A space satellite could be mistaken for a ballistic missile." And when many countries possess nuclear weapons, if a warhead were set off in one city, it might be impossible to identify the aggressor and therefore even to threaten retaliation. Thus the major assumption of the mutual-deterrence strategy falls to the ground.

The best possible system of inspection techniques has weaknesses. A "foolproof" inspection system is a politician's catchword. The Melman report recognizes this truth and suggests a design to compensate for the inevitable gaps in inspection. This design, called "inspection by the people," counts upon the help of plain citizens of every country to enforce international agreements. Secret rearmament requires the participation of a large number of persons; among them there

are certain to be some who are not in sympathy with the evasion efforts and who might therefore be expected to report them to the international inspectorate. Such persons must be encouraged and protected. If channels of communication from the population to the inspecting organization are always kept open, news of clandestine violations—the use of certain machines, the production of materials, the operation of prohibited processes—will almost certainly trickle through. A constant appeal urging the theme that "the international agreement is mankind's shield against mutual extermination and that a violation of this agreement is thereby a crime against humanity" would, in Melman's view, evoke a cooperative response in every country and make untenable the position of any government, or group of officials, found guilty of breaking the law.

Can it not be said that this is the most important conclusion of the Columbia study? In a sense it links together the different approaches of Mills, Pauling, Schweitzer and others who warn and strive to educate the world before it is too late. For after assessing the causes of war, analyzing the various strategies, designing meticulous disarmament and inspection schemes, one faces the irreducible truth that we can live together or die together. It is too much to expect men all at once to throw away their weapons and embrace. But a beginning must be made, and that beginning depends, as the authors of the Melman report tell us, on conceding to each other what moral capacity we have, on having faith even in the enemy's awareness of his humanity.

"We cannot," says Schweitzer in a moving appeal broadcast from Norway last year and reprinted in his little book, "continue in this paralyzing mistrust. If we want to work our way out of the desperate situation in which we find ourselves, another spirit must enter into the people. It can only come if the awareness of its necessity suffices to give us strength to believe in its coming. We must presuppose the awareness of this need in all the peoples who have suffered along with us. We must approach them in the spirit that we are human beings, all of us, and that we feel ourselves fitted to feel with each other; to think and to will together in the same way. . . ."

# 39

# The Use and Misuse of Game Theory

ANATOL RAPOPORT

*December 1962*

We live in an age of belief—belief in the omnipotence of science. This belief is bolstered by the fact that the problems scientists are called on to solve are for the most part selected by the scientists themselves. For example, our Department of Defense did not one day decide that it wanted an atomic bomb and then order the scientists to make one. On the contrary, it was Albert Einstein, a scientist, who told Franklin D. Roosevelt, a decision maker, that such a bomb was possible. Today, in greater measure than ever before, scientists sit at the decision makers' elbows and guide the formulation of problems in such a way that scientific solutions are feasible. Problems that do not promise scientific solutions generally tend to go unformulated. Hence the faith in the omnipotence of science.

The self-amplifying prestige of science among decision makers has been further amplified in this period by the popularization of a scientific aid to the task of decision making itself. This is game theory—a mathematical technique for the analysis of conflict first propounded by the late John Von Neumann in 1927 and brought to wide notice by Von Neumann and Oskar Morgenstern in 1944 in a book entitled *Theory of Games and Economic Behavior*. Now, game theory is an intellectual achievement of superlative originality and has opened a large new field of research. Unfortunately this is not the way game theory has been embraced in certain quarters where Francis Bacon's dictum "Knowledge is power" is interpreted in its primitive, brutal sense. The decision makers in our society are overwhelmingly preoccupied with power conflict, be it in business, in politics or in the military. Game theory is a "science of conflict." What could this new science be but a reservoir of power for those who get there fastest with the mostest?

A thorough understanding of game theory should dim these greedy hopes. Knowledge of game theory does not make any one a better card player, businessman or military strategist, because game theory is not primarily concerned with disclosing the optimum strategy for any particular conflict situation. It is concerned with the logic of conflict, that is, with the theory of strategy. In this lies both the strength and the limitation of the technique. Its strength derives from the powerful and intricate mathematical apparatus that it can bring to bear on the strategic analysis of certain conflict situations. The limitations are those inherent in the range of conflicts to which this analysis can be successfully applied.

No one will doubt that the logic of strategy does not apply to certain conflicts. For example, there are no strategic considerations in a dogfight. Such a conflict is better thought of as being a sequence of events, each of which triggers the next. A growl is a stimulus for a countergrowl, which in turn stimulates the baring of teeth, sudden thrusts and so on. Signals stimulate postures; postures stimulate actions. Human quarrels, where symbolic rather than physical injuries are mutually stimulated, are frequently also of this sort. Conflicts of this kind can be called fights. The motivation in a fight is hostility. The goal is to eliminate the opponent, who appears as a noxious stimulus, not as another ego, whose goals and strategies, even though hostile, must be taken into account. Intellect, in the sense of calculating capacity, foresight and comparison of alternative courses of action, need not and usually does not play any part in a fight.

Game theory applies to a very different type of conflict, now technically called a game. The well-known games such as poker, chess, ticktacktoe and so forth are games in the strict technical sense. But what makes parlor games games is not their entertainment value or detachment from real life. They are games because they are instances of formalized conflict: there is conflict of interest between two or more parties; each party has at certain specified times a range of choices of what to do prescribed by the rules; and the outcome representing the sum total of choices made by all parties, and in each case involving consideration of the choice made by or open to the other parties, determines an assignment of pay-offs to each party. By extension, any conflict so conducted falls into the category of games, as defined in game theory. Nor does it matter whether the rules are results of common agreement, as in parlor games, or simply of restraints imposed by the situation. Even if no rules of warfare are recognized, a military situation can still be considered as a game if the range of choices open to each opponent at any given stage can be exactly specified.

Let us see how chess and poker each fulfill these requirements. In chess the conflict of interest is, of course, implied

*Bark and counterbark*

in each player's desire to win. The range of choices consists for each player of all the legal moves open to him when it is his turn to move. The outcome is determined by all the choices of both players. The pay-offs are usually in psychological satisfaction or dissatisfaction. In poker the situation is much, but not entirely, the same. The choices are (at specified times) whether or not to stay in; which cards, if any, to throw off; whether or not to raise and by how much and so on. The outcome of each round is the designation of one of the players as the winner. Pay-offs are usually in money.

Poker differs from chess in one important respect. In a poker game there is an extra (invisible) player, who makes just one choice at the beginning of each round. This choice is important in determining the outcome, but the player who makes it has no interest in the game and does not get any pay-off. The player's name is Chance, and his choice is among the nearly 100 million trillion trillion trillion trillion trillion ($10^{68}$) arrangements of the deck at the beginning of each round. Chance makes no further choices during the round; the rest is up to the players. One can argue that Chance continues to interfere, for example by causing lapses of memory, directing or misdirecting the attention of the players and so forth. But game theory is concerned only with what perfect players would do.

Although Chance may thus play a part, the game as defined by game theory is clearly distinguished from gambling as treated by the much older and better-known mathematics of gambling. The latter has considerable historical importance, since it is in the context of gambling theory that the mathematical theory of probability was first developed some 300 years ago. This theory has since been incorporated into all branches of science where laws of chance must be taken into account, as in the physics of small particles, genetics, actuarial science, economics, experimental psychology and the psychology of mass be-

havior. For the gambler the mathematical theory of probability makes possible a precise calculation of the odds. This often calls for considerable mathematical sophistication. It is irrelevant, however, to the playing of the game; it is relevant only in deciding whether or not to play. The gambling problem is solved when the odds of the possible outcomes have been calculated. If there are several such outcomes, the gains or losses associated with each are multiplied by the corresponding probabilities and the products are added (with proper signs attached). The resulting number is the expected gain; that is, what can be reasonably expected over a long series of bets when the bets are placed according to the odds offered. A rational gambler is one who accepts or offers the gambles in such a way as to maximize his expected gain. All gambling houses are rational gamblers. That is why they stay in business.

The inadequacy of gambling theory as a guide in a true game is shown clearly in the well-known fact that the rational gambler is likely to meet with disaster in a poker game. The rational gambler will make his decisions strictly in accordance with the odds. He will never bluff, and he will bet in proportion to the strength of his hand. As a result he will betray his hand to his opponents, and they will use the information to his disadvantage.

Gambling theory is of even less use to the ticktacktoe player. Ticktacktoe is a game in which there is a best move in every conceivable situation. Chance, we know, is not involved at all in some games. To be sure, chance is involved in all card games but, as the example of poker shows, something else is involved, namely a strategic skill that is not part of gambling theory at all.

Consider what goes on in the mind of a chess player: If I play Knight to Queen's Bishop's 4, thus threatening his rook, he can reply Rook to King's 2, check. In that case I have the choice of either interposing the Bishop or King

*The omnipotent scientist*

to Queen's 1. On the other hand, he can ignore the threat to the rook and reply with a counterthreat by Bishop to Knight's 5, in which case I have the following choices . . .

The stronger the player, the longer this chain of reasoning is likely to be. But because of the limitations on how much we can hold in our minds at one time, the chain of reasoning must stop somewhere. For the chess player it stops a few moves ahead of the situation at hand, at a set of possible new situations among which he must choose. The one situation that will actually occur depends partly on his own choices and partly on the choices available to the opponent (over whom the first player has no control). Two decisions are involved in the choice of action: first, which situations may actually occur? Second, which of all those situations is to be preferred?

Now, these questions can be answered without ambiguity if the game is thought out to the end. In a game such as chess, however, it is out of the question to foresee all the alternatives to the end (except where checkmates or clear wins are foreseen as forced). The good chess player then does the next best thing: he calculates the relative values of the various possible future positions according to his experience in evaluating such positions. How then does he know which position will be actually arrived at, seeing that he controls only his own moves, not those of the opponent? Chess players recognize two chess philosophies.

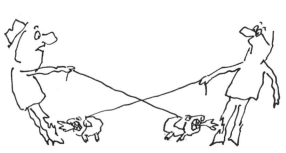

*Bite and counterbite*

*Escalated conflict*

*Playing the stock market or a slot machine involves no game theory*

One is "playing the board," the other is "playing the opponent."

Playing the opponent makes chess akin to psychological warfare. The great chess master José Capablanca tells in his memoirs of an incident that illustrates the drama of such conflicts. In a tournament in 1918 he was matched with Frank J. Marshall, the U.S. champion. Marshall offered an unexpected response to Capablanca's accustomed opening attack, and the play proceeded not at all in line with the usual variations of this opening. Capablanca suspected that Marshall had discovered a new variation in the attack and had kept this knowledge as a secret weapon, to be used only at the most propitious time, namely in an international tournament with the eyes of the chess world on his play against a truly formidable opponent. Capablanca had been picked as the victim of the new strategy.

"The lust of battle, however," Capablanca continues, "had been aroused within me. I felt that my judgment and skill were being challenged by a player who had every reason to fear both (as shown by the records of our previous encounters), but who wanted to take advantage of the element of surprise and of the fact of my being unfamiliar with a

*Psychological warfare in chess*

*Advanced psychological warfare in chess*

thing to which he had devoted many nights of toil.... I considered the position then and decided that I was in honor bound ... to accept the challenge."

He did and went on to win the game. Capablanca's decision was based on taking into account his opponent's thought processes, not only those pertaining to the game but also Marshall's ambitions, his opinion of Capablanca's prowess, his single-mindedness and so on. Capablanca was playing the opponent.

Although the drama of games of strategy is strongly linked with the psychological aspects of the conflict, game theory is not concerned with these aspects. Game theory, so to speak, plays the board. It is concerned only with the logical aspects of strategy. It prescribes the same line of play against a master as it does against a beginner. When a stragetic game is completely analyzed by game-theory methods, nothing is left of the game. Ticktacktoe is a good example. This game is not played by adults because it has been completely analyzed. Analysis shows that every game of ticktacktoe must end in a draw. Checkers is in almost the same state, although only exceptionally good players know all the relevant strategies. A generation ago it was thought that chess too was approaching the "draw death." But new discoveries and particularly the introduction of psychological warfare into chess, notably by the Russian masters, has given the game a reprieve. Nevertheless H. A. Simon and Allen Newell of the Carnegie Institute of Technology have seriously predicted that within 10 years the world's chess champion will be an electronic computer. The prediction was made more than three years ago. There is still a good chance that it will come true.

Is the aim of game theory, then, to reveal the logic of every formalized game so that each player's best strategy is discovered and the game as a whole is killed because its outcome in every instance will be known in advance? This is by no means the case. The class of games for which such an analysis can

be carried through even in principle, let alone the prodigious difficulty of doing it in practice, is only a very small class.

Games of this class are known as games of perfect information. They are games in which it is impossible to have military secrets. Chess is such a game. Whatever the surprise Marshall thought he had prepared for Capablanca, he was not hiding something that could not be discovered by any chess player. He only hoped that it would be overlooked because of human limitations.

Not all games are games of perfect information. Poker is definitely not such a game. The essence of poker is in the circumstance that no player knows the entire situation and must be guided by guesses of what the situation is and what the others will do. Both chess and poker are "zero-sum" games in the sense that what one player wins the other or others necessarily lose. Not all games are of this sort either.

To understand the differences among these various classes of games, let us look at some examples from each class. The essential idea to be demonstrated is that each type of situation requires a different type of reasoning.

An improbably elementary situation in business competition will serve to illustrate the class of games of perfect information. The situation is otherwise a two-person zero-sum game. The Castor Company, an old, established firm, is being squeezed by Pollux, Incorporated, an aggressive newcomer. The Castor people guide their policies by the balance sheet, which is projected one year ahead. The Pollux people also guide their policies by a balance sheet, not their own but the Castor Company's. Their aim is to put Castor out of business, so they consider Castor's losses their gains and vice versa, regardless of what their own balance shows. Both are faced with a decision, namely whether or not to undertake an extensive advertising campaign. The outcome depends on what both firms do, each having control over only its own decision. Assume, however, that both firms have enough information to know what the outcomes will be, given both decisions [*see matrix at left in bottom illustration on page 290*].

From Castor's point of view, a better or a worse outcome corresponds to each of its decisions, depending on what Pollux does. Of the two worse outcomes associated with Castor's two possible decisions, $3 million in the red and $1 million in the red (both occurring if Pollux advertises), clearly the second is preferred. Castor's manager now puts himself into the shoes of Pollux' manager and asks what Pollux would do if Castor chose the lesser of the two evils. Clearly Pollux would choose to advertise to prevent the outcome that would be better for Castor ($1 million in the black). Getting back into his own shoes, Castor's manager now asks what he would do knowing that this was Pollux' decision. Again the answer is advertise. Exactly similar reasoning leads Pollux to its decision, which is advertise. Each has chosen the better of the two worse alternatives. In the language of game theory this is called the minimax (the maximum of the minima). This solution is always prescribed no matter how many alternatives there are, provided that the gains of one are the losses of the other and provided that what is the "best of the worst" for one is also the "best of the worst" for the other. In this case the game has a saddle point (named after the position on the saddle that is lowest with respect to front and back and highest with respect to right and left). Game theory shows that whenever a saddle point exists, neither party can improve the outcome for itself (or worsen it for the other). The outcome is forced, as it is in ticktacktoe.

The next situation is quite different. It is a two-person zero-sum game, again involving the choice of two strategies on each side. In this case, however, the choices must be made in the absence of the information that guides the opponent's decision. Appropriately this is a military situation enveloped in the fog of battle.

A commander of a division must decide which of two sectors to attack. A breakthrough would be more valuable in one than in the other, but the more valuable sector is also likely to be more strongly defended. The defending commander also has a problem: which sector to reinforce. It would seem obvious that the more critical sector should be reinforced at the expense of the secondary one. But it is clear to the defending commander that the problem is more complicated. Secrecy is of the essence. If he does exactly what the enemy expects him to do, which is to reinforce the critical sector, will this not be to the enemy's advantage? Will not the attacker, knowing that the important sector is more strongly defended, attack the weaker one, where a breakthrough, even though less valuable, is more certain? Should the defender therefore not do the opposite of what the enemy expects and reinforce the secondary sector, since that

is where the enemy, wishing to avoid the stronger sector, will probably attack? But then is not the enemy smart enough to figure this out and so attack the primary center and achieve a breakthrough where it counts?

The attacking commander is going through the same tortuous calculations. Should he attack the secondary sector because the primary one is more likely to be strongly defended or should he attack the primary one because the enemy expects him to avoid it?

In despair the attacking commander calls in a game theorist for consultation. If the game theorist is to help him, the

general must assign numerical values to each of the four outcomes; that is, he must estimate (in relative units) how much each outcome is "worth" to him. He assigns the values shown in the top illustration on the next page. Working with these figures, the game theorist will advise the general as follows: "Roll a die. If ace or six comes up, attack sector 1, otherwise attack sector 2."

If the defending commander assigns the same values (but with opposite signs, since he is the enemy) to the four outcomes, his game theorist will advise him to throw two pennies and reinforce sector 1 if they both come up heads,

*Game theory in "Tosca": Tosca double-crosses Scarpia*

*Scarpia derives satisfaction from the thought of what is going to happen*

*Tosca and Cavaradossi discover the double double cross*

ATTACKER'S PAY-OFF        DEFENDER'S PAY-OFF        MINIMAX SOLUTION

| A \ D | $D_1$ | $D_2$ |
|---|---|---|
| $A_1$ | −10 | +30 |
| $A_2$ | +5 | −15 |

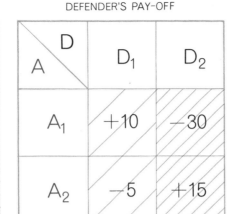

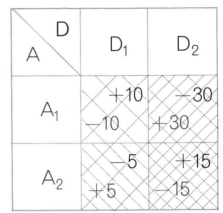

**TWO-PERSON ZERO-SUM GAME** of an attacking ($A$) and a defending commander ($D$), in which neither possesses the information that guides his opponent's decisions, is summarized in these three matrices. The first commander has the choice of attacking a primary sector ($A_1$) or a secondary sector ($A_2$). The matrix at left shows the values he assigns to the four possibilities. The second commander has the choice of defending either sector. The matrix at center shows his assigned values. As the number of diagonal lines in each matrix square indicates, the first commander should decide by chance, using two-to-one odds in favor of the secondary sector; likewise for the defending commander, except that the odds are three to one. These results are combined in the matrix at right.

otherwise he should reinforce sector 2.

The solutions seem bizarre, because we think of tossing coins to make decisions only in matters of complete indifference. To be sure, a tossed-coin decision is sometimes used to settle an argument, but we do not think of such decisions as being rational and do not hire experts to figure them out. Nevertheless, the game theorists' decisions are offered not only as rational decisions but also as the best possible ones under the circumstances.

To see why this is so, imagine playing the game of button-button. You hide a button in one hand and your opponent tries to guess which. He wins a penny if he guesses right and loses a penny if he guesses wrong. What is your best pattern of choices of where to hide the button in a series of successive plays? You will certainly not choose the same hand every time; your opponent will quickly find this out. Nor will you alternate between the two hands; he will find this out too. It is reasonable to conclude (and it can be proved mathematically) that the best pattern is no pattern. The best way to ensure this is to abdicate your role as decision maker and let chance decide for you. Coin tossing as a guide to strategy is in this case not an act of desperation but a rational policy.

In the button-button game the pay-offs are exactly symmetrical. This is why decisions should be made by a toss of a fair coin. If the pay-offs were not symmetrical—for example, if there were more advantage in guessing when the coin was in the right hand—this bias would have to be taken into account. It would be reflected in letting some biased chance device make the decision. Game theory provides the method of computing the bias that maximizes the long-run expected gain.

CASTOR AND POLLUX

| C \ P | $P_Y$ | $P_N$ |
|---|---|---|
| $C_Y$ | −1 | +1 |
| $C_N$ | −3 | +2 |

TOSCA'S PAY-OFF

| T \ S | $S_K$ | $S_D$ |
|---|---|---|
| $T_K$ | +5 | −10 |
| $T_D$ | +10 | −5 |

SCARPIA'S PAY-OFF

| T \ S | $S_K$ | $S_D$ |
|---|---|---|
| $T_K$ | +5 | +10 |
| $T_D$ | −10 | −5 |

**ZERO-SUM AND NONZERO-SUM GAMES** are represented in these three game-theory matrices. The matrix at left is that of the two-person zero-sum game of perfect information discussed in the text. The matrix tabulates the results for Castor Company (in millions of dollars) of any combination of decisions; e.g., if Castor and Pollux, Incorporated, both advertise ($Cy$ and $Py$), Castor loses $1 million. For Pollux, which will decide on the basis of the effect on Castor, this is a positive pay-off. Tosca and Scarpia are involved in a nonzero-sum game (also discussed in the text), that is, a gain for one does not imply a loss for the other. Tosca's line of reasoning can be determined from the matrix at center: if she keeps her bargain with Scarpia ($Tk$), then she loses everything if he double-crosses her ($Sd$); her gain is greatest and her loss least if she double-crosses him ($Td$). Scarpia, as the matrix at right indicates, reasons along the same line, in reverse. They both lose equally; if they had trusted each other, they would have gained equally.

The attacker's game theorist, then, has figured out that the attacker stands the best chance if he allows chance to decide, using two-to-one odds in favor of sector 2. This is the meaning of rolling a die and allowing four sides out of six to determine the second sector. This is the best the attacker can do against the best the defender can do. The defender's best is to let chance decide, using three-to-one odds in favor of sector 2. Game theory here prescribes not the one best strategy for the specific occasion but the best mixture of strategies for this kind of occasion. If the two commanders were confronted with the same situation many times, these decisions would give each of them the maximum pay-offs they can get in these circumstances if both play rationally.

At this point one may protest that it is difficult, if not impossible, to assign numerical values to the outcome of real situations. Moreover, identical situations do not recur, and so the long-run expected gain has no meaning. There is much force in these objections. We can only say that game theory has gone just so far in baring the essentials of strategic conflict. What it has left undone should not be charged against it. In what follows some further inadequacies of game theory will become apparent. Paradoxically, in these inadequacies lies most of the value of the theory. The shortcomings show clearly how far strategic thinking can go.

In the next class of games to be illustrated there are choices open to the two parties where the gain of one does not imply loss for the other and vice versa. Our "nonzero-sum" game is a tale of lust and betrayal. In Puccini's opera *Tosca* the chief of police Scarpia has condemned Tosca's lover Cavaradossi to death but offers to save him in exchange for Tosca's favors. Tosca consents, the agreement being that Cavaradossi will go through a pretended execution. Scarpia and Tosca double-cross each other. She stabs him as he is about to embrace her, and he has not given the order to the firing squad to use blank cartridges.

The problem is to decide whether or not it was to the best advantage of each party to double-cross the other. Again we must assign numerical values to the outcome, taking into account what each outcome is worth both to Tosca and to Scarpia [see two matrices at right in bottom illustration on opposite page].

The values, although arbitrary, present the situation reasonably. If the bargain is kept, Tosca's satisfaction of getting her lover back is marred by her surrender to the chief of police. Scarpia's satisfaction in possessing Tosca will be marred by having had to reprieve a hated rival. If Tosca double-crosses Scarpia and gets away with it, she will win most (+ 10) and he will lose most (− 10), and vice versa. When both double-cross each other, both lose, but not so much as each would have lost had he or she been the sucker. For example, the dying Scarpia (we assume) derives some satisfaction from the thought of what is going to happen just before the final curtain, when Tosca rushes to her fallen lover and finds him riddled with bullets.

Let us now arrive at a decision from Tosca's point of view: whether to keep the bargain or to kill Scarpia. Tosca has no illusions about Scarpia's integrity. But she is not sure of what he will do, so she considers both possibilities: If he keeps the bargain, I am better off double-crossing him, since I will get Cavaradossi without Scarpia if I do and Cavaradossi with Scarpia if I don't. If he double-crosses me, I am certainly better off double-crossing him. It stands to reason that I should kill him whatever he does.

Scarpia reasons in exactly the same way: If she keeps the bargain, I am bet-ter off double-crossing her, since I will get rid of Cavaradossi if I do and have to put up with him if I don't. If she double-crosses me, I certainly should see to it that I am avenged. The execution, therefore, must go on.

The result is the denouement we know. Tosca and Scarpia both get − 5. If they had trusted each other and had kept the trust, each would have got + 5.

The shortcoming of strategic thinking becomes obvious in this example. Evidently more is required than the calculation of one's own pay-offs if the best decisions are to be made in conflict situations. Game theory can still treat the foregoing case satisfactorily by introducing the notion of a coalition. If Tosca and Scarpia realize that the interests of both will be best served if both keep the bargain, they need not both be losers. Coalitions, however, bring headaches of their own, as will be seen in the next example.

Abe, Bob and Charlie are to divide a dollar. The decision as to how to divide it is to be by majority vote. Abe and Bob form a coalition and agree to split the dollar evenly between them and so freeze Charlie out. The rules of the game allow bargaining. Charlie approaches Bob

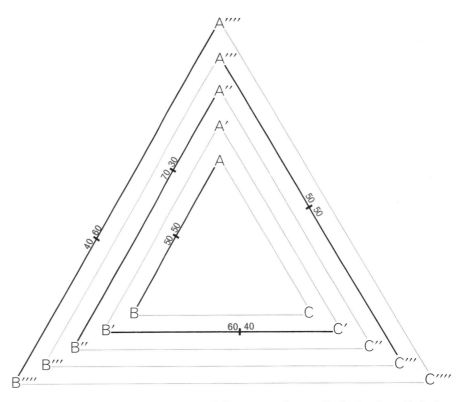

COALITION GAME involves splitting a dollar among three individuals; the split is decided by majority vote. Abe and Bob (*A and B*) form a coalition that excludes Charlie (*C*). Charlie then (*C'*) offers Bob (*B'*) 60 cents of the dollar, and so on. Any division is inherently unstable because two can always do better for themselves than can three, and two can enforce any division. No game-theory strategy will guarantee a division satisfactory to all.

*Maximum in the military application of game theory*

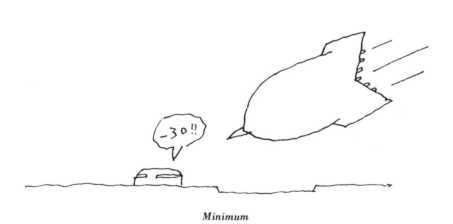

*Minimum*

with a proposition. He offers Bob 60 cents of the dollar if Bob will shift his vote to freeze Abe out. Abe does not like this arrangement, so he offers Bob 70 cents to shift his vote again to freeze Charlie out. Bob is about to rejoice in his good fortune, which he attributes to his bargaining shrewdness, when he notices that Abe and Charlie are off in a corner. Bob is shrewd enough to guess what they are discussing, and he is right. They are discussing the folly of respectively getting 30 cents and nothing when they have the power to freeze Bob out and split the dollar between them. In fact, they do this. Bob now approaches Abe hat in hand and offers him 60 cents if he will come back. The question is: Should Abe accept the offer?

The game-theory solutions to prob-

lems of this sort are extremely involved and need not be pursued here. Instead let us try to summarize in general terms the values and limitations of the game-theory approach to human conflict.

The value of game theory is not in the specific solutions it offers in highly simplified and idealized situations, which may occur in formalized games but hardly ever do in real life. Rather, the prime value of the theory is that it lays bare the different kinds of reasoning that apply in different kinds of conflict.

Let us go back to our examples and compare them. The decisions made by Castor and Pollux were clear-cut, and they were the best decisions on the basis of the knowledge at hand. As we have seen, both firms were guided by the principle of the minimax, choosing the best

of the worst outcomes. When both choose the minimax, neither firm can improve its position. Had one of the generals used such a decision, he would have been clearly at a disadvantage. Military secrecy introduces an element of randomness to confound the enemy and brings in a different kind of reasoning. Such reasoning would have been useless in the Castor and Pollux example, because in their case each knew what the other's best decision had to be, and this knowledge made no difference to either. The difference between the two situations is immediately apparent to the game theorist. In the first case the minimax choice of one player is also the minimax choice of the other, in the second case it is not.

Consider the Tosca-Scarpia game. Here both parties have the same minimax choice, which, in fact, they choose. The outcome is bad for both. Why is this? Again the answer is clear to the game theorist. Tosca and Scarpia were playing the game as if it were a zero-sum game, a game in which what one party wins the other necessarily loses. If we examine the pay-offs, we find that this is not the case. Both parties could have improved their pay-offs by moving from the minimax solution to the coalition solution (keeping the bargain and getting + 5 each). Life would be simple if advantage in conflicts could always be obtained by forming and keeping proper coalitions. But the dilemma plaguing Abe, Bob and Charlie deprives us of that hope also. Moreover, both the Tosca-Scarpia game and the divide-the-dollar game reveal that decisions based on calculated self-interest can lead to disaster.

Whether game theory leads to clear-cut solutions, to vague solutions or to impasses, it does achieve one thing. In bringing techniques of logical and mathematical analysis to bear on problems involving conflicts of interest, game theory gives men an opportunity to bring conflicts up from the level of fights, where the intellect is beclouded by passions, to the level of games, where the intellect has a chance to operate. This is in itself no mean achievement, but it is not the most important one. The most important achievement of game theory, in my opinion, is that game-theory analysis reveals its own limitations. Because this negative aspect is far less understood than the positive aspect, it will be useful to delve somewhat deeper into the matter.

The importance of game theory for decision making and for social science can be best understood in the light of the history of science. Scientists have been able to avoid much futile squandering of effort because the very foundations of science rest on categorical statements about what cannot be done. For example, thermodynamics shows that perpetual-motion machines are impossible. The principles of biology assert the impossibility of a spontaneous generation of life and of the transmission of acquired characteristics; the uncertainty principle places absolute limits on the precision of certain measurements conducted simultaneously; great mathematical discoveries have revealed the impossibility of solving certain problems.

Absolute as these impossibilities are, they are not absolutely absolute but are so only in certain specific contexts. Progress in science is the generalization of contexts. Thus the conservation of mechanical energy can be circumvented by converting other forms of energy into mechanical energy. The simpler conservation law is violated, but it is re-established in a more general thermodynamic context. In this form it can again be seemingly violated, but it is again re-established in the still broader context of $E = mc^2$. Angles can be mechanically trisected by instruments more complicated than the straightedge and the compass. Life can probably be synthesized, but not in the form of maggots springing from rotting meat; acquired characteristics can probably be genetically transmitted, but not by exercising muscles.

The negative verdicts of science have often been accompanied by positive codicils. The power conferred by science, then, resides in the knowledge of what cannot be done and, by implication, of what can be done and of what it takes to do it.

The knowledge we derive from game theory is of the same kind. Starting with the simplest type of game, for example two-person zero-sum games with saddle points, we learn from game-theory analysis that the outcome of such games is predetermined. This leads to a verdict of impossibility: neither player can do better than his best. Once these bests are discovered, it is useless to play such a game. If war were a two-person zero-sum game with a saddle point, the outcome of each war could conceivably be calculated in advance and the war would not need to be fought. (The conclusion that wars need to be fought because they are not two-person zero-sum games with saddle points is not warranted!)

Examining now the two-person zero-sum game without a saddle point, we

*Strategic thinking in a two-person zero-sum situation with a saddle point*

*Strategic thinking in a two-person zero-sum situation without a saddle point*

*Strategic thinking in a two-person nonzero-sum situation*

*Strategic thinking in a three-person constant-sum situation with a coalition*

294

*Communication is a prerequisite for the resolution of conflict*

arrive at another verdict of impossibility: It is impossible to prescribe a best strategy in such a game. It is still possible, however, to prescribe a best mixture of strategies. The meaning of a strategy mixture and the advantage of using it can be understood only in a certain context, namely in the context of an expected gain. This in turn requires that our concept of preference be defined with a certain degree of specificity. To choose the best strategy in a saddle-point game it is necessary only to rank-order the preferences for the possible outcomes. To choose the best strategy mixture an interval scale (like that of temperature) must be assigned to our preferences. Unless this more precise quantification of preferences can be made, rational decisions cannot be made in the context of a game without a saddle point.

I have often wondered to what extent decision makers who have been "sold" on game theory have understood this last verdict of impossibility, which is no less categorical than the verdict on squaring the circle with classical tools. I have seen many research proposals and listened to long discussions of how hot and cold wars can be "gamed." Allowing for the moment that hot and cold wars are zero-sum games (which they are not!), the assignment of "utilities" to outcomes must be made on an interval scale. There is the problem. Of course, this problem can be bypassed, and the utilities can be assigned one way or another, so that we can get on with

the gaming, which is the most fun. But of what practical use are results based on arbitrary assumptions?

That is not all. By far the most important conflicts that plague the human race do not fit into the two-person zero-sum category at all. The Tosca-Scarpia game and the Abe-Bob-Charlie game are much more realistic models of human conflicts, namely dramas, in which individuals strive for advantage and come to grief. In these games there are neither pure nor mixed strategies that are best in the sense of guaranteeing the biggest pay-offs under the constraints of the game. No argument addressed individually to Tosca or to Scarpia will convince either that it is better to keep the bargain than to double-cross the other. Only an argument addressed to both at once has this force. Only collective rationality will help them to avoid the trap of the double double cross.

Similarly we can tell nothing to Abe, Bob or Charlie about how to behave to best advantage. We can only tell them collectively to settle the matter in accordance with some pre-existing social norm. (For example, they can take 33 cents apiece and donate one to charity.) This solution is based on an ethical principle and not on strategic considerations.

The role of social norms in games with more than two players was not missed by Von Neumann and Morgenstern. The importance of honesty, social responsibility and kindred virtues has been pointed out by sages since the dawn of history. Game theory, however,

gives us another perspective on these matters. It shows how the "hardheaded" analysis of conflicts (with which game theory starts) comes to an impasse, how paradoxical conclusions cannot be avoided unless the situation is reformulated in another context and unless other, extra-game-theory concepts are invoked. Thus acquaintance with these deeper aspects of game theory reveals that the poker game is not the most general or the most sophisticated model of conflict, nor the most relevant in application, as professional strategists often implicitly assume.

Game theory, when it is pursued beyond its elementary paradox-free formulations, teaches us what we must be able to do in order to bring the intellect to bear on a science of human conflict. To analyze a conflict scientifically, we must be able to agree on relative values (to assign utilities). We must learn to be perceptive (evaluate the other's assignment of utilities). Furthermore, in order to engage in a conflict thus formalized, we must be able to communicate (give a credible indication to the other of how we assign utilities to outcomes). At times we must learn the meaning of trust, or else both we and our opponents will invariably lose in games of the Tosca-Scarpia type. At times we must be able to convince the other that he ought to play according to certain rules or even that he ought to play a different game. To convince the other we must get him to listen to us, and this cannot usually be done if we ourselves do not listen. Therefore we must learn to listen in the broadest sense of listening, in the sense of assuming for a while the other's world outlook, because only in this way will we make sense of what he is saying.

All these skills are related not to know-how but to wisdom. It may happen that if we acquire the necessary wisdom, many of the conflicts that the strategy experts in their professional zeal insist on formulating as battles of wits (or, worse, as battles of wills) will be resolved of their own accord.

*Another prerequisite is the assignment of utilities to outcomes*

Bleckman

# 40

# Thermonuclear War

JAMES R. NEWMAN

*March 1961*

A review of *On Thermonuclear War*, by Herman Kahn. Princeton, N. J.: Princeton University Press, 1961; and *Arms Control*. Special issue of *Daedalus* (Journal of the American Academy of Arts and Sciences), Fall, 1960.

Is there really a Herman Kahn? It is hard to believe. Doubts cross one's mind almost from the first page of this deplorable book: no one could write like this; no one could think like this. Perhaps the whole thing is a staff hoax in bad taste.

The evidence as to Kahn's existence is meager. The biographical note states that he was born in Bayonne, N.J., in 1922, that he studied at the University of California at Los Angeles and the California Institute of Technology, that he has worked for 12 years for the Rand Corporation as a "military planner." An autobiographical footnote states that he was trained as a physicist and a mathematician. This explains, he says, why he finds it "both satisfying and illuminating to distinguish the three kinds of [military] deterrence by I, II, and III, [despite the fact that] many people have been distressed with the nonsuggestive nature of the ordinal numbers." I find this passionate attachment to the ordinal numbers implausible. One more personal note. In his preface Kahn says he carried the manuscript around for a year "on airplanes and railroads." This has the ring of truth. Every now and then the reader gains the impression that he is getting a collection of airport and station jottings; at times it almost seems as though the author is suffering from motion sickness.

Kahn may be the Rand Corporation's General Bourbaki, the imaginary individual used by a school of French mathematicians to test outrageous ideas. The style of the book certainly suggests teamwork. It is by turns waggish, pomp-ous, chummy, coy, brutal, arch, rude, man-to-man, Air Force crisp, energetic, tongue-tied, pretentious, ingenuous, spastic, ironical, savage, malapropos, square-bashing and moralistic. Solecisms, pleonasms and jargon abound; the clichés and fused participles are spectacular; there are many sad examples of what Fowler calls cannibalism—words devouring their own kind. How could a single person produce such a caricature?

No less remarkable is the substance of the book. An ecstatic foreword by Klaus Knorr of Princeton University's Center of International Studies states that this is "not a book about the moral aspects of military problems." The disclaimer is much to the point; it is exactly wrong. This is a moral tract on mass murder: how to plan it, how to commit it, how to get away with it, how to justify it.

The argument of "On Thermonuclear War," so far as it attains coherence, runs like this. Kahn says he is concerned with "alternative national postures" to deter war and to survive it if it comes. It is quite possible, he believes, that we shall have another world war; in fact, several. But one war at a time. What should be done to reduce the threat? World government? Disarmament? These, he says, are utopian. Can we rely on the uncertain balance of terror to postpone the date of mankind's final war? In Kahn's view it is dangerous to hold that an all-out war is "rationally infeasible." The "survival-conscious person" has to think more boldly. We must be ready to fight as well as deter. And if we do fight "we have to 'prevail' in some meaningful sense if we cannot win."

We must therefore be equipped to erase cities, especially control centers ("Finite Deterrence"); we must have "Counterforce as Insurance," "Preattack Mobilization Base," "Limited War Capability" and "Long War (2–30 Days) Capability." Kahn defines the last concept elegantly: "Almost no matter how well one does on the first day of the war, if he has no capability on the second day—and the enemy does have some capability on that day—he is going to lose the war."

Do we need civil defense? The important thing is to fit civil defense into the large strategic program: "Counterforce" and "Credible First Strike Capability," to make sure we gain the most effective "posture" for "Preattack and Postattack Coercion." Three types of deterrence (*i.e.*, I, II and III); all kinds of weapons; readiness for all kinds of wars; a habituation to "tense situations"; "keeping our conceptual doctrinal and linguistic framework up to the moment"—these are some of the elements in the Kahn program of preparation for *Der Tag*.

Kahn summarizes his general notion of the most desirable "posture." We should have, he says, "at least, enough capability to launch a first strike in the kind of tense situation that would result from an outrageous Soviet provocation, so as to induce uncertainty in the enemy as to whether it would not be safer to attack us directly rather than provoke us. The posture should have enough of a retaliatory capacity to make this direct attack unattractive." The Higher Incoherence, otherwise known as the game-theory approach to nuclear-age strategy (which is much admired and fostered by the Rand Corporation) characterizes the argument. There is a Jewish anecdote which runs:

"Where are you going?"

"To Minsk."

"Shame on you! You say this to make me think you are going to Pinsk. But I happen to know you *are* going to Minsk."

What Bertrand Russell's paradox of the class of all classes is to the foundations of mathematics, this anecdote is to the game of international out-think. Kahn is a Minsk-to-Pinsk out-thinker.

When the war is ended (2–30 days), what then? (Do we all join up again?) Some persons have said that after a thermonuclear war the world will be a graveyard and the rats will inherit the earth. Nonsense, says Kahn. This is the "layman's view," although there are many military planners, scientists, "intellectuals" and even generals who hold it. This shows they have not thought hard enough about the question. The 52 Nobel prize winners who in 1955 issued the Mainau Declaration (" 'All nations must come to the decision to renounce force as a final resort of policy. If they are not prepared to do this they will *cease to exist*' ") are well-meaning chaps but they are guilty of "rhetoric." The facts, adduced by "homework" and "sober study," are otherwise. Kahn has "researched" the matter and is in a position to assure us that, while a thermonuclear war "is quite likely to be an *unprecedented catastrophe* for the defender" [his italics], this is "a far cry from an 'unlimited' one." The limits on the magnitude of the catastrophe "seem to be closely dependent on what kinds of preparations have been made, and on how the war is started and fought."

In Kahn's view we must distinguish between 100 million dead and 50 million dead. We must face the task in assessing "postwar states . . . of distinguishing among the possible degrees of awfulness." After all, it would be better to have "a country which survives a war with, say, 150 million people and a gross national product (GNP) of $300 billion a year, [than] a nation

which emerges with only 50 million people and a GNP of $10 billion. The former would [still] be the richest and the fourth largest nation in the world [while] the latter would be a pitiful remnant. . . ."

To clear the mind "for deliberations in this field" Kahn gives us the table reproduced below. (The cryptic caption is from the book.) "Here," says Kahn, "I have tried to make the point that if we have a posture which might result in 40 million dead in a general war, and as a result of poor planning, apathy, or other causes, our posture deteriorates and a war occurs with 80 million dead, we have suffered an additional disaster, an *unnecessary* additional disaster that is almost as bad as the original disaster." Eliminating the *unnecessary* dead is, of course, "something vastly worth doing." And yet, Kahn complains, "it is very difficult to get this point across to laymen or experts with enough intensity to move them to action. The average citizen has a dour attitude toward planners who say that if we do thus and so it will not be 40 million dead—it will be 20 million dead." I suggest the "dour attitude" may be due to the fact that, unlike Kahn, we have not been mathematically trained, and big numbers are apt to be confusing.

Taking 40 million or 80 million dead as a round figure, we might ask whether the postwar "environment" would be so "hostile" that "we or our descendants would prefer being dead than [sic] alive?" Not at all, says Kahn. "Objective studies [made by Kahn and his colleagues] indicate that even though the amount of human tragedy would be

greatly increased in the postwar world, the increase would not preclude normal and happy lives for the majority of survivors and their descendants." "Would the survivors live as Americans are accustomed to living—with automobiles, television, ranch houses, freezers and so on?" Kahn is optimistic. "No one can say, but I believe there is every likelihood that even if we make almost no preparations for recuperation except to buy radiation meters, write and distribute manuals, train some cadres for decontamination and the like, and make some other minimal plans, the country would recover rather rapidly . . . from the small attack."

Kahn admits it may take a little time to get back to normalcy. A number of cities may have disappeared, and the economic engine would require retuning. The economy is sometimes compared to a living organism, which may die even if 99 per cent of its cells are undamaged, but the analogy "seems to be completely wrong as far as long-term recuperation is concerned." The economy is "even more flexible than a salamander (which can grow new parts when old ones are destroyed) in that large sections of it can operate independently (with some degradation, of course). In addition, no matter how much destruction is done, if there are survivors, they will put *something* together. The creating (or recreating) of a society is an art rather than a science; even though empirical and analytic 'laws' have been worked out, we do not really know how it is done, but almost everybody (Ph.D. or savage) can do it."

Parts of the land may become uninhabitable due to fallout. But in general, according to Kahn, the fallout and contamination danger has been exaggerated. Still, if the Strategic Air Command should follow the suggestion "that some people have made" and move "into the Rocky Mountains or the Great American Desert—then some wars might easily result in the creation of large areas that one would not wish to live in, even by industrial standards. It is very unlikely that areas such as the Rocky Mountains would ever be decontaminated. Some people might be willing to visit and perhaps hunt or fish for a few weeks (the game would be edible) but, unless they had a very good reason to stay, it would be unwise to live there and even more unwise to raise a family there."

Kahn favors us with a lengthy analysis of genetic damage. It is not so easily repaired as a ranch house; on the other hand, the damage is likely to be "spread out," and on the installment plan we

## TRAGIC BUT DISTINGUISHABLE POSTWAR STATES

| DEAD | ECONOMIC RECUPERATION |
|---|---|
| 2,000,000 | 1 YEAR |
| 5,000,000 | 2 YEARS |
| 10,000,000 | 5 YEARS |
| 20,000,000 | 10 YEARS |
| 40,000,000 | 20 YEARS |
| 80,000,000 | 50 YEARS |
| 160,000,000 | 100 YEARS |

*WILL THE SURVIVORS ENVY THE DEAD?*

could afford it. We might have to pay for a war through "20 or 30 or 40 generations. But even this is a long way from annihilation. It might well turn out, for example, that U. S. decision makers would be willing, among other things, to accept the high risk of an additional 1 per cent of our children being born deformed *if that meant not giving up Europe to Soviet Russia.*" Kahn is of the opinion that if genetic damage is "borne by our descendants and not by our own generation," we must not take it too much to heart. ("While I believe that this statement is a defendable one, it is not one I would care to defend in the give and take of a public debate") Embryonic deaths are "of limited significance.... These are conceptions which would have been successful if it had not been for radiation that damaged the germ cell and thus made the potential conception result in a failure. There will probably be five million of these in the first generation, and one hundred million in future generations. I do not think of this last number as too important, except for the small fraction that involves detectable miscarriages or stillbirths. On the whole, the human race is so fecund that a small reduction in fecundity should not be a serious matter even to individuals."

Leaving aside the question of genetic deaths, which lie within a price range Kahn feels we should be prepared to pay, how many of the living—to wit, us—should we be prepared to throw into the pot? Kahn says that 180 million "is too high a price to pay for punishing the Soviets for their aggression." But there remains the "hard and unpleasant question": If not 180 million, then how many? Maybe even 100 million is too high. "Almost nobody," Kahn observes, "wants to go down in history as the first man to kill 100,000,000 people." "I have discussed this question," he says, "with many Americans, and after about fifteen minutes of discussion their estimates of an acceptable price generally fall between 10 and 60 million, clustering toward the upper number.... The way one seems to arrive at the upper limit of 60 million is rather interesting. He takes one-third of a country's population, in other words somewhat less than half." It is gratifying to learn that "no American that I have spoken to who was at all serious about the matter believed that any U S. action, limited or unlimited, would be justified...if more than half of our population would be killed in retaliation."

One small brush stroke may be permitted to fill out this portrait of the mind

of Herman Kahn. We are asked to imagine ourselves in the "postwar situation." We will have been exposed to "extremes of anxiety, unfamiliar environment, strange foods, minimum toilet facilities, inadequate shelters, and the like. Under these conditions some high percentage of the population is going to become nauseated, and nausea is very catching. If one man vomits, everybody vomits. It would not be surprising if almost everybody vomits. Almost everyone is likely to think he has received too much radiation. Morale may be so affected that many survivors may refuse to participate in constructive activities, but would content themselves with sitting down and waiting to die—some may even become violent and destructive. However, the situation would be quite different if radiation meters were distributed. Assume now that a man gets sick from a cause other than radiation. Not believing this, his morale begins to drop. You look at his meter and say, 'You have received only ten roentgens, why are you vomiting? Pull yourself together and get to work.'"

Herman Kahn, we are told, is "one of the very few who have managed to avoid the 'mental block' so characteristic of writers on nuclear warfare." The mental block consists, if I am not mistaken, of a scruple for life. This evil and tenebrous book, with its loose-lipped pieties and its hayfoot-strawfoot logic, is permeated with a bloodthirsty irrationality such as I have not seen in my years of reading. We are now in a position to comprehend the noble Houyhnhnm's horror at Gulliver's account of the condition of man:

"He said, whoever understood the Nature of *Yahoos* might easily believe it possible for so vile an Animal, to be capable of every action I had named, if their Strength and Cunning equalled their Malice.... That, although he hated the *Yahoos* of this Country, yet he no more blamed them for their odious Qualities, than he did a Gnnayh (A Bird of Prey) for its Cruelty, or a sharp Stone for cutting his Hoof. But, when a Creature pretending to Reason, could be capable of such Enormities, he dreaded lest the Corruption of that Faculty might be worse than Brutality itself."

After the unsavory experience of the Kahn book I had looked forward to the arms-control issue of *Dædalus.* This is a quarterly journal published by the American Academy of Arts and Sciences. The Academy is an organization of 1,800 members, one of whose major activities is to elect new members an-

nually. Its net being widely flung, I had assumed I would encounter the opinions of educators, scientists and other serious-minded men who had thought fruitfully about some of the steps which have to be taken if human life is to continue on this planet. I am bound to say that I was deeply disappointed. In addition to giving the floor to Herman Kahn—a 37-page dose on "Doomsday Machines," "Doomsday-in-a-Hurry Machine" and other obscene lunacies—the issue contains a medley of pieces scored more or less in the Kahn key, and an assortment of legal, political and economic oddments which offer neither light nor warmth nor shelter to the anxious reader. A few bright interludes are to be found in this dismal repertory (Erich Fromm and Kenneth E. Boulding are among those who have some penetrating things to say), but if this issue of *Dædalus* is a fair sample of the high thoughts of American academicians on the "potentially feasible routes as well as the obstacles to arms control," then American arts and sciences are in a bad way, and unless a new crop of sensible politicians comes to the rescue we can kiss ourselves and posterity good-by. It is not only the fact that *Dædalus* gave Kahn houseroom that shocked me but also that so many of the contributors accept his inhumanity as the basis for their own speculations.

The contents of the issue are divided into six main groups: "Background," "Major Issues and Problems," "The Implementation of Arms Control," "The Formation of United States Arms Control Policy," "Related Techniques and Issues," "Beyond the Cold War." I shall sample, which is quite enough.

Donald G. Brennan, a mathematician and "communication theorist," discussing the "setting and goals of arms control," tells us that "there is an increasing recognition of the fact that the simple form of the 'balance of terror' theory to implement Type A deterrence is inadequate, and that the balance, as was aptly noted by Wohlstetter, is 'delicate'"; that there are "pro" and "con" hazards to arms control: "it may improve some component of our security, either in the short or the long term, and it may degrade some other—again, either in the short or the long term"; that "smaller" nuclear weapons having a yield as low as that of 55 tons of TNT can be developed and "probably would be advantageous for the United States, *provided that they did not lead to the use of much larger weapons*" (a point which may have escaped you); that it is important for us "to educate the Soviets in mutually desirable strategies [a Marquis of Queensberry touch] and armament policies. For this purpose we would first have to educate ourselves in some detail as to what these were—which hardly prevails at the present time." "We have men today," says Brennan, "as capable as those who drafted our own Constitution; it is not necessary to wait for the once-in-a-century appearance of an Abraham Lincoln." I must assume from all this that Brennan is better in the theory than in the practice of communication.

Edward Teller coaxes us to accept the concept of limited war. "Limited warfare can very well stay limited." (After all, half a loaf is better than none.) "All-out war will never be in our interest, and we should never start it." If the Russians should start, "they will probably pick a time when our guard is down. While a limited nuclear war is in progress, we shall be much better prepared than in times of peace. The time of a limited nuclear conflict, therefore, would be the worst time for the Russians to launch an all-out attack." If I understand this sequence, it implies that the U. S. would be well advised to get into a limited war as quickly as possible and keep it going indefinitely.

From Henry A. Kissinger comes a "reappraisal" of limited war. With that waffling judiciousness so characteristic of his writings, he weighs the question of whether the limited war we must be ready to fight should be "conventional or nuclear" (taking it for granted, of course, that we are also prepared to wage a "general nuclear war"). His answer is yes and no and both and neither. A substantial build-up of conventional forces and a greater reliance on a conventional strategy are "essential," he says; on the other hand, "it is equally vital not to press the conclusions too far. . . . Conventional forces should not be considered a substitute for a capability of waging a limited nuclear war, but a complement to it. It would be suicidal to rely entirely on conventional arms against an opponent equipped with nuclear weapons. . . . A conventional war can be kept within limits only if nuclear war seems more unattractive. . . . The aggressor must understand that we are in a position to match any increment of force, nuclear or conventional, that he may add"—an interesting form of potlatch.

In a thoughtful essay on the domestic implications of arms control, Kenneth Boulding disposes effectively of the myth that U. S. prosperity depends on arms expenditures. If we had money to spend sensibly, we might in time become sensible enough to know how to spend it. The economic problem is really trivial: re-duce armaments and family life will go on, as will business, industry, schools, milk deliveries. The nation would thrive even if the Pentagon were turned into a garage. The Pentagonians, to be sure, might not thrive until they learned another trade. A specter, says Boulding, is haunting the chancelleries and the general staffs, "more frightening perhaps than that which Karl Marx invoked in 1848; it is the specter of Peace—that drab girl with the olive-branch corsage whom no red-blooded American (or Russian) could conceivably warm up to. She haunts us because we cannot go back to Napoleon, or to Lee, or even to Mac-Arthur: the military are caught in an implacable dynamic of technical change which makes them increasingly less capable of defending the countries which support them, except at an increasingly intolerable cost. The grotesque irony of national defense in the nuclear age is that, after having had the inestimable privilege of losing half (or is it three-quarters, or all?) our population, we are supposed to set up again the whole system which gave rise to this holocaust."

Jerome B. Wiesner, in his analysis of arms-limitation systems, makes some obvious but very useful points. In this tortuous business it is the obvious which is usually scanted. He points out that until recently neither the U. S. nor Russia was sincerely attempting to reach an agreement on arms limitation; moreover, that the U. S. delegations to the disarmament discussions, the nuclear-test-ban conferences and the surprise-attack conferences "had very inadequate technical preparation." Defy the foul fiend, yes, but how? The objective should be to find security systems which are less dangerous than the arms bolero rather than to achieve a system capable of providing absolute security, "an obviously unattainable goal." Wiesner makes it clear that an unconscionable amount of time has been spent on the essentially trivial issue of detecting underground tests. "Ironically," he says, "an inspection system for monitoring a truly comprehensive disarmament agreement would probably have no need at all for a system to detect underground nuclear tests." The pother about nuclear-stockpile control does not impress him. Admittedly it is impossible to determine exactly the size of the stockpiles. But the range of uncertainty is not so great as we have been led to believe, and it is likely that "an intensive study of the physical means of estimating past nuclear production could greatly reduce this uncertainty." Wiesner makes an excellent point about establishing a stable deterrence system using only a relatively small number of ballistic missiles. Assume the deterrent force consists of a number of Minuteman missiles installed in underground concrete emplacements. Such emplacements can be made pretty secure against very heavy shock waves, so that if guided missiles are used to deliver nuclear weapons in attacking hardened targets, the accuracy of the missiles would be very important. "If a nuclear weapon had to make impact within one-half mile of a target to destroy it, a missile having a median accuracy of half a mile would have a 0.5 probability of doing so, two missiles would have a 0.75 probability of doing so, three missiles a 0.875 probability and four missiles would have approximately a 0.94 probability of destroying the target. When the number of targets to be attacked is large and the number of survivors that can be tolerated is small, the certainty with which each individual target must be destroyed becomes extreme, and the number of attacking missiles required can become quite large. . . . To demonstrate how difficult it is to destroy a hardened missile force, an example will be given. If it is agreed that each side is to have 200 missiles in its deterrent force and if the missiles were protected for 300 pound/sq. inch overpressure, 1,000 missiles having a median accuracy of one mile would be required to have a 0.9 probability of reducing the attacked force to 10 missiles. It obviously would not require a very intensive inspection effort to detect an attempted build-up of this magnitude."

Two more pieces are worth mentioning. Saville R. Davis, managing editor of *The Christian Science Monitor,* has a tidily lethal piece called "Recent Policy Making in the United States Government." It will not be celebrated by admirers of John Foster Dulles, Admiral Arthur Radford, Lewis Strauss and Edward Teller. Radford's and Strauss's torpedoing of Harold Stassen's efforts to negotiate a disarmament agreement makes a distressing tableau; so does the pressure that Strauss, with the help of Teller and the late Ernest O. Lawrence, brought to bear on the President on behalf of continued nuclear testing in order to perfect "clean" weapons. "Evidence is available to the writer [says Davis] which clearly indicates that the President and Mr. Dulles were unwitting prisoners, in their lonely isolation at the top of the government pyramid, of the special selection of knowledge and attitudes which came to them through official channels and especially through Mr. Strauss. They had no alternative against which to measure the partisan quality of this advice or its scientific inadequacies."

Suppose every person had the means of killing himself and his family, knowing that this would result in the death of two Russian families. This is the exemplar of contemporary lunacy, and in discussing it psychiatrists should have a principal role. The profession is ably represented in this symposium by Erich Fromm. He puts the case for unilateral disarmament. Among the concrete steps which could be taken unilaterally so as to induce the other side to reciprocate are the sharing of scientific information, the stopping of atomic tests, troop reductions, evacuation of certain military bases, the discontinuation of German rearmament. Risks are involved in these steps, to be sure, but they are not crippling risks, they do not invite a major assault and they are risks any sane man would prefer to the risks of the arms race.

Fromm emphasizes the need for breaking through the "thought barrier," the "frozen stereotypes" which prevent us from seeking peace by any means other than threat and counterthreat. We do nothing more than circle the forest. It is absurd to imagine that a policy based on deterrence can indefinitely keep the peace, but even if it did, what kind of peace would it be? Under the constant threat of destruction, most human beings begin to come apart. They grow callous, hostile, increasingly indifferent to the values we cherish. Freedom is lost, the individual becomes nothing. "Things are in the saddle, and ride mankind," Emerson said a century ago. In the computer and missile age they ride us harder.

Fromm examines some of the popular psychological arguments advanced against disarmament, for example "The Russians cannot be trusted." If this is meant in the moral sense, it is true: political leaders of any nationality are rarely trustworthy. They may be good to their mothers, kind to their children and honorable in dealing with the grocer, but public and private faces are not the same. The state, which is an idol, can and does commit immoral acts, which the community applauds—acts which the community would deprecate or punish if committed by individuals.

Yet the phrase "Trust the Russians" has a meaning that is relevant to politics. We must have faith in their being sane men whose conduct is therefore to some extent predictable. We must not assume that they will destroy themselves for the pleasure of destroying us. A rational policy of disarmament must have its ultimate roots in two simple convictions: that we want to live, that the Russians want to live. This conviction is no less applicable to those who make the decisions than to those who are compelled to obey them. This issue of sanity, as Fromm points out, leads to another consideration which affects us as much as it does the Russians. "In the current discussion on armament control, many arguments are based on the question of what is *possible*, rather than what is *probable*. The difference between these two modes of thinking is precisely the difference between *paranoid* and *sane* thinking." It is *possible* my wife and children are planning to poison me; it is possible a meteorite will hit me on the head when I leave the house this afternoon; it is possible the manuscript of this review will disintegrate by molecular action before my secretary types it; but I do not live by these possibilities. One lives by what is likely or one is mad. A certain faith in life, in oneself and in others is necessary in order to operate. In the twilight of our probationership here on earth—to use Locke's phrase—we learn to depend on probabilities. Arms control is hopeless if we insist on covering every possibility.

"At midwinter in the year 1085 William the Conqueror wore his crown at Gloucester and there he had deep speech with his wise men." Thence, as Frederic Maitland tells us, missions went throughout England bringing back a *descriptio* of the new realm. These are preserved in the famous *Domesday Book*. We have need of a *descriptio* of man's realm; of what he has made of the world and of himself in 20,000 years; of the treasures he has stored, of the ideas he has borne and nurtured. We have need, appropriately, of a new *Domesday Book* to enable us to reckon whether we wish to preserve what we have and what we are or whether we are preparing to risk all in the war of every man against every man. For this we need deep speech with our wise men, a few of whom, but too few, have thus far been called into council. That there is some wisdom in the pages of this issue of *Dædalus*, some sane counsel, is good; that they can scarcely be heard over the strident clamor is tragic.

# 41

# Steps Toward Disarmament

P. M. S. BLACKETT

*April 1962*

The representatives of 17 nations—the two main nuclear powers, seven nations allied with one or the other of them, and eight uncommitted nations—have convened at Geneva for the third formal, full-dress attempt since the end of World War II to negotiate disarmament. It must be conceded that the circumstances are not entirely favorable to agreement. During 1961 the U.S. and the U.S.S.R. reversed the trend of nearly a decade and increased their military expenditures by something on the order of 25 per cent. The three-year moratorium on the testing of nuclear weapons was terminated by the series of Soviet tests in the fall; on the eve of the Geneva meeting the U.S. announced its intention to move its present series of underground tests into the atmosphere if the U.S.S.R. did not immediately agree to a test ban.

On the other hand, both the Soviet and the Western bloc are committed by categorical public statements to the objective of complete and general disarmament under strict inspection and control. What is more, practical military considerations, arising from the nature of nuclear weapons, commend substantial reduction in armaments to the great powers as a measure that will increase their security in the first step toward disarmament.

In considering possible first steps that would lead to increased security for both sides, partisans of each side should try to understand how the present military situation must look to the other. A military commander, in planning a campaign or a battle, attempts to do this as a matter of course. He has first to find out all he can about the material facts of his opponent's military deployment and secondly to assess the probable intentions of his opponent for its use. This is the process that has been described as "guessing what is happening on the other side of the hill." A similar obligation rests on those who plan a disarmament negotiation. A military planner, it is true, can much more easily put himself mentally in the position of his military opponent than a statesman can think himself into the position of his opposite number, because a statesman must enter imaginatively into the political as well as the military thought processes of his opponent. This is hard to do at a time of acute ideological struggle. It is nonetheless essential that the military and political leaders of both sides do just this. No small part of the present crisis, concerning armaments in general and nuclear weapons in particular, has been due to a tendency in the West to attribute to ideological motives actions by the U.S.S.R. that seem to have been motivated mainly by military considerations. Conversely, much of the West's defense policy appears to have been influenced by political and economic factors.

It may be useful to start by describing the most important elements in the military capabilities of the Soviet bloc and the Western alliance. In recent months there have been significant disclosures about the nuclear weapons and their means of delivery possessed by both sides. On November 12 of last year Robert S. McNamara, Secretary of Defense of the U.S., said that the U.S. nuclear-strike force consists of 1,700 intercontinental bombers, including 630 B-52's, 55 B-58's and 1,000 B-47's. He said that the U.S. possesses in addition several dozen operational intercontinental ballistic missiles (ICBM's), some 80 Polaris missiles in nuclear-powered submarines, about the same number of Thor and Jupiter intermediate-range missiles, some 300 carrier-borne aircraft armed with megaton war heads and nearly 1,000 supersonic land-based fighters with nuclear war heads. According to his deputy, Roswell L. Gilpatric, "the total number of our nuclear delivery vehicles, tactical as well as strategic, is in the tens of thousands, and of course we have more than one war head for each vehicle.... We have a second-strike capability that is at least as extensive as what the Soviets can deliver by striking first, therefore we can be confident that the Soviets will not provoke a major conflict." The U.S. stockpile of nuclear weapons is most often estimated as around 30,000 megatons, that is, enough for some 30,000 one-megaton bombs.

Naturally no such precise figures for Soviet strength are available. I have seen no reliable estimates of the U.S.S.R.'s nuclear stockpile, nor of its possible nuclear-armed submarine strength, nor of its nuclear-armed fighter-bomber strength (the last, of course, would not have sufficient range to contribute to the Soviet strike power against the U.S.). But recent semiofficial estimates from Washington give the U.S.S.R. some 50 ICBM's, some 150 intercontinental bombers and some 400 medium-range missiles (the last able to cover Europe but not the U.S.). The same sources indicate that the U.S. may have a small lead over the U.S.S.R. in the number of ICBM's. That such estimates should issue from Washington may seem surprising in view of the role that an alleged "missile gap" played in the 1960 presidential election campaign. That the estimates are realistic, however, is indicated by the statement of Senator Stuart Symington that the U.S. intelligence estimate of the missile force available to the U.S.S.R. at the middle of 1961 was only 3.5 per cent of the number predicted a few years ago. The corresponding estimate of Soviet bomber strength, he revealed, was 19 per cent of the number predicted in 1956 [see illustrations at right]. Mr. Symington explained that the new figures are predicated on intelligence about Soviet "intentions" as well as "capability" and expressed his own disquiet at "the tentativeness at best of our intelligence estimates." It is one of the purposes of this article to attempt to elucidate some of these Soviet intentions.

At first sight there appears to be a contradiction between Washington's claim of a marked over-all nuclear superiority and the recent statement by Marshal Rodion Y. Malinovsky, the Soviet Minister of Defense, that the U.S.S.R. has the power to destroy all the important industrial, administrative and political centers of the U.S. and "whole countries that have provided their territories for the siting of American war bases." The explanation may be as follows. To carry out such destruction would require not more than 1,000 megatons of nuclear destructive power, say five megatons for each of 100 key targets in the U.S. and another 500 megatons for Western Europe and U.S. bases overseas. At only 100,000 dead per megaton such an attack would kill 100 million people. The U.S. stockpile, estimated at 30,000 megatons, is 30 times greater than the U.S.S.R. would need to carry out the retaliatory blow described by Malinovsky.

There is, of course, the possibility that the new U.S. estimates of Soviet nuclear strength are too low. After all, firm

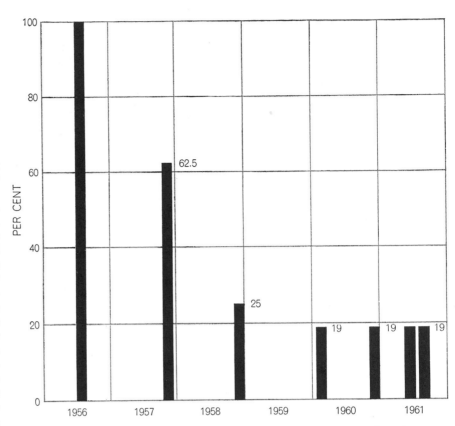

U.S. ESTIMATE OF SOVIET HEAVY-BOMBER STRENGTH by the middle of 1961, according to an article by Senator Stuart Symington in *The Reporter*, decreased by 81 per cent between August, 1956 (*bar at left*), and August, 1961 (*right*). Senator Symington's figures were given in percentages, rather than absolute numbers, for security reasons.

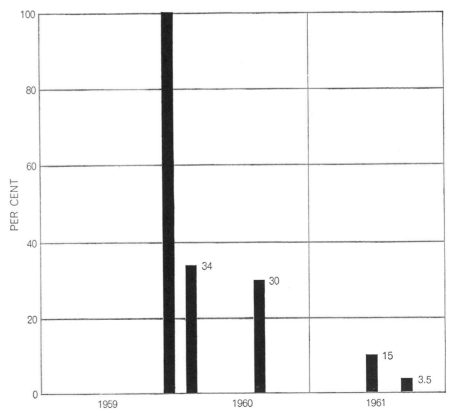

U.S. ESTIMATE OF SOVIET OPERATIONAL ICBM STRENGTH similarly decreased, according to Senator Symington, by 96.5 per cent between December, 1959, and September, 1961.

information about Soviet military preparations is notoriously hard to come by. It seems certain, however, that the U.S. Department of Defense must believe the estimates to be roughly correct. It would be politically disastrous for the Administration to be found guilty of underestimating Soviet nuclear strength. But even assuming that the estimates of the relative strength of the two sides are only approximately correct, they show that the possibility of a rationally planned surprise nuclear attack by the U.S.S.R. on the nuclear delivery system of the West must be quite negligible. The question of why the U.S.S.R. has built such a small nuclear delivery system should perhaps be replaced by the question of why the U.S. has built such an enormous striking capacity.

In order to understand the possible motives behind Soviet defense policy, it is necessary to consider the history of the growth of nuclear-weapon power. During the period of U.S. atomic monopoly or overwhelming numerical superiority, say from 1947 to 1954, the role of the U.S. Strategic Air Command was to attack and destroy Soviet cities in case of war. This countercity policy, like most traditional military doctrines, had both an offensive and a defensive aspect. From the Western viewpoint, under the doctrine of "massive retaliation," this nuclear striking power was seen to be both a deterrent to the possibility of attack by Soviet land forces and, in the extreme "roll back," or "liberation," statement of the doctrine, an offensive weapon to obtain political concessions by threat of its use. By 1954 the threat was implemented by more than 1,000 intercontinental B-47 bombers, plus larger numbers of shorter range vehicles deployed around the U.S.S.R.

From the U.S.S.R.'s point of view, its land forces were the only available counter to the Western nuclear monopoly during this period. The answer to the threat of nuclear attack was the threat of taking over Europe on the ground. In retrospect the military reaction of the U.S.S.R. seems understandable. It started a crash program to produce its own nuclear weapons. It also embarked on a huge air defense program; by 1953 it was credited with an operational fighter strength of some 10,000 aircraft. As Western nuclear strength grew, the U.S.S.R. gradually built up its land forces so as to be able to invade Europe, even after a U.S. nuclear attack. At the political level the U.S.S.R. consolidated its forward military line by the political coup in 1948 in Czechoslovakia and integrated the other satellite countries more closely into the Soviet defense system. Since the main military threat then to the U.S.S.R. was from manned nuclear bombers, the greatest possible depth for air defense was vital. During World War II it was found that the efficacy of a fighter defense system increased steeply with the depth of the defense zone. Finally, the U.S.S.R. maintained strict geographical secrecy over its land area so as to deny target information to the U.S. Strategic Air Command.

The doctrine of massive retaliation became less and less plausible as the Soviet nuclear stockpile grew. It had to be abandoned after 1954, when hydrogen bombs became available to both East and West. When the U.S.S.R. proceeded to build up a fleet of long-range bombers to deliver its hydrogen bombs, the U.S. became vulnerable to nuclear counterattack. Some form of nuclear stalemate by balance of terror seemed to have arrived.

This balance seemed still further strengthened about 1957, when rapid progress in the technology of nuclear weapons and missiles made it possible to carry multimegaton hydrogen bombs in ICBM's. Because such missiles are most difficult, if not impossible, to destroy in flight, a nuclear aggressor would have to leave no enemy missiles undestroyed if it wanted to keep its own major cities from being wiped out by a retaliatory attack. The advent of long-range missiles therefore made the balance of terror more stable.

Two contrasting systems of military theory evolved in response to this new situation. The first led off from the premise that a rather stable kind of military balance had been reached, in which neither side could make use of its strategic nuclear power without ensuring its own destruction. In other words, the balance of terror was likely to be rather stable against rational action, even though the actual nuclear strengths of the two sides were markedly different, as indeed they were in the middle 1950's, when the U.S. was already vastly stronger in over-all deployed nuclear strength. This view rested on the assumption that neither side could hope to knock out the other's nuclear system entirely. Since some power to retaliate would survive attack, a rational government would be nearly as much, if not just as much, deterred from a first strike by the expectation that it would suffer, say, 10 million deaths as it would be if the expectation were 100 million.

This view led to the practical conclusion that "enough is enough." In today's jargon this is the policy of the minimum deterrent—that is, the possession of a nuclear force adequate only for a retaliatory attack on enemy cities but incapable of successful attack on the enemy's nuclear delivery system. It is clear that only a small nuclear delivery system is necessary for a minimum deterrent. One big hydrogen bomb dropped on a big city could kill several millions. The small delivery system must, however, be highly invulnerable. Otherwise the enemy might think it possible to bring off a

**MINIMUM DETERRENT** strategy of a nuclear opponent of the U.S. could logically

successful "counterforce" first strike, aimed at the destruction of the system. Little operational intelligence is needed for such a minimum deterrent policy because this involves attack on cities, whose locations are known, and does not involve surprise attack on nuclear bases, whose locations therefore do not need to be known.

On the political plane, it was thought, the resulting period of relative stability would be favorable for a serious attempt to negotiate a substantial measure of disarmament, both nuclear and conventional. Far-reaching disarmament was seen to be highly desirable, if only because such a balance of terror is stable

solely against rational acts of responsible governments. It is not stable against irresponsible actions of individuals or dissident groups or technical accidents. A few suitably placed individuals—a missile crew or the crew of a nuclear bomber on a routine flight—could kill a few million enemy city dwellers on their own initiative. The best way to reduce this danger is to reduce drastically the number of nuclear weapons on both sides.

The second and quite different doctrine was that the balance of terror was not even stable against rational acts of responsible governments. This was based on the view that a determined

nuclear power might be able to launch a surprise counterforce attack on the enemy's nuclear delivery system of such strength that the enemy would not be able to retaliate. The aggressor, without suffering unacceptable casualties, would then have the enemy at its mercy. The practical consequence of this doctrine is to strive for maximum superiority in number of weapons, maximum invulnerability of one's own nuclear delivery system and maximum intelligence about the enemy's nuclear system.

Plainly a successful counterforce attack would require knowledge of the location of all the enemy's nuclear missile and air bases and the power to dispatch

be based on an attack on the U.S. population rather than on U.S. airfields and missile bases. The colored dots on this map represent the 25 largest U.S. cities. In the 1960 census the combined population of the metropolitan areas of these cities was 60.8 million.

several weapons against each, so as to ensure that at least one reached its target. A counterforce strategy thus implies the necessity for a many-fold nuclear superiority over the enemy. Moreover, to have the slightest chance of success such an attack must come as a complete surprise to the enemy: it must be a first strike. This policy has various pseudonyms: maximum deterrent posture, first-counterforce-strike capability, or, in plain English, preparation for nuclear aggression.

Since the possession of nuclear armament raises the possibility that either side could adopt either one of these strategies, both of them must have been discussed in military circles in Moscow and Washington during the years after the explosion of the first hydrogen bombs in 1954. Let us try to find out how the discussions went by studying what shape the nuclear-defense policies of the U.S.S.R. and the U.S. took in the subsequent years.

If the Washington figures for Soviet nuclear strength are valid, it is clear that the U.S.S.R. has planned for a purely retaliatory nuclear role and has definitely not planned for a surprise attack on the U.S. delivery system. As long ago as 1956 the U.S.S.R. was believed to have the capability of making 25 long-range bombers a month. It appears today to have only some 150, compared with the 1,700 U.S. long-range bombers able to reach the U.S.S.R. Even though Soviet medium-range bombers could reach the U.S. on a one-way flight, this is much more than counterbalanced by the 1,500 or so Western fighter bombers, carrier-borne aircraft and medium-range missiles able to reach the U.S.S.R. It is also probable that the U.S.S.R. could have made many more than the 50 or so ICBM's with which it is now credited, since its space program indicates substantial industrial resources for making missiles. The evidence is that the U.S.S.R. has based its safety on the retaliatory power of a small number of missiles and aircraft operating from bases whose exact locations are kept as secret as possible. The deterrent value of its missiles is certainly enhanced by the prestige of its space program.

That the U.S.S.R. believed the danger of a major war, intentionally initiated, had been reduced by the advent of hydrogen bombs seems indicated by the fact that it reduced the total number of men in its armed forces from 5.8 million in 1955 to 3.6 million in 1959. In January, 1960, Premier Khrushchev announced the U.S.S.R.'s intention to re-

duce this to 2.4 million by the end of 1961. The U.S.S.R. needed fewer troops because it no longer had to rely on a retaliatory land blow in Europe to counter a Western nuclear attack. Its concern about the danger of accidental, irresponsible or escalated war is probably one of the reasons for its strong espousal in 1955 of a drastic measure of comprehensive and general disarmament.

Turning to the history of U.S. defense policy over this period, it is to be noted that the total service manpower fell slowly from 2.9 million in 1955 to 2.6 million in 1960. The development of improved nuclear weapons, missiles and aircraft continued, but not at a great rate, even after the Soviet launching of an artificial satellite in 1957 and much boasting by the U.S.S.R. of its missile prowess. Although subjected to considerable public pressure to engage in a crash program to close the alleged missile gap, President Eisenhower maintained that the existing program was adequate for the safety of the nation. In his last State of the Union Message in January, 1961, he declared: "The 'bomber gap' of several years ago was always a fiction and the 'missile gap' shows every sign of being the same."

As 1954 was the year of the hydrogen bomb, so 1961 was for both sides in the cold war the year of the Great Rearmament. In the U.S.S.R. the decrease of total armed forces to 2.4 million projected for 1961 was deferred and the arms budget was markedly increased. In July the Soviet Government went on the diplomatic offensive to bring about changes in the status of Berlin and to get the division of Germany recognized. In August it began testing nuclear weapons again, in spite of a promise in January, 1960, by Premier Khrushchev that the U.S.S.R. would not be the first to do so. No doubt there were some political motives behind these drastic moves. Possibly heavy pressure was put on Khrushchev from China and from the opposition elements in the U.S.S.R. to admit that his policy of coexistence had not produced political gains commensurate with its possible military risks. But such drastic changes, with the inevitable adverse reaction of much of world opinion, would hardly have been made unless there were strong military reasons for them. To get at these reasons it is necessary to recall in more detail the circumstances in which the changes took place.

In the first place the flights of the U.S. reconnaissance U-2 aircraft must have had decisive importance in shaping the

attitudes of Soviet military leaders. Although the over-all nuclear strength of the U.S. is now, and was then, much greater than that of the U.S.S.R., Soviet leaders could reckon that one vital factor would make a U.S. nuclear attack on the U.S.S.R. exceedingly risky: the secrecy as to the location of the Soviet nuclear bases. Obviously one of the main objectives of the U-2 flights was to locate those nuclear bases. The Soviet command knew that the U-2 flights had been going on for some years before the first aircraft was shot down in the spring of 1960; presumably they reacted by greater dispersal and camouflage. What must have disturbed the Soviet military staff was President Eisenhower's justification of the flights as essential for U.S. security. This implied that U.S. security could only be maintained if the U.S. had sufficient information as to the location of Soviet nuclear sites to make possible a successful surprise attack on the Soviet retaliatory force.

If these were the Soviet fears, the rejection by the U.S.S.R. early in 1961 of the British-American draft of a treaty to ban the testing of nuclear weapons finds explanation in the same jealous military concern to protect the country's geographical security. A detailed study of this document makes it clear that the elaborate international inspection system proposed for the prevention of underground tests could conceivably have served to reveal the location of at least some of the Soviet missile sites. It would be hard to convince a military staff officer of any nationality that this possibility was negligible. If the West had been content to monitor only the atmosphere against test violations, a much less comprehensive inspection system would have sufficed and a test-ban treaty might well have been signed. The Soviet fear of inspection may have been the more acute because there was so little in the U.S.S.R. to inspect.

The resumption of testing by the U.S.S.R. in September, 1961, would seem to fall into the same pattern of motivation. Although its timing may have been influenced by the Berlin crisis, which Khrushchev himself brought to a head, the testing of war heads with an explosive force of up to 60 megatons and the simultaneous well-publicized success of putting seven ICBM's on their target in the Pacific at a range of some 7,000 miles was an effective way of re-establishing the U.S.S.R.'s confidence in the few deployed ICBM's that formed its main retaliatory force. Soviet spokesmen were at pains to promote the credibility of the U.S.S.R.'s deterrent by emphasizing to the U.S. the accuracy of its missiles and the possible power of the war heads demonstrated in these tests.

In the redirection of Soviet military policy considerable weight must also have been carried by the fear that if the NATO rearmament continued, the time could not be far distant when West Germany would get *de facto* control of its own nuclear weapons. In Soviet eyes the refusal of the West to take disarmament seriously at the "Committee of Ten" conference in 1960 was evidently decisive. As early as November, 1960, the Russians stated that if the West continued to temporize on disarmament, the U.S.S.R. would be forced into massive rearmament.

Sometime in the latter half of 1960 or early in 1961 it seems probable that the Soviet military staff began to have doubts as to the adequacy of the minimum deterrent posture in relation to the near-maximum deterrent posture of the U.S. It must have been later than January of 1960, for in that month Khrushchev announced a drastic cutback of both long-range bombers and conventional forces. Since the effectiveness of the Soviet minimum deterrent rested so heavily on geographical secrecy, the U.S.S.R. command may have feared that the U.S., by further air or satellite reconnaissance, or by espionage or defections, would ultimately acquire the intelligence necessary to make a successful nuclear attack on Soviet nuclear bases. Probably the main fear of the Soviet Government was that circumstances might arise in which the U.S. Government would be pushed by irresponsible or fanatical groups into reckless action. The Russians certainly noted the doctrine of some civilian analysts that it would be quite rational to make a "preemptive first strike" even at the cost of 10 million deaths to the attacking side, and the doctrine of others that the U.S. should prepare itself mentally and materially to suffer such casualties.

In the U.S. the program for the Great Rearmament was projected as early as 1959 by the Democratic National Committee. In preparation for the impending presidential election the party leadership published a detailed study of defense problems and recommended a $7 billion increase (16 per cent) in the $43 billion defense budget proposed by President Eisenhower. The funds were to go partly for increased conventional forces and partly to increase the strength and reduce the vulnerability of the U.S. nuclear striking power. In January, 1961, almost immediately after taking office, the Administration authorized an increase of $3 billion and later in the year another $4 billion, thus carrying out the program in full. The present plans include the provision of up to 800 ICBM's of the solid-fuel Minuteman type in underground "hardened" bases by 1965.

The Democratic Party's campaign for increased nuclear armaments was closely linked with the theoretical doctrine of the instability of the balance of terror, derived from the alleged overwhelming advantage accruing to the nuclear aggressor. This was ably argued by civilian analysts closely associated with the U.S. Air Force. The U.S.S.R. was said to have both the capability and the intention to launch a surprise nuclear attack on the U.S. In retrospect, it would seem that these "looking-glass strategists" endowed the U.S.S.R. with a capability that it did not have and that the U.S. had once had and had now lost.

That the Soviet military staff had reason to take this element in U.S. opinion seriously may be judged by the fact that President Kennedy himself found it necessary to launch in the fall of 1961 a vigorous campaign against all those in the U.S. who urge "total war and total victory over communism . . . who seek to find an American solution for all problems"—against those who were living in the long-past era of the U.S. nuclear monopoly. In this campaign President Kennedy has been vigorously supported by ex-President Eisenhower. Very possibly the U.S.S.R. may have overestimated the potential influence of the proponents of aggressive nuclear strategy and the ultra-right-wing groups that yearn "to get it over with." Nonetheless, the fact that both Kennedy and Eisenhower have felt it necessary to combat them must also imply that the Soviet military planners could not afford to ignore their existence.

The Kennedy Administration's recent vigorous emphasis on the overwhelming nuclear superiority of the U.S. over the U.S.S.R., and the assertion that the U.S. possesses a second strike that is as strong as the Soviet first strike might perhaps be held in the U.S.S.R. to suggest a move by the U.S. Administration toward a preventive war posture. Undoubtedly the exact reverse is the case. The Administration's statements are designed to bury officially the fear of a Soviet first strike, sedulously propagated by those who believe that the U.S.S.R. has planned for, and in fact now has, a first-counterforce capability, and so at a time of crisis might use it. If this were in truth the situation, the argument that the U.S. must forestall the Soviet blow might seem strong. The Kennedy Administration evidently foresaw this danger arising and effectively removed it by denying that the U.S.S.R. has ever had an effective first-strike capacity; thus there would be no reason for a forestalling blow in a crisis. The President, by emphasizing U.S. nuclear superiority over the U.S.S.R., has forestalled the potential forestallers, or, in the current jargon, has pre-empted the potential pre-empters. At the same time he has refuted many of the arguments on which the Democratic Party based much of its election campaign, and indeed many of the arguments for his own present rearmament program.

It is, for instance, hard to see the military justification for the program of up to 800 Minuteman ICBM's in the next few years. If these are, as claimed, reasonably invulnerable, this number is at least 10 times larger than is necessary for an effective retaliatory force to attack Soviet cities.

The only military circumstance that could justify such a continuous build-up of nuclear striking force would be that the other party could adequately protect its cities or succeed in perfecting an anti-missile defense system. Recently Soviet generals have boasted that "the complex and important problems of destroying enemy rockets in flight have been solved." This must refer to the scientific and technical problems; these have also been solved in the U.S. A complete anti-missile defense system that is of any operational significance certainly does not exist today and, in my view, will not exist in the foreseeable future. Suppose, however, that I am wrong and that a system can eventually be constructed capable of destroying, say, 50 per cent of a retaliatory missile attack by 50 ICBM's, so reducing the number reaching the target to 25. Even this reduced blow would kill tens of millions of people. Moreover, it would only be necessary to increase the strength of the retaliatory force from 50 to 100 missiles to cancel out the antimissile missile. This illustrates the general conclusion that since a purely retaliatory nuclear force can be quite small, any possible defense system, either active or passive, can be canceled out by a small number of additional missiles. The fact that a purely retaliatory posture is little affected by technological innovation, whereas a

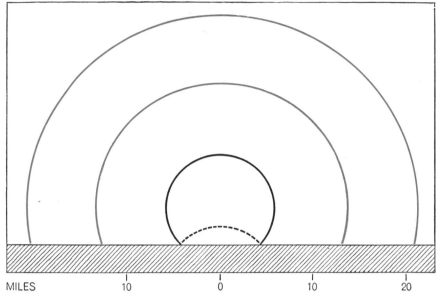

MILES        10        0        10        20

AIR BURST of a nuclear bomb would maximize its effects on a city, the most widespread of which would be due to heat. This drawing outlines the effects of a 10-megaton bomb set off at 20,000 feet. At 12 miles (*inner colored circle*) from "ground zero" the fireball, 3.4 miles in diameter, would deliver 30 calories per square centimeter at a rate sufficient to ignite virtually all flammable building materials. At 20 miles (*outer colored circle*) from ground zero the heat would be 12 calories per square centimeter, enough to cause third-degree burns and start many fires. Arc extending upward from ground below the burst is a reflected shock wave that would amplify blast effects of the explosion (*see drawing below*).

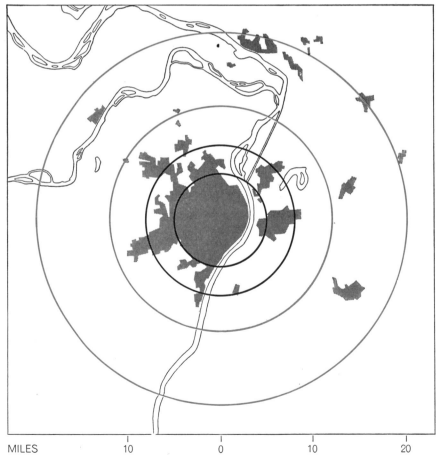

MILES        10        0        10        20

RADII OF EFFECTS of a 10-megaton air burst are superimposed on a map of St. Louis and the surrounding area. The two colored circles correspond to the colored circles in the drawing at the top of the page. The black circles concern effects due to blast. At a distance of five miles (*inner black circle*) from ground zero virtually all buildings would be destroyed. At eight miles (*outer black circle*) virtually all wooden buildings would be destroyed.

counterforce posture is very much affected, may prove a vital factor in disarmament negotiations.

It cannot be seriously believed now that the U.S.S.R. has either the capability or the intention of making an all-out attack on U.S. missile sites and bomber bases. Much genuine alarm in the West might have been allayed if the U.S.S.R. had been more successful in making clearer its disbelief in the military possibility of a successful first-counterforce strike and its intention not to plan for such a possibility. After the brutality of Soviet action in Hungary in 1956 and the technological triumph of the artificial satellite the following year, there may have been legitimate grounds in the West for fearing that the U.S.S.R. might adopt the Western policy of massive retaliation, which, against a nuclear power, requires a counterforce capability. In January, 1960, however, Khrushchev explicitly declared the Soviet commitment to a purely retaliatory strategy. The Soviet second-strike force was strong enough, he said, "to wipe the country or countries which attack us off the face of the earth." To his own rhetorical question, "Will they not, possibly, show perfidy and attack us first...and thus have an advantage to achieve victory?" he replied: "No. Contemporary means of waging war do not give any country such advantages." In addition to freeing resources for capital development, the Soviet minimum-deterrent strategy has avoided the greatest military danger: that the U.S. might attack the U.S.S.R. because of a belief that the U.S.S.R. was about to attack the U.S.

If the analysis given here is approximately correct, what are the prospects of progress toward disarmament at the present meeting in Geneva? Both blocs are fully committed by official pronouncements to the goal of complete and general disarmament under strict control and inspection—notably by the British Commonwealth Prime Ministers' statement in the spring of 1961, by President Kennedy's speech to the General Assembly of the United Nations and by the Soviet-American Joint Statement of Principles, both in September of 1961. Moreover, both sides are committed to attempting to work out first steps of the disarmament process that do not impair the present strategic balance.

Clearly, conventional and nuclear disarmament must go in parallel. The fear of the West of Soviet superiority in trained and deployed land forces must be met by a drastic reduction during the first stage to low levels such as those suggested by the Anglo-French memorandum of 1954: one million or at most 1.5 million men each for the U.S., the U.S.S.R. and China. When the correspondingly limited contributions to the land forces of NATO from Great Britain, France and West Germany are taken into account, the armies of the Soviet bloc would not have the capability of overrunning Europe in a surprise land attack.

The number of nuclear weapons in existence on both sides, their explosive power and the diversity of the delivery systems are so overwhelming that no small step in nuclear disarmament can have much significance. In a situation in which the U.S. has 10,000 delivery vehicles and a stockpile of 30,000 megatons of explosive (which is said to be increasing at the fastest rate in its history), a first disarmament step involving only a small percentage reduction is not worth negotiating. To justify the labor of negotiating any agreed reduction, and to offset the undoubted strains and disputes that will inevitably arise from the operation of any inspection and control system, the negotiated reduction must be a major one; in fact, of such magnitude as to change qualitatively the nature of the relative nuclear postures of the two giant powers.

The simplest big first step, and the one most consistent with realistic military considerations, is that both giant powers should reduce their nuclear forces to a very low and purely retaliatory role. That is, each should retain only enough invulnerable long-range vehicles to attack the other's cities if it is itself attacked, say less than 100 ICBM's with one-megaton war heads. This is still an enormous force, capable of killing tens of millions of people. A reduction to a level of 20 ICBM's or less would be much preferable. Such a reduction would at once prevent nuclear weapons from being used by sane governments as weapons of aggression or coercion. It would not, of course, prevent them from being used by irresponsible groups who do not calculate the cost. It is only at a later stage in disarmament, when nuclear weapons are completely destroyed, that this danger will be excluded. It has always been clear that the ever present danger of accidental or irresponsible war is a cogent reason for big and rapid steps in the disarmament process.

Detailed studies are needed of possible ways in which both the U.S.S.R. and the U.S. could take such an impor-

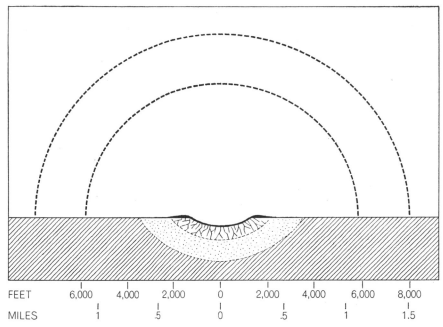

GROUND BURST of a nuclear bomb would be required to neutralize a "hardened" (i.e., buried) missile site. Diameter of the crater dug by a 10-megaton ground burst in dry soil would be 2,600 feet; the depth of the crater would be 250 feet. Radius of the underground "plastic zone" (*outer line below ground*) would be 3,250 feet; the radius of the "rupture zone" (*inner line below ground*) would be 2,000 feet. At a distance of 1.1 miles from ground zero the blast would exert an air pressure of some 300 pounds per square inch (*inner circle above ground*); at a distance of 1.5 miles (*outer circle above ground*), 100 pounds per square inch.

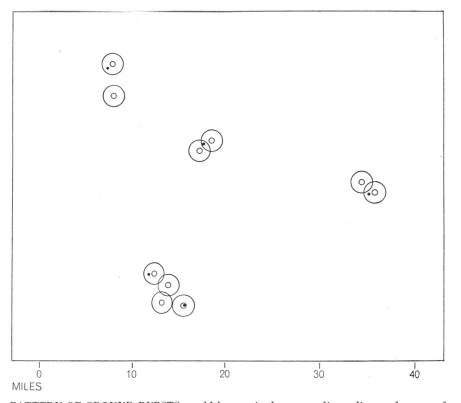

PATTERN OF GROUND BURSTS would be required to neutralize a dispersed group of hardened missile sites. In this schematic drawing a "circle of probable error" of one mile is assumed for each of the attacking missiles; this implies that at least two missiles would be directed at each of the sites. There are five sites, represented by dots. The smaller of each of the 10 pairs of concentric circles represents the 2,600-foot diameter of a 10-megaton bomb crater; the larger of the circles, the 1.1-mile radius at which the air pressure is 300 pounds per square inch. The total weight of the attack on the five bases is 100 megatons. The scale of the drawing is the same as that of the map of St. Louis at the bottom of page 306.

tant first step without upsetting the present strategic balance. A major problem is how to phase the building up of a system of general inspection while at the same time making a drastic reduction in nuclear delivery systems by their actual destruction under international verification. Taking military considerations only into account, I believe that a procedure acceptable to both blocs could be devised.

The difference hitherto between the proposed Western and Soviet first steps in relation to nuclear weapons has been often simplified to the statement that the U.S.S.R. wants disarmament without control and the West wants control without disarmament. It would be more accurate to say that the clash is on the phasing of the stages of disarmament and the stages of control.

In its 1960 proposals the U.S.S.R. suggested that, in the first step, international teams should be dispatched to inspect the destruction of all rocket weapons, military aircraft and other carriers of nuclear weapons. It did not propose the inspection or control of those that remain waiting to be destroyed. Full inspection of a country was to be undertaken only when all weapons had been destroyed. It is clear that the U.S.S.R.'s first steps of disarmament are consistent with its presumed military policy of relying for its safety from nuclear attack on a relatively small force of purely retaliatory nuclear weapons in secret sites.

On the other hand, the U.S. proposals in 1960 envisaged widespread inspection in the first stages and no actual disarmament until the second stage. This proposal might make military sense if put by a weak nuclear power to a much stronger one. But when put by a strong power to a weaker one, rejection must have been expected. If the U.S.S.R. had accepted the proposal, the geographical secrecy of its nuclear sites would have been lost and it would have been vulnerable to nuclear attack from the much stronger West.

Any realistic first stage must start from the fact that the present nuclear balance, such as it is, has a highly asymmetric character: the West's much greater nuclear power is balanced by Soviet geographical secrecy. Since the military balance is asymmetric, so must be any mutually acceptable first step. Concessions must be made by both sides and these must be based on the realities of the military postures of the two blocs.

The U.S.S.R. should accept general inspection not, as in their proposals hitherto, when disarmament is complete but at some intermediate stage on the road to disarmament. Reciprocally, the West should not demand widespread inspection before any disarmament has taken place, as it has done hitherto, but only after substantial destruction of nuclear armaments has taken place under international verification.

In the first stage, therefore, all parties might supply to one another a list of nuclear weapons and their delivery systems, together with research and production facilities. The exact location of sites would not be included at this stage. An agreed number of weapons would then be destroyed and their destruction would be verified by on-site inspection by the international control organization. When this destruction has been verified, a general inspection, using some sampling technique, would begin. The object would then be to verify the correctness of the original declared inventories by checking the numbers remaining after the agreed reductions had been verified, and to proceed to the elimination of the armament remaining.

A word must be said about the place of a test-ban agreement in the stages of a disarmament plan. If this agreement did not involve a type of inspection that might reveal the Soviet nuclear sites, it would be advantageous for it to be included in the first stage, or preferably agreed to at once. If, however, it involved widespread inspection that might reveal these sites, Soviet military planners would certainly advise its rejection. It would then have to wait for the second stage of disarmament, when general inspection starts after the destruction of agreed numbers of nuclear weapons in the first stage.

Some such compromise between Western and Soviet proposals would seem to meet many of the reciprocal criticisms made by the two parties of their respective 1960 proposals without compromising the military security of either. The problem becomes more difficult, however, when nonmilitary considerations are taken into account. Since nonmilitary considerations have played a major role in shaping the defense policies of the great powers, they must inevitably also affect their disarmament policies. For example, if it is difficult to find legitimate military reasons for the vast number of U.S. nuclear weapons and delivery vehicles, it is clear that military arguments alone are not likely to be dominant in U.S. discussion of a possible drastic first step toward nuclear disarmament. This is widely admitted in the U.S., where the impediments to disarmament are being seen more and more as economic, political and emotional in origin rather than as based on operational military considerations. A vital aspect of the problem for the U.S. is the effect that drastic disarmament steps would have not only on the economy as a whole but also on those special sections of high-grade, science-based and highly localized industries that are now so overwhelmingly involved in defense work. A valuable step would be for both the U.S. and Soviet governments to produce and publish detailed and politically realistic economic plans for the transition to a purely retaliatory capacity.

It is fair to conclude that a realistic military basis for an agreed drastic first step in disarmament may not be impossible to find. The urgency of the situation was declared with eloquence by President Kennedy in his speech to the United Nations in September:

"Today, every inhabitant of this planet must contemplate the day when this planet may no longer be habitable. Every man, woman and child lives under a nuclear sword of Damocles, hanging by the slenderest of threads, capable of being cut at any moment by accident or miscalculation or by madness.... The risks inherent in disarmament pale in comparison to the risks inherent in an unlimited arms race."

This great goal of disarmament will be achieved only if the real nature of the arguments against disarmament are clearly identified and frankly faced. The problems of disarmament must not be obscured, as they sometimes have been in the past, by ingenious but fallacious military doctrine applied to false intelligence estimates.

The growing power of China, and the evidence of an ideological rift between it and Russia, provide an added reason for urgency in the drive for disarmament. The U.S.S.R. and the U.S. will be wise to limit drastically their nuclear arms before China becomes a major nuclear power. It is to be observed that whatever influence China may now be exerting on the U.S.S.R. to adopt a harder policy with the West certainly arises in part from the failure of Premier Khrushchev's campaign for disarmament. This failure greatly weakens Khrushchev's argument for the feasibility of peaceful coexistence of the Soviet and the Western worlds. It would seem urgently necessary to attempt to bring China into the disarmament negotiations as soon as possible.

# 42

# The Detection of
# Underground Explosions

SIR EDWARD BULLARD

*July 1966*

Serious interest in the detection of underground explosions dates from 1958. In that year a conference of "experts" was held in Geneva to study methods of detecting violations of a possible agreement on the suspension of nuclear tests. At that time the interest in an agreement to stop testing came mostly from a wish to slow the arms race and to moderate the "cold war"; it was thought that it might be relatively easy to reach an agreement in this field and that it might then be possible to make progress in related fields of disarmament and arms control. The wish to halt the contamination of the atmosphere was also a powerful motive, but this is not relevant to underground testing. Since the conclusion of the treaty prohibiting atmospheric tests both France and China have exploded nuclear devices, thus increasing the membership of the "nuclear club" from three to five. The main reason today for wishing to stop underground testing is the desire to hinder additional nations from developing nuclear weapons by persuading them to join in a test-ban treaty.

The 1958 conference showed how

SUBSIDENCE CRATER was left by a nuclear explosion set off 1,620 feet belowground at the Nevada test site of the U.S. Atomic Energy Commission. The explosion, of "intermediate yield," took place in a formation of volcanic tuff. The "chimney" produced by the explosion, however, extended upward into alluvium, a loosely consolidated material, with the result that the chimney collapsed.

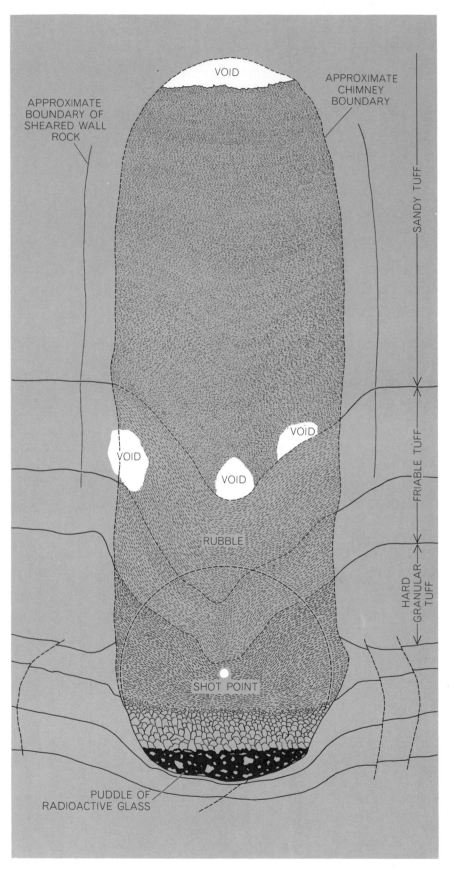

UNDERGROUND NUCLEAR EXPLOSION leaves a rubble-filled chimney above a puddle of radioactive glass. The explosion that produced this result at the Nevada test site was
named "Rainier." The shot was fired at a depth of about 800 feet. Because the yield of the
Rainier explosion was so low (1.7 kilotons) its effects did not extend upward into the alluvium above the chimney and thus the chimney did not collapse to produce a crater at
the surface. The explosion created a temporary cavity (*white dot*) 19.8 meters in radius.

little anyone knew about underground
explosions. To remedy this situation
large programs of investigation were
started in both the U.S. and Britain. My
purpose is not to describe the renewal
of the whole field of seismology that has
followed from this work but to summarize the present state of affairs as it
relates to the detection of clandestine
tests of nuclear weapons below the
surface of the earth. The detection of
such tests has three aspects: detection,
location and identification. Has anything happened? If it has, where did it
happen? Was it a nuclear explosion?

When a nuclear explosion occurs in
a cavity within the earth, the explosive
device is converted to a very hot gas
that reaches the wall of the cavity as
a shock wave. The pressure exerted on
the wall produces stress differences
much exceeding the strength of the
rock, and the temperature near the wall
rises above the melting point. A "plastic"
wave travels through the rock, leaving
it permanently distorted. The wave
spreads for a few hundred meters until
it is so attenuated by spreading and absorption that it can no longer permanently distort the rock. Beyond this
point the plastic wave becomes an
elastic wave and is propagated much as
if it had been produced by an earthquake. Some seconds or minutes after
the explosion the roof of the cavity collapses, leaving a chimney-like hole filled
with debris and with a puddle of molten
rock at the bottom. Much work has
been done on the processes occurring in
the immediate neighborhood of the explosion, and the main features of these
processes are understood.

From the point of view of detection
the important result of an underground explosion is the motion of the
ground produced by the seismic wave at
distant points. It is customary to use
the measured ground motions at such
points to calculate a quantity known as
the magnitude of the earthquake or explosion causing the seismic waves. The
magnitude is intended to be a measure
of the energy in these waves. Owing to
the large range of energies to be covered the scale is a logarithmic one; each
increase of one unit in the magnitude
corresponds to an increase of 250-fold
in the total seismic energy. The relation
of magnitude to seismic energy is very
uncertain, perhaps by as much as a factor of four. Our interest, however, is
not in the seismic energy but in the
ground motions. In spite of the uncertainty in the relation of the magnitude
to energy, the magnitude does constitute

a useful summary of the observed ground motions.

The energy in the seismic waves is of course only a part of all the energy released by an explosion; in smaller explosions it is commonly a few parts per thousand of the entire energy. The principal matter of practical significance is the energy of the explosion—the "yield"—needed to produce seismic signals of a given magnitude. This energy depends on the nature of the material in which the explosion occurs. The explosive energy needed to produce a given seismic effect in alluvium—unconsolidated material of the kind deposited by a river—is about 10 times the energy needed in granite. The volcanic tuff of the test site in Nevada requires twice the explosive energy that is needed in granite. It is unlikely that a clandestine explosion would be made in alluvium; the unconsolidated material above the cavity made by such an explosion usually collapses, leaving a crater at the surface that could be photographed by a reconnaissance satellite [see illustration on page 309].

There has been considerable difference of opinion as to the true relation between explosive energy and seismic magnitude. From the study of a large number of explosions it appears that the amplitude of the first few cycles of ground motion at distant stations is proportional to the energy released by the explosion. From this it may be deduced that the magnitude increases by one for a tenfold increase in the explosive energy [see illustration at right]. Most relations previously proposed gave a slower increase in magnitude with explosive energy. The revision is due to the results from the 80-kiloton explosion ("Longshot") recently set off in hard rock in the Aleutian Islands, which gave records of amplitude comparable to a magnitude-six earthquake.

There is little room for difference of opinion concerning smaller magnitudes, for which data are available from numerous test explosions: an explosion of one kiloton in hard rock produces a magnitude-four seismic signal. The size of the smallest underground explosion that would be important in verifying a test-ban agreement is not well defined; in most discussions the lower limit has been considered to be a yield of a few kilotons, corresponding to a seismic signal of about magnitude four.

Because the surface of the earth is in continuous motion all seismic signals have to be detected against a background of seismic noise. The main ad-

vances in recent years have resulted from improvements in the ratio of signal to noise. These improvements are of three kinds: the choice of quiet sites, the use of arrays of instruments and the processing of the recorded signals.

The background noise has many causes, among which are storms at sea, waves breaking on the shore, the rocking of trees and buildings by the wind, vibrations from traffic and machinery and the effects of innumerable earthquakes too small to be detected individually. The amount of disturbance varies greatly from place to place, and one of the most useful seismological discoveries of recent years is that there exist sites where the noise level is exceptionally low. As might be expected, such sites are on hard rock, in the interiors of continents and remote from industrial activity. Excellent sites have been found at Yellowknife in northwestern Canada, in the mountain states of the U.S., in the Sahara, in India and in Australia; doubtless others could be found in the U.S.S.R., Africa and South America.

The background noise has a spectrum with a broad maximum at periods of

oscillation between five and 10 seconds, whereas the bulk of the energy in the initial part of a seismic record (the .P wave, or compressional wave) has periods between .5 and 1.5 seconds. A considerable improvement in the ratio of signal to noise can therefore be obtained by passing the signal through a filter that removes signals with periods greater than about one second. It is also desirable to remove signals whose frequencies are above 10 cycles per second; there is appreciable noise and almost no signal in this frequency range. When this filtering has been done, the ground motion at the best sites is less than $10^{-7}$ centimeter—only a few times the diameter of an atom—for a large part of the time. A poor site may give values 10 or 100 times greater.

Background noise is not the only impediment to obtaining a clear record of an explosion or earthquake. As a seismic wave travels through the earth it is reflected and scattered by discontinuities. Such reflected and scattered waves complicate the signal and increase the difficulty of extracting information about the source. A rather similar effect is produced by scattering

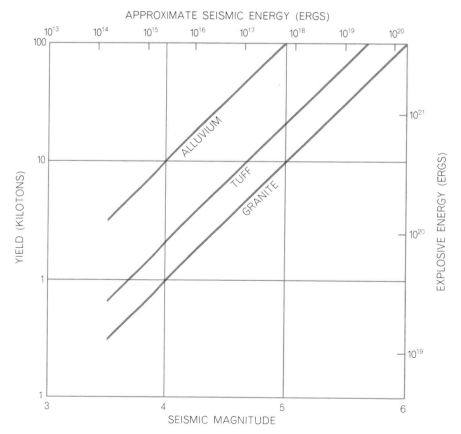

EXPLOSIVE ENERGY required to produce a seismic disturbance of a given magnitude is shown for three kinds of material. There is some evidence that the curves for alluvium and tuff should bend upward to give lower magnitudes for large explosions. A kiloton is the energy released by the explosion of 1,000 tons of TNT, equivalent to about $4.2 \times 10^{19}$ ergs.

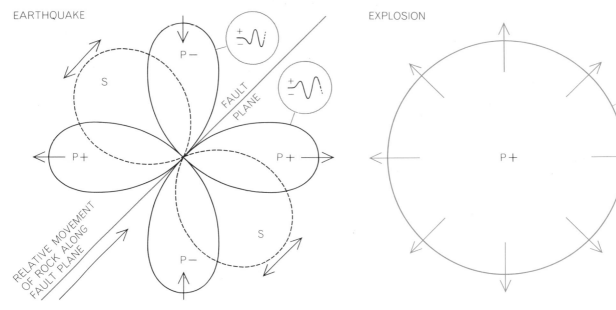

EARTHQUAKE

EXPLOSION

**COMPARISON OF EXPLOSION WITH EARTHQUAKE** shows that an earthquake (*left*) characteristically produces compressional *P* waves and distortional *S* waves, both of which vary with direc-tion. In contrast, an explosion (*right*) produces an initial outward compressional motion in all directions. An earthquake normally originates in a fault where two bodies of rock slip past each other.

from topographic irregularities at the surface of the earth near the seismo-graph. This "signal-generated noise" is particularly troublesome when a large part of the path of the signal from source to seismograph runs through the outermost part of the earth. At greater distances the seismic ray goes steeply down from the source and steeply up to the seismograph [*see illustration below*].

There is then only a small part of its path over which it is affected by the complications of geology; the greater part is deep in the earth's mantle, where the material is more uniform than it is near the surface. Accordingly if one wishes to study the structure of the earth's crust, one should make measure-ments within, say, 1,000 kilometers of the seismic source, but if one wishes to

study the nature of seismic sources or to detect clandestine explosions, one should observe at distances beyond 3,000 kilometers.

Between 3,000 and 10,000 kilometers the amplitudes of the first few cycles of the seismic signals are as great as or greater than they are between 1,000 and 2,000 kilometers. Beyond 10,000 kilometers the earth's core casts a shad-

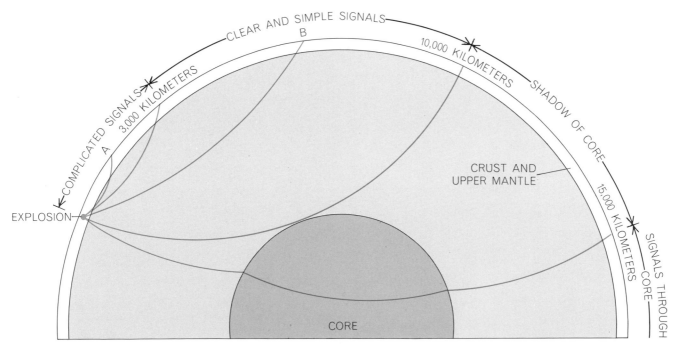

**SEISMIC WAVES,** whether produced by an earthquake or an ex-plosion, can be recorded at great distances. In fact, the signals re-ceived at distances between 3,000 and 10,000 kilometers from an event are usually clearer than signals received within 3,000 kilo-meters. A ray received at *A* has its entire path in the crust and up-per mantle of the earth, where irregularities in the rocks scatter the energy and complicate the record. The ray to *B* has most of its path deep in the earth, where the irregularities are few.

ow, making satisfactory observations of the earlier part of the seismogram impossible. Between 14,000 kilometers and the antipodes (20,000 kilometers) waves can be received through the core and could probably be used in the study of seismic sources, but this range of distances has been little studied for purposes of bomb-test detection. The range from 3,000 to 10,000 kilometers is convenient because the whole of the U.S.S.R. lies within this range from stations in the U.S.

A large improvement in the ratio of signal to noise can be obtained by the use of an array of seismographs instead of a single instrument. The idea of an array is not a new one; arrays have been employed for many years in the reception of short radio waves and in radio astronomy. If a row of *n* detectors receives a signal, and if the outputs from all the detectors are combined (after delaying the outputs to allow for different times of arrival at the different instruments), all the signals will add up to give a signal with an amplitude *n* times the amplitude of that from a single detector [*see illustration on next page*]. The array can be regarded as a telescope that selects signals coming from a given direction or, what is the same thing, as a device that selects signals arriving with a given apparent velocity of propagation over the ground. By adjusting the correction for the time lag it is possible to accentuate different parts of the seismogram, for example the compressional *P* waves or the distortional *S* waves. If the instruments of the array are not all on a single line but are spread out in two dimensions on the surface of the earth, the time lags can be adjusted so as to pick out signals coming from a given compass direction as well as those coming at a given angle to the vertical.

Obviously there would be no advantage in using an array if it enhanced the noise as much as it enhanced the signal. Provided that the noise at the different detectors is uncorrelated, an array of *n* detectors will yield a total value for noise that is roughly the square root of *n* times the noise received by a single detector. As a consequence the ratio of signal amplitude to noise amplitude will be increased in the ratio of *n* divided by the square root of *n*, that is, by a factor equal to the square root of *n*. For a real array the noise, particularly the signal-generated noise, will be partially correlated at the different detectors, and the improvement attainable may be greater or less than the

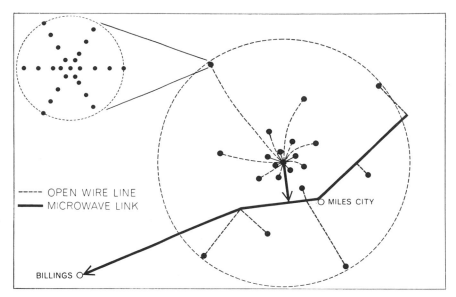

OPEN WIRE LINE
MICROWAVE LINK
MILES CITY
BILLINGS

LARGE SEISMIC ARRAY recently installed by the AEC near Billings, Mont., consists of 21 clusters, each containing 25 seismometers. The clusters are seven kilometers in diameter. The entire seismic array is spread over a circular area 200 kilometers in diameter.

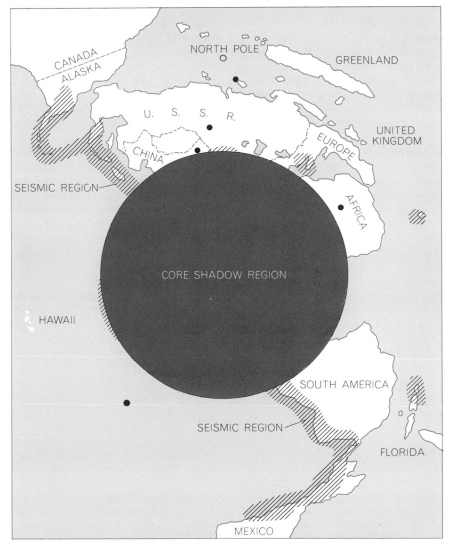

SEISMIC VIEW OF THE EARTH from the center of the Montana array shows the region in which seismic events can be detected and that in which they cannot be. Earthquake regions are hatched. Black dots indicate nuclear test sites in the U.S.S.R. and elsewhere.

314

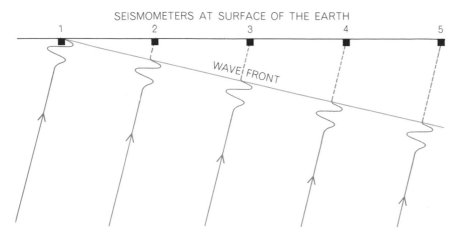

**FUNCTION OF SEISMIC ARRAY is to enhance the signal from a seismic event while suppressing noise. The diagram shows a seismic wave traveling upward to a line of five seismometers near the surface of the earth. The signal arrives first at the detector at the far left. To reach the next detector the wave must travel an extra distance. If the signals from the detectors are suitably delayed before they are added together, the combined output from the five detectors will have an amplitude five times that from a single detector.**

square root of $n$. In practice it is usually possible to reach an improvement equal to the square root of $n$ by using an array whose diameter is 20 kilometers or more. Since the number of detectors, $n$, in an array can exceed 100, the factor represented by the square root of $n$ may represent a very large improvement.

An array can be organized in many ways to meet particular requirements, and much work has been done on the properties of different layouts. A common arrangement is to put two lines of instruments on the arms of a cross each about 20 kilometers in length. In the large-aperture seismic array recently installed in Montana the array consists of 21 clusters of seismometers spread over a circle 200 kilometers in diameter; each cluster contains 25 detectors spread over a circle seven kilometers in diameter. All 525 detectors are at the bottom of holes 200 feet deep to reduce noise generated at the surface by wind and other causes [see top illustration on preceding page].

Filtering and adding the outputs from a number of instruments is a linear process, and is especially suitable for improving the ratio of signal to noise when the signal is smaller than the noise, or not much larger. When the signal has been raised well above the noise level by these methods, it is advantageous to use nonlinear methods to "clean up" the record.

A particularly useful technique is to take the two records produced by summing all the instruments on each arm of a cross array, to accentuate the parts of these records that are similar and to reduce the parts that are dissimilar. This can be done by feeding the two records into a multiplier, multiplying them point by point and smoothing the product over time intervals of 1.5 or two seconds [see illustration on opposite page]. Such a record is called a correlogram.

With the aid of correlograms it has been possible to separate the signal of a small (400-pound) underground chemical explosion from the seismic disturbance produced by a small earthquake, both recorded by an array 2,400 kilometers distant [see top illustration on page 318]. If the array had been located several hundred kilometers farther away from the disturbances, the record of the explosion would probably have been even simpler. The detection of so small an explosion even though it was obscured by a preceding earthquake is remarkable. It shows that an explosion of the size of significance for bomb-test detection will usually be detected; the main difficulty is to identify it as an explosion.

Information on the accuracy with which a seismic event can be located is not as complete as could be wished. The onset of the signals from an explosion can be determined to within a few tenths of a second. If a network of 10 or 20 stations could achieve this accuracy, the source would be located within a few kilometers. In practice the accuracy is degraded by local variations in the velocity of seismic waves, about which information is largely lacking. If data from stations distant from the event are used, it seems realistic to

estimate that the site can be located within a circular area whose radius is about eight kilometers. Stations that are 500 to 2,000 kilometers from the event may give much larger errors, owing to irregularities in seismic velocities in the crust and upper mantle.

Since the great majority of events recorded at seismic stations are earthquakes, the critical problem in attempting to detect clandestine explosions is to distinguish a very few explosions among a great many earthquakes. Not surprisingly, small earthquakes are much more numerous than large ones. Roughly speaking, the number decreases by a factor of 10 for each increase of one unit in seismic magnitude. Throughout the world each year there are about 10,000 earthquakes of magnitude between four and five and only 10 or 20 of magnitude seven or greater [see bottom illustration on page 318].

About 1.5 percent of the world's earthquakes occur in the U.S.S.R. In an average year some 170 of these will be above magnitude four and only two or three will be above magnitude six. These numbers are only estimates of mean rates; the actual numbers fluctuate considerably from year to year. The fluctuations exceed the statistical expectancy for random events because in some years there may be 100 or more aftershocks following a single large earthquake. In a given year the total number of earthquakes above magnitude four in the U.S.S.R. might be as high as 300 or as low as 100.

The distribution of earthquakes over the world is extremely nonuniform; the great bulk of them occur along the Alpine-Himalayan chain, on the margins of the Pacific Ocean and along the mid-ocean ridges. In the U.S.S.R. earthquakes are practically confined to a belt along the southern border of the country and to the Kamchatka Peninsula, with a few on a north-south line to the east of the Lena River [see illustration on pages 316 and 317]. The areas of older rocks that constitute the greater part of the country appear to be completely free from earthquakes. The Russian testing ground near Semipalatinsk is just north of the earthquake belt and the testing ground in Novaya Zemlya is in an almost earthquake-free region.

Some earthquakes cause fractures at the surface of the earth; others cause fracturing entirely below the surface, commonly at depths of from 10 to 60 kilometers. The deepest earthquakes are about 750 kilometers below the surface.

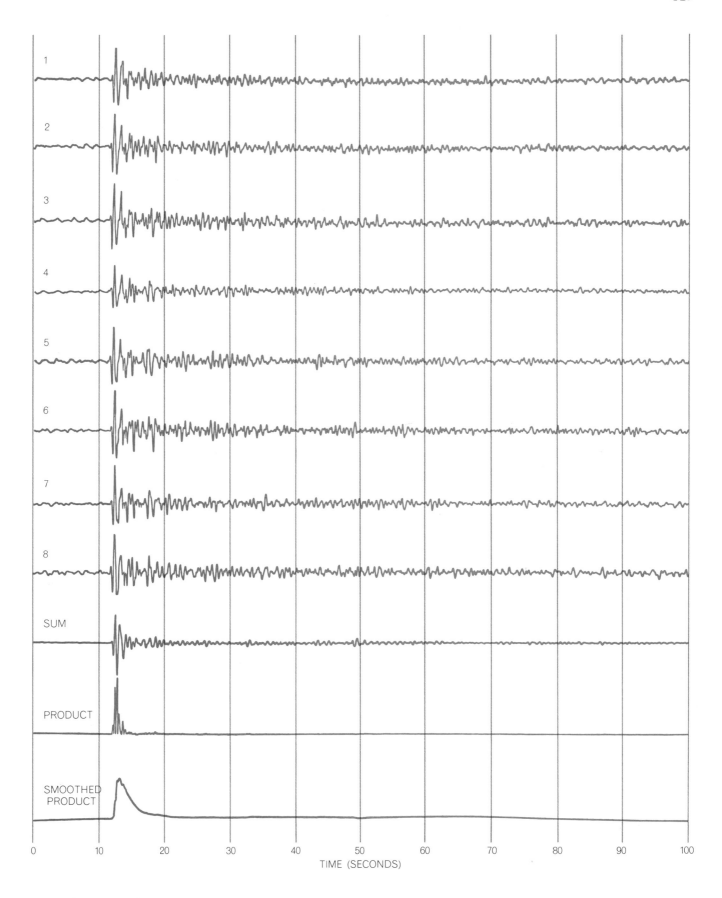

EXPLOSION NEAR SEMIPALATINSK, a test site in the U.S.S.R., produced these records at a seismic array at Eskdalemuir in Scotland, 5,300 kilometers from the event. The first eight traces are from single instruments on one arm of the cross array. The next trace shows the sum of all instruments on both arms of the array. The next trace gives the product of the records from the two arms; the bottom trace shows the product after smoothing. The first wave to arrive is the compressional $P$ wave of the explosion. The oscillations that are removed in the last two traces are noise generated by the signal itself. The explosion took place on May 16, 1964.

The deep earthquakes occur only in certain places, particularly on the inner side of the earthquake belt surrounding the Pacific. This can be seen in the map below: in the Kamchatka region the earthquakes just offshore are shallow and most of those inland are deep.

At first it seems that it should not be too difficult to distinguish an earthquake from an explosion. The mechanism at the source is quite different. An earthquake normally has its origin in a fault where two bodies of rock slip past each other, whereas an explosion pro-

duces a sudden and spherically symmetrical pressure pulse. The first motion from an explosion is always an outward-moving compressional pulse, which is similar in all directions. A fault, on the other hand, generates both compressional $P$ waves and distortional $S$ waves whose motion differs in different directions. The first motion is inward in some places and outward in others [see top illustration on page 312]. Moreover, the motion produced by an explosion is much simpler than that from an earthquake. This is not unexpected; an explosion is localized in both space and

time, whereas a fault extends for some distance and may take some seconds to complete its movement.

These criteria have been considerably sharpened by the development of seismic arrays. The initial wave form can now be seen with great clarity, largely freed from the effects of the transmission path. For example, an underground explosion near Semipalatinsk on May 16, 1964, produced closely similar records at stations at Eskdalemuir in Scotland and Yellowknife in Canada [see top illustration on page 319]. The first few seconds of the records are almost

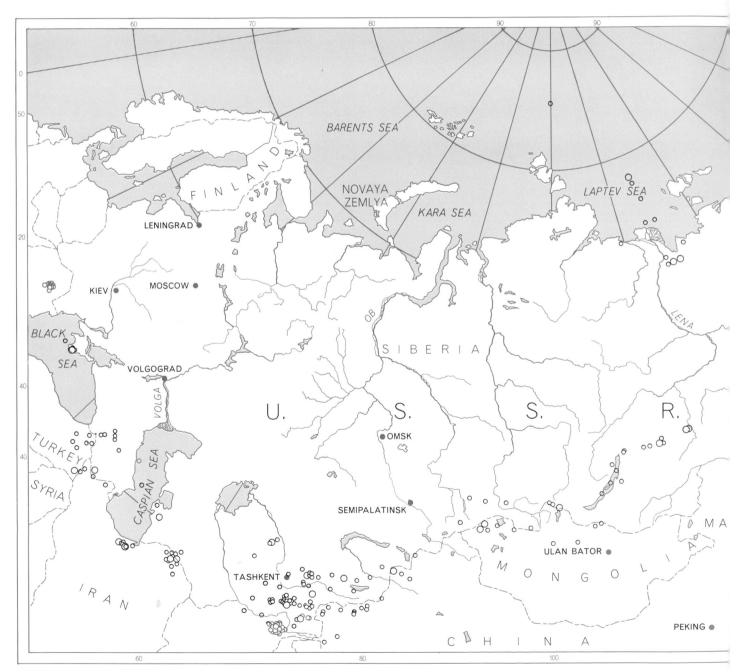

**SEISMIC MAP OF U.S.S.R.** shows earthquakes above magnitude 5¼ that occurred from 1911 to 1959. The shocks above magnitude 6½ are represented by larger circles, those of 5¼ to 6½ by smaller circles. Black circles indicate shallow earthquakes: depth less than 35 kilometers. Open colored circles indicate depths of between 35 and 300 kilometers; colored circles with dots indicate depths exceeding 300 kilometers. The map is based on one published by the Soviet Academy of Sciences. Novaya Zemlya and Semipalatinsk,

identical at the two stations in spite of the widely separated paths followed by the seismic waves and the completely different geological setting of the two arrays. The wave form must therefore be determined largely by the events at the source of the energy.

Another striking example is given by comparing the records of the seismic waves from an explosion at the French testing ground in Algeria with those from an earthquake in Libya. The arrays at Eskdalemuir, Yellowknife and a station at Pole Mountain, Wyo., gave very similar records of the explosion and

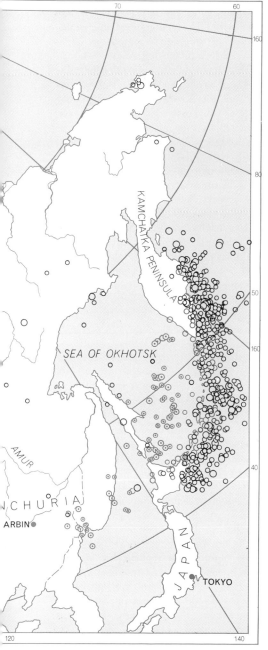

used as nuclear-test sites, are in regions that are free from earthquakes. In the Kamchatka area the earthquakes just offshore are shallow and most of those inland are deep.

much more complicated and dissimilar records of the earthquake [see bottom illustration on page 319].

It has been suggested that the complexity in the wave form of earthquakes is due not to the source mechanism but to the tendency of earthquakes to occur in more complex geological settings than those in which explosions are fired. The recent Longshot explosion in the Aleutians has demonstrated that this is not so. This shot gave characteristic explosion records in spite of being fired in the circum-Pacific earthquake belt. The suggestion that it is only earthquakes of magnitude above 4.5 that have complicated records also appears to have no foundation.

The determination of depth is another criterion that enables a large number of earthquakes to be eliminated. An event occurring at a depth exceeding eight kilometers is certainly an earthquake. About 30 percent of all earthquakes occur at depths exceeding 50 kilometers. The number with depths between eight kilometers and 50 kilometers is not easy to determine because the complexity of the wave form makes it difficult to detect the wave reflected from the surface above the site of the fault. If an earthquake with a simple wave form occurred in this depth range, it should be possible to determine its depth and thus demonstrate that it is not an explosion. By a careful examination of records from arrays it is sometimes possible to estimate the depth of an explosion fired at a depth of as little as one kilometer below the surface of the earth. The determination of depth is of particular importance for detecting explosions in Kamchatka, where most of the shallow earthquakes occur in the sea just offshore and most of those on land are deep.

Attempts to use criteria depending on S waves and surface waves have not been very successful. The horizontally polarized S waves from an explosion are smaller than those from an earthquake, but there are many small earthquakes without detectable S waves or surface waves. The probable explanation is that the period of these waves falls near the peak in the frequency distribution of background noise. It is also true that less effort has been put into such methods than has been devoted to the study of P waves; moreover, the instruments used in the arrays are designed to respond to vertical motions and thus are unsuitable for detecting horizontally polarized waves. It may be that the matter deserves closer attention. It would be particularly in-

teresting to see what results could be obtained from a cross array equipped with instruments capable of responding along three coordinate axes.

The critical question is whether, when all these criteria have been used, there is a residuum of earthquakes that are indistinguishable from explosions. The results of an examination of a worldwide sample of 161 earthquakes with depths of less than 50 kilometers have been published by the United Kingdom Atomic Energy Authority. There were seven earthquakes, or 4.5 percent, that could not be distinguished from explosions. This would give an average of eight "suspicious events" per year in the U.S.S.R. Estimates by some other authorities are about twice as high. The discrepancy is not surprising. "Suspicious" is not a precisely defined concept; how many events one regards as suspicious will depend to some extent on how suspicious one is.

The number of arrays used in the Atomic Energy Authority's investigation was three, but for a large part of the time only one was in operation. Five of the seven explosion-like earthquakes were recorded while only one array was in operation. Some improvement is to be expected from the use of more arrays, particularly from those recently established in Australia and India, which receive signals from the U.S.S.R. in directions differing widely from the directions of signals received in Britain, Canada and the U.S. The arrays themselves are also susceptible to improvement; those employed in gathering the sample had only 20 detectors, compared with the total of 525 in the array in Montana. The use of more refined determinations of focal depth would also probably remove some ambiguous events.

It is clearly possible to chip away at the number of earthquakes within the U.S.S.R. that might be mistaken for explosions, but there is no reason to suppose that this number can be reduced to zero. Even if there are no earthquakes that virtually duplicate explosions, there may be earthquakes whose fault planes lie at such an angle that they give an initial outward motion at all distances beyond 3,000 kilometers. If one of these happened to give a simple wave form and to be shallow, it might not be distinguished from an explosion. With a system as complicated as the earth almost anything can happen occasionally. It seems prudent to assume that we shall not get a system that never mistakes an earthquake for an explosion.

In a sample of 35 underground explo-

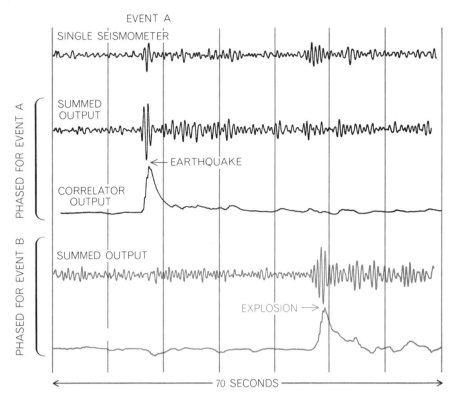

EVENT A
SINGLE SEISMOMETER

PHASED FOR EVENT A

SUMMED OUTPUT

←— EARTHQUAKE

CORRELATOR OUTPUT

PHASED FOR EVENT B

SUMMED OUTPUT

EXPLOSION →

←——————— 70 SECONDS ———————→

**EARTHQUAKE FOLLOWED BY SMALL EXPLOSION** at a distance of some 2,400 kilometers were clearly distinguished by the seismic array at Yellowknife in Canada. The explosion was produced by .2 kiloton of chemical explosive set off belowground at Climax, Colo. The top trace shows the output of a single seismometer. The second trace shows the effect of adding the traces from 20 detectors with time lags chosen to emphasize the earthquake. In the fourth trace the lags emphasize the explosion. The remaining traces are "correlograms" obtained by multiplying and smoothing the summed output from the two arms of the array. Had the explosion been farther away the record probably would have been simpler.

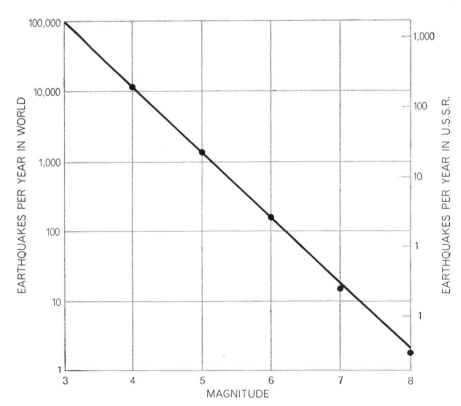

**ANNUAL NUMBER OF EARTHQUAKES** above a given magnitude are shown for the entire world and for the U.S.S.R. It is assumed that 1.4 percent of all earthquakes occur inside Russian territory. There are sizable annual fluctuations around these mean rates.

sions studied by seismological means, one on Novaya Zemlya would have been regarded as an earthquake if it had occurred in a seismic region. Such misinterpretation is not, however, of much significance. In order to take advantage of it one would have to know beforehand that the conditions were such that an explosion would be misinterpreted as an earthquake.

Much attention has been given to the possibility that the seismic effects of an explosion could be reduced or modified so that it was undetectable or appeared to be an earthquake. The most promising method is to set off the explosion in a very large cavity. If this were done, theory suggests that the amplitude of the seismic signal would be reduced by a factor of 100.

The theory has been verified for chemical explosions of up to about one ton and there is no reason to suppose that it would not apply to nuclear explosions. In spite of the importance of the matter no experiments have been made with kiloton explosions. This is probably because of the great engineering difficulties involved in making cavities several hundred feet in diameter and also because of the possibility that such a cavity would collapse after an explosion and form a detectable crater at the surface. If the method were to be used for concealing a clandestine explosion, there would also be great difficulty in removing millions of cubic feet of material from the cavity without leaving conspicuous signs on the surface.

Other methods of concealment that have been proposed are to set off several explosions in a short interval or to set off an explosion just after an earthquake has occurred. Neither method is likely to be effective. The second entails a wait of years before a suitable earthquake occurs near the place where the test has been prepared.

It may be that it is possible to devise a method of concealing a nuclear explosion. As far as is known, however, this has not been done. To do it would itself involve an extensive series of tests. One of the reasons for wishing that a comprehensive test ban be not too long delayed is that the ban would prevent the development of methods of concealment that, if they were discovered before tests stop, would make agreement more difficult.

This article is not the place to discuss at length the influence that the facts presented here should have on international negotiations. The main political problem is to decide what means of verification, if any, are needed to ensure that a "suspicious event" is not a clandestine explosion. The only effective means of verification is to inspect the neighborhood of the event. The exact nature of the inspection is not of the first importance; a violator of a test ban would be unlikely to allow an inspection of a real explosion. Rather than be caught red-handed he would incur the odium of refusing access to the site or of abrogating the treaty. The object of having the right of inspection is to demonstrate to ourselves and to the world in general that suspicious events are not explosions, and to get a strong indication of a violation from the obstruction that would occur if a proposal were made to inspect the site of an actual explosion.

A formal theory can be set up: If there are $N$ suspicious events in country $A$ per year, and country $B$ is allowed $n$ inspections, what is the probability of $A$'s conducting $M$ tests in a year without $B$'s asking to inspect one or more of them? $A$ will require this probability to be quite high, say .95, since the consequences of being caught cheating on a matter concerned with nuclear weapons are serious. $B$, on the other hand, requires a high probability of detecting $A$'s violation before too many tests have been carried out. In other words, $B$ wants the probability formula to produce a low number. The following figures show for various numbers of inspections the probabilities that $B$ will *not* ask to inspect one or more of the tests when there are five tests among 15 suspicious events:

| Number of inspections | 1 | 2 | 3 | 4 |
|---|---|---|---|---|
| Probabilities | .67 | .43 | .26 | .15 |

In summary, one can say that the technical facts are as clear as can be expected for a complicated subject and that there are now no very serious differences of opinion about them. Some further improvements in equipment are possible, particularly in the provision of more seismographic arrays. These, however, can hardly affect the political decisions, which must depend on a balance between the desirability of a test ban and the disadvantages of knowing that a few ambiguous events will occur in the U.S.S.R. each year, any of which could be an explosion but all or most of which are earthquakes. In practice the choice is among the following three courses:

First, to say that the real risks exceed the hypothetical gains and that we

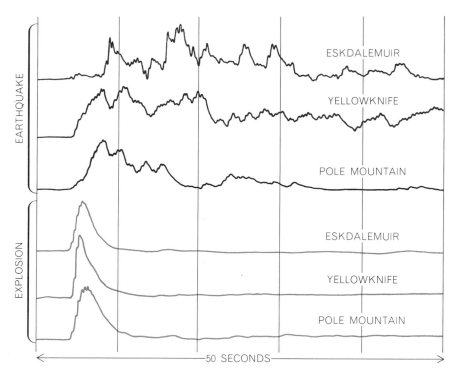

EARTHQUAKE AND EXPLOSION are compared in the correlograms obtained from seismic arrays at Eskdalemuir, Yellowknife and Pole Mountain, Wyo. The earthquake (*top*), which occurred in Libya, gives a long-drawn-out, complicated record that is visibly different at different stations. The nuclear explosion (*bottom*), which was conducted by the French government at its test site in Algeria, gives simple and similar records at all stations.

therefore do not want to seek a test ban.

Second, to accept something like the Russian offer (since withdrawn) of "two or three" inspections a year.

Third, to say that no nation is likely to conduct clandestine tests in circumstances where there is an appreciable probability of their opponent's knowing that they are doing so, even if the violation cannot be proved in a court of law, and that a treaty without inspection is therefore acceptable.

Most informed opinion would accept the second course. The difficulty is that the Russians have stated categorically and recently that it is not acceptable to them. It is therefore not a possible choice for the U.S. The acceptance of the third course would be made easier if there were little chance of making important improvements in weapons by underground testing. In the absence of detailed knowledge one can only say that it is difficult to imagine what the improvements could be. It might be objected that acceptance of a test ban without inspection would leave us on a slippery slope where, if we were sufficiently lacking in judgment or firmness, we might find it hard to refuse unverified agreements about matters vital to our security.

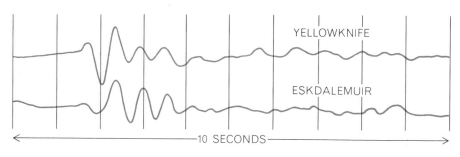

FRENCH UNDERGROUND EXPLOSION produced these records at Yellowknife and Eskdalemuir, respectively 9,000 and 3,500 kilometers from the event. The two traces are the summed output of all the instruments at each site. It can be seen that the initial wave form depends on the character of the original signal and is largely independent of the distance and the path from the source to the seismic arrays and of the geology near the arrays.

# 43

# National Security and the Nuclear-Test Ban

JEROME B. WIESNER and HERBERT F. YORK

*October 1964*

The partial nuclear-test ban—the international treaty that prohibits nuclear explosions in the atmosphere, in the oceans and in outer space—has been in effect for a little more than a year. From July, 1945, when the first atomic bomb was set off in New Mexico, until August, 1963, when the U.S. completed its last series of atmospheric bomb tests in the Pacific, the accumulated tonnage of nuclear explosions had been doubling every three years [*see top illustration on page 322*]. Contamination of the atmosphere by fission products and by the secondary products of irradiation (notably the long-lived carbon 14) was approaching a level (nearly 10 percent of the natural background radiation) that alarmed many biologists. A chart plotting the accumulation of radioactive products can also be read as a chart of the acceleration in the arms race.

Now, for a year, the curve has flattened out. From the objective record it can be said that the improvement of both the physical and the political atmosphere of the world has fulfilled at least the short-range expectations of those who advocated and worked for the test ban. In and of itself the treaty does no more than moderate the continuing arms race. It is nonetheless, as President Kennedy said, "an important first step—a step toward peace, a step

toward reason, a step away from war."

The passage of a year also makes it possible to place in perspective and evaluate certain misgivings that have been expressed about the effect on U.S. national security of the suspension of the testing of nuclear weapons in the atmosphere. These misgivings principally involve the technology of nuclear armament. National security, of course, involves moral questions and human values—political, social, economic and psychological questions as well as technological ones. Since no one is an expert in all the disciplines of knowledge concerned, it is necessary to consider one class of such questions at a time, always with the caution that such consideration is incomplete. As scientists who have been engaged for most of our professional lifetimes in consultation on this country's military policy and in the active development of the weapons themselves, we shall devote the present discussion primarily to the technological questions.

The discussion will necessarily rest on unclassified information. It is unfortunate that so many of the facts concerning this most important problem are classified, but that is the situation at this time. Since we have access to classified information, however, we can assure the reader that we would not have to modify any of the arguments we present

here if we were able to cite such information. Nor do we know of any military considerations excluded from open discussion by military secrecy that would weaken any of our conclusions. We shall discuss the matter from the point of view of our country's national interest. We believe, however, that a Soviet military technologist, writing from the point of view of the U.S.S.R., could write an almost identical paper.

Today as never before national security involves technical questions. The past two decades have seen a historic revolution in the technology of war. From the blockbuster of World War II to the thermonuclear bomb the violence of military explosives has been scaled upward a million times. The time required for the interhemispheric transport of weapons of mass destruction has shrunk from 20 hours for the 300-mile-per-hour B-29 to the 30-minute flight time of the ballistic missile. Moreover, the installation of the computer in command and control systems has increased their information-processing capacity by as much as six orders of magnitude compared with organizations manned at corresponding points by human nervous systems.

It has been suggested by some that technological surprise presents the primary danger to national security. Yet recognition of the facts of the present state of military technology must lead to the opposite conclusion. Intercontinental delivery time cannot be reduced to secure any significant improvement in the effectiveness of the attack. Improvement by another order of magnitude in the information-processing capacity of the defending system will not make nearly as large a difference in its operational effectiveness.

The point is well illustrated by the 100-megaton nuclear bomb. Whether or not it is necessary, in the interests of national security, to test and deploy a bomb with a yield in the range of 100 megatons was much discussed during the test-ban debates. The bomb was frequently referred to as the "big" bomb, as if the bombs now in the U.S. arsenal were somehow not big. The absurdity of this notion is almost enough by itself to settle the argument. A one-megaton bomb is already about 50 times bigger than the bomb that produced 100,000 casualties at Hiroshima, and 10 megatons is of the same order of magnitude as the grand total of all high explosives used in all wars to date. Other technical considerations that surround this question are nonetheless illuminating and worth exploring.

There is, first of all, the "tactics" of the missile race. The purpose of a missile system is to be able to destroy or, perhaps more accurately, able to threaten to destroy enemy targets. No matter what the statesmen, military men and moralists on each side may think of the national characteristics, capabilities and morality of the other side, no matter what arguments may be made about who is aggressive and who is not or who is rational and who is not, the military planners on each side must reckon with the possibility that the other side will attack first. This means that above all else the planner must assure the survival of a sufficient proportion of his own force, following the heaviest surprise attack the other side might mount, to launch a retaliatory attack. Moreover, if the force is to be effective as a deterrent to a first strike, its capacity to survive and wreak revenge and even win, whatever that may mean, must be apparent to the other side.

Several approaches, in fact, can be taken to assure the survival of a sufficient missile force after a first attack on it. The most practical of these are: (1) "hardening," that is, direct protection against physical damage; (2) concealment, including subterfuge and, as in the case of the Polaris submarine missiles, mobility, and (3) numbers, that is,

presenting more targets than the attacker can possibly cope with. The most straightforward and certain of these is the last: numbers. For the wealthier adversary it is also the easiest, because he can attain absolute superiority in numbers. A large number of weapons is also a good tactic for the poorer adversary, because numbers even in the absence of absolute superiority can hopelessly frustrate efforts to locate all targets.

There is an unavoidable trade-off, however, between the number and the size of weapons. The cost of a missile depends on many factors, one of the most important being gross size or weight. Unless one stretches "the state of the art" too far in the direction of sophistication and miniaturization, the cost of a missile turns out to be roughly proportional to its weight, if otherwise identical design criteria are used. The protective structures needed for hardening or the capacity of submarines needed to carry the missile also have a cost roughly proportional to the volume of the missile. Some of the ancillary equipment has a cost proportional to the size of the missile and some does not; some operational expenditures vary directly with size or weight and some do not. The cost of the warhead generally does not, although the more powerful warhead requires the larger missile. It is not possible to put all these factors together in precise bookkeeping form, but it is correct to say that the cost of a missile, complete and ready for firing, increases somewhat more slowly than linearly with its size.

On the other hand—considering "hard" targets only—the effectiveness of a missile increases more slowly than cost as the size of the missile goes up. The reason is that the radius of blast damage, which is the primary effect employed against a hard target, increases only as the cube root of the yield and because yield has a more or less direct relation to weight. Against "soft" targets, meaning population centers and conventional military bases, even "small" bombs are completely effective, and nothing is gained by increasing yield. Given finite resources, even in the wealthiest economy, it would seem prudent to accept smaller size in order to get larger numbers. On any scale of investment, in fact, the combination of larger numbers and smaller size results in greater effectiveness for the missile system as a whole, as contrasted to the effectiveness of a single missile.

This line of reasoning has, for some years, formed the basis of U.S. mis-

sile policy. The administration of President Eisenhower, when faced with the choice of bigger missiles (the liquid-fueled Atlas and Titan rockets) as against smaller missiles (the solid-fueled Minuteman and Polaris rockets), decided to produce many more of the smaller missiles. The administration of President Kennedy independently confirmed this decision and increased the ratio of smaller to larger missiles in the nation's armament. During the test-ban hearings it was revealed that the U.S. nuclear armament included bombs of 23-megaton yield and higher, carried by bombers. Recently Cyrus R. Vance, Under Secretary of Defense, indicated that the Air Force has been retiring these large bombs in favor of smaller ones. There are presumably no targets that call for the use of such enormous explosions.

The argument that says it is now critical for U.S. national security to build very big bombs and missiles fails completely when it is examined in terms of the strictly technical factors that determine the effectiveness of a missile attack. In addition to explosive yield the principal factors are the number of missiles, the overall reliability of each missile and the accuracy with which it can be delivered to its target. The effectiveness of the attack—the likelihood that a given target will be destroyed—can be described by a number called the "kill probability" ($P_k$). This number depends on the number of missiles ($N$) launched at the target, the reliability ($r$) of each missile and the ratio of the radius of damage ($R_k$) effected by each missile to the accuracy with which the missiles are delivered to the target (CEP). The term "CEP," which stands for "circular error probable," implies that the distribution of a large number of hits around a given target will follow a standard error curve; actually, for a variety of reasons (which include the presence of systematic errors, coupling between certain causes of error and the sporadic nature of the larger error factors) the distribution does not really follow a standard error curve. The term "CEP" is still useful, however, and can be defined simply as the circle within which half of a large number of identical missiles would fall.

Now, in the case of a soft target, $R_k$ is very large for the present range of warhead yields in the U.S. arsenal. The reason is that soft targets are so highly vulnerable to all the "prompt" effects (particularly the incendiary effects) of thermonuclear weapons. The range of these effects, modified by various attenuation factors, increases approximately as the square root or the cube root of the yield at large distances. Under these circumstances, given the accuracy of existing fire-control systems, the ratio $R_k$/CEP is large and the likelihood that the target will be destroyed becomes practically independent of this ratio. Instead $P_k$ depends primarily on $r$, the reliability of the missile. If $r$ is near unity, then a single missile ($N = 1$) will do the job; if $r$ is not near unity, then success in the attack calls for an offsetting increase in the number of missiles $[P_k = 1 - (1 - r)^N]$. In either case changes in $R_k$ make little difference. That is to say, a "big" bomb cannot destroy a soft target any more surely than a "small" one can.

When it comes to hard targets, the ratio $R_k$/CEP becomes much smaller even for bombs of high yield. The blast effects—including the ground rupture, deformation and shock surrounding the crater of a surface burst—have comparatively small radii at intensities sufficient to overcome hardening. Moreover, as mentioned above, the radii of these effects increase only as the cube root of the yield. This rule of thumb is modified somewhat in both directions by the duration of the blast pulse, local variations in geology and other factors, but it is sustained by a voluminous record from weapons tests. Since the radius of blast damage is of the same order of size as the circular error probable, or smaller, the ratio $R_k$/CEP must be reckoned with in an attack on a hard target. Yet even in this situation the cube root of a given increase in yield would contribute much less to success than a comparable investment in numbers, reliability or accuracy.

Yield is of course a product of the yield-to-weight ratio of the nuclear explosive employed in the warhead multiplied by the weight of the warhead. In order to gain significant increases in the first of these two quantities further nuclear tests would be necessary. Increase in the weight of the warhead, on the other hand, calls for bigger and more efficient missiles. In the present state of the art, efforts to improve CEP and reliability as well as weight-carrying capacity hold out more promise than efforts to improve the yield-to-weight ratio. The reason is that missile design and control involve less mature and less fully exploited technologies than the technology of nuclear warheads. Finally, an increase in the number of missiles, although not necessarily cheap, promises more straightforward and assured

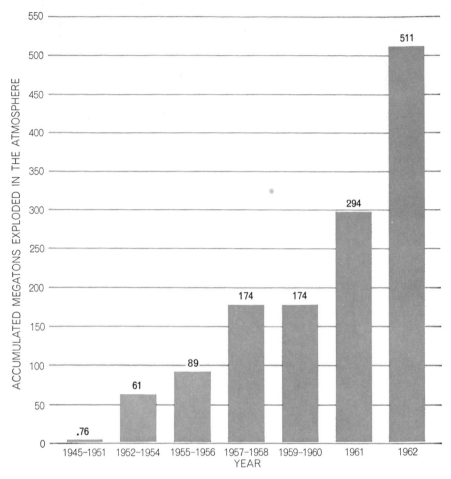

NUCLEAR EXPLOSIONS IN THE ATMOSPHERE from 1945 to 1962, the last full year in which the U.S. and the U.S.S.R. set off such explosions, are presented on the basis of accumulated megatons. The bars of equal height for the periods 1957–1958 and 1959–1960 reflect the informal moratorium on testing. The overall increase in megatons has doubled every three years. The data for this chart are from *Federal Radiation Council Report No. 4.*

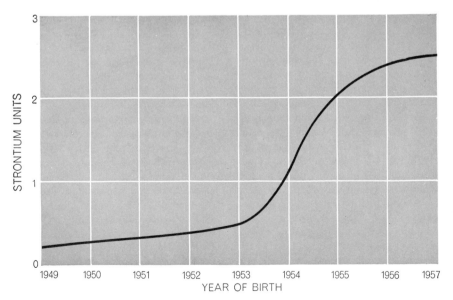

STRONTIUM 90 IN THE TEETH OF INFANTS between 1949 and 1957 was measured in a cooperative project of the Committee for Nuclear Information and the Washington University and St. Louis University schools of dentistry. This curve shows the strontium-90 activity in the deciduous teeth of bottle-fed infants. Because such teeth can be collected only some six years after the birth of the child, the curve does not come up to the present. The sharp rise in strontium-90 activity coincides with the period of extensive nuclear testing beginning in 1953. In the course of the project 110,000 teeth were collected.

results than a fractional increase in yield-to-weight ratio. Of all the various possible technical approaches to improving the military effectiveness of an offensive missile force, therefore, the only one that calls for testing (whether underground or in the atmosphere) is the one that offers the smallest prospect of return.

Suppose, however, a new analysis, based on information not previously considered, should show that it is in fact necessary to incorporate the 100-megaton bomb in the U.S. arsenal. Can this be done without further weapons tests? The answer is yes. Because the U.S.S.R. has pushed development in this yield range and the U.S. has not, the U.S. 100-megaton bomb might not be as elegant as the Soviet model. It would perhaps weigh somewhat more or at the same weight would produce a somewhat lower yield. It could be made, however, and the basic techniques for making it have been known since the late 1950's. The warhead for such a bomb would require a big missile, but not so big as some being developed by the National Aeronautics and Space Administration for the U.S. space-exploration program. Such a weapon would be expensive, particularly on a per-unit basis; under any imaginable circumstances it would be of limited use and not many of its kind would be built.

The extensive series of weapons tests carried out by the U.S.—involving the detonation of several hundred nuclear bombs and devices—have yielded two important bodies of information. They have shown how to bring the country's nuclear striking force to its present state of high effectiveness. And they have demonstrated the effects of nuclear weapons over a wide range of yields. Among the many questions that call for soundly based knowledge of weapons effects perhaps none is more important in a discussion of the technical aspects of national security than: What would be the result of a surprise attack by missiles on the country's own missile forces? Obviously if the huge U.S. investment in its nuclear armament is to succeed in deterring an attacker, that armament must be capable of surviving a first strike.

A reliable knowledge of weapons effects is crucial to the making of rational decisions about the number of missiles needed, the hardening of missile emplacements, the degree of dispersal, the proportion that should be made mobile and so on. The military planner must bear in mind, however, that such decisions take time—years—to carry out and require large investments of finite physical and human resources. The inertia of the systems is such that the design engineer at work today must be concerned not with the surprise attack that might be launched today but rather with the kind and size of forces that might be launched against them years in the future. In addition to blast, shock and other physical effects, therefore, the planner must contend with a vast range of other considerations. These include the yields of the various bombs the attacker would use against each target; the reliability and accuracy of his missiles; the number and kind of weapons systems he would have available for attack; the tactics of the attacker, meaning the number of missiles he would commit to a first strike, the fractions he would allocate to military as against civilian targets and the relative importance he would assign to various kinds of military targets, the effects of chaos on the defender's capacity to respond, and so on. In all cases the planner must project his thinking forward to some hypothetical future time, making what he can of the available intelligence about the prospective attacker's present capabilities and intentions. Plainly all these "other considerations" involve inherently greater uncertainties than the knowledge of weapons effects.

The extensive classified and unclassified literature accumulated in two decades of weapons tests and available to U.S. military planners contains at least some observations on all important effects for weapons with a large range of yields. These observations are more or less well understood in terms of physical theories; they can be expressed in numerical or algebraic form, and they can be extrapolated into areas not fully explored in the weapons tests conducted by the U.S., for example into the 100-megaton range. As one departs from the precise circumstances of past experiments, of course, extrapolation becomes less and less reliable. Nonetheless, some sort of estimate can be made about what the prompt and direct effects will be under any conceivable set of circumstances.

Consider, in contrast, the degree of uncertainty implicit in predicting the number and kind of weapons systems that might be available to the prospective attacker. Such an uncertainty manifested itself in the famous "missile gap" controversy. The remarkable difference between the dire predictions made in the late 1950's—based as they were on the best available intelligence—and the actual situation that developed in the early 1960's can be taken as indicating the magnitude of the uncertainties that surround the variables other than weapons effects with which the military planner must contend. Moreover, these factors, as they concern a future attack, are uncertain not only to the defender; they are almost as uncertain to the attacker.

Uncertainties of this order and kind defy reduction to mathematical expression. A human activity as complex as modern war cannot be computed with the precision possible in manipulation of the data that concern weapons effects. What is more, the uncertainties about this single aspect of the total problem are not, as is sometimes assumed, multiplicative in estimation of the overall uncertainty. Most, but not all, of the uncertainties are independent of one another. The total uncertainty is therefore, crudely speaking, the square root of the sum of the squares of the individual uncertainties.

In our view further refinement of the remaining uncertainties in the data concerning prompt direct physical effects can contribute virtually nothing more to management of the real military and political problems, even though it would produce neater graphs. Furthermore, if new effects should be discovered either experimentally or theoretically in the future, or if, in certain peculiar environments, some of the now known effects should be excessively uncertain, it will be almost certainly possible to "overdesign" the protection against them. Thus, although renewed atmospheric testing would contribute some refinement to the data on weapons effects, the information would be, at best, of marginal value.

Such refinements continue to be sought in the underground tests that are countenanced under the partial test ban. From this work may also come some reductions in the cost of weapons, modest improvements in yield-to-weight ratios, devices to fill in the spectrum of tactical nuclear weapons and so on. There is little else to justify the effort and expenditure. The program is said by some to be necessary, for example, to the development of a pure fusion bomb, sometimes referred to as the "neutron bomb." It is fortunate that this theoretically possible (stars are pure fusion systems) device has turned out to be so highly difficult to create; if it were relatively simple, its development might open the way to thermonuclear armament for the smallest and poorest powers in the world. The U.S., with its heavy investment in fission-to-fusion technology, would be the last nation to welcome this development and ought to be the last to encourage it. Underground testing is also justified for its contribution to the potential peaceful uses of nuclear explosives. Promising as these may be, the world could forgo them for a time in exchange for cessation of the arms race. Perhaps the best rationale for the underground-test program is that it helps to keep the scientific laboratories of the military establishment intact and in readiness—in

readiness, however, for a full-scale resumption of the arms race.

Paradoxically one of the potential destabilizing elements in the present nuclear standoff is the possibility that one of the rival powers might develop a successful antimissile defense. Such a system, truly airtight and in the exclusive possession of one of the powers, would effectively nullify the deterrent force of the other, exposing the latter to a first attack against which it could not retaliate. The possibilities in this quarter have often been cited in rationalization of the need for resuming nuclear tests in the atmosphere. Here two

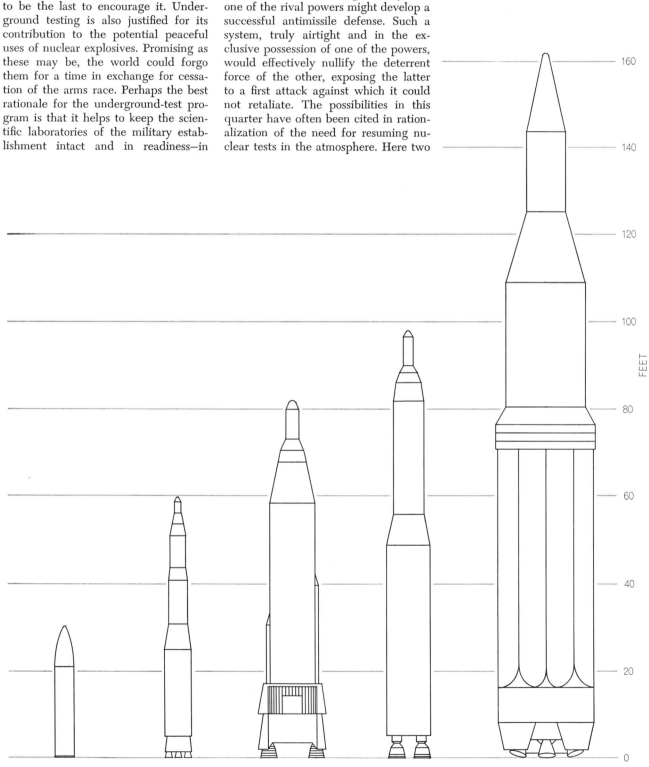

**PAYLOAD OF EXISTING ROCKETS** sets a limit on the size of nuclear weapons that can be used in a rocket attack. The five U.S. rockets shown here are drawn to scale. At left is the Polaris Type A-3, designed for launching from submarines; it weighs 30,000 pounds, has a range of 2,500 nautical miles and can carry a nuclear warhead of about one megaton. Second from left is Minuteman II; it weighs 65,000 pounds, has a range of 6,300 nautical miles and can carry a warhead of about one megaton. Third is Atlas; it weighs 269,000 pounds, has a range of 9,000 nautical miles and can carry a warhead of about five megatons. Fourth is Titan II; it weighs 303,000 pounds, has a range of 6,300 nautical miles and can carry a warhead of about 20 megatons. To lift a larger warhead would require a rocket such as Saturn I (*right*), which weighs 1,138,000 pounds. Data are from the journal *Missiles and Rockets*.

questions must be examined. One must first ask if it is possible to develop a successful antimissile defense system. It then becomes appropriate to consider whether or not nuclear weapons tests can make a significant contribution to such a development.

Any nation that commits itself to large-scale defense of its civilian population in the thermonuclear age must necessarily reckon with passive modes of defense (shelters) as well as active ones (antimissile missiles). It is in the active mode, however, that the hazard of technological surprise most often lurks. The hazard invites consideration if only for the deeper insight it provides into the contemporary revolution in the technology of war.

The primary strategic result of that revolution has been to overbalance the scales in favor of the attacker rather than the defender. During World War II interception of no more than 10 percent of the attacking force gave victory to the defending force in the Battle of Britain. Attrition of this magnitude was enough to halt the German attack because it meant that a given weapons-delivery system (bomber and crew) could deliver on the average only 10 payloads of high explosive; such a delivery rate was not sufficient to produce backbreaking damage. In warfare by thermonuclear missiles the situation is quantitatively and qualitatively different. It is easily possible for the offense to have in its possession and ready to launch a number of missiles that exceeds the number of important industrial targets to be attacked by, let us say, a factor of 10. Yet the successful delivery of only one warhead against each such target would result in what most people would consider an effective attack. Thus where an attrition rate of only 10 percent formerly crowned the defense with success, a penetration rate of only 10 percent (corresponding to an attrition rate of 90 percent) would give complete success to the offense. The ratio of these two ratios is 100 to one; in this sense the task of defense can be said to have become two orders of magnitude more difficult.

Beyond this summary statement of the situation there are many general reasons for believing that defense against thermonuclear attack is impossible. On the eve of attack the offense can take time to get ready and to "point up" its forces; the defense, meanwhile, must stay on the alert over periods of years, perpetually ready and able to fire within the very few minutes available after the first early warning. The attacker can pick its targets and can choose to concentrate its forces on some and ignore others; the defense must be prepared to defend all possible important targets. The offense may attack the defense itself; then, as soon as one weapon gets through, the rest have a free ride.

The hopelessness of the task of defense is apparent even now in the stalemate of the arms race. A considerable inertia drags against the movement of modern, large-scale, unitary weapons systems from the stage of research and development to operational deployment. The duration and magnitude of these enterprises, whether defensive or offensive, practically assure that no system can reach full deployment under the mantle of secrecy. The designer of the defensive system, however, cannot begin until he has learned something about the properties and capabilities of the offensive system. Inevitably the defense must start the race a lap behind. In recent years, it seems, the offense has even gained somewhat in the speed with which it can put into operation stratagems and devices that nullify the most extraordinary achievements in the technology of defense. These general observations are expensively illustrated in the development and obsolescence of two major U.S. defense systems.

Early in the 1950's the U.S. set out to erect an impenetrable defense against a thermonuclear attack by bombers. The North American continent was to be ringed with a system of detectors that would flash information back through the communications network to a number of computers. The computers were to figure out from this data what was going on and what ought to be done about it and then flash a series of commands to the various interceptor systems. In addition to piloted aircraft, these included the Bomarc (a guided airborne missile) and the Nike-Hercules (a ballistic rocket). By the early 1960's this "Sage" system was to be ready to detect, intercept and destroy the heaviest attack that could be launched against it.

The early 1960's have come and yet nothing like the capability planned in the 1950's has been attained. Why not? Time scales stretched out, subsystems failed to attain their planned capabilities and costs increased. Most important, the offense against which the system was designed is not the offense that actually exists in the early 1960's. Today the offensive system on both sides is a mixture of missiles and bombers.

The Sage system has a relatively small number of soft but vital organs completely vulnerable to missiles—a successful missile attack on them would give a free ride to the bombers. As early as 1958 the Department of Defense came to realize that this would be the situation, and the original grand plan was steadily cut back. In other words, the Sage system that could have been available, say, in 1963 and that should have remained useful at least through the 1960's would in principle have worked quite well against the offense that existed in the 1950's.

To answer the intercontinental ballistic missile, the Department of Defense launched the development of the Nike-Zeus system. Nike-Zeus was intended to provide not a defense of the continent at its perimeter but a point defense of specific targets. To be sure, the "points" were fairly large—the regions of population concentration around 50 to 70 of the country's biggest cities. The system was to detect incoming warheads, feeding the radar returns directly into its computers, and launch and guide an interceptor missile carrying a nuclear warhead into intersection with the trajectory of each of the incoming warheads.

Nike-Zeus was not designed to defend the 1,000 or so smaller centers outside the metropolitan areas simply because there are too many of these to be covered by the resources available for a system so huge and complicated. Nor was the system designed to defend the retaliatory missiles, the security of these forces being entrusted to the more reliable protection of dispersal, concealment, mobility and number. In principle, the defense of a hardened missile silo would have presented by far the simplest case for proof of the effectiveness of Nike-Zeus as advanced by those who contend that such a system can be made to "work." There would be no ambiguity about the location of the target of the incoming warhead. By the same token Nike-Zeus might have been considered for the defense of a few special defense posts, such as the headquarters of the Air Defense Command of the Strategic Air Command. These special cases are so few in number, however, that it had to be concluded that the attacker would either blast his way through to them by a concentration of firepower or ignore them altogether.

At the time of the conception of the Nike-Zeus system its designers were confronted with a comparatively simple problem, namely that of shooting down

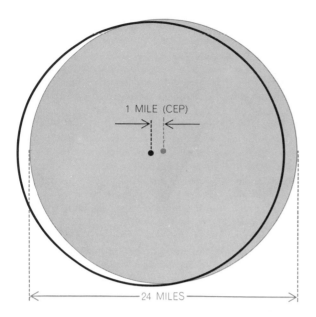

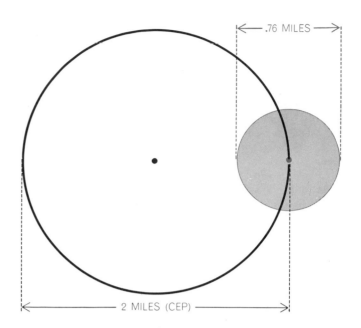

ACCURACY OF MISSILES has far more relevance for hard targets than for soft. It has been estimated that at maximum range both U.S. and Soviet rockets have a "circular error probable" of about a mile, that is, there is a probability of 50 percent that a rocket will hit within a mile of the target. The radius of fire damage for a one-megaton bomb will not be greatly affected by a near miss (*cir-* *cles at left*). The radius of kill for a one-megaton bomb aimed at a target hardened to 300 pounds per square inch, however, is so much smaller than the circular error probable (*circles at right*) that a number of weapons would have to be used to assure a hit. The illustrations in this article and the figures on which they are based are not the responsibility of the authors but of the editors.

the warheads one by one as they presented themselves to the detectors. Even this simple problem had to be regarded as essentially unsolvable, in view of the fact that a 90 per-cent success in interception constitutes failure in the inverted terms of thermonuclear warfare. At first, therefore, the designers of the offensive system did not take the prospect of an antimissile system seriously. Then the possibility that the problem of missile interception might be solved in principle gave them pause. Thereupon the designers of the offense began to invent a family of "penetration aids," that is, decoys and confusion techniques. The details of these and the plans for their use are classified, but the underlying principles are obvious. They include light decoys that can be provided in large numbers but that soon betray their character as "atmospheric sorting" separates them from the heavier decoys (and actual warheads) that can be provided in smaller numbers to confuse the defending detectors down to the last minute. Single rockets can also eject multiple warheads. Both the decoys and the warheads can be made to present ambiguous cross sections to the radar systems. These devices and stratagems overwhelmed the designed capability of the Nike-Zeus system and compelled its recent abandonment.

If the installation of the system had proceeded according to plan, the first Nike-Zeus units would have been operational within the next year or two. This could have been celebrated as a technical milestone. As a means of defense of a substantial percentage of the population, however, the system would not have reached full operational deployment until the end of the decade. In view of its huge cost the system should then have looked forward to a decade of useful life until, say, the late 1970's. Thus, in inexorable accordance with the phase-lag of the defense, the U.S. population was to be defended a decade too late by a system that might have been effective in principle (although most probably not in practice) against the missiles of the early 1960's.

The race of the tortoise and the hare has now entered the next lap with the development of the Nike-X system as successor to Nike-Zeus. The Advanced Research Projects Agency of the Department of Defense has been spending something on the order of $200 million a year on its so-called Defender Program, exploring on the broadest front the principles and techniques that might prove useful in the attempt to solve the antimissile problem. Although nothing on the horizon suggests that there is a solution, this kind of work must go forward. It not only serves the forlorn hope of developing an active antimissile defense but also promotes the continued development of offensive weapons. The practical fact is that work on defensive systems turns out to be the best way to promote invention of the penetration aids that nullify them.

As the foregoing discussion makes clear, the problems of antimissile development are problems in radar, computer technology, missile propulsion, guidance and control. The nuclear warheads for the antimissile missile have been ready for a long time for delivery to the right place at the right time. Although it is argued that certain refinements in the existing data about weapons effects are needed, the other uncertainties all loom much larger than the marginal uncertainties in these physical effects. The antimissile defense problem, then, is one in which nuclear testing can play no really significant part.

The pursuit of an active defense system demands parallel effort on the passive defense, or shelter, front because the nature of the defense system strongly conditions the tactics of the offense that is likely to be mounted against it. To take a perhaps farfetched example, a Nike-Zeus system that provided protection for the major population centers might invite the attacker to concentrate the weight of his assault in ground bursts on remote military installations and unprotected areas adjacent to cities, relying on massive fallout to imperil the population centers. This example serves also to suggest how heavily the effectiveness of any program for sheltering the civilian population depends on the tactics of the attacker. Fallout shelters by themselves are of no avail if the attacker chooses to assault the population centers directly.

In any speculation about the kind of attack to which this country might be exposed it is useful to note where the military targets are located. Most of the missile bases are, in fact, far from the largest cities. Other key military installations, however, are not so located. Boston, New York, Philadelphia, Seattle, San Francisco, Los Angeles (Long Beach) and San Diego all have important naval bases. Essential command and control centers are located in and near Denver, Omaha and Washington, D.C. The roll call could be extended to include other major cities containing military installations that would almost certainly have to be attacked in any major assault on this country. The list does not stop with these; it is only prudent to suppose still other cities would come under attack, because there is no way to know in advance what the strategy may be.

The only kind of shelter that is being seriously considered these days, for other than certain key military installations, is the fallout shelter. By definition fallout shelters offer protection against nothing but fallout and provide virtually no protection against blast, fire storms and other direct effects. Some people have tried to calculate the percentage of the population that would be saved by fallout shelters in the event of massive attack. Such calculations always involve predictions about the form of the attack, but since the form is unknowable the calculations are nonsensical. Even for the people protected by fallout shelters the big problem is not a problem in the physical theory of gamma-ray attenuation, which can be neatly computed, but rather the sociological problem of the sudden initiation of general chaos, which is not subject to numerical analysis.

Suppose, in spite of all this, the country were to take fallout shelters seriously and build them in every city and town. The people living in metropolitan areas that qualify as targets because they contain essential military installations and the people living in metropolitan areas that might be targeted as a matter of deliberate policy would soon recognize that fallout shelters are inadequate. That conclusion would be reinforced by the inevitable reaction from the other side, whose military planners would be compelled to consider a massive civilian-shelter program as portending a first strike against them. Certainly the military planners of the U.S. would be remiss if they did not take similar note of a civilian-shelter program in the U.S.S.R. As a step in the escalation of the arms race toward the ultimate outbreak of war, the fallout shelter would lead inevitably to the blast shelter. Even with large numbers of blast shelters built and evenly distributed throughout the metropolitan community, people would soon realize that shelters alone are not enough. Accidental alarms, even in tautly disciplined military installations, have shown that people do not always take early warnings seriously. Even if they did, a 15-minute "early" warning provides less than enough time to seal the population into shelters. Accordingly, the logical next step is the live-in and work-in blast shelter leading to still further disruption and distortion of civilization. There is no logical termination of the line of reasoning that starts with belief in the usefulness of fallout shelters; the logic of this attempt to solve the problem of national security leads to a diverging series of ever more grotesque measures. This is to say, in so many words, that if the arms race continues and resumes its former accelerating tempo, 1984 is more than just a date on the calendar 20 years hence.

Ever since shortly after World War II the military power of the U.S. has been steadily increasing. Throughout this same period the national security of the U.S. has been rapidly and inexorably diminishing. In the early 1950's the U.S.S.R., on the basis of its own unilateral decision and determination to accept the inevitable retaliation, could have launched an attack against the U.S. with bombers carrying fission bombs. Some of these bombers would have penetrated our defenses and the American casualties would have numbered in the millions. In the later 1950's, again on its own sole decision and determination to accept the inevitable massive retaliation, the U.S.S.R. could have launched an attack against the U.S. using more and better bombers, this time carrying thermonuclear bombs. Some of these bombers would have penetrated our defenses and the American casualties could have numbered in the tens of millions.

Today the U.S.S.R., again on the basis of its own decision and determination to accept the inevitable retaliation, could launch an attack on the U.S. using intercontinental missiles and bombers carrying thermonuclear weapons. This time the number of American casualties could very well be on the order of 100 million.

The steady decrease in national security did not result from any inaction on the part of responsible U.S. military and civilian authorities. It resulted from the systematic exploitation of the products of modern science and technology by the U.S.S.R. The air defenses deployed by the U.S. during the 1950's would have reduced the number of casualties the country might have otherwise sustained, but their existence did not substantively modify this picture. Nor could it have been altered by any other defense measures that might have been taken but that for one reason or another were not taken.

From the Soviet point of view the picture is similar but much worse. The military power of the U.S.S.R. has been steadily increasing since it became an atomic power in 1949. Soviet national security, however, has been steadily decreasing. Hypothetically the U.S. could unilaterally decide to destroy the U.S.S.R. and the U.S.S.R. would be absolutely powerless to prevent it. That country could only, at best, seek to wreak revenge through whatever retaliatory capability it might then have left.

Both sides in the arms race are thus confronted by the dilemma of steadily increasing military power and steadily decreasing national security. *It is our considered professional judgment that this dilemma has no technical solution.* If the great powers continue to look for solutions in the area of science and technology only, the result will be to worsen the situation. The clearly predictable course of the arms race is a steady open spiral downward into oblivion.

We are optimistic, on the other hand, that there is a solution to this dilemma. The partial nuclear-test ban, we hope and believe, is truly an important first step toward finding a solution in an area where a solution may exist. A next logical step would be the conclusion of a comprehensive test ban such as that on which the great powers came close to agreement more than once during 10 long years of negotiation at Geneva. The policing and inspection procedures so nearly agreed on in those parleys would set significant precedents and lay the foundations of mutual confidence for proceeding thereafter to actual disarmament.

# Commentary

Although we scientists often pride ourselves that our familiarity with the "scientific method" enables us to form more rational political views and to raise the level of political discussions, most of our writings on the problems that confront our nation today resemble the monologues that men in politics use. We state our views and desires but do not point out the areas of disagreement with our opponents, or the extent, or the reasons therefor. No fruitful scientific discussion could proceed on such a basis, and in the following paragraphs, in which I comment on the article "National Security and the Nuclear-Test Ban," by Jerome B. Wiesner and Herbert F. York, I hope to come closer to a dialogue. Thus I shall attempt to specify both the areas in which I agree with the authors and those in which I disagree, and to give my reasons for disagreeing.

In the early stages of writing this letter it appeared that only the views expressed by Wiesner and York on policies, attitudes and technical questions would have to be discussed. It soon became apparent, however, that there is a third subject that could not be disregarded: the inferences the daily press drew from the article and the extrapolations it attached thereto.

Turning first to *questions of broad policy expressed in the article itself*, there is much with which it would seem a vast majority of our colleagues can agree. The principal area of agreement concerns the success of the test-ban treaty. One would have to be blind not to see that the tensions between the U.S.S.R. and our country have much relaxed since this treaty has been in effect. It would be stretching a point to say that the cessation of testing is only the consequence, and not at the very least partially a *cause*, of the relaxation of tensions. As to the delay that peaceful uses of atomic explosives suffer as a result of the test cessation, the article says, "Promising as peaceful uses of nuclear explosives may be, the world could forgo them for a time" in exchange for a quieter international atmosphere, and I can only concur.

On the other hand, it would be a mistake to overlook the fact that the test-ban treaty is the result of extensive negotiations in which *both* parties made

significant concessions. There is no evidence that generous acts of the U.S. through which it unilaterally weakens itself have any but adverse effects on the policy of the U.S.S.R. Nor does the insistence on *mutual* concessions have to create an unfriendly atmosphere. On the contrary, the constant pressure on our government, by means of public statements, to give in, raises false hopes in the negotiators of the U.S.S.R. The thwarting of these hopes, and the irritation of our own negotiators because of these pressures, do create· an unhappy atmosphere. It is to be hoped that the article by Wiesner and York, with its insistence on a comprehensive test-ban treaty but without equal insistence on policing and inspection, will not have such an effect. It certainly counsels moderation not only to our government but also to that of the U.S.S.R. (An apt description of the adverse circumstances under which our negotiators often labor was given by R. Gilpin in his *American Scientists and Nuclear Weapons Policy*.)

Another statement of the article with which few will quarrel is that "if the great powers continue to look for solutions in the area of science and technology only, the result will be to worsen the situation." It is unfortunate that the subtitle of the article abbreviates this to "there can be no technical solution to the problem of national security." This subtitle, whether written by the authors or by a somewhat careless editor, printed as it is in large italics, could give the impression, and has given the impression to some of the daily press, that technical problems will play no further role in the future. This is, of course, not the meaning of the statement quoted. In fact, the sentence "Today as never before national security involves technical questions" stands just three inches below the subtitle.

Actually the great powers have never confined themselves to looking for solutions in the area of science and technology only but have initiated extensive negotiations toward easing tensions. As we have seen, some of these were successful.

If one is asked whether one agrees or disagrees with the policies recommended by the article, one soon discovers that the article does not recommend any policies. It leaves its reader with a sense of frustrated disorientation cou-

pled with the impression that the past policies of this country were fundamentally wrong and something fundamentally new has to be tried. Some passages even carry the implication that our defense preparations have aggravated the danger to our freedom, independence and survival. "Ever since shortly after World War II the military power of the U.S. has been steadily increasing. Throughout the same period the national security of the U.S. has been rapidly and inexorably diminishing." Does this suggest that the decrease of our security is a *result* of the increase of our military power? To some who wish to think so it apparently does; to the people of countries whose military power did not grow adequately—to the people of Czechoslovakia, Hungary and Tibet—it would not. When I maintained in a discussion with one of the top scientific negotiators of the U.S.S.R. that the U.S. used its early atomic monopoly with great restraint, he answered, "I don't know. We wanted to do many things that [as a result of your atomic strength] we [the U.S.S.R.] could not." I am afraid that the borders of Stalinist Russia would have moved *much* farther to the West in Europe had the military power of the U.S. not been "steadily increasing."

Furthermore, is it really true that the national security of the U.S. has been "rapidly and inexorably diminishing"? If one thinks only in terms of physical possibilities, in terms of a fanatical enemy who takes seriously Lenin's dictum "Better only one-third of the world's population surviving if those are then good communists," the security has decreased. But is that a valid picture? Have we not spoken of the relaxation of tensions before? True, we still hear the threats of burying us, alone or in collaboration with the Chinese brothers, the praise of the "irreconcilable class hatred that exposes and strikes the enemies of our social system," and the glorification of the sparks flying from the sabers of cavalrymen (both in Khrushchev's speech on culture of March 8, 1963). But one also hears, with an increasing volume, the realization of the need for coexistence and of peaceful competition. One does not have to shut one's ears to the threats in order to hear also the voice of reason and adjustment. Furthermore, is it not clear that the realization of the need for coexistence is in large measure the result of the understanding that the sparks flying from the sabers of cavalrymen can ignite other fires? A U.S. that is not only strong but *evidently* strong is in the interests of all: it is reassuring to the West and should turn the interest of the rulers of the East away from domination and toward the true welfare of their people. Are there no signs that it is at least beginning to do so?

To put the preceding point more pragmatically: Although the worst conceivable alternative may have become worse with the progress of time, the probability of such an alternative has decreased sharply. As a result, from the point of view of the most likely turn of events, the security of the U.S. has probably greatly increased, particularly in the course of the past few years.

Let us now turn to the *technical points of the article.* I am sorry to say that I find it more difficult to agree with them than with the general statements discussed before. The first remark that comes to mind is that the alleged need for developing the 100-megaton bomb was not the only, in fact not the principal, argument against the test-ban treaty. Personally I feel that this treaty was worth what we paid for it, and that it benefits both the U.S. and the U.S.S.R. It would be only fair toward those who opposed the treaty, however, to state that they were chiefly concerned with the testing of certain defense measures against antiballistic missiles, not with the development of the 100-megaton bomb.

Nevertheless, we must recall in connection with this bomb that the U.S.S.R. found it worthwhile to break the test moratorium in order to test it. In addition, the statement of the article that "on any scale of investment, the combination of larger numbers and smaller size results in greater effectiveness of the missile system" cannot be maintained. The cost of a missile is approximately proportional to the square root of its explosive yield. The illustration in the article that shows a linear relation is incorrect (and is contradicted in the text). Hence the yield is proportional to the square of the cost. The range of destruction is proportional to the cube root of the yield and hence to the 2/3 power of the cost. Finally, the area of destruction is proportional to the *square* of the range of destruction and hence to the 4/3 power of the cost. It follows that if one disregards soft targets that can be destroyed by a single smaller bomb, the cost effectiveness of weapons actually *increases* somewhat with their

yield. Even this is not the complete picture. A hardened defense installation can be destroyed with a smaller explosion only if this takes place closer by and at lower altitudes. It is easier to prevent such an explosion by antimissile measures, or otherwise, than an explosion at the larger distances and high altitudes at which a very large bomb can still destroy the installation.

Drs. Wiesner and York state that we do not need tests in order to design a 100-megaton bomb. This is true but disregards the time element. The time schedules for the production of a bomb with the characteristics that exploit the inherent advantages of size are extremely long in the absence of tests. If the bomb should be needed, it would be meager comfort to know that, given only a few more years, we could have had it.

The cases for and against the big bomb remain unproved and I personally cannot become enthusiastic about it. I do feel strongly, however, that the rejection of defense measures, in particular the rejection of civil defense, is unjustified. Drs. Wiesner and York recognize that the instability that has to be overcome is due to the "overbalance [of] the scales in favor of the attacker rather than defender." If this is so, a wholehearted effort should be made to redress the overbalance. It does not seem, however, that they have explored the possibilities of civil defense even halfheartedly. Most readers will be struck by the contradiction between the postulates that, on the one hand, the retaliatory installations are invulnerable and, on the other hand, that shelters are useless. In fact, when discussing civil defense, the authors say: "The only kind of shelter that is being seriously considered these days, for other than certain key military installations, is the fallout shelter." They then proceed to show that fallout shelters by themselves do not suffice to render the position of the defender strong enough. Although they do not state this explicitly, they give the impression that they would not be opposed to abandoning altogether the fallout shelter program as insufficient. The opposite alternative, to strengthen the civil defense program by the installation of blast shelters at important locations, is dismissed all too easily with arguments that are in no way convincing. Thus the writers mention the danger of a short warning time in a surprise attack but do not mention that a complete surprise is difficult to achieve, and in fact the two world wars did not break out without warning. In addition, the shel-

ters could well be located in such a way that they could be reached by most people in the 15-minute warning time the writers concede. Similarly, the writers mention the possibility of chaos and disorientation in shelters but fail to mention that these dangers are much greater if no shelters exist. They do not mention either that the history of past disasters does not bear out their fears of antisocial behavior as long as proper leadership is provided. During the siege of Budapest people stayed in shelters for many weeks but continued to cooperate and help each other. The authors emphasize throughout the article that, in contrast to an attack of World War II, a nuclear attack that is 10 percent effective would be considered successful. This would hardly be the case if the population were well sheltered. Hence a combination of antiballistic missiles and shelters seems to hold more promise of reducing the "overbalance" of offense over defense than any other measure known to me.

Finally, and perhaps most importantly, the authors do not mention that no nation will dare to disarm if its population remains exposed to the awesome dangers the authors so well depict. Hence civil defense is also a prerequisite to disarmament.

Let us come finally to the "extrapolations" contained in the *reports of the daily press.* Many of these were crude exaggerations that may have served a useful purpose, however, by attracting attention to the article. They were encouraged by the mode of communication of the article, which made most of its points by implication. However, the result often approached the bizarre. Even *The New York Times* headline reads "Disarmament Is Called the Answer to 'Stalemate,'" as if it were desirable to have a checkmate ending to the game. The words "stalemate" and "disarmament" occur once each in the article, the former toward the middle and the latter as the last word.

A more nearly justified "extrapolation" made from the article is that no further methods of offense or defense need be explored, that is, that military science is a complete and closed book. Even this is only implied by the article. To evaluate it, it may be useful to recall similar statements about other areas, in particular about physics. These were made around the turn of the century, before the advent of atomic theory, before virtually any knowledge of the nucleus, before quantum and relativity theories, before any inkling of the results of almost all areas that are at present at the center of interest of research in physics. Statements of this sort mean partly that those who make them have, at the time of making the statements, no promising ideas in the field about which they speak. Others may have such ideas, and those making the statements may themselves conceive such ideas at a later time. The statements in question also appear to herald the impending initiation of new lines of endeavor. In the case of physics this was the turn toward microscopic phenomena; in the area of "weaponry" it may well be the exploration of a more effective defense.

Having stated the areas of disagreement, it would be well to reemphasize the agreement with what appears to be the main thesis of Wiesner and York: the importance of not relying on physical power alone. It is a truism that the purpose of power is only the achievement of certain goals, called national objectives. However, military power, like police power, works best when it works through its presence rather than by active involvement, and when it is supporting persuasion to follow rules of conduct that are just and reasonable. Certainly included is the rule to leave our country free to follow its own path of independence and individual freedom.

EUGENE P. WIGNER

Princeton University
Princeton, N.J.
December, 1964

# 44

# The Shelter-Centered Society

ARTHUR I. WASKOW
*May 1962*

Civil defense has been a topic of increasing public concern since July 25 of last year, when President Kennedy, in an address to the nation on the Berlin crisis, called for the preparation of shelters against the local fallout that results from a nuclear explosion on the ground. Most public discussion—for example, the hearings and debates on civil defense legislation in Congress—has focused on whether fallout shelters would be useful during and after a nuclear attack. There has been little examination of the possible effects that the creation and operation of a shelter-centered civil defense system might have on U.S. society during peacetime. It is apparent, however, that these effects must be reckoned with, whether or not the usefulness of the system is ever tested.

In January the Peace Research Institute, a nonprofit organization headed by James J. Wadsworth, former Permanent U.S. Representative to the United Nations, brought together a group of social scientists to consider this aspect of the proposed civil defense system. The conference is one of a number of projects recently undertaken by the Peace Research Institute to stimulate scientific research concerning ways to further the cause of peace, in the same fashion that research has been effectively used in recent years in support of the arms race.

The conferees [*see Editor's Note at the bottom of page 333*] prepared themselves by studying a number of background papers and documents; they were briefed, in the course of their deliberations, by officials of the National Security Council and the Office of Civil Defense and by experts from the Institute for Defense Analyses. In addition to their basic knowledge of social theory and of experimental findings in their fields of learning, the conferees brought to their study such special data as recent observations of overseas reactions to the U.S. civil defense program, scientific surveys of U.S. public opinion and locally observed reactions among students, patients and the general public.

On one conclusion the conferees felt they could confidently agree: The existence of a shelter-centered civil defense would be a wholly new departure in U.S. history. Because the prospect is without precedent they did not attempt to produce ironclad predictions of what would happen. They sought rather to define the problems that are likely to develop. As the product of their work together, the conferees issued not conclusions but a series of questions. They shared unanimously the sentiment that the questions are urgent and that action taken without careful consideration of these questions might lead to irreversible and disastrous consequences.

The civil defense program, the conferees agreed, portends an unprecedented departure in U.S. life because it implies major effects on our society as a whole. In official statements the program is described as "minimal insurance" against the "unlikely" event of a nuclear war. Because of its uniquely potent psychological and social appeal to survival instincts, however, it would be extremely difficult to limit the program to any predetermined minimum. A drive for continuous expansion of the program—a drive far more powerful than the usual pressures to expand, for example, Social Security—threatens to press constantly on the decision makers. Once having promised survival to some, they would soon have to meet all objections of inadequacy or noncoverage by broadening the program geographically and by improving its quality. To inspire any hope of effectiveness in war, civil defense must be able to call forth virtually universal teamwork. Since failure in any of a number of crucial tasks could gravely impair the operation of the system, the program must instill in all Americans a wholehearted willingness to carry out difficult orders on short notice. It would require the training of a large cadre of men and women to a fine pitch of elaborate knowledge and total dedication, and the training of the rest of the population for unquestioning obedience in a crisis.

All the evidence from the experience of the armed forces indicates that such "training" comes not from reading textbooks or instruction posters but from actually rehearsing crisis behavior. Civil defense would require whole detachments of civil defense workers to go into and stay in shelters, whole populations to

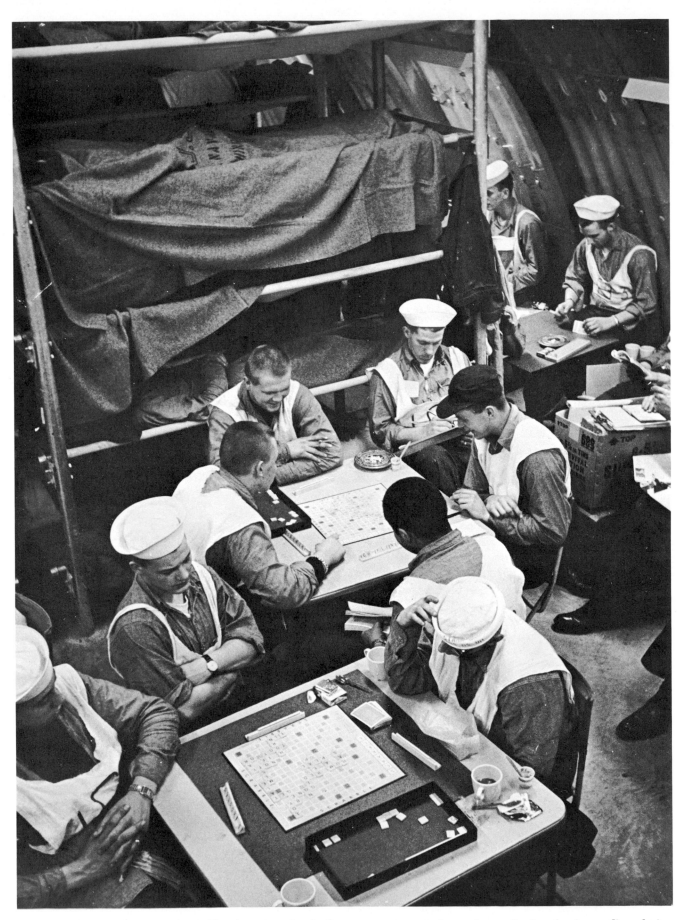

**FALLOUT-SHELTER TEST** run by the Navy last winter involved 100 men, who stayed in an underground steel and concrete shelter for two weeks. Ventilating fans supplied filtered air. The sailors ate emergency rations and spent the time sleeping, reading, playing cards and Scrabble. The test was designed to study the physical effects of long confinement underground in cramped quarters.

practice responding to emergency signals, whole cities to drill on a winter night. The demand for disciplined obedience to authority extended to the entire population would be entirely new in U.S. life. Indeed, in virtually no society is there any precedent for maintaining a large portion of a civilian population over a long time in trained readiness for a threatening event with a low probability of occurrence.

Proponents of civil defense have made a virtue of the need for such co-ordination of sentiment and action and have argued that by making the danger to survival obvious to all, the shelter program would enhance a national sense of community. But the conferees, on the basis of sociological evidence from the past, were unanimous in the doubt that feelings of community would be thereby reinforced. People working together to face danger perceived as equally threatening to all, in a civil defense program perceived as equally protective of all, might well have their community solidarity strengthened. But the danger of attack weighs differently on different Americans, the prospective usefulness of shelters is vastly different in different situations, and the work of building and operating shelters would actually be done by different agencies working along different lines at different levels of expense and with different chances of success or failure. In these conditions the evidence suggests that existing stresses and strains in the community would be amplified.

Already the civil defense effort has strained the web of community. Some people have concluded that shelters (private or public) would be useless to them unless they were prepared to limit the number of occupants to those whom the shelter could physiologically support. They have therefore announced their intention of excluding neighbors, or people from the next block, or strangers from the next county, or casual visitors to town, from the family or community shelter. Suburbia has been pitted against city, one state against another. These strains cannot be expected to disappear. It is indeed likely that they would worsen as cities realize how vulnerable they are to attack, as racial and ethnic groups compete for space in and access to community shelters, as farmers realize that refugees from the cities will deplete their food stocks.

From the point of view of the individual, the conferees agreed, the announcement of the civil defense program represented simultaneously a highly authoritative threat of personal death and social destruction and a promise that there is a way to meet this threat. Even a program announced as "minimal insurance" against an "unlikely war" signals to people that the danger is high ("Otherwise, why the program?") and that safety is possible ("Maybe not for other people, but for me").

It might seem that the threat is actually posed by the existence of nuclear weapons rather than by a shelter program. In people's minds, however, it is the civil defense program that spells out the danger. The call for civil defense is seen as a warning that war is highly possible and imminent; the physical trappings of civil defense make a visible and immediate impact on local and family affairs. In comparison the distant and half-realized military arsenal is a much less potent symbol of danger. At a still more distant remove is the outside military threat, which can be ignored, denied or suppressed in one's mind. Civil defense, however, is immediately visible, tangible and unavoidable.

The probable effect of this powerful threat and promise is to bring about three distinct reactions in the population. First, the threat generates anxiety in almost the entire population. Second, the promise of some protection provides a considerable amount of relief from the anxiety. It goes without saying, however, that the relief cannot fully or permanently offset the anxiety. The relief depends on sustained conviction that the shelter program is adequate, whereas the anxiety can disappear only if the threat disappears. Contradictory as they are, these two reactions are likely to be confounded by a third reaction. Among some people—an unknown proportion of the population—the civil defense program is bound to stir a dark attraction to the world of which civil defense is a warning: a world wiped clean of complications, ambiguities and dissension. The coexistence in the population of deep anxiety, precarious hope and an obsessive concern with violence and death would constitute a new situation for U.S. society.

This analysis of the impending impact of civil defense on American life suggests a number of urgent questions. First among them is the question of whether or not a commitment to the proposed program will tend to restrict the U.S. Government's freedom to negotiate with Communist governments. In the long run such a development would make it more likely that conflicts between the U.S.S.R. and the U.S. would lead to the use of force.

It is conceivable that public opinion might come under the sway of intense and uncontrolled hostility to the idea of negotiation with Communist states. Such hostility could result from the public's interpreting civil defense to mean that an enemy threatens imminent death to home, neighborhood and nation. Against this prospect it is sometimes argued that the shock and immediacy of a civil defense program might bring home to Americans the possibility and the peril of a nuclear war and thereby increase interest in and devotion to negotiations toward such goals as mutual disarmament. It was the unanimous judgment of the conferees, however, that a reaction in favor of disarmament is extremely

EDITOR'S NOTE

This article is a condensed version of the report of a conference on the potential implications of a national civil defense program, held by the Peace Research Institute in Washington, D.C., on January 13 and 14, 1962. Partial support for the conference was given by the National Institute of Mental Health and the American Psychological Association. The full report can be obtained from the Peace Research Institute.

The conferees were Raymond A. Bauer of the Harvard University Graduate School of Business Administration, Urie Bronfenbrenner of the Department of Psychology at Cornell University; Morton Deutsch of the Bell Telephone Laboratories; Herbert H. Hyman of the Department of Sociology at Columbia University; Erich Lindemann, professor of psychiatry at the Harvard Medical School and Psychiatrist-in-Chief of the Massachusetts General Hospital; David Riesman of the Department of Social Relations at Harvard University; Stephen B. Withey, director of Public Affairs Studies at the Survey Research Center of the University of Michigan; and Donald N. Michael, director of Planning and Programs for the Peace Research Institute, who was chairman of the conference. All the conferees are in essential agreement on the substance of the report.

unlikely. They agreed that those people who were committed to supporting disarmament before the call for civil defense might take the call as a signal for desperately intensifying their previous efforts. For almost everyone else civil defense and disarmament are what is known in social psychology as "dissonant"; that is, civil defense fits into a view of the world in which negotiation has failed and war is looming, whereas disarmament fits into a view of the world in which negotiation seems possible and war seems avoidable. Confronted by the physical reality of shelters such popular support as there is for disarmament might weaken and wane. In other words, the shelters themselves might be symbolically even more threatening to hopes of disarmament than the call for civil defense.

Meanwhile there is some evidence to suggest that the civil defense program might have the unintended effect of restricting the area open to the U.S. Government for negotiation. Public opinion surveys, conducted by the Survey Research Center of the University of Michigan, have already shown that the President's call for civil defense, regardless of what he intended, was widely accepted as a warning of intense and immediate danger of war ("Why else should he want us to do this?"), a warning that negotiations with the U.S.S.R. were not working. Might not the popular anxiety aroused by this "signal" lead to popular belief that negotiations cannot succeed? Since a real civil defense program is not just paper but underground buildings and training programs that persist over long periods, might the hostility to negotiation also persist into periods when the Government would see a possibility of resolving previous crises? If so, the ability of the Government to negotiate might be severely impaired by anxiety among the people.

Urie Bronfenbrenner of Cornell University was able to report, on the basis of his recent trip to the U.S.S.R., that many Russians, plain citizens as well as officials, see in the U.S. civil defense program a threat of war. Thus, regardless of the carefully qualified remarks of the President in his call for civil defense, the call was seen in both the U.S. and the U.S.S.R. as a signal that war was near, that negotiations were failing, and in each country as evidence that the other was threatening aggression. The dangers of such an atmosphere are obvious.

The creation of civil defense will bring to life in every nook and cranny of the nation special institutions economically dependent on and deriving their power and prestige from a civil defense program: Government agencies, private builders and suppliers, a cadre of trained shelter managers, and so on. Even if not a single individual in these groups directly or deliberately attacked the notion of negotiating with an adversary, might not the mere existence of these groups immensely complicate the task of working out a plan for general disarmament that would not disrupt U.S. society?

Investigators in social psychology suggest that the existence of an omnipresent, persistent and highly visible symbol of one line of action might distract attention from other, parallel and alternative lines of action that lack such a symbol. Thus fallout shelters on a national scale might constantly call attention to nuclear war as the technique for conducting international conflict. Both proponents and opponents of civil defense might tend to concentrate on nuclear dangers and their mitigation instead of trying to discover alternatives to the use of nuclear weapons, such alternatives as "conventional war," political and economic pressure, tension-reducing initiatives or the invention of international institutions for conflict control.

The proposed civil defense program raises corresponding questions about this country's relations with its allies and neutral nations. Through civil defense the American people would in effect be searching for a way of survival for themselves and would tacitly be abandoning non-Americans to die if nuclear war should come. Among U.S. citizens the civil defense program might produce feelings of isolation from and lack of interest in the rest of the world. Among even the most friendly peoples the program might be taken to mean that the U.S. had withdrawn from its undertaking to promote the interests and the defense of its allies. From a recent visit to Japan, David Riesman of Harvard University was able to report that many people in that country feel considerable uneasiness over the U.S. shelter program and fear that it may indeed symbolize a turning inward of U.S. interests and policy. If this response should become more widespread, it might be seized on by some Americans as evidence of the unreliability of allies who could be so easily annoyed by American attempts at self-protection. Therefore the possibility exists that feelings of isolation in the U.S. could feed on feelings of isolation from the U.S., the two processes being constantly reinforced by progress in the civil defense program and culminating in substantial alienation of the U.S. from its friends and allies.

Just as civil defense might affect the course of American foreign policy, so it might change the ways in which the country's traditional democratic processes work at home. If mobilization of total support and participation should become a goal of the program, would not the civil defense organization be impelled to invade the privacy and liberty of individuals? There is the great danger that teachers, clergymen, editors, civil servants and other leaders of opinion would be required to become enforcers of official policy on civil defense—in the assumed best interests of those coerced. What if some teachers, clergymen, editors and civil servants should disagree with the policy? Would they be punished for encouraging "shelter dodgers" as they would be if they incited draft dodgers? Although no such official pressures have yet been exerted, it is disquieting to hear reliable reports from New York City that some high school students who refused to participate in shelter drills have been refused recommendations for admission to college.

It is possible that the "confusion" that has marked the civil defense effort so far is in reality a kind of unconscious civilian resistance to the half-understood possibilities of enforced conformity. Civilian doubts, hesitancies and ambivalences may be having a "last fling" in anxious anticipation of the absolute unanimity and centralization that might

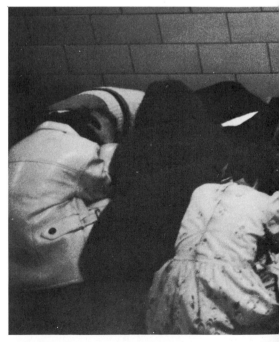

SHELTER DRILL is conducted from time to time in the Norwalk, Conn., elementary

be required if civil defense became a paramilitary organization of the entire population. Since the doubts and confusions themselves interfere with civil defense, they may themselves bring about more effort to eliminate the doubts and to control their expression more tightly. Thus the fears of centralized control could help to bring such control into force. To prevent this kind of repressive process from beginning or continuing—if it can be prevented at all—would take an understanding of the problem on the part of citizens and Government officials and a careful re-examination of the whole situation to locate the points at which the process could most effectively be halted.

These problems might be compounded if the peculiar imagery and symbolism involved in civil defense attracts particular personality types into leadership of civil defense organizations. Such special groups of people might be benign; for example, they might well include protective individuals, committed to saving, helping and nursing their fellow human beings in time of disaster. But the image of a world of death and destruction might act as a kind of "pornography of violence" to attract people to civil defense work who "want to get it over with"—who see nuclear war as a relief of intolerable tensions and as a way of "ending" international conflict—or who see themselves as survivors and rulers in a world where affluence and internal bickering had given way to pioneer exertions and tightly controlled or-

der. In fact, civil defense publications in certain localities suggest that people with these kinds of personality structure have already begun to dominate some local civil defense organizations.

The combined promise of life and warning of death put forward by civil defense involves such strains in individual hopes and fears that any failure, mistake, corruption or delay in the program might have far more basic consequences than even total collapse of an ordinary political proposal. There is a disturbing prospect that widespread disillusionment with the political leader, the scientist and the expert and even democratic government itself could grow out of a civil defense program.

Since there is a considerable "cultural lag" in translating new knowledge of weapons effects into new civil defense procedures, in transmitting the new procedures to local civil defense officials, and in putting the new procedures to work, there might exist at any time a publicly known gap between the need and the program. Such a gap might well provoke anger and disenchantment in citizens who knew that it might mean their death. The anger and disenchantment might be deep enough and broad enough to exceed the normal amount of disapproval felt by free peoples toward democratic governments.

For example, a leading businessman could be asked by his local civil defense director to make expensive alterations in his building at the very moment when the same businessman hears from his

Washington sources that these alterations are out of date and that specifications may soon be changed. Should he go ahead with the alterations or not? In either case, what would he and the local civil defense director think of the Federal Government? Since Federal officials will have recognized this difficulty, would they hesitate to keep the civil defense program up to date for fear of destroying morale? If they should decide to maintain an out-of-date program, what would happen to their own morale and self-respect?

Finally, it should be pointed out that in civil defense, as in every program in any society with any form of government, sooner or later there are sure to be mistakes, instances of corruption and so on. In most public programs people take such occasions in their stride (although in the U.S. they have sometimes led to dangerous contempt for politics and politicians). In a program that would be as deeply tied to national and personal life or death as civil defense, any fall from grace might provoke a much more serious revulsion against political leaders and possibly even against democratic politics. Similarly, the image of the scientist and the expert might suffer badly if the presumed experts disagree profoundly on what nuclear war could mean or urge a program of civil defense that is later shown to be ineffective. Since the scientist is perceived as the ranking "expert," disenchantment with him might generalize to disbelief in Government experts of all kinds. And cynicism

schools. These pupils are first-graders at the Tracey School. The children march out of their classrooms and line up along the walls of an interior corridor. They sit on the floor, head down and hands clasped behind neck for protection against a nuclear explosion.

among Government officials, if they privately see mistakes, failures, inadequacies and delays in so crucial a program, might still further sap belief in democracy and might make corruption or apathy more likely.

Of all the questions raised by civil defense, the conferees agreed, the most troublesome and dangerous is the question of how the commitment, once made, could ever be limited or reversed. A shelter-oriented civil defense system will of necessity create a large, highly organized institution with crucial connections in every area of American life. The civil defense hierarchy—Federal, state and local—will be carrying the heavy burden of training a large cadre of shelter managers and survival specialists. The trained cadres in turn will have the task of training large groups of people to follow orders in a hurry at a time of catastrophe and crisis, in spite of their anxieties, fears and lack of information. The civil defense organization will have to insist on the economic adjustments and controls necessary to implement and operate a shelter program without unsettling either long-range business investments or day-to-day business where shelters are being dug or buildings modified. If the civil defense organization should attempt to supply only shelter structures and other physical necessities, fearful of involvement in educational, psychological or economic problems, it is likely that local demands for drills, training and controls will quickly arise as the shelters themselves appear. The very existence of the civil defense system will make it a social force of great importance, powerful even if it abjures power.

As the shelter program grows, proponents and administrators of civil defense might become more preoccupied with imagined attack situations. They might feel that in the overwhelming emergency of attack it will be necessary to have fully co-ordinated action in order to use the shelters efficiently. Individuals who have not learned their places and tasks—who in panic block shelter entrances, who forget to take along essential personal medicines or who bring along too much luggage for the available space—will be endangering not only their own lives but also the lives of others. For this reason the civil defense organization will find itself more and more pressed to demand virtually universal acceptance of civil defense and preparation for it.

The social, psychological and political momentum generated by an operating civil defense system would therefore make it increasingly impossible for the nation to change its mind once the program was under way. The program would be much harder to reverse than most political decisions. One well-established social and psychological mechanism that might make civil defense irreversible is the common transformation of conflict and confusion into rigid and overwhelming commitment. The puzzlement and disagreement that will have preceded adoption of a civil defense program might bring about such a commitment to the program as it goes into effect. Having cast its doubts away and invested huge amounts of labor, capital, energy and imagination in civil defense, the nation might find it difficult to surrender the hope of survival through the new program. Once it acts for civil defense, the nation would find itself seeking constantly for new reasons to be so acting. This phenomenon—the bringing of ideas and wishes into line with action already being taken—is well known to behavioral scientists and politicians.

With the continued escalation of the destructiveness and accuracy of nuclear weapons and their delivery vehicles, it would be difficult to resist demands for constant expansion and intensification of civil defense. At any extreme—even underground cities—civil defense would be inadequate against many conceivable weapons (some already in existence) that could be brought against it. Thus at any existing level of civil defense various groups (for instance, the political opposition) might be able to demand a larger and more effective program. A system of escalating political blackmail could well develop. Such pressure would be difficult to resist if the Government had already implied that it could protect at least some people. With the pressure always in one direction, the civil defense program might be always expanding, never contracting.

The difficulty of reversing policy on civil defense can be made clearer by examination of the conditions in the past under which some major political decisions have actually been reversed by the U.S. political system. The abandonment of the National Industrial Recovery Act in the early days of the Roosevelt administration seems to have been one such reversal. But two crucial conditions accompanied that episode: it was legitimized by an institution specially assigned to make such reversals, the Supreme Court; and it was made palatable by the adoption of a series of alternative New Deal programs. The fulfillment of these two conditions in the case of civil defense would be difficult. In matters of defense policy it is hard to see how any outside institutionalized body, such as the courts, could reverse the decisions made by the President and Congress. In fact, the decisions of the President in the field of defense have since 1940 been given such weight that it is difficult to tell whether a real option for reversal would exist even in the hands of Congress. As to alternatives, it has already been suggested that disarmament, for example, could probably not mobilize as much support within a shelter-centered society as it could before shelters were begun. Alternatives may therefore be hard to promote even should it seem wise at some later date to dismantle the civil defense program.

In thus detailing the questions raised by civil defense, the conferees did not suggest that a policy of no civil defense would be without problems. The anxiety aroused by the cold war and its crises, by nuclear weapons and the images of the destruction they would deliver already exists in the U.S. To do nothing would not reduce the anxiety, and to do nothing might well lead to problems other than those herein described as possible results of implementing civil defense.

Whether the appropriate action should be a search for alternatives to civil defense or a search for ways of so managing civil defense as to lessen or eliminate its difficulties is for the nation to decide. What is essential is that the problems that might very well grow out of civil defense be examined carefully and that policy be re-evaluated so that the nation need not be confronted by a world in which the possible difficulties have become real disasters.

The conferees therefore put to the American people and its leadership these questions:

Are we prepared to accept the possibility that unhappy social and political consequences will occur if the proposed civil defense program is implemented?

If we reject the possibility that adverse consequences will develop, on what grounds do we do so and with what assurance that we are right?

If we are not prepared to accept unhappy consequences, are we prepared to recognize the difficulties consciously and apply the arduous study that would be necessary to discover if ways exist of avoiding those consequences?

If, after study, there seems to be no way of avoiding the unhappy results, do we have alternative policies in mind?

# 45

# A "Light" ABM System

ROBERT S. McNAMARA

*September 1967*

An address delivered at the Conference of United Press International Editors and Publishers, San Francisco, 18 September, 1967. Not published in *Scientific American.*

Man has lived now for more than twenty years in what we have come to call the Atomic Age. If then, man is to have a future at all, it will have to be a future overshadowed with the permanent possibility of thermonuclear holocaust. About that fact, we are no longer free. Our freedom in this question consists rather in facing the matter rationally and realistically and discussing actions to minimize the danger.

No sane citizen, no sane political leader, no sane nation wants thermonuclear war. But merely not wanting it is not enough. We must understand the difference between actions which increase its risk, those which reduce it, and those which, while costly, have little influence one way or another.

Now this whole subject matter tends to be psychologically unpleasant. But there is an even greater difficulty standing in the way of constructive and profitable debate over the issues. And that is that nuclear strategy is exceptionally complex in its technical aspects. Unless these complexities are well understood, rational discussion and decision making are simply not possible. What I want to do this afternoon is deal with these complexities and clarify them

with as much precision and detail as time and security permit.

One must begin with precise definitions. The cornerstone of our strategic policy continues to be to deter deliberate nuclear attack upon the United States, or its allies, by maintaining a highly reliable ability to inflict an unacceptable degree of damage upon any single aggressor, or combination of aggressors, at any time during the course of a strategic nuclear exchange—even after our absorbing a surprise first strike. This can be defined as our "assured destruction capability."

Now it is imperative to understand that assured destruction is the very essence of the whole deterrence concept. We must possess an actual assured destruction capability. And that actual assured destruction capability must also be credible. Conceivably, our assured destruction capability could be actual, without being credible—in which case, it might fail to deter an aggressor. The point is that a potential aggressor must himself believe that our assured destruction capability is in fact actual, and that our will to use it in retaliation to an attack is in fact unwavering. The conclusion, then, is clear: if the United States is to deter a nuclear attack on itself or on our allies, it must possess an actual, and a credible assured destruction capability.

When calculating the force we require, we must be "conservative" in all our estimates of both a potential aggres-

sor's capabilities, and his intentions. Security depends upon taking a "worst plausible case"—and having the ability to cope with that eventuality. In that eventuality, we must be able to absorb the total weight of nuclear attack on our country—on our strike-back forces; on our command and control apparatus; on our industrial capacity; on our cities; and on our population—and still, be fully capable of destroying the aggressor to the point that his society is simply no longer viable in any meaningful twentieth-century sense. That is what deterrence to nuclear aggression means. It means the certainty of suicide to the aggressor—not merely to his military forces, but to his society as a whole.

Now let us consider another term: "first-strike capability." This, in itself, is an ambiguous term, since it could mean simply the ability of one nation to attack another nation with nuclear forces first. But as it is normally used, it connotes much more: the substantial elimination of the attacked nation's retaliatory second-strike forces. This is the sense in which "first-strike capability" should be understood.

Now, clearly, such a first-strike capability is an important strategic concept. The United States cannot—and will not—ever permit itself to get into the position in which another nation, or combination of nations, would possess such a first-strike capability, which could be effectively used against it. To get into such a position vis-à-vis any other nation or nations would not only constitute an intolerable threat to our security, but it would obviously remove our ability to deter nuclear aggression—both against ourselves and against our allies.

Now, we are not in that position today—and there is no foreseeable danger of our ever getting into that position. Our strategic offensive forces are immense: 1000 Minutemen missile launchers, carefully protected below ground, 41 Polaris submarines, carrying 656 missile launchers—with the majority of these hidden beneath the seas at all times; and about 600 long-range bombers, approximately forty percent of which are kept always in a high state of alert. Our alert forces alone carry more than 2200 weapons, averaging more than one megaton each. A mere 400 one-megaton weapons, if delivered on the Soviet Union, would be sufficient to destroy over one-third of her population, and one-half of her industry. And all of these flexible and highly reliable forces are equipped with devices that insure their penetration of Soviet defenses.

Now what about the Soviet Union? Does it today possess a powerful nuclear arsenal? The answer is that it does.

Does it possess a first-strike capability against the United States? The answer is that it does not.

Can the Soviet Union in the foreseeable future, acquire such a first-strike capability against the United States? The answer is that it cannot. It cannot because we are determined to remain fully alert, and we will never permit our own assured destruction capability to be at a point where a Soviet first strike capability is even remotely feasible.

Is the Soviet Union seriously attempting to acquire a first-strike capability against the United States? Although this is a question we cannot answer with absolute certainty, we believe the answer is no. In any event, the question itself is—in a sense—irrelevant. It is irrelevant since the United States will so continue to maintain—and where necessary strengthen—our retaliatory forces, that whatever the Soviet Union's intentions or actions, we will continue to have an assured destruction capability vis-à-vis their society in which we are completely confident.

But there is another question that is most relevant. And that is, do we—the United States—possess a first-strike capability against the Soviet Union? The answer is that we do not. And we do not, not because we have neglected our nuclear strength. On the contrary, we have increased it to the point that we possess a clear superiority over the Soviet Union. We do not possess first-strike capability against the Soviet Union for precisely the same reason that they do not possess it against us. And that is that we have both built up our "second-strike capability"* to the point that a first-strike capability on either side has become unattainable.

There is, of course, no way in which the United States could have prevented the Soviet Union from acquiring its present second strike capability—short of a massive pre-emptive strike on the Soviet Union in the 1950s. The blunt fact is, then, that neither the Soviet Union nor the United States can attack the other without being destroyed in retaliation; nor can either of us attain a first-strike capability in the foreseeable future. The further fact is that both the Soviet Union and the United States presently possess an actual and credible second-strike capability against one another—and it is precisely this mutual capability that provides us both with the strongest possible motive to avoid a nuclear war.

The more frequent question that arises in this connection is whether or not the United States possesses nuclear superiority over the Soviet Union. The answer is that we do. But the answer is—like everything else in this matter—technically complex.

The complexity arises in part out of what measurement of superiority is most meaningful and realistic. Many commentators on the matter tend to define nuclear superiority in terms of gross megatonnage, or in terms of the number of missile launchers available. Now, by both these two standards of measurement, the United States does have a substantial superiority over the Soviet Union in the weapons targeted against each other. But it is precisely these two standards of measurement that are themselves misleading. For the most meaningful and realistic measurement of nuclear capability is neither gross megatonnage, nor the number of available missile launchers; but rather the number of separate warheads that are capable of being *delivered* with accuracy on individual high-priority targets with sufficient power to destroy them.

Gross megatonnage in itself is an inadequate indicator of assured destruction capability, since it is unrelated to survivability, accuracy, or penetrability, and poorly related to effective elimination of multiple high-priority targets. There is manifestly no advantage in over-destroying one target, at the expense of leaving undamaged other targets of equal importance. Further, the number of missile launchers available is also an inadequate indicator of assured destruction capability, since the fact is that many of our launchers will carry multiple warheads. But by using the realistic measurement of the number of warheads available, capable of being reliably delivered with accuracy and effectiveness on the appropriate targets in the United States or Soviet Union, I can tell you that the United States currently possesses a superiority over the Soviet Union of at least three or four to one.

Furthermore, we will maintain a superiority—by these same realistic criteria—over the Soviet Union for as far ahead in the future as we can realistically plan. I want, however, to make one point patently clear: our current numerical superiority over the Soviet Union in reliable, accurate, and effective warheads is both greater than we had originally planned, and is in fact more than we require.

Moreover, in the larger equation of security, our "superiority" is of limited significance—since even with our current superiority, or indeed with any numerical superiority realistically attainable, the blunt, inescapable fact remains that the Soviet Union could still—with its present forces—effectively destroy the United States, even after absorbing the full weight of an American first strike. I have noted that our present superiority is greater than we had planned. Let me explain to you how this came about, for I think it is a significant illustration of the intrinsic dynamics of the nuclear arms race.

In 1961, when I became Secretary of Defense, the Soviet Union possessed a very small operational arsenal of intercontinental missiles. However, they did possess the technological and industrial capacity to enlarge that arsenal very substantially over the succeeding several years. Now, we had no evidence that the Soviets did in fact plan to fully use that capability.

But as I have pointed out, a strategic planner must be "conservative" in his calculations; that is, he must prepare for the worst plausible case and not be content to hope and prepare merely for the most probable. Since we could not be certain of Soviet intentions—since we could not be sure that they would not undertake a massive build-up—we had

---

* A "second-strike capability" is the capability to absorb a surprise nuclear attack, and survive with sufficient power to inflict unacceptable damage on the aggressor.

to insure against such an eventuality by undertaking ourselves a major build-up of the Minuteman and Polaris forces. Thus, in the course of hedging against what was then only a theoretically possible Soviet build-up, we took decisions which have resulted in our current superiority in numbers of warheads and deliverable megatons.

But the blunt fact remains that if we had had more accurate information about planned Soviet strategic forces, we simply would not have needed to build as large a nuclear arsenal as we have today.

Now let me be absolutely clear. I am not saying that our decision in 1961 was unjustified. I am simply saying that it was necessitated by a lack of accurate information. Furthermore, that decision in itself—as justified as it was—in the end, could not possibly have left unaffected the Soviet Union's future nuclear plans. What is essential to understand here is that the Soviet Union and the United States mutually influence one another's strategic plans.

Whatever be their intentions, whatever be our intentions, actions—or even realistically potential actions—on either side relating to the build-up of nuclear forces, be they either offensive or defensive weapons, necessarily trigger reactions on the other side. It is precisely this action-reaction phenomenon that fuels an arms race.

Now, in strategic nuclear weaponry, the arms race involves a particular irony. Unlike any other era in military history, today a substantial numerical superiority of weapons does not effectively translate into political control, or diplomatic leverage. While thermonuclear power is almost inconceivably awesome, and represents virtually unlimited potential destructiveness, it has proven to be a limited diplomatic instrument. Its uniqueness lies in the fact that it is at one and the same time, an all powerful weapon—and a very inadequate weapon. The fact that the Soviet Union and the United States can mutually destroy one another—regardless of who strikes first—narrows the range of Soviet aggression which our nuclear forces can effectively deter. Even with our nuclear monopoly in the early postwar period, we were unable to deter the Soviet pressures against Berlin, or their support of aggression in Korea. Today, our nuclear superiority does not deter all forms of Soviet support of communist insurgency in Southeast Asia.

What all of this has meant is that we, and our allies as well, require substantial non-nuclear forces in order to cope with levels of aggression that massive strategic forces do not in fact deter. This has been a difficult lesson both for us and for our allies to accept, since there is a strong psychological tendency to regard superior nuclear forces as a simple and unfailing solution to security, and an assurance of victory under any set of circumstances.

What is important to understand is that our nuclear strategic forces play a vital and absolutely necessary role in our security and that of our allies, but it is an intrinsically limited role. Thus, we and our allies must maintain substantial conventional forces, fully capable of dealing with a wide spectrum of lesser forms of political and military aggression—a level of aggression against which the use of strategic nuclear forces would not be to our advantage, and thus a level of aggression which these strategic nuclear forces by themselves cannot effectively deter. One cannot fashion a credible deterrent out of an incredible action. Therefore security for the United States and its allies can only arise from the possession of a whole range of graduated deterrents, each of them fully credible in its own context.

Now I have pointed out that in strategic nuclear matters, the Soviet Union and the United States mutually influence one another's plans. In recent years the Soviets have substantially increased their offensive forces. We have, of course, been watching and evaluating this very carefully. Clearly, the Soviet build-up is in part a reaction to our own build-up since the beginning of this decade. Soviet strategic planners undoubtedly reasoned that if our build-up were to continue at its accelerated pace, we might conceivably reach, in time, a credible first-strike capability against the Soviet Union.

That was not in fact our intention. Our intention was to assure that they—with their theoretical capacity to reach such a first-strike capability—would not in fact outdistance us. But they could not read our intentions with any greater accuracy than we could read theirs. And thus the result has been that we have both built up our forces to a point that far exceeds credible second-strike capability against the forces we each started with. In doing so, neither of us has reached a first-strike capability. And the realities of the situation being what they are—whatever we believe their intentions to be, and whatever they believe our intentions to be—each of us can deny the other a first-strike capability in the foreseeable future.

Now, how can we be so confident that this is the case? How can we be so certain that the Soviets cannot gradually outdistance us—either by some dramatic technological break-through, or simply through our imperceptibly lagging behind, for whatever reason: reluctance to spend the requisite funds; distraction with military problems elsewhere; faulty intelligence; or simple negligence and naivete? All of these reasons—and others —have been suggested by some commentators in this country, who fear that we are in fact falling behind to a dangerous degree.

The answer to all of this is simple and straightforward. We are not going to permit the Soviets to outdistance us, because to do so would be to jeopardize our very viability as a nation. No President, no Secretary of Defense, no Congress of the United States—of whatever political party, and of whatever political persuasion—is going to permit this nation to take that risk.

We do not want a nuclear arms race with the Soviet Union—primarily because the action-reaction phenomenon makes it foolish and futile. But if the only way to prevent the Soviet Union from obtaining first-strike capability over us is to engage in such a race, the United States possesses in ample abundance the resources, the technology, and the will to run faster in that race for whatever distance is required.

But what we would much prefer to do is to come to a realistic and reasonably riskless agreement with the Soviet Union, which would effectively prevent such an arms race. We both have strategic nuclear arsenals greatly in excess of a credible assured destruction capability. These arsenals have reached that point of excess in each case for precisely the same reason: we each have reacted to the other's build-up with very conservative calculations. We have, that is, each built a greater arsenal than either of us needed for a second-strike capability, simply because we each wanted to be able to cope with the "worst plausible case."

But since we now each possess a deterrent in excess of our individual needs, both of our nations would benefit from a properly safe-guarded agreement first to limit, and later to reduce, both our offensive and defensive strategic nuclear forces. We may, or we may not, be able to achieve such an agreement. We hope we can. And we believe such an agreement is fully feasible, since it is clearly in both our nations' interests.

But reach the formal agreement or not, we can be sure that neither the Soviets nor we are going to risk the other obtaining a first-strike capability. On the contrary, we can be sure that we are both going to maintain a maximum effort to preserve an assured destruction capability. It would not be sensible for either side to launch a maximum effort to achieve a first-strike capability. It would not be sensible because the intelligence-gathering capability of each side being what it is, and the realities of lead-time from technological breakthrough to operational readiness being what they are, neither of us would be able to acquire a first-strike capability in secret.

Now, let me take a specific case in point. The Soviets are now deploying an anti-ballistic missile system. If we react to this deployment intelligently, we have no reason for alarm. The system does not impose any threat to our ability to penetrate and inflict massive and unacceptable damage on the Soviet Union. In other words, it does not presently affect in any significant manner our assured destruction capability. It does not impose such a threat because we have already taken the steps necessary to assure that our land-based Minuteman missiles, our nuclear submarine-launched new Poseidon missiles, and our strategic bomber forces have the requisite penetration aids—and in the sum, constitute a force of such magnitude, that they guarantee us a force strong enough to survive a Soviet attack and penetrate the Soviet ABM deployment.

Now let me come to the issue that has received so much attention recently: the question of whether or not we should deploy an ABM system against the Soviet nuclear threat. To begin with, this is not in any sense a new issue. We have had both the technical possibility and the strategic desirability of an American ABM deployment under constant review since the late 1950s. While we have substantially improved our technology in the field, it is important to understand that none of the systems at the present or foreseeable state of the art would provide an impenetrable shield over the United States. Were such a shield possible, we would certainly want it—and we would certainly build it.

And at this point, let me dispose of an objection that is totally irrelevant to this issue. It has been alleged that we are opposed to deploying a large-scale ABM system because it would carry the heavy price tag of $40 billion. Let me make it very clear that the $40 billion is not the issue. If we could build and deploy a genuinely impenetrable shield over the United States, we would be willing to spend not $40 billion, but any reasonable multiple of that amount that was necessary.

The money in itself is not the problem: the penetrability of the proposed shield is the problem. There is clearly no point, however, in spending $40 billion if it is not going to buy us a significant improvement in our security. If it is not, then we should use the substantial resources it represents on something that will.

Every ABM system that is now feasible involves firing defensive missiles at incoming offensive warheads in an effort to destroy them. But what many commentators on this issue overlook is that any such system can rather obviously be defeated by an enemy simply sending more offensive warheads, or dummy warheads, than there are defensive missiles capable of disposing of them. And this is the whole crux of the nuclear action-reaction phenomenon. Were we to deploy a heavy ABM system throughout the United States, the Soviets would clearly be strongly motivated to so increase their offensive capability as to cancel out our defensive advantage.

It is futile for each of us to spend $4 billion, $40 billion, or $400 billion—and at the end of all the spending, and at the end of all the deployment, and at the end of all the effort, to be relatively at the same point of balance on the security scale that we are now.

In point of fact, we have already initiated offensive weapons programs costing several billions in order to offset the small present Soviet ABM deployment, and the possibly more extensive future Soviet ABM deployments. That is money well spent; and it is necessary. But we should bear in mind that it is money spent because of the action-reaction phenomenon. If we in turn opt for heavy ABM deployment—at whatever price—we can be certain that the Soviets will react to offset the advantage we would hope to gain.

It is precisely because of this certainty of a corresponding Soviet reaction that the four prominent scientists—men who have served with distinction as the Science Advisors to Presidents Eisenhower, Kennedy, and Johnson, and the three outstanding men who have served as Directors of Research and Engineering to three Secretaries of Defense—have unanimously recommended aginst the deployment of an ABM system designed to protect our population against a Soviet attack. These men are Doctors Killian, Kistiakowsky, Wiesner, Hornig, York, Brown, and Foster.

The plain fact of the matter is that we are now facing a situation analogous to the one we faced in 1961: we are uncertain of the Soviets' intentions. At that time we were concerned about their potential offensive capabilities; now we are concerned about their potential defensive capabilities. But the dynamics of the concern are the same. We must continue to be cautious and conservative in our estimates—leaving no room in our calculations for unnecessary risk. And at the same time, we must measure our own response in such a manner that it does not trigger a senseless spiral upward of nuclear arms.

Now, as I have emphasized, we have already taken the necessary steps to guarantee that our offensive strategic weapons will be able to penetrate future, more advanced, Soviet defenses. Keeping in mind the careful clockwork of lead-time, we will be forced to continue that effort over the next few years if the evidence is that the Soviets intend to turn what is now a light and modest ABM deployment into a massive one. Should they elect to do so, we have both the lead-time and the technology available to so increase both the quality and quantity of our offensive strategic forces—with particular attention to highly reliable penetration aids—that their expensive defensive efforts will give them no edge in the nuclear balance whatever.

But we would prefer not to have to do that. For it is a profitless waste of resources, provided we and the Soviets can come to a realistic strategic arms-limitation agreement. As you know, we have proposed U.S.-Soviet talks on this matter. Should these talks fail, we are fully prepared to take the appropriate measures that such a failure would make necessary. The point for us to keep in mind is that should the talks fail—and the Soviets decide to expand their present modest ABM deployment into a massive one—our response must be realistic. There is no point whatever in our responding by going to a massive ABM deployment to protect our population, when such a system would be ineffective against a sophisticated Soviet offense.

Instead, realism dictates that if the Soviets elect to deploy a heavy ABM system, we must further expand our sophisticated offensive forces, and thus preserve our overwhelming assured destruction capability. But the intractable fact is that should the talks fail, both the Soviets and ourselves would be forced to continue on a foolish and reckless course. It would be foolish and reckless because—in the end—it would provide neither the Soviets, nor us, with any greater relative nuclear capability. The time has come for us both to realize that, and to act reasonably. It is clearly in our own mutual interest to do so.

Having said that, it is important to distinguish between an ABM system designed to protect against a Soviet attack on our cities, and ABM systems which have other objectives. One of the other uses of an ABM system which we should seriously consider is the greater protection of our strategic offensive forces. Another is in relation to the emerging nuclear capability of Communist China.

There is evidence that the Chinese are devoting very substantial resources to the development of both nuclear warheads, and missile delivery systems. As I stated last January, indications are that they will have medium-range ballistic missiles within a year or so, an initial intercontinental ballistic missile capability in the early 1970s, and a modest force in the mid-70s. Up to now, the lead-time factor has allowed us to postpone a decision on whether or not a light ABM deployment might be advantageous as a countermeasure to Communist China's nuclear development. But the time will shortly be right for us to initiate production if we desire such a system.

China at the moment is caught up in internal strife, but it seems likely that her basic motivation is developing a strategic nuclear capability is an attempt to provide a basis for threatening her neighbors, and to clothe herself with the dubious prestige that the world pays to nuclear weaponry. We deplore her development of these weapons, just as we deplore it in other countries. We oppose nuclear proliferation because we believe that in the end it only increases the risk of a common and cataclysmic holocaust.

President Johnson has made it clear that the United States will oppose any efforts of China to employ nuclear blackmail against her neighbors. We possess now, and will continue to possess for as far ahead as we can foresee, an overwhelming first-strike capability with respect to China. And despite the shrill and raucous propaganda directed at her own people that "the atomic bomb is a paper tiger," there is ample evidence that China well appreciates the destructive power of nuclear weapons. China has been cautious to avoid any action that might end in a nuclear clash with the United States—however wild her words—and understandably so. We have the power not only to destroy completely her entire nuclear offensive forces, but to devastate her society as well.

Is there any possibility, then, that by the mid-1970s China might become so incautious as to attempt a nuclear attack on the United States or our allies? It would be insane and suicidal for her to do so, but one can conceive conditions under which China might miscalculate. We wish to reduce such possibilities to a minimum. And since, as I have noted, our strategic planning must always be conservative, and take into consideration even the possible irrational behavior of potential adversaries, there are marginal grounds for concluding that a light deployment of U.S. ABMs against this possibility is prudent.

The system would be relatively inexpensive—preliminary estimates place the cost at about $5 billion—and would have a much higher degree of reliability against a Chinese attack, than the much more massive and complicated system that some have recommended against a possible Soviet attack. Moreover, such an ABM deployment designed against a possible Chinese attack would have a number of other advantages. It would provide an additional indication to Asians that we intend to deter China from nuclear blackmail, and thus would contribute toward our goal of discouraging nuclear weapon proliferation among the present non-nuclear countries.

Further, the Chinese-oriented ABM deployment would enable us to add—as a concurrent benefit—a further defense of our Minuteman sites against Soviet attack, which means that at modest cost we would in fact be adding even greater effectiveness to our offensive missile force and avoiding a much more costly expansion of that force.

Finally, such a reasonably reliable ABM system would add protection of our population against the improbable but possible accidental launch of an intercontinental missile by any one of the nuclear powers.

After a detailed review of all these considerations, we have decided to go forward with this Chinese-oriented ABM deployment, and we will begin actual production of such a system at the end of this year. In reaching this decision, I want to emphasize that it contains two possible dangers—and we should guard carefully against each.

The first danger is that we may psychologically lapse into the old oversimplification about the adequacy of nuclear power. The simple truth is that nuclear weapons can serve to deter only a narrow range of threats. This ABM deployment will strengthen our defensive posture—and will enhance the effectiveness of our land-based ICBM offensive forces. But the independent nations of Asia must realize that these benefits are no substitute for their maintaining, and where necessary strengthening, their own conventional forces in order to deal with the more likely threats to the security of the region.

The second danger is also psychological. There is a kind of mad momentum intrinsic to the development of all new nuclear weaponry. If a weapon system works—and works well—there is strong pressure from many directions to procure and deploy the weapon out of all proportion to the prudent level required.

The danger in deploying this relatively light and reliable Chinese-oriented ABM system is going to be that pressures will develop to expand it into a heavy Soviet-oriented ABM system.

We must resist that temptation firmly —not because we can for a moment afford to relax our vigilance against a possible Soviet first-strike—but precisely because our greatest deterrent against such a strike is not a massive, costly, but highly penetrable ABM shield, but rather a fully credible offensive assured destruction capability.

The so-called heavy ABM shield—at the present state of technology—would in effect be no adequate shield at all against a Soviet attack, but rather a strong inducement for the Soviets to vastly increase their own offensive forces. That, as I have pointed out, would make it necessary for us to respond in turn—and so the arms race would rush hopelessly on to no sensible purpose on either side.

Let me emphasize—and I cannot do so too strongly—that our decision to go ahead with a *limited* ABM deployment in no way indicates that we feel an agreement with the Soviet Union on the limitation of strategic nuclear offensive and defensive forces is any the less urgent or desirable.

The road leading from the stone axe to the ICBM—though it may have been more than a million years in the building—seems to have run in a single direction. If one is inclined to be cynical, one might conclude that man's history seems to be characterized not so much by consistent periods of peace, occasionally punctuated by warfare; but rather by persistent outbreaks of warfare, wearily put aside from time to time by periods of exhaustion and recovery—that parade under the name of peace.

I do not view man's history with that degree of cynicism, but I do believe that man's wisdom in avoiding was is often surpassed by his folly in promoting it. However foolish unlimited war may have been in the past, it is now no longer merely foolish, but suicidal as well.

It is said that nothing can prevent a man from suicide, if he is sufficiently determined to commit it. The question is what is our determination in an era when unlimited war will mean the death of hundreds of millions—and the possible genetic impairment of a million generations to follow? Man is clearly a compound of folly and wisdom—and history is clearly a consequence of the admixture of those two contradictory traits. History has placed our particular lives in an era when the consequences of human folly are waxing more and more catastrophic in the matters of war and peace.

In the end, the root of man's security does not lie in his weaponry. In the end, the root of man's security lies in his mind.

What the world requires in its 22nd Year of the Atomic Age is not a new race towards armament. What the world requires in its 22nd Year of the Atomic Age is a new race towards reasonableness. We had better all run that race.

Not merely we the administrators. But we the people.

# 46

# Anti-Ballistic-Missile Systems

RICHARD L. GARWIN and HANS A. BETHE

*March 1968*

Last September, Secretary of Defense McNamara announced that the U.S. would build "a relatively light and reliable Chinese-oriented ABM system." With this statement he apparently ended a long and complex debate on the merits of any kind of anti-ballistic-missile system in an age of intercontinental ballistic missiles carrying multimegaton thermonuclear warheads. Secretary McNamara added that the U.S. would "begin actual production of such a system at the end of this year," meaning the end of 1967.

As two physicists who have been concerned for many years with the development and deployment of modern nuclear weapons we wish to offer some comments on this important matter. On examining the capabilities of ABM systems of various types, and on considering the stratagems available to a determined enemy who sought to nullify the effectiveness of such a system, we have come to the conclusion that the "light" system described by Secretary McNamara will add little, if anything, to the influences that should restrain China indefinitely from an attack on the U.S. First among these factors is China's certain knowledge that, in McNamara's words, "we have the power not only to destroy completely her entire nuclear offensive forces but to devastate her society as well."

An even more pertinent argument against the proposed ABM system, in our view, is that it will nourish the illusion that an effective defense against ballistic missiles is possible and will lead almost inevitably to demands that the light system, the estimated cost of which exceeds $5 billion, be expanded into a heavy system that could cost upward of $40 billion. The folly of undertaking to build such a system was vigorously stated by Secretary McNamara. "It is important to understand," he said, "that none of the [ABM] systems at the present or foreseeable state of the art would provide an impenetrable shield over the United States.... Let me make it very clear that the [cost] in itself is not the problem: the penetrability of the proposed shield is the problem."

In our view the penetrability of the light, Chinese-oriented shield is also a problem. It does not seem credible to us that, even if the Chinese succumbed to the "insane and suicidal" impulse to launch a nuclear attack on the U.S. within the next decade, they would also be foolish enough to have built complex and expensive missiles and nuclear warheads peculiarly vulnerable to the light ABM system now presumably under construction (a system whose characteristics and capabilities have been well publicized). In the area of strategic weapons a common understanding of the major elements and technical possibilities is essential to an informed and reasoned choice by the people, through their government, of a proper course of action. In this article we shall outline in general terms, using nonsecret information, the techniques an enemy could employ at no great cost to reduce the effectiveness of an ABM system even more elaborate than the one the Chinese will face. First, however, let us describe that system.

Known as the Sentinel system, it will provide for long-range interception by Spartan antimissile missiles and short-range interception by Sprint antimissile missiles. Both types of missile will be armed with thermonuclear warheads for the purpose of destroying or inactivating the attacker's thermonuclear weapons, which will be borne through the atmosphere and to their targets by reentry vehicles (RV's). The Spartan missiles, whose range is a few hundred kilometers, will be fired when an attacker's reentry vehicles are first detected rising above the horizon by perimeter acquisition radar (PAR).

If the attacker is using his available propulsion to deliver maximum payload, his reentry vehicles will follow a normal minimum-energy trajectory, and they will first be sighted by one of the PAR's when they are about 4,000 kilometers, or about 10 minutes, away [*see illustration, page 348*]. If the attacker chooses to launch his rockets with less than maximum payload, he can put them either in a lofted trajectory or in a depressed one. The lofted trajectory has certain advantages against a terminal defense system. The most extreme example of a depressed trajectory is the path followed by a low-orbit satellite. On such a trajectory a reentry vehicle could remain below an altitude of 160 kilometers and would not

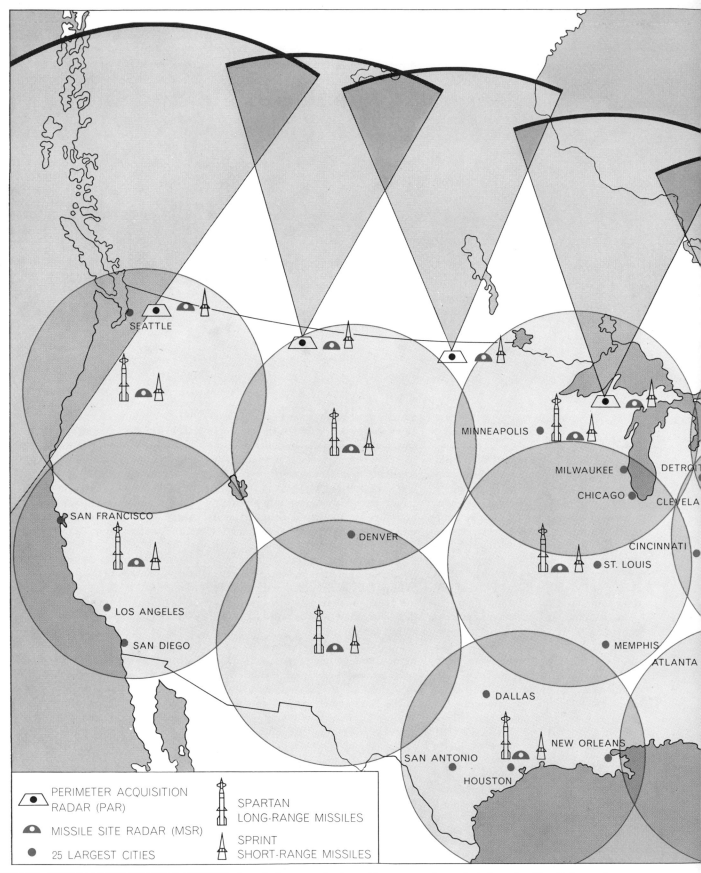

PERIMETER ACQUISITION RADAR (PAR)

MISSILE SITE RADAR (MSR)

25 LARGEST CITIES

SPARTAN LONG-RANGE MISSILES

SPRINT SHORT-RANGE MISSILES

**SENTINEL ANTI-BALLISTIC-MISSILE SYSTEM,** described as a "relatively light and reliable Chinese-oriented ABM system," is now under construction at an estimated cost exceeding $5 billion. Designed to defend the entire U.S., Sentinel will depend on perhaps six perimeter acquisition radars (PAR's) along the country's borders to detect enemy missiles as they come over the northern horizon. The arcs at the end of the radar "fans" show where an enemy reentry vehicle (RV) would be detected if it were in a low, satellite-like orbit. The PAR's will alert Spartan interceptors located at some 10 or a dozen sites around the U.S. The sites shown on the map are not actual ones but indicate how the U.S. could be covered by a pattern of 10 Spartan sites, assuming that the Spar-

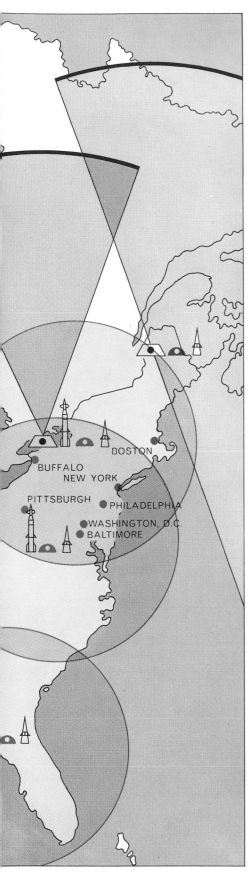

tans have an effective range of 600 kilometers. Each Spartan site will be protected by short-range Sprint missiles and will include missile-site radar (MSR) to help guide both types of missiles. Sprints and MSR's will also guard the PAR installations.

be visible to the horizon-search radar until it was some 1,400 kilometers, or about three minutes, away. This is FOBS: the fractional-orbit bombardment system, which allows intercontinental ballistic missiles to deliver perhaps 50 to 75 percent of their normal payload.

In the Sentinel system Spartans will be launched when PAR has sighted an incoming missile; they will be capable of intercepting the missile at a distance of several hundred kilometers. To provide a light shield for the entire U.S. about half a dozen PAR units will be deployed along the northern border of the country to detect missiles approaching from the general direction of the North Pole [see illustration at left]. Each PAR will be linked to several "farms" of long-range Spartan missiles, which can be hundreds of kilometers away. Next to each Spartan farm will be a farm of Sprint missiles together with missile-site radar (MSR), whose function is to help guide both the Spartans and the shorter-range Sprints to their targets. The task of the Sprints is to provide terminal protection for the important Spartans and MSR's. The PAR's will also be protected by Sprints and thus will require MSR's nearby.

Whereas the Spartans are expected to intercept an enemy missile well above the upper atmosphere, the Sprints are designed to be effective within the atmosphere, at altitudes below 35 kilometers. The explosion of an ABM missile's thermonuclear warhead will produce a huge flux of X rays, neutrons and other particles, and within the atmosphere a powerful blast wave as well. We shall describe later how X rays, particles and blast can incapacitate a reentry vehicle.

Before we consider in detail the capabilities and limitations of ABM systems, one of us (Garwin) will briefly summarize the present strategic position of the U.S. The primary fact is that the U.S. and the U.S.S.R. can annihilate each other as viable civilizations within a day and perhaps within an hour. Each can at will inflict on the other more than 120 million immediate deaths, to which must be added deaths that will be caused by fire, fallout, disease and starvation. In addition more than 75 percent of the productive capacity of each country would be destroyed, regardless of who strikes first. At present, therefore, each of the two countries has an assured destruction capability with respect to the other. It is usually assumed that a nation faced with the assured destruction of 30 percent of its population and pro-

ductive capacity will be deterred from destroying another nation, no matter how serious the grievance. Assured destruction is therefore not a very flexible political or military tool. It serves only to preserve a nation from complete destruction. More conventional military forces are needed to fill the more conventional military role.

Assured destruction was not possible until the advent of thermonuclear weapons in the middle 1950's. At first, when one had to depend on aircraft to deliver such weapons, destruction was not really assured because a strategic air force is subject to surprise attack, to problems of command and control and to attrition by the air defenses of the other side. All of this was changed by the development of the intercontinental ballistic missile and also, although to a lesser extent, by modifications of our B-52 force that would enable it to penetrate enemy defenses at low altitude. There is no doubt today that the U.S.S.R. and the U.S. have achieved mutual assured destruction.

The U.S. has 1,000 Minuteman missiles in hardened "silos" and 54 much larger Titan II missiles. In addition we have 656 Polaris missiles in 41 submarines and nearly 700 long-range bombers. The Minutemen alone could survive a surprise attack and achieve assured destruction of the attacker. In his recent annual report the Secretary of Defense estimated that as of October, 1967, the U.S.S.R. had some 720 intercontinental ballistic missiles, about 30 submarine-launched ballistic missiles (excluding many that are airborne rather than ballistic) and about 155 long-range bombers. This force provides assured destruction of the U.S.

Secretary McNamara has also stated that U.S. forces can deliver more than 2,000 thermonuclear weapons with an average yield of one megaton, and that fewer than 400 such weapons would be needed for assured destruction of a third of the U.S.S.R.'s population and three-fourths of its industry. The U.S.S.R. would need somewhat fewer weapons to achieve the same results against the U.S.

It is worth remembering that intercontinental missiles and nuclear weapons are not the only means of mass destruction. They are, however, among the most reliable, as they were even when they were first made in the 1940's and 1950's. One might build a strategic force somewhat differently today, but the U.S. and the U.S.S.R. have no incentive for doing so. In fact, the chief virtue of assured destruction may be that it removes the need to race—there is no reward for

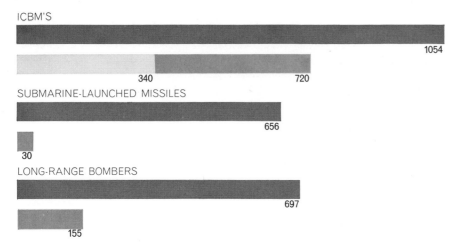

ICBM'S

1054

340    720

SUBMARINE-LAUNCHED MISSILES

656

30

LONG-RANGE BOMBERS

697

155

COMPARISON OF NUCLEAR WEAPON CARRIERS shows that the U.S. (*gray bars*) out-weighs the U.S.S.R. in every category. These figures, representing U.S. intelligence esti-mates as of October 1, 1967, appear in the Secretary of Defense's latest annual report. The report notes that the number of Soviet ICBM's had increased from 340 a year earlier. Of the 1,054 U.S. ICBM's, 1,000 are Minutemen and 54 are Titan II's. The 30 submarine-launched ballistic missiles credited to the U.S.S.R. are believed to have a range considerably less than the range of some 2,500 kilometers for the 656 Polaris missiles aboard 41 nuclear-powered U.S. submarines. The report states that the combined U.S. force could deliver up to 4,500 nuclear warheads compared with about 1,000 larger warheads for the U.S.S.R. force.

getting ahead. One really should not worry too much about new means for delivering nuclear weapons (such as bombs in orbit or fractional-orbit sys-tems) or about advances in chemical or biological warfare. A single thermonu-clear assured-destruction force can de-ter such novel kinds of attack as well.

Now, as Secretary McNamara stated in his September speech, our defense ex-perts reckoned conservatively six to 10 years ago, when our present strategic-force levels were planned. The result is that we have right now many more mis-siles than we need for assured destruc-tion of the U.S.S.R. If war comes, there-fore, the U.S. will use the excess force in a "damage-limiting" role, which means firing the excess at those elements of the Russian strategic force that would do the most damage to the U.S. Inas-much as the U.S.S.R. has achieved the level of assured destruction, this action will not preserve the U.S., but it should reduce the damage, perhaps sparing a small city here or there or reducing somewhat the forces the U.S.S.R. can use against our allies. To the extent that this damage-limiting use of our forces reduces the damage done to the U.S.S.R. it may slightly reduce the deterrent ef-fect resulting from assured destruction. It must be clear that only surplus forces will be used in this way. It should be said, however, that the exact level of casualties and industrial damage re-quired to destroy a nation as a viable society has been the subject of surpris-ingly little research or even argument.

One can conceive of three threats to the present rather comforting situation of mutual assured destruction. The first would be an effective counterforce sys-tem: a system that would enable the U.S. (or the U.S.S.R.) to incapacitate the other side's strategic forces before they could be used. The second would be an effective ballistic-missile defense com-bined with an effective antiaircraft sys-tem. The third would be a transition from a bipolar world, in which the U.S. and the U.S.S.R. alone possess over-whelming power, to a multipolar world including, for instance, China. Such threats are of course more worrisome in combination than individually.

American and Russian defense plan-ners are constantly evaluating less-than-perfect intelligence to see if any or all of these threats are developing. For pur-poses of discussion let us ask what re-sponses a White side might make to various moves made by a Black side. Assume that Black has threatened to negate White's capability of assured de-struction by doing one of the following things: (1) it has procured more inter-continental missiles, (2) it has installed some missile defense or (3) it has built up a large operational force of missiles each of which can attack several targets, using "multiple independently target-able reentry vehicles" (MIRV's).

White's goal is to maintain assured destruction. He is now worried that Black may be able to reduce to a dan-gerous level the number of White war-heads that will reach their target.

White's simplest response to all three threats—but not necessarily the most ef-fective or the cheapest—is to provide himself with more launch vehicles. In addition, in order to meet the first and third threats White will try to make his launchers more difficult to destroy by one or more of the following means: by making them mobile (for example by placing them in submarines or on rail-road cars), by further hardening their permanent sites or by defending them with an ABM system.

Another possibility that is less often discussed would be for White to arrange to fire the bulk of his warheads on "eval-uation of threat." In other words, White could fire his land-based ballistic mis-siles when some fraction of them had al-ready been destroyed by enemy war-heads, or when an overwhelming attack is about to destroy them. To implement such a capability responsibly requires excellent communications, and the deci-sion to fire would have to be made with-in minutes, leading to the execution of a prearranged firing plan. As a complete alternative to hardening and mobility, this fire-now-or-never capability would lead to tension and even, in the event of an accident, to catastrophe. Still, as a supplemental capability to ease fears of effective counterforce action, it may have some merit.

White's response to the second threat —an increase in Black's ABM defenses— might be limited to deploying more launchers, with the simple goal of satu-rating and exhausting Black's defenses. But White would also want to consider the cost and effectiveness of the follow-ing: penetration aids, concentrating on undefended or lightly defended targets, maneuvering reentry vehicles or multi-ple reentry vehicles. The last refers to several reentry vehicles carried by the same missile; the defense would have to destroy all of them to avoid damage. Finally, White could reopen the ques-tion of whether he should seek assured destruction solely by means of missiles. For example, he might reexamine the effectiveness of low-altitude bombers or he might turn his attention to chemical or biological weapons. It does not much matter how assured destruction is achieved. The important thing, as Sec-retary McNamara has emphasized, is that the other side find it credible. ("The point is that a potential aggressor must himself believe that our assured destruc-tion capability is in fact actual, and that our will to use it in retaliation to an at-tack is in fact unwavering.")

It is clear that White has many op-tions, and that he will choose those that are most reliable or those that are cheapest for a given level of assured destruction. Although relative costs do depend on the level of destruction required, the important technical conclusion is that for conventional levels of assured destruction it is considerably cheaper for White to provide more offensive capability than it is for Black to defend his people and industry against a concerted strike.

As an aside, it might be mentioned that scientists newly engaged in the evaluation of military systems often have trouble grasping that large systems of the type created by or for the military are divided quite rigidly into several chronological stages, namely, in reverse order: operation, deployment, development and research. An operational system is not threatened by a system that is still in development; the threat is not real until the new system is in fact deployed, shaken down and fully operative. This is particularly true for an ABM system, which is obliged to operate against large numbers of relatively independent intercontinental ballistic missiles. It is equally true, however, for counterforce reentry vehicles, which can be ignored unless they are built by the hundreds or thousands. The same goes for MIRV's, a development of the multiple reentry vehicle in which each reentry vehicle is independently directed to a separate target. One must distinguish clearly between the *possibility* of development and the development itself, and similarly between development and actual operation. One must refrain from attributing to a specific defense system, such as Sentinel, those capabilities that *might* be obtained by further development of a different system.

It follows that the Sentinel light ABM system, to be built now and to be operational in the early 1970's against a possible Chinese intercontinental ballistic missile threat, will have to reckon with a missile force unlike either the Russian or the American force, both of which were, after all, built when there was no ballistic-missile defense. The Chinese will probably build even their first operational intercontinental ballistic missiles so that they will have a chance to penetrate. Moreover, we believe it is well within China's capabilities to do a good job at this without intensive testing or tremendous sacrifice in payload.

Temporarily leaving aside penetration aids, there are two pure strategies for attack against a ballistic-missile defense. The first is an all-warhead attack in which one uses large booster rockets to transport many small (that is, fractional-megaton) warheads. These warheads are

separated at some instant between the time the missile leaves the atmosphere and the time of reentry. The warheads from one missile can all be directed against the same large target (such as a city); these multiple reentry vehicles (MRV's) are purely a penetration aid. Alternatively each of the reentry vehicles can be given an independent boost to a different target, thus making them into MIRV's. MIRV is not a penetration aid but is rather a counterforce weapon: if each of the reentry vehicles has very high accuracy, then it is conceivable that each of them may destroy an enemy missile silo. The Titan II liquid-fuel rocket, designed more than 10 years ago,

could carry 20 or more thermonuclear weapons. If these were employed simply as MRV's, the 54 Titans could provide more than 1,000 reentry vehicles for the defense to deal with.

Since the Spartan interceptors will each cost $1 million to $2 million, including their thermonuclear warheads, it is reasonable to believe thermonuclear warheads can be delivered for less than it will cost the defender to intercept them. The attacker can make a further relative saving by concentrating his strike so that most of the interceptors, all bought and paid for, have nothing to shoot at. This is a high-reliability penetration strategy open to any country that

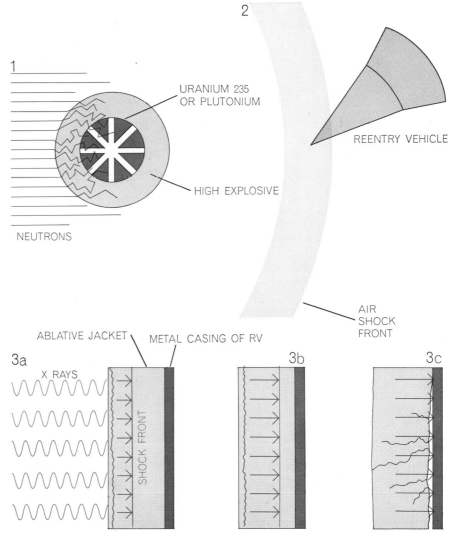

MECHANISMS FOR KILLING REENTRY VEHICLES include the neutrons, blast and X radiation from a thermonuclear explosion. Neutrons (1) can penetrate the fission trigger of an enemy warhead, causing the uranium 235 or plutonium to melt and lose its shape. It can then no longer be assembled for firing. If the defensive warhead is fired inside the atmosphere, the resulting shock front of air (2) can cause the incoming reentry vehicle (RV) to decelerate with a force equivalent to several hundred times the force of gravity, thereby leading to its destruction or malfunction. If the explosion is outside the atmosphere, the X rays travel unimpeded to their target. On striking an RV (3a) they are absorbed by and intensely heat a thin layer of the RV's heat jacket. This creates a shock front that travels through the jacket (3b, 3c) and may cause the jacket to break up or detach from the RV.

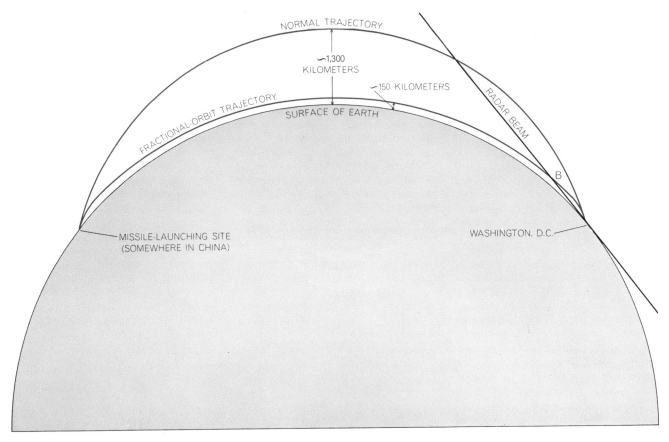

**MISSILE TRAJECTORIES** can be chosen by the attacker to reduce the effectiveness of the defender's radar. The normal trajectory, which requires the least fuel, carries an ICBM to an altitude of about 1,300 kilometers. On its return to the earth the missile would intersect the path of a horizon-search radar at a distance of about 4,000 kilometers ($A$), when the missile was about 10 minutes away. Longer-range but less precise radars may be able to detect the missile earlier. On a fractional-orbit trajectory the missile would stay so close to the earth that it would not cross the radar horizon ($B$) until it was about 1,400 kilometers, or about three minutes, away.

can afford to spend a reasonable fraction of the amount its opponent can spend for defense.

The second pure strategy for attack against an ABM defense is to precede the actual attack with an all-decoy attack or to mix real warheads with decoys. This can be achieved rather cheaply by firing large rockets from unhardened sites to send light, unguided decoys more or less in the direction of plausible city targets. If the ABM defense is an area defense like the Sentinel system, it must fire against these threatening objects at very long range before they reenter the atmosphere, where because of their lightness they would behave differently from real warheads. Several hundred to several thousand such decoys launched by a few large vehicles could readily exhaust a Sentinel-like system. The attack with real warheads would then follow.

The key point is that since the putative Chinese intercontinental-ballistic-missile force is still in the early research and development stage, it can and will be designed to deal with the Sentinel system, whose interceptors and sensors

are nearing production and are rather well publicized. It is much easier to design a missile force to counter a defense that is already being deployed than to design one for any of the possible defense systems that might or might not be deployed sometime in the future.

One of us (Bethe) will now describe (1) the physical mechanisms by which an ABM missile can destroy or damage an incoming warhead and (2) some of the penetration aids available to an attacker who is determined to have his warheads reach their targets.

Much study has been given to the possibility of using conventional explosives rather than a thermonuclear explosive in the warhead of a defensive missile. The answer is that the "kill" radius of a conventional explosive is much too small to be practical in a likely tactical engagement. We shall consider here only the more important effects of the defensive thermonuclear weapon: the emission of neutrons, the emission of X rays and, when the weapon is exploded in the atmosphere, blast.

Neutrons have the ability to penetrate

matter of any kind. Those released by defensive weapons could penetrate the heat shield and outer jacket of an offensive warhead and enter the fissile material itself, causing the atoms to fission and generating large amounts of heat. If sufficient heat is generated, the fissile material will melt and lose its carefully designed shape. Thereafter it can no longer be detonated.

The kill radius for neutrons depends on the design of the offensive weapon and the yield, or energy release, of the defensive weapon. The miss distance, or distance of closest approach between the defensive and the offensive missiles, can be made small enough to achieve a kill by the neutron mechanism. This is particularly true if the defensive missile and radar have high performance and the interception is made no more than a few tens of kilometers from the ABM launch site. The neutron-kill mechanism is therefore practical for the short-range defense of a city or other important target. It is highly desirable that the yield of the defensive warhead be kept low to minimize the effects of blast and heat on the city being defended.

The attacker can, of course, attempt to shield the fissile material in the offensive warhead from neutron damage, but the mass of shielding needed is substantial. Witness the massive shield required to keep neutrons from escaping from nuclear reactors. The size of the reentry vehicle will enable the defense to make a rough estimate of the amount of shielding that can be carried and thus to estimate the intensity of neutrons required to melt the warhead's fissile material.

Let us consider next the effect of X rays. These rays carry off most of the energy emitted by nuclear weapons, especially those in the megaton range. If sufficient X-ray energy falls on a reentry vehicle, it will cause the surface layer of the vehicle's heat shield to evaporate. This in itself may not be too damaging, but the vapor leaves the surface at high velocity in a very brief time and the recoil sets up a powerful shock wave in the heat shield. The shock may destroy the heat shield material or the underlying structure.

X rays are particularly effective above the upper atmosphere, where they can travel to their target without being absorbed by air molecules. The defense can therefore use megaton weapons without endangering the population below; it is protected by the intervening atmosphere. The kill radius can then be many kilometers. This reduces the accuracy required of the defensive missile and allows successful interception at ranges of hundreds of kilometers from the ABM launch site. Thus X rays make possible an area defense and provide the key to the Sentinel system.

On the other hand, the reentry vehicle can be hardened against X-ray damage to a considerable extent. And in general the defender will not know if the vehicle has been damaged until it reenters the atmosphere. If it has been severely damaged, it may break up or burn up. If this does not happen, however, the defender is helpless unless he has also constructed an effective terminal, or short-range, defense system.

The third kill mechanism—blast—can operate only in the atmosphere and requires little comment. Ordinarily when an offensive warhead reenters the atmosphere it is decelerated by a force that, at maximum, is on the order of 100 g. (One g is the acceleration due to the earth's gravity.) The increased atmospheric density reached within a shock wave from a nuclear explosion in air can produce a deceleration several times greater. But just as one can shield against neutrons and X rays one can

shield against blast by designing the reentry vehicle to have great structural strength. Moreover, the defense, not knowing the detailed design of the reentry vehicle, has little way of knowing if it has destroyed a given vehicle by blast until the warhead either goes off or fails to do so.

The main difficulty for the defense is the fact that in all probability the offensive reentry vehicle will not arrive as a single object that can be tracked and fired on but will be accompanied by many other objects deliberately placed there by the offense. These objects come under the heading of penetration aids. We shall discuss only a few of the many types of such aids. They include fragments of the booster rocket, decoys, fine metal wires called chaff, electronic countermeasures and blackout mechanisms of several kinds.

The last stage of the booster that has

propelled the offensive missile may disintegrate into fragments or it can be fragmented deliberately. Some of the pieces will have a radar cross section comparable to or larger than the cross section of the reentry vehicle itself. The defensive radar therefore has the task of discriminating between a mass of debris and the warhead. Although various means of discrimination are effective to some extent, radar and data processing must be specifically set up for this purpose. In any case the radar must deal with tens of objects for each genuine target, and this imposes considerable complexity on the system.

There is, of course, an easy way to discriminate among such objects: let the whole swarm reenter the atmosphere. The lighter booster fragments will soon be slowed down, whereas the heavier reentry vehicle will continue to fall with essentially undiminished speed. If a swarm of objects is allowed to reenter,

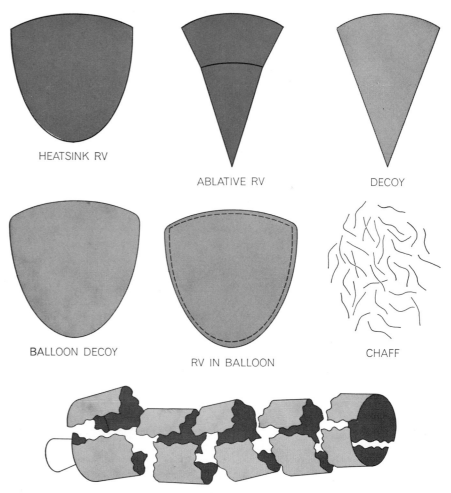

HEATSINK RV            ABLATIVE RV            DECOY

BALLOON DECOY          RV IN BALLOON          CHAFF

BOOSTER FRAGMENTS

**PENETRATION AIDS** include objects that will reflect radar signals and thus simulate or conceal actual reentry vehicles (*color*). A decoy might be a simple conical structure or even a metallized balloon. RV's could be placed inside the same kind of balloon. Fragments of the launching vehicle and its fuel tank provide radar reflectors at no cost. Short bits of metal wire, called chaff, also make a cheap and lightweight reflector of radar signals.

however, one must abandon the concept of area defense and construct a terminal defense system. If a nation insists on retaining a pure area defense, it must be prepared to shoot at every threatening object. Not only is this extremely costly but also it can quickly exhaust the supply of antimissile missiles.

Instead of relying on the accidental targets provided by booster fragments, the offense will almost certainly want to employ decoys that closely imitate the radar reflectivity of the reentry vehicle. One cheap and simple decoy is a balloon with the same shape as the reentry vehicle. It can be made of thin plastic covered with metal in the form of foil,

strips or wire mesh. A considerable number of such balloons can be carried uninflated by a single offensive missile and released when the missile has risen above the atmosphere.

The chief difficulty with balloons is putting them on a "credible" trajectory, that is, a trajectory aimed at a city or some other plausible target. Nonetheless, if the defending force employs an area defense and really seeks to protect the entire country, it must try to intercept every suspicious object, including balloon decoys. The defense may, however, decide not to shoot at incoming objects that seem to be directed against nonvital targets; thus it may choose to

limit possible damage to the country rather than to avoid all damage. The offense could then take the option of directing live warheads against points on the outskirts of cities, where a nuclear explosion would still produce radioactivity and possibly severe fallout over densely populated regions. Worse, the possibility that reentry vehicles can be built to maneuver makes it dangerous to ignore objects even 100 kilometers off target.

Balloon decoys, even more than booster fragments, will be rapidly slowed by the atmosphere and will tend to burn up when they reenter it. Here again a terminal ABM system has a far better

RADAR BLACKOUTS can be created if enough free electrons are released in a sizable volume of space. An attacker can use thermonuclear explosions to generate the electrons required. A fireball blackout results when the heat from a nuclear explosion strips electrons from atoms and molecules of air. In this diagram a high-altitude fireball has been created by an enemy missile launched in a low orbit. The beta rays (electrons) released in the decay of fission products can create another type of blackout. If the beta rays are released at high altitude, they travel along the lines of force in the earth's magnetic field (*parallel colored lines*) and remove electrons from molecules in the atmosphere below. An effective beta blackout could be produced by spreading the fission products of a one-chance than an area defense system to discriminate between decoys and warheads. One possibility for an area system is "active" discrimination. If a defensive nuclear missile is exploded somewhere in the cloud of balloon decoys traveling with a reentry vehicle, the balloons will either be destroyed by radiation from the explosion or will be blown far off course. The reentry vehicle presumably will survive. If the remaining set of objects is examined by radar, the reentry vehicle may stand out clearly. It can then be killed by a second interceptor shot. Such a shoot-look-shoot tactic may be effective, but it obviously places severe demands on the ABM missiles and

SPRINT
MISSILE    MSR
FARM

megaton explosion over an area 200 kilometers in radius. For several minutes the electron cloud would heavily absorb the long waves emitted by a PAR unit, here aimed to the north. The shorter MSR waves, however, would be attenuated only briefly.

the radar tracking system. Moreover, it can be countered by the use of small, dense decoys within the balloon swarms.

Moreover, it may be possible to develop decoys that are as resistant to X rays as the reentry vehicle and also are simple and compact. Their radar reflectivity could be made to simulate that of a reentry vehicle over a wide range of frequencies. The decoys could also be made to reenter the atmosphere—at least down to a fairly low altitude—in a way that closely mimicked an actual reentry vehicle. The design of such decoys, however, would require considerable experimentation and development.

Another way to confuse the defensive radar is to scatter the fine metal wires of chaff. If such wires are cut to about half the wavelength of the defensive radar, each wire will act as a reflecting dipole with a radar cross section approximately equal to the wavelength squared divided by $2\pi$. The actual length of the wires is not critical; a wire of a given length is also effective against radar of shorter wavelength. Assuming that the radar wavelength is one meter and that one-mil copper wire is cut to half-meter lengths, one can easily calculate that 100 million chaff wires will weigh only 200 kilograms (440 pounds).

The chaff wires could be dispersed over a large volume of space; the chaff could be so dense and provide such large radar reflection that the reentry vehicle could not be seen against the background noise. The defense would then not know where in the large reflecting cloud the reentry vehicle is concealed. The defense would be induced to spend several interceptors to cover the entire cloud, with no certainty, even so, that the hidden reentry vehicle will be killed. How much of the chaff would survive the defensive nuclear explosion is another difficult question. The main problem for the attacker is to develop a way to disperse chaff more or less uniformly.

An active alternative to the use of chaff is to equip some decoys with electronic devices that generate radio noise at frequencies selected to jam the defensive radar. There are many variations on such electronic countermeasures, among them the use of jammers on the reentry vehicles themselves.

The last of the penetration aids that will be mentioned here is the radar blackout caused by the large number of free electrons released by a nuclear explosion. These electrons, except for a few, are removed from atoms or molecules of air, which thereby become ions.

There are two main causes for the formation of ions: the fireball of the explosion, which produces ions because of its high temperature, and the radioactive debris of the explosion, which releases beta rays (high-energy electrons) that ionize the air they traverse. The second mechanism is important only at high altitude.

The electrons in an ionized cloud of gas have the property of bending and absorbing electromagnetic waves, particularly those of low frequency. Attenuation can reach such high values that the defensive radar is prevented from seeing any object behind the ionized cloud (unlike chaff, which confuses the radar only at the chaff range and not beyond).

Blackout is a severe problem for an area defense designed to intercept missiles above the upper atmosphere. The problem is aggravated because area-defense radar is likely to employ low-frequency (long) waves, which are the most suitable for detecting enemy missiles at long range. In some recent popular articles long-wave radar has been hailed as the cure for the problems of the ABM missile. It is not. Even though it increases the capability of the radar in some ways, it makes the system more vulnerable to blackout.

Blackout can be caused in two ways: by the defensive nuclear explosions themselves and by deliberate explosions set off at high altitude by the attacker. Although the former are unavoidable, the defense has the choice of setting them off at altitudes and in locations that will cause the minimum blackout of its radar. The offense can sacrifice a few early missiles to cause blackout at strategic locations. In what follows we shall assume for purposes of discussion that the radar wavelength is one meter. Translation to other wavelengths is not difficult.

In order to totally reflect the one-meter waves from our hypothetical radar it is necessary for the attacker to create an ionized cloud containing $10^9$ electrons per cubic centimeter. Much smaller electron densities, however, will suffice for considerable attenuation. For the benefit of technically minded readers, the equation for attenuation in decibels per kilometer is

$$\alpha = \frac{4.34}{3 \times 10^5} \frac{\omega_p^2}{\omega^2 + \gamma_e^2} \gamma_e .$$

Here $\omega_p$ is the plasma frequency for the given electron density, $\omega$ is the radar frequency in radians per second and $\gamma_e$ is the frequency of collisions of an electron with atoms of air. At normal tempera-

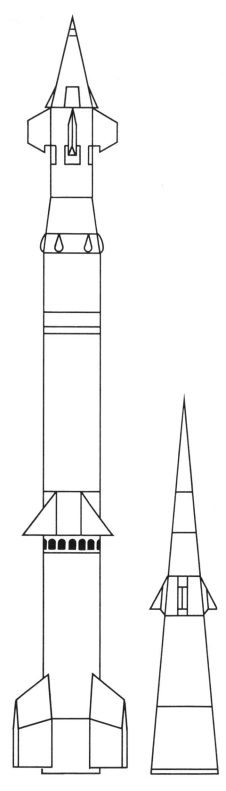

**SENTINEL MISSILES** are the long-range Spartan (*left*), with a reported range of several hundred kilometers, and the Sprint (*right*), which will be used for terminal defense, usually below 35 kilometers. Both will be equipped with thermonuclear warheads designed to destroy or disable the attacker's bomb-carrying reentry vehicles. The Spartan, about 55 feet long, is a three-stage solid-fuel rocket. The smaller and faster Sprint missile uses two stages of solid fuel.

tures this frequency $\gamma_e$ is the number $2 \times 10^{11}$ multiplied by the density of the air ($\rho$) compared with sea-level density ($\rho_0$), or $\gamma_e = 2 \times 10^{11} \rho/\rho_0$. At altitudes above 30 kilometers, where an area-defense system will have to make most of its interceptions, the density of air is less than .01 of the density at sea level. Under these conditions the electron collision frequency $\gamma_e$ is less than the value of $\omega = (2\pi \times 3 \times 10^8)$ and therefore can be neglected in the denominator of the equation. Using that equation, we can then specify the number of electrons, $N_e$, needed to attenuate one-meter radar waves by a factor of more than one decibel per kilometer: $N_e > 350\rho_0/\rho$. At an altitude of 30 kilometers, where $\rho_0/\rho$ is about 100, $N_e$ is about $3 \times 10^4$, and at 60 kilometers $N_e$ is still only about $3 \times 10^6$. Thus the electron densities needed for the substantial attenuation of a radar signal are well under the $10^9$ electrons per cubic centimeter required for total reflection. The ionized cloud created by the fireball of a nuclear explosion is typically 10 kilometers thick; if the attenuation is one decibel per kilometer, such a cloud would produce a total attenuation of 10 decibels. This implies a tenfold reduction of the outgoing radar signal and another tenfold reduction of the reflected signal, which amounts to effective blackout.

The temperature of the fireball created by a nuclear explosion in the atmosphere is initially hundreds of thousands of degrees centigrade. It quickly cools by radiation to about 5,000 degrees C. Thereafter cooling is produced primarily by the cold air entrained by the fireball as it rises slowly through the atmosphere, a process that takes several minutes.

When air is heated to 5,000 degrees C., it is strongly ionized. To produce a radar attenuation of one decibel per kilometer at an altitude of 90 kilometers the fireball temperature need be only 3,000 degrees, and at 50 kilometers a temperature of 2,000 degrees will suffice. Ionization may be enhanced by the presence in the fireball of iron, uranium and other metals, which are normally present in the debris of nuclear explosion.

The size of the fireball can easily be estimated. Its diameter is about one kilometer for a one-megaton explosion at sea level. For other altitudes and yields there is a simple scaling law: the fireball diameter is equal to $(Y\rho_0/\rho)^{1/3}$, where $Y$ is the yield in megatons. Thus a fireball one kilometer in diameter can be produced at an altitude of 30 kilometers (where $\rho_0/\rho = 100$) by an explosion of only 10 kilotons. At an altitude of 50 kilometers (where $\rho_0/\rho = 1,000$), a one-megaton explosion will produce a fireball 10 kilometers in diameter. At still higher altitudes matters become complicated because the density of the atmosphere falls off so sharply and the mechanism of heating the atmosphere changes. Nevertheless, fireballs of very large diameter can be expected when megaton weapons are exploded above 100 kilometers. These could well black out areas of the sky measured in thousands of square kilometers.

For explosions at very high altitudes (between 100 and 200 kilometers) other phenomena become significant. Collisions between electrons and air molecules are now unimportant. The condition for blackout is simply that there be more than $10^9$ electrons per cubic centimeter.

At the same time very little mass of air is available to cool the fireball. If the air is at first fully ionized by the explosion, the air molecules will be dissociated into atoms. The atomic ions combine very slowly with electrons. When the density is low enough, as it is at high altitude, the recombination can take place only by radiation. The radiative recombination constant (call it $C_R$) is about $10^{-12}$ cubic centimeter per second. When the initial electron density is well above $10^9$ per cubic centimeter, the number of electrons remaining after time $t$ is roughly equal to $1/C_R t$. Thus if the initial electron density is $10^{12}$ per cubic centimeter, the density will remain above $10^9$ for 1,000 seconds, or some 17 minutes. The conclusion is that nuclear explosions at very high altitude can produce long-lasting blackouts over large areas.

The second of the two mechanisms for producing an ionized cloud, the beta rays issuing from the radioactive debris of a nuclear explosion, can be even more effective than the fireball mechanism. If the debris is at high altitude, the beta rays will follow the lines of force in the earth's magnetic field, with about half of the beta rays going immediately down into the atmosphere and the other half traveling out into space before returning earthward. These beta rays have an average energy of about 500,000 electron volts, and when they strike the atmosphere, they ionize air molecules. Beta rays of average energy penetrate to an altitude of about 60 kilometers; some of the more energetic rays go down to about 50 kilometers. At these levels, then, a high-altitude explosion will give

**PROTOTYPE OF MISSILE-SITE RADAR is** the multifunction array radar (MAR) at the Army's White Sands Missile Range in New Mexico. MSR, a smaller version of MAR, will be used to guide Spartans and Sprints to their targets. MAR provided an early demonstration that the fastest way to aim a radar beam at different parts of the sky is to switch the beam electronically.

rise to sustained ionization as long as the debris of the explosion stays in the vicinity.

One can show that blackout will occur if $y \times t^{-1.2} > 10^{-2}$, where $t$ is the time after the explosion in seconds and $y$ is the fission yield deposited per unit horizontal area of the debris cloud, measured in tons of TNT equivalent per square kilometer. The factor $t^{-1.2}$ expresses the rate of decay of the radioactive debris. If the attacker wishes to cause a blackout lasting five minutes ($t = 300$), he can achieve it with a debris level $y$ equal to 10 tons of fission yield per square kilometer. This could be attained by spreading one megaton of fission products over a circular area about 400 kilometers in diameter at an altitude of, say, 60 kilometers. Very little could be seen by an area-defense radar attempting to look out from under such a blackout disk. Whether or not such a disk could actually be produced is another question. Terminal defense would not, of course, be greatly disturbed by a beta ray blackout.

The foregoing discussion has concentrated mainly on the penetration aids that can be devised against an area-defense system. By this we do not mean to suggest that a terminal-defense system can be effective, and we certainly do not wish to imply that we favor the development and deployment of such a system.

Terminal defense has a vulnerability all its own. Since it defends only a small area, it can easily be bypassed. Suppose that the 20 largest American cities were provided with terminal defense. It would be easy for an enemy to attack the 21st largest city and as many other undefended cities as he chose. Although the population per target would be less than if the largest cities were attacked, casualties would still be heavy. Alternatively the offense could concentrate on just a few of the 20 largest cities and exhaust their supply of antimissile missiles, which could readily be done by the use of multiple warheads even without decoys.

It was pointed out by Charles M. Herzfeld in *The Bulletin of the Atomic Scientists* a few years ago that a judicious employment of ABM defenses could equalize the risks of living in cities of various sizes. Suppose New York, with a population of about 10 million, were defended well enough to require 50 enemy warheads to penetrate the defenses, plus a few more to destroy the city. If cities of 200,000 inhabitants were left undefended, it would be equally "attractive" for an enemy to attack New York and penetrate its defenses as to attack an undefended city.

Even if such a "logical" pattern of ABM defense were to be seriously proposed, it is hard to believe that people in the undefended cities would accept their statistical security. To satisfy everyone would require a terminal system of enormous extent. The highest cost estimate made in public discussions, $50 billion, cannot be far wrong.

Although such a massive system would afford some protection against the U.S.S.R.'s present armament, it is virtually certain that the Russians would react to the deployment of the system. It would be easy for them to increase the number of their offensive warheads and thereby raise the level of expected damage back to the one now estimated. In his recent forecast of defense needs for the next five years, Secretary McNamara estimated the relative cost of ABM defenses and the cost of countermeasures that the offense can take. He finds invariably that the offense, by spending considerably less money than the defense, can restore casualties and destruction to the original level before defenses were installed. Since the offense is likely to be "conservative," it is our belief that the actual casualty figures in a nuclear exchange, after both sides had deployed ABM systems and simultaneously increased offensive forces, would be worse than these estimates suggest.

Any such massive escalation of offensive and defensive armaments could hardly be accomplished in a democracy without strong social and psychological effects. The nation would think more of war, prepare more for war, hate the potential enemy and thereby make war more likely. The policy of both the U.S. and the U.S.S.R. in the past decade has been to reduce tensions to provide more understanding, and to devise weapon systems that make war less likely. It seems to us that this should remain our policy.

# Commentary

As one who has been intimately associated with the U.S. Anti-Ballistic-Missile (ABM) Defense System Program for the past five years and who has participated in many constructive dialogues on the program with Drs. Garwin and Bethe, I should like to offer some comments regarding their article "Anti-Ballistic-Missile Systems."

As a general comment, let me note that neither the authors nor myself are arguing about technical facts, since in this complex area we are dealing at most with technical judgments, and more often with political opinion. Further, there are limits placed on our discussion by the necessary security classification that exists regarding the details of the program. Drs. Garwin and Bethe make three primary points, and I should like to comment on each of these in turn.

1. An ABM system designed to counter a potential Chinese ICBM threat is unnecessary because our overwhelming nuclear offensive power "should restrain China indefinitely from an attack on the U.S."

An opposing political judgment to that presented by the authors would state that deterrent power only has meaning if the potential enemy believes it will be used. I believe it is well within the realm of possibility that some future Chinese leaders might not believe we would use our nuclear power in a crisis, when we would know that we could be destroyed with certainty by a much less potent force. I readily admit that this is only opinion, but I believe it to be no more or no less credible than that expressed by the authors. The fact is that neither they nor I nor our present government leadership will be making the decisions on either side at the time of some future crisis. I believe the main purpose of our light anti-China defense system should be to give our future leadership more options in order to lend credibility to our deterrent.

2. The Sentinel system will "nourish the illusion that an effective defense against ballistic missiles is possible and will lead almost inevitably to demands that the light system...be expanded into a heavy system."

This argument has been repeated time and again, and I believe it is a toothless old saw. This country has resisted the expansion of weapon systems, both offensive and defensive, time and again by sound arguments against the necessity of such expansion. There is no reason why this pressure, if it exists at all, cannot be countered by effective leadership. The point is further clouded by the lack of precise definition of the word "effective." Somehow we only pay attention to the extremes; either the system is so "ineffective" that it does not work at all or it is so "effective" that there must be an insatiable demand for more of this good thing. Let me first state unequivocally that I agree with the authors that I cannot foresee a system so effective that by expanding the system we would have any real chance of negating the deterrent power of a sophisticated enemy such as the U.S.S.R. To attempt such an objective would truly be a waste of resources; but I should like to define an effective system as one that is able to successfully counter light attacks, and I would contend that this is a very useful military objective and a very necessary one in the situation when the U.S.S.R. has some ballistic-missile defense. I do not believe it is adequate for us to only have the power to use our force en masse to overwhelm the defense and produce "assured destruction." Suppose, for example, for any reason a potential enemy sent over a couple of missiles, perhaps to take out a military target with very few civilian casualties. Would we respond by an overwhelming attack against their cities, knowing full well that the counterresponse would lead to our destruction? The light defense goes a long way toward removing the military credibility of any light attacks and thus introduces a firebreak that makes the ballistic missile useful solely as a deterrent, which is a situation we have grown to accept.

3. Our Sentinel Chinese-oriented system is, or will be, penetrable by the Chinese.

No discussion of this subject is possible in any depth without getting into classified material. Certainly there are possible penetration aids, as the authors have pointed out, that could be attempted to defeat the Sentinel system, but the uncertainties are not all on the side of the defense. The decision to deploy the Sentinel system was a very conscious one, with a great deal of analysis done to provide reasonable assurance that the system could evolve to handle future penetration aids that could be adopted by the Chinese. This does not mean an expansion of the system to the $40-billion variety but a continued technical upgrading as necessary to include and counter advances in technology. Certainly there is no guarantee that this is possible, but there were adequate possibilities to warrant the deployment decision.

While the subject of ballistic-missile defense may be an unusual one for your journal, it is appropriate since it is primarily the technical community that has expressed opinions on the subject. I would only hope that future treatment of ABM would present both sides of this very complicated and important issue.

DANIEL J. FINK

General Manager
Space Systems
Valley Forge Space Technology Center
Missile and Space Division
General Electric Company
Philadelphia, Pa.
May, 1968

It is always helpful to know what is being said by eminent scientists about our major weapons system developments. I deeply regret that military classification prevents the scientists and engineers who know most about ballistic-missile defense capabilities from publication of a better-informed treatment of this whole subject.

What concerns me about the Garwin-Bethe approach to ballistic-missile defense is to read again about all those clever tricks that a theoretical physicist can invent on paper for an enemy use to defeat a defensive system—ideas that have been around for 10 years to my knowledge and possibly longer. The inference is that these clever tricks are easy to do with high confidence of real-world effectiveness, and that defense is helpless to react. Nothing could be further from the truth. The U.S. Air Force and the U.S. Navy are able to draw on 20 years of ballistic-missile technology and 10 years of space technology, and they still find that achievement of effective penetration aids takes years and hundreds of millions of dollars of development and testing to have *high confidence* in penetration aids. Surely no nation, not even the Chinese, would be so foolish as to plan to produce and deploy a weapons system as expensive as the ICBM without the years of painstaking and costly testing effort needed to give them high confidence that their systems would work if called on.

The fact of the matter is that the Army-proposed level of defense and our estimates of its probable effectiveness in the 1975-to-1980 time frame were carefully reviewed by technically competent and well-informed groups outside of the Army before the Secretary of Defense recommended approval of production and deployment of the Nike Sentinel system. There is nothing in the Garwin-Bethe analysis that was not known and adequately considered at that time.

A. W. BETTS

Lieutenant General, GS
Chief of Research and Development
Department of the Army
Washington, D.C.
May, 1968

---

We agree with Dr. Fink that competent proponents and opponents of ABM systems differ primarily not on technical facts but in technical judgments and political opinions.

Dr. Fink's Point 2 is the crux of the matter. Within a few weeks after the decision to deploy Sentinel, demands were made by influential persons to expand the system as soon as possible to the $40-billion variety. Certainly effective leadership has in the past sometimes successfully resisted the expansion of weapons systems or of other programs, but the effectiveness of governments varies, and a moment of weakness or of political horse trading could saddle us with a costly, useless and dangerous system. We think the existence of a light ABM will make it more difficult to resist pressure for its expansion.

General Betts and Dr. Fink note that successful penetration aids require more than invention, and take a lot of effort on development and testing. They also point out that a defense can be modified technically to deal with any of them. Security classification precludes meaningful technical discussion here of these moves and countermoves, but it is our experience, based on more than 10 years of involvement in these and similar military-technical problems, that in general the required reaction by the defense is slower, much more costly and less certain than the action of the offense. We do not agree that a system on the scale of Sentinel can long stay ahead of Chinese penetration aids.

The elements of the Sentinel system and the production capacity we build to produce it, together with the avowed possibility of technical upgrading of the system, may lead the U.S.S.R. to consider it as directed against themselves. Whether warranted or not, such a conclusion on their part would lead to a revived and intensified arms race. In the words of former Secretary McNamara, "the Soviets and ourselves would be forced to continue on a foolish and feckless course."

The other half of the arms race deserves mention as well. Should the U.S. become convinced, rightly or wrongly, of an expansion and increased capability of the Soviet ABM system around Moscow, our present deterrent policy would require high assurance as to the adequacy of our penetration aids and perhaps even an increase in numbers of missiles as well. As a consequence of our maintaining a high-confidence assured-destruction capability, if war actually comes, the U.S.S.R. would very likely suffer even more complete destruction than if they had not built an ABM system. Like the U.S., the U.S.S.R. would be well advised not to build an ABM system that could present even the appearance of effectiveness against a deterrent force.

RICHARD L. GARWIN

IBM-Watson Laboratory
    at Columbia University
New York, N.Y.

HANS A. BETHE

Laboratory of Nuclear Studies
Cornell University
Ithaca, N.Y.
May, 1968

# 47

# The Dynamics of the Arms Race

GEORGE W. RATHJENS

*April 1969*

The world stands at a critical juncture in the history of the strategic arms race. Within the past two years both the U.S. and the U.S.S.R. have decided to deploy new generations of offensive and defensive nuclear weapons systems. These developments, stimulated in part by the emergence of China as a nuclear power, threaten to upset the qualitatively stable "balance of terror" that has prevailed between the two superpowers during most of the 1960's. The new weapons programs portend for the 1970's a decade of greatly increased military budgets, with all the concomitant social and political costs these entail for both countries. Moreover, it appears virtually certain that at the end of all this effort and all this spending neither nation will have significantly advanced its own security. On the contrary, it seems likely that another upward spiral in the arms race would simply make a nuclear exchange more probable, more damaging or both.

As an alternative to this prospect, the expectation of serious arms-limitation talks between the U.S. and the U.S.S.R. holds forth the possibility of at least preventing an acceleration of the arms race. In the circumstances it seems worthwhile to inquire into the nature of the forces that impel an arms race. In doing so we may determine how best to damp this newest cycle of military competition, either by mutual agreement or by unilateral restraint, before it is beyond control.

There are a number of new weapons systems under development in both the U.S. and the U.S.S.R., but the possibilities that are likely to be at the center of discussion not only in the forthcoming negotiations but also in the current Congressional debate are the anti-ballistic-missile (ABM) concept and the multiple-independently-targeted-reentry-vehicle (MIRV) concept. These systems, one defensive and the other offensive, can usefully be discussed together because of the way they interact. In fact, the intrinsic dynamics of the arms race can be effectively illustrated by concentrating on these two developments.

It is now 18 months since former Secretary of Defense McNamara announced the decision of the Johnson Administration to proceed with the deployment of the Sentinel system: a "thin" ABM system originally described as being intended to cope with a hypothetical Chinese missile attack during the 1970's. The technology of the Sentinel system and some of the means a determined adversary might employ to defeat it were discussed in some detail a year ago in SCIENTIFIC AMERICAN [see "Anti-Ballistic-Missile Systems," by Richard L. Garwin and Hans A. Bethe, beginning on page 343]. At this point I should like to review some of the background of the ABM problem.

Before the Sentinel decision most of the interest in a ballistic-missile defense for the U.S. was focused on the Nike-X program. This concept involved the use of two kinds of interceptor to protect the population and industry of the country against a hypothetical Russian missile attack. Interception would first be attempted outside the earth's atmosphere with Spartans, long-range missiles with nuclear warheads in the megaton range. The effectiveness of the defense, however, would depend primarily on the use of Sprints, short-range missiles with kiloton-yield warheads designed to intercept incoming missiles after they have reentered the atmosphere. The system also envisaged suitable radars and computers to control the engagement.

The Spartans could in principle defend large areas; indeed, about a dozen sites could defend the entire country. A defense based solely on them could be rendered ineffective, however, by fairly simple countermeasures, in particular by large numbers of lightweight decoys (which would be indistinguishable to a radar from an actual reentry vehicle containing a warhead) or by measures that would make the radar ineffective, for example the use of nuclear explosions, electronic jammers or light, widely dispersed metal "chaff."

The effectiveness of a Sprint defense would be less degraded by such countermeasures. Light decoys could be distinguished from actual reentry vehicles because they would be disproportionately slowed by the atmosphere and possibly because their wake in the atmo-

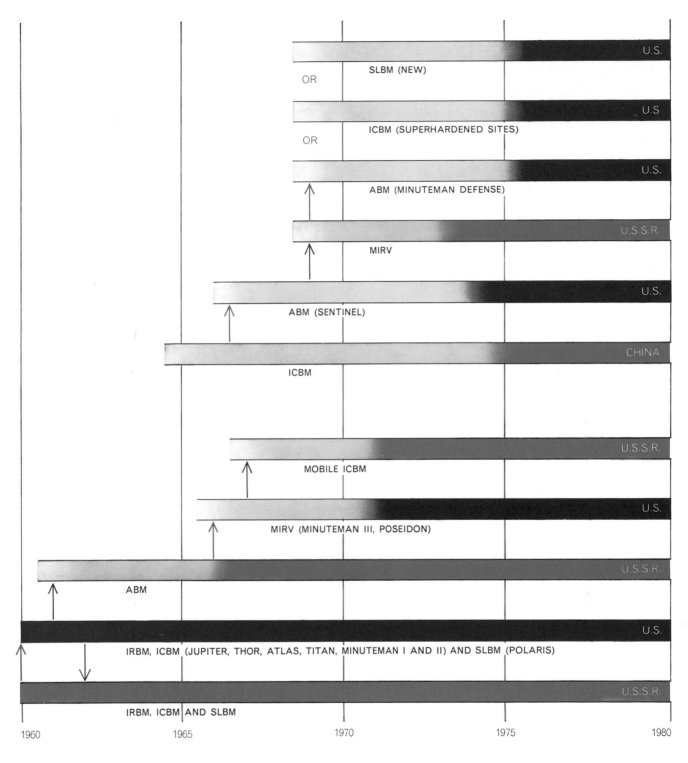

SLBM (NEW) — U.S.

OR

ICBM (SUPERHARDENED SITES) — U.S

OR

ABM (MINUTEMAN DEFENSE) — U.S.

MIRV — U.S.S.R.

ABM (SENTINEL) — U.S.

ICBM — CHINA

MOBILE ICBM — U.S.S.R.

MIRV (MINUTEMAN III, POSEIDON) — U.S.

ABM — U.S.S.R.

IRBM, ICBM (JUPITER, THOR, ATLAS, TITAN, MINUTEMAN I AND II) AND SLBM (POLARIS) — U.S.

IRBM, ICBM AND SLBM — U.S.S.R.

1960    1965    1970    1975    1980

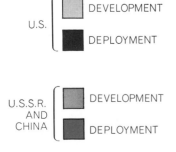

U.S.
DEVELOPMENT
DEPLOYMENT

U.S.S.R.
AND
CHINA
DEVELOPMENT
DEPLOYMENT

ACTION-REACTION PHENOMENON, stimulated in most cases by uncertainty about an adversary's intentions and capabilities, characterizes the dynamics of the arms race. Starting at bottom left, American overreaction to uncertainty at the time of the erroneous "missile gap" in 1960 led to the massive growth of U.S. missile forces during the 1960's. The scale of this deployment may have led in turn to the recent large Russian buildup in strategic offensive forces and also to the deployment of a limited ABM system around Moscow. The U.S. response to the possible extension of the Moscow ABM system into a countrywide system (and to the deployment of a Russian anti-aircraft system, which until recently was thought to be a countrywide ABM system) was to equip its Minuteman III and Poseidon missiles with MIRV warheads. A likely Russian reaction to the potential counterforce threat posed by the MIRV's is the development of land-mobile ICBM's. Another action-reaction chain may have been triggered by the emergence of China as a nuclear power. The resulting deployment of the U.S. Sentinel system, particularly in its expanded versions, seems certain to have an effect on Russian planners, who may push for the development of their own MIRV systems, provoking a variety of American counterresponses. In the author's view, breaking the action-reaction cycle by limiting ABM defenses should be given first priority in any forthcoming arms-control talks.

sphere would be different. Radar blackout would also be much less of a problem. Because of their short range, however, Sprints could defend only those targets in their immediate vicinity. Thus an adversary could choose to attack some cities with enough weapons to overwhelm the defense while leaving others untargeted. Heavy radioactive fallout could also be produced over large parts of the country by an adversary's delivering large-yield weapons outside the areas covered by Sprint defenses. A nationwide defense of the Sprint type would therefore require a nationwide fallout-shelter program.

Although combining Sprints and Spartans in a single system, as was proposed for the Nike-X system, would complicate an adversary's penetration problem,

in a competition with a determined and resourceful adversary the advantage in an offense-defense duel would still lie with the offense. As a result, in spite of strong advocacy by the Army and support from the other branches of the military and from members of Congress, the decision to deploy the Nike-X system was never made.

At the heart of the debate about whether or not to deploy the Nike-X system was the question of what the Russian reaction to such a decision would be. It was generally conceded that the system might well save large numbers of lives in the event of war, if the U.S.S.R. were simply to employ the forces projected in the available intelligence estimates. On that basis propo-

nents argued in favor of deployment in spite of the high costs, variously estimated as being from $13 billion to $50 billion. Such deployment was opposed, particularly by Secretary McNamara, because of the belief that the U.S.S.R. could and would improve its offensive capabilities in order to negate whatever effectiveness the system might have had. Indeed, because the deployment of a U.S. ABM system would introduce large uncertainties into the calculus of the strategic balance, there were occasional expressions of concern that the U.S.S.R. might overreact. Hence the damage inflicted on us in the event of war might even be greater than it would be if the Nike-X system were not deployed.

The Sentinel system announced in 1967 would have far less capability than

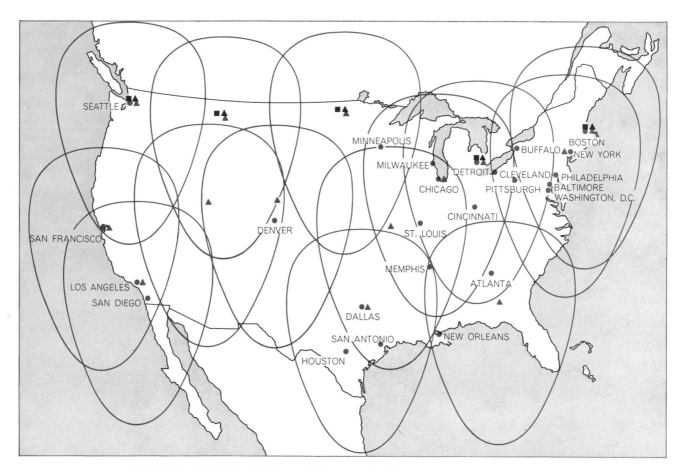

**▲ SPARTAN SITE**

**▲ SPRINT SITE**

**■ PAR SITE**

**● 25 LARGEST CITIES**

SENTINEL SYSTEM, a "thin" anti-ballistic-missile (ABM) system described by the Johnson Administration as being intended to defend the U.S. against a hypothetical Chinese missile attack in the 1970's, is depicted on this map in its original form. The main defense would be provided by Spartan missiles, long-range ABM missiles with nuclear warheads in the megaton range, designed to intercept incoming missiles outside the earth's atmosphere. The Spartans would be deployed at about 14 locations in order to provide a "thin" or "light" area defense of the whole country. The range of each "farm" of Spartans is indicated by the egg-shaped area around it; for missiles attacking over the northern horizon, the intercept range of the Spartan is elongated somewhat to the south. The Sentinel system would also include some Sprint missiles, short-range ABM missiles with much smaller warheads, designed to intercept incoming missiles after they have reentered the atmosphere. The Sprints were originally to be deployed to defend only the five or six perimeter acquisition radars, or PAR's, which were to be deployed across the northern part of the country. In President Nixon's proposed modification of the Sentinel scheme Spartans, Sprints and PAR sites would be deployed in a somewhat different array to provide additional protection for our land-based retaliatory forces against a hypothetical surprise attack by the Russians.

the Nike-X system. It would include some Sprint missiles to defend key radars (five or six perimeter acquisition radars, or PAR's, to be deployed across the northern part of the country), but the main defense would be provided by Spartan missiles located to provide a "thin" or "light" defense for the entire country [*see illustration on opposite page*]. Spokesmen for the Johnson Administration argued that such a deployment would be almost completely effective in dealing with a possible Chinese missile attack during the 1970's, but that it would be so ineffective against a possible Russian attack that the U.S.S.R. would not feel obliged to improve its strategic offensive forces as a response to the decision. Both arguments were seriously questioned.

Garwin and Bethe, for example, contended that even the first-generation Chinese missiles might well be equipped with penetration aids that would defeat the Sentinel system. Other experts pointed out that the system, like the Nike-X system, could never be tested adequately short of actual war, and that in view of its complexity there would be a high probability of a catastrophic failure.

The contention that the U.S.S.R. would not react to the Sentinel decision seemed at least as questionable as the assertions of great effectiveness against the Chinese. Whatever the initial capability of the Sentinel system, it seemed clear that the Sentinel decision would at least shorten the lead time for the deployment of a system of the Nike-X type. Moreover, the fact that Sentinel was strongly and publicly supported as a first step toward an "anti-Soviet" system could hardly escape the attention of Russian decision-makers.

Since the announcement of the Sentinel decision, and particularly since the change in the Administration, the arguments in favor of the decision have become confused. It has been variously suggested by Administration spokesmen that the primary purpose would be (1) to defend the American population and industry against a possible Chinese attack, (2) to provide at least some protection for population and industry against a possible Russian attack, (3) to defend Minuteman missile sites against a possible Russian attack and (4) to serve as a bargaining counter in strategic-arms-limitation talks with the U.S.S.R. It might be noted that no one in recent months has seriously suggested that a Russian reaction to the decision is unlikely. In fact, all but the first

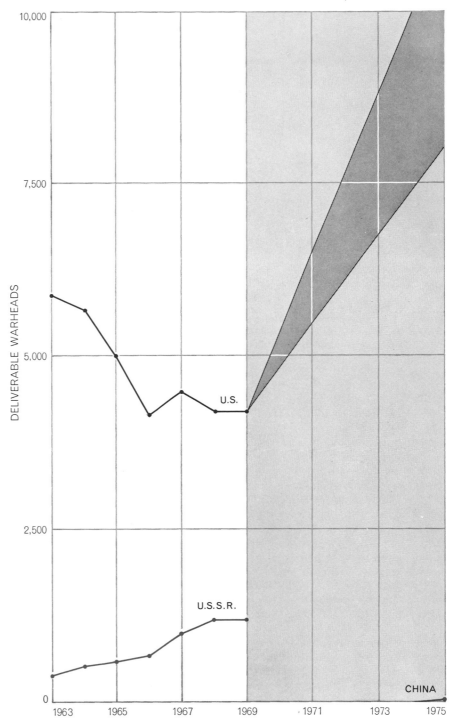

STRATEGIC OFFENSIVE FORCES of the U.S., the U.S.S.R. and China are compared here in terms of the number of megaton-range nuclear warheads the U.S. could deliver against either of the two other powers as of a given date, and vice versa. The U.S. missile force grew rapidly during the 1960's, offsetting a partial phasing-out of intercontinental bombers. The Russian bomber force, meanwhile, has remained at a constant level, while their missile force has grown steadily and shows no sign of leveling off. Thus at present the U.S. maintains a superiority over the U.S.S.R. of about four to one in numbers of deliverable warheads. The hypothetical Chinese strategic force was recently estimated by Secretary of Defense Laird to be 20 to 30 deliverable warheads by the mid-1970's. The effect of the present U.S. program to equip its Minuteman III and its Poseidon missiles with multiple independently targeted reentry vehicles (MIRV's) would be to increase greatly the number of U.S. deliverable warheads by 1975 (*dark gray area*). It is not known what compensating actions, if any, will be taken by the other powers in response to this development. It has been estimated by former Secretary of Defense McNamara that 400 one-megaton warheads would suffice to destroy 75 percent of the industry and 33 percent of the population of the U.S.S.R.

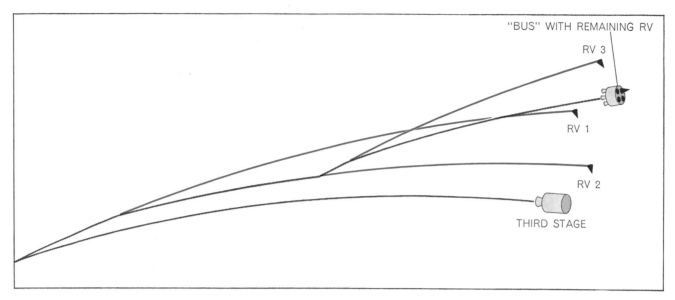

MIRV CONCEPT on which the current U.S. Minuteman III and Poseidon programs are based is illustrated by the idealized drawings on these two pages. Each offensive missile will carry aloft a "bus," containing a number of individual reentry vehicles, or RV's (in this example four are shown). A single guidance and propulsion system will control the orientation and velocity of the bus, from which the reentry vehicles will be released sequentially (left). After each release there will be a further adjustment in the velocity

of the arguments cited above imply the likelihood of a Russian response.

President Nixon's reaffirmation, albeit with some modification, of the Sentinel decision was presumably made on the basis of his judgment that the first and third of the aforementioned arguments justify the costs of such a system, not only the direct dollar cost but also the cost in terms of the impact on Russian decision-making and any other costs that may be imputed to the system. Whether or not his decision is correct depends strongly on how serious the possibility of a Russian reaction is. Before dealing further with that question it will be useful to bring MIRV's into the picture.

The problem of simulating an actual warhead reentry vehicle is a comparatively easy one, provided that the attacker need not be concerned with differences in the interaction of decoys and warheads with the atmosphere during reentry. If one wishes to build decoys and warheads that will be indistinguishable down to low altitudes, however, the problem is a formidable one, particularly if one demands high confidence in the indistinguishability of the two types of object. Improved radar resolution and increased traffic-handling and data-processing capability make the problem of effective decoy design increasingly difficult. The development of interceptors capable of high acceleration will also complicate the offense's problem. With such interceptors the decision to engage reentering objects can be de-

ferred until they are well down into the atmosphere; the longer the defense can wait, the more stringent are the demands of decoy simulation on the offense.

As the problem becomes more difficult, the ratio of decoy weight to warhead weight increases. There comes a point at which, if one wants really high confidence of penetration, one might just as well use several warheads on each missile rather than a single warhead and several decoys, each of which may be as heavy, or nearly as heavy, as a warhead. Hence multiple warheads are in a sense the ultimate in high-confidence penetration aids (assuming that one relies on exhaustion or saturation of defense capabilities as the preferred tactic for defeating the defense). To be effective, however, multiple warheads must be sufficiently separated so that a single interceptor burst will not destroy more than one incoming warhead. Moreover, the utility of multiple warheads for destroying targets, particularly small ones that would not justify attack by more than one or two small warheads, will be greatly enhanced if they can be individually guided.

In principle each reentry vehicle could have its own "post-boost" guidance and propulsion system. That, however, is not the concept of the MIRV's in our Poseidon and Minuteman III missile systems, which are now under development. Rather, a single guidance and propulsion system will control the orientation and velocity of a "bus" from

which reentry vehicles will be released sequentially [see illustration on these two pages]. After each release there will be a further adjustment in the velocity and direction of the bus. Thus each reentry vehicle can be directed to a separate target. The targets can be rather widely separated, the actual separation depending on how much energy (and therefore weight) one is willing to expend in the post-boost maneuvers of the bus. It is an ingenious—and demanding —concept.

Two rationales have been advanced for the decision to proceed with the U.S. MIRV programs. One is that with MIRV's the U.S. can have a high confidence of being able to penetrate an adversary's ABM defenses. The apparent deployment of a limited Russian ABM system in the vicinity of Moscow and U.S. concern about a possibly more widespread Russian ABM-system deployment have been important considerations in the decision to go ahead with the U.S. MIRV programs.

The second rationale is that a MIRV system enables one to strike more targets with a given number of boosters than would be the case if one were using one warhead per missile. This rationale has been important for two reasons.

First, it enabled spokesmen for the Johnson Administration to argue against expanding the size of our strategic missile force during a period when Russian forces were growing rapidly. They were able to contend in the face of political opposition on both flanks that, whereas

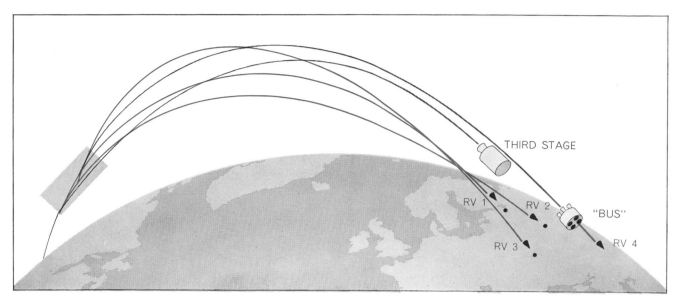

**and the direction of the bus. Thus each reentry vehicle can be directed to a separate target (*right*). The actual separation of the targets depends on how much energy (and therefore weight) one is willing to expend in the post-boost maneuvers of the bus. Besides** being a potentially attractive means of penetrating an adversary's ABM defenses, MIRV's could conceivably be effective some day as a "counterforce" weapon, that is, a system capable of destroying the adversary's strategic offensive forces in a preemptive attack.

we did not contemplate expanding the number of our offensive missiles, the number of warheads we could deliver would increase rapidly.

Second, it raised the prospect of a missile force that could be used as a very effective "counterforce" weapon. This means that with MIRV's a limited number of missiles might be capable of destroying a larger intercontinental-ballistic-missile (ICBM) force in a preemptive attack. To achieve this performance, however, particularly against hardened offensive missile sites, would require a substantial improvement in accuracy and a high post-boost reliability—no mean feats with a device as complicated as the MIRV bus.

What bearing will the deployment of the ABM and the MIRV systems have on the future of the arms race? In attempting to answer this difficult question it is instructive to consider the extent to which the choices of each of the superpowers regarding strategic weapons have been influenced by the other's decisions.

The actual role of this action-reaction phenomenon is a matter of considerable debate in American defense circles. Indeed, the differences in views on this question account for most of the dispute of the past few years regarding the objectives to be served by strategic forces and their desired size and qualities. Thus whether the U.S. should be content with an adequate retaliatory, or "assured destruction," capability or go further and

try to build a capability that would permit us to reduce damage to ourselves in the event of war must clearly depend on a judgment on whether Russian defense decisions could be influenced significantly by our decisions. Those who have felt that Russian defense planning would be responsive to our actions have held that for the most part any attempt by us to develop such "damage-limiting" capabilities with respect to the U.S.S.R. would be an effort doomed to failure. The U.S.S.R. would simply improve its offensive capabilities to offset the effects of any measures we might take. This was the basis for the rejection by the American leadership of the requests by the Army for large-scale ABM-system deployment and for the rejection of requests by the Air Force for much larger ICBM forces.

Although there is considerable evidence to support the claim that the action-reaction phenomenon does apply to defense decision-making, to explain all the major decisions of the superpowers in terms of an action-reaction hypothesis is an obvious oversimplification. The American MIRV deployment has been rationalized as a logical response to a possible Russian ABM-system deployment, but there were also other motivations that were important: the desire to keep our total missile force constant while increasing the number of warheads we could deploy, the long-term possibility of MIRV's giving us an effective counterforce capability, and finally the simple desire to bring to fruition an

interesting and elegant technological concept.

Nevertheless, the action-reaction phenomenon, with the reaction often premature and/or exaggerated, has clearly been a major stimulant of the strategic arms race. Examples from the past can be cited to support this point: (1) the American reaction, indeed overreaction, to uncertainty at the time of the "missile gap," which played a central role in the 1960 Presidential election but was soon afterward shown by improved intelligence to be, if anything, in favor of the U.S.; (2) the Russian decision to deploy the "Tallinn" air-defense system, possibly made in the mistaken expectation that the U.S. would go ahead with the deployment of B-70 bombers or SR-71 strike-reconnaissance aircraft; (3) the U.S. response to the Tallinn system (which until recently was thought to be an ABM system) and to the possible extension of the Moscow ABM system into a countrywide system. It was in order to have high assurance of its ability to get through these possible Russian ABM defenses that the U.S. embarked on the development of various penetration aids and even of new missiles: Minuteman III and Poseidon.

These examples have in common the fact that if doubt exists about the capabilities or intentions of an adversary, prudence normally requires that one respond not on the basis of what one expects but on a considerably more pessimistic projection. The U.S. generally

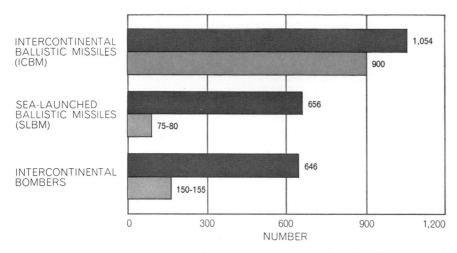

INVENTORY of the intercontinental delivery vehicles possessed by the U.S. (*gray*) and the U.S.S.R. (*color*) as of October, 1968, shows that in this category also the U.S. continues to hold a clear superiority over the U.S.S.R. This measure is not entirely satisfactory, however, in that it does not take into account qualitative differences in the various systems, the expected interactions during a nuclear exchange or the match of the weapons with the targets. When such factors are taken into account, the differences between the U.S. and the U.S.S.R. are not significant. In his final "posture" statement for fiscal 1969, for example, former Secretary McNamara estimated that, barring drastic changes in the strategic balance, in an all-out nuclear exchange in the mid-1970's each country could inflict a minimum of 120 million fatalities on the other—regardless of which country struck first.

bases its plans—and makes much of the fact—on what has become known as the "greater-than-expected threat." In so doing, the Americans (and presumably the Russians) have often overreacted. The extent of the overreaction is directly dependent on the degree of uncertainty about any adversary's intentions and capabilities.

The problem is compounded by lead-time requirements for response. According to the Johnson Administration, the decisions to go ahead with Minuteman III, Poseidon and Sentinel had to be made when they were because of the possibility that in the mid-1970's the Russians might have a reasonably effective ABM system and the Chinese an ICBM capability. The Russians had to make a decision to develop the Tallinn system (if the decision was made because of the B-70 program) long before we ourselves knew whether or not we would deploy an operational B-70 force.

Once the decisions to respond to ambiguous indications of adversary activity were made it often proved impossible to modify the response, even when new intelligence became available. For example, between the time the Sentinel decision was announced and the first Congressional debate on the appropriation took place during the summer of 1968, evidence became available that the Chinese threat was not developing as rapidly as had been feared. Yet in spite of this information those in Congress who attempted at that time to defer

the appropriation for Sentinel failed. Similarly, at this writing, as the Poseidon and Minuteman III programs begin to gain momentum, it seems much less likely than it did at the time of their conception that the U.S.S.R. will deploy the kind of ABM system that was the Johnson Administration's main rationale for these programs. On the Russian side, the Tallinn deployment continued long after it became clear that no operational B-70 force would ever be built.

Of the kinds of weapons development that can stimulate overresponse on the part of one's adversary, it is hard to imagine one more troublesome than ABM defenses. In addition to uncertainty about adversary intentions and the need (because of lead-time requirements) for early response to what the adversary might do, there is the added fact that the uncertainties about how well an ABM system might perform are far larger than they are for strategic offensive systems. The conservative defense planner will design his ABM system on the assumption that it may not work as well as he hopes, that is, he will overdesign it to take into account as fully as he can all imaginable modes of failure and enemy offensive threats. The offensive planner, on the other hand, will assume that the defense might perform much better than he expects and will overdesign his response. Thus there is overreaction on both sides. These uncertainties result in a divergent process: an arms race with no apparent limits other

than economic ones, each round being more expensive than the last. Moreover, because of overreaction on the part of the offense there may be an increase in the ability of each side to inflict damage on the other.

All one needs to make this possibility a reality is a triggering mechanism. The Russian ABM program, by stimulating the Minuteman III and Poseidon programs, may have served that purpose. The Chinese nuclear program may also have triggered an action-reaction chain, of which the Sentinel response is the second link [*see illustration on page 356*].

It can be assumed that there will be considerable pressure and effort to make Sentinel highly effective against a "greater than expected" Chinese threat. Such a system will undoubtedly have some capability against Russian ICBM's. Russian decision-makers, who must assume that Sentinel might perform better than they expect, will at least have to consider this possibility as they plan their offensive capabilities. More important, they will have to respond on the assumption that the Sentinel decision may foreshadow a decision to build an anti-Russian ABM system. Hence it is probably not a question of whether the U.S.S.R. will respond to Sentinel but rather of whether the U.S.S.R. will limit its response to one that does not require a U.S. counterresponse, and of whether it is too late to stop the Sentinel deployment.

It is apparent that reduction in uncertainty about adversary intentions and capabilities is a *sine qua non* to curtailing the strategic arms race. There are a number of ways to accomplish this (in addition to the gathering of intelligence, which obviously makes a great contribution).

First, there is unilateral disclosure. In the case of the U.S. there has been a conscious effort to inform both the American public and the Russian leadership of the rationale for many American decisions regarding strategic systems and, to the extent consistent with security, of U.S. capabilities. This has been done particularly through the release by the Secretary of Defense of an annual "posture" statement, a practice that, it is hoped, will be continued by the U.S. and will be emulated someday by the U.S.S.R. This would be in the interest of both countries. Because there has been no corresponding effort by the Russians the U.S. probably overreacts to Russian decisions more than the U.S.S.R. does to American decisions. (At least it is easier to trace a causal relationship between Russian decisions and U.S. reactions than it is between U.S. decisions and Russian reactions.)

Second, negotiations to curtail the arms race (even if abortive) or any other dialogue may be very useful if such efforts result in a reduction of uncertainty about the policies, capabilities or intentions of the parties.

Third, some weapons systems may be less productive of uncertainty than others that might be chosen instead. For example, it is likely to be less difficult to measure the size of a force of submarine-launched or fixed missiles than it is to measure the size of a mobile land force. Similarly, it would be easier to persuade an adversary that a small missile carried only a single warhead than would be the case with a large vehicle. Such considerations must be borne in mind in evaluating alternative weapons systems.

In short, although uncertainty about adversary capabilities and intentions may not always be bad (in some instances the existence of uncertainty has contributed to deterrence), the U.S. and the U.S.S.R. would seem well advised to make great efforts to avoid giving each other cause for overreacting to decisions because of inadequate understanding of their meaning.

The importance of somehow breaking the action-reaction chains that seem to drive the arms race is obvious when one considers the enormous resources involved that could otherwise be used to meet pressing social needs. In addition, there is particular importance in doing so at present because the concurrent deployment of MIRV's and ABM systems is likely to have drastic destabiliz-

ing consequences. It is conceivable that one of the superpowers with an ABM system might develop MIRV's to the point where it could use them to destroy the bulk of its adversary's ICBM force in a preemptive attack. Its air and ABM defenses would then have to deal with a much degraded retaliatory blow, consisting of the sea-launched forces and any ICBM's and aircraft that might have survived the preemptive attack. The problems of defense in such a contingency would remain formidable. They would be significantly less difficult, however, than if the adversary's ICBM force had not been seriously depleted. In fact, the defense problem would be relatively simple if a large fraction of the adversary's retaliatory capability were, as is true for the U.S. and to a far greater degree for the U.S.S.R., in its land-based ICBM's, most of which would presumably have been destroyed.

It may seem unlikely that either superpower would initiate such a preemptive attack, in view of the great uncertainties in effectiveness (particularly with respect to defenses) and the disastrous consequences if even a comparatively small fraction of the adversary's retaliatory force should get through. With both MIRV's and an ABM system, however, such a preemptive attack would not seem as unlikely as it does

now. It might not appear irrational to some, for example, if an uncontrollable nuclear exchange seemed almost certain, and if by striking first one could limit damage to a significantly lower level than if the adversary were to strike the first blow. In short, if one or both of the two superpowers had such capabilities, the world would be a much more unstable place than it is now.

Obviously neither superpower would permit its adversary to develop such capabilities without responding, if it could, by strengthening its retaliatory forces. The response problem becomes more

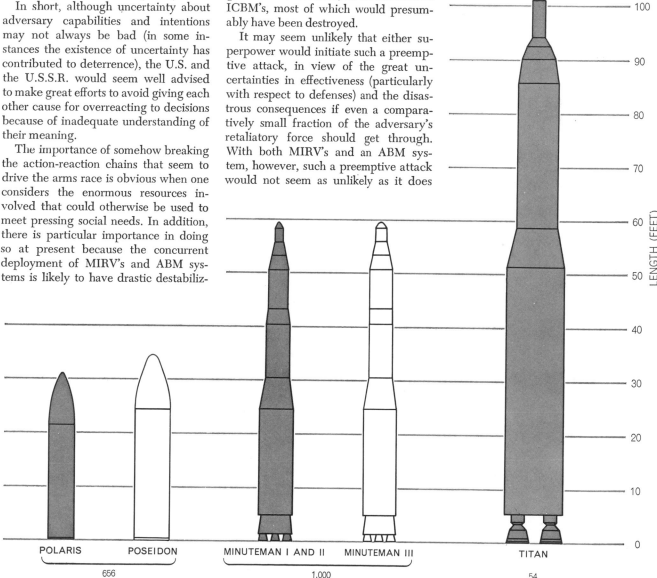

U.S. OFFENSIVE MISSILES currently deployed or under development are drawn here to scale. The sea-based Polaris and Poseidon and the land-based Minutemen are capable of carrying warheads with a total explosive yield of about one megaton each. The land-based Titan II can carry a warhead of more than five megatons. Poseidon and Minuteman III, which are under development, are designed to carry MIRV warheads. The total number of missiles scheduled for deployment in each category is indicated at the bottom.

difficult, however, if the adversary develops both MIRV's and an ABM system than if only one is developed.

Against a MIRV threat alone there are such obvious responses as defense of ICBM sites or greater reliance on sea-launched or other mobile systems. Such responses are likely to be acceptable because, whereas the costs of highly invulnerable systems are large (perhaps several times larger than the costs of simple undefended ICBM's), only relatively small numbers of such secure retaliatory weapons would be required to provide an adequate "assured destruction" capability. Indeed, a force the size of the present Polaris submarine fleet would seem to be more than adequate. The response to an ABM system alone might also be kept within acceptable limits because the expenditures required to offset the effects of defense are likely to be small compared with the costs of the defense.

If it is necessary to acquire retaliatory capabilities that are comparatively invulnerable to MIRV attack in numbers sufficient to saturate or exhaust ABM defenses, however, the total cost could be very great. In fact, if one continued to rely heavily on exhaustion of defenses as the preferred technique for penetration, the offense might no longer have a significant cost-effectiveness advantage over the defense. Thus the concurrent

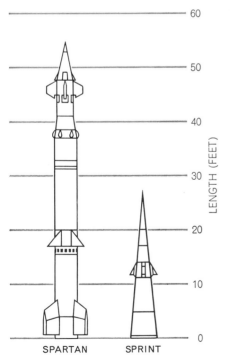

U.S. DEFENSIVE MISSILES currently being deployed as part of the modified Sentinel system are drawn here to scale. The Spartan and Sprint carry warheads in the megaton range and kiloton range respectively.

development of MIRV's and ABM systems raises the specter of a more precarious balance of terror a few years hence, a rapidly escalating arms race in the attempt to prevent the instabilities from getting out of hand, or quite possibly both.

With this background about the roles of uncertainty and the action-reaction phenomenon in stimulating the arms race, one can draw some general conclusions about the functions and qualities of future strategic forces. We must first recognize that two kinds of instability must be considered: crisis instability (the possibility that when war seems imminent, one side or the other will be motivated to attack preemptively in the hope of limiting damage to itself) and arms-race instability (the possibility that the development or deployment decisions of one country, or even the possibility of such decisions, may trigger new development or deployment decisions by another country).

The first kind of instability is illustrated in the chart on the opposite page, which is based on former Secretary McNamara's posture statement for fiscal 1967. This shows that—assuming two possible expanded Russian threats, various damage-limiting efforts by the U.S. and failure of the U.S.S.R. to react to extensive U.S. damage-limiting efforts by improving its retaliatory capability—American fatalities in 1975 would be only about a third as great in the event of a U.S. first strike as they would be in the case of a Russian first strike. (In the present situation the advantage of the attacker is negligible.) Obviously if war seemed imminent, with the strategic balances assumed in this example, there would be tremendous pressure on the U.S. to strike first. There would be corresponding pressure on the U.S.S.R. to do likewise if a Russian first strike could result not only in a much higher level of damage to the U.S. but also in a diminution in damage to the U.S.S.R. The incentives would be mutually reinforcing.

To minimize the chance of a failure of deterrence in a time of crisis, it seems important for both the U.S. and the U.S.S.R. to develop strategic postures such that preemptive attack would have as small an effect as possible on the anticipated outcome of a thermonuclear exchange. Actually, of course, it is extremely unlikely that the Russians would passively watch the U.S. develop the extensive damage-limiting postures assumed in the foregoing example. Instead they would probably react by modify-

ing their posture so that the advantage to the U.S. of attacking preemptively would be less than is indicated in the chart. Thus the example can also be used to illustrate the second kind of instability.

To the extent that one accepts the action-reaction view of the arms race, one is forced to conclude that virtually anything we might attempt in order to reduce damage to ourselves in the event of war is likely to provoke an escalation in the race. Moreover, many of the choices we might make with damage-limitation in mind are likely to make preemptive attack more attractive and war therefore more probable. The concurrent development of MIRV's and ABM systems is a particularly good example of this.

One is struck by the fact that there is an inherent inconsonance in the objectives spelled out in our basic military policy, namely "to deter aggression at any level and, should deterrence fail, to terminate hostilities in concert with our allies under conditions of relative advantage while limiting damage to the U.S. and allied interests." Hard choices must be made between attempting to minimize the chance of war's occurring in a time of crisis and attempting to minimize the consequences if it does occur.

The decisions made by U.S. planners in recent years with respect to new weapons development and deployment reflect a somewhat inconsistent philosophy on this point. The U.S. has generally avoided actions whose primary rationale was to limit damage that the U.S.S.R. might inflict on it, actions to which the Russians would probably respond. Accordingly the U.S. has not deployed an anti-Russian ABM system and has given air defense a low priority.

On the other hand, where there were reasons other than a desire to improve American damage-limiting capability with respect to the U.S.S.R., the U.S. has proceeded with programs in spite of their probably escalating effect on the arms race or their effect on first-strike incentives. This was true in the case of the MIRV's and Sentinel.

The U.S. will face more such decisions. For example, it may appear necessary to change the U.S. strategic offensive posture in order to make American forces less vulnerable to possible Russian MIRV attack. The nature of these decisions will depend on the importance attached to the action-reaction phenomenon and to the effect of improved counterforce capabilities on the probability of war. Emphasis on these two factors

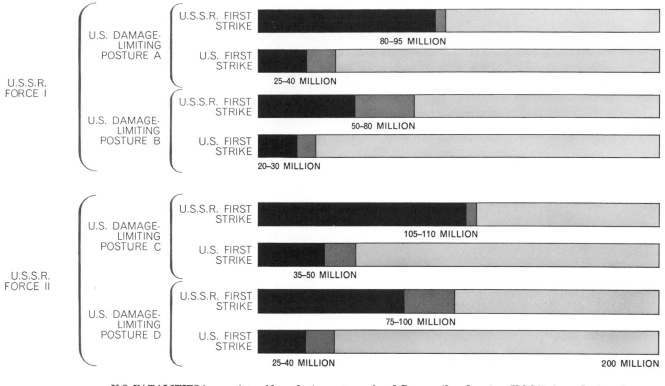

DEATHS

PROBABLE
DEATHS

SURVIVORS

U.S. FATALITIES in a variety of hypothetical nuclear exchanges in the mid-1970's are rounded off to the nearest five million in this bar chart. U.S.S.R. force *I* is basically an extrapolation of current Russian forces reflecting some future growth in both offensive and defensive capability; force *II* assumes a major Russian response to our deployment of an ABM system. Two of the four U.S. damage-limiting programs, postures *A* and *B*, are tailored against U.S.S.R. force *I*; the other two, postures *C* and *D*, are tailored against U.S.S.R. force *II*. The chart illustrates the basic incompatibility between a policy of attempting to minimize the consequences of nuclear war and a policy of attempting to minimize the probability of nuclear war. If war seemed imminent, with the strategic balances as hypothesized in this chart, there would be tremendous pressure on both sides to strike first, and as a result of this added incentive the chances of escalation would be enhanced. The chart is based on information contained in former Secretary McNamara's posture statement for fiscal 1967.

implies discounting options that would increase U.S. counterforce capability against Russian strategic forces, which in turn might provoke an expansion of Russian offensive forces. Options requiring long lead times would also be discounted, since decisions regarding them might have to be made while there was still uncertainty about whether the U.S.S.R. was developing MIRV's.

Should more weight be given in the future to developing damage-limiting capabilities? Or should more weight be given to minimizing the probability of a thermonuclear exchange and curtailing the strategic arms race? It is hard to see how one can have it both ways.

In spite of some changes in technology, there is little to indicate that the U.S. could get very far with damage-limiting efforts, considering the determination of the Russians and the options available to them for denying the attainment of such U.S. capabilities. The emergence of new nuclear powers, the rapid pace of technological advance and the other important demands on American resources suggest that a clear first priority should be assigned to moderating the action-reaction cycle. Moving toward greater emphasis on damage-limitation would seem justified only if the U.S. can persuade itself that the Russians will not react to American moves as the U.S. would to theirs, and if means can be chosen that will not increase the probability of war.

No treatment of the dynamics of the strategic arms race would be complete without some discussion of the possibility of ending it, or at least curtailing it, through negotiations. Both the urgency and the opportunity are great, but the latter may be waning. This opportunity is in part a consequence of the present military balance, as well as of somewhat changed views in both the U.S. and the U.S.S.R. about strategic capabilities and objectives.

With the rapid growth in its strategic offensive forces during the past few years, the U.S.S.R. can at long last enter negotiations without conceding inferiority or (which is worse from the Russian point of view) exposing itself to the possibility of being frozen in such a position. Moreover, the U.S.S.R. may at long last be prepared to accept the prevailing American view about the action-reaction phenomenon, and about the intrinsic advantage of the offense and the futility of defense. The apparent decision of the Russians not to proceed with a nationwide ABM system at present, and their professed willingness to enter into negotiations to control both offensive and defensive systems, may be evidence of this convergence of viewpoints.

On the American side there is at long last a quite general, if not yet universal, acceptance of the concept of nuclear "sufficiency": the idea that beyond a certain point increased nuclear force cannot be translated into useful political power. Acceptance of this concept is an almost necessary condition to termination of the arms race.

In considering negotiations with the U.S.S.R. on the strategic arms problem, the first factor to be kept in mind is the objectives to be sought. It would be a mistake to expect too much or to aspire to too little. One obvious aim is to reduce strategic armaments in order to lessen significantly the damage that would be sustained by the U.S. (and the U.S.S.R.) in the event of a nuclear exchange. Regrettably this goal is not likely to be realized in the near future. In the first place, any initial understandings will probably not involve reductions in strategic forces. Even if they did, the reductions would be limited. One cannot expect potential damage levels to be lowered by more than a few percent, even with fairly substantial cuts in strategic forces, because the capabilities of the superpowers are already so great.

Other objectives have been considered: reducing the incentives to strike preemptively in time of crisis, reducing the probability of accident or miscalculation, and increasing the time available for decision-making in the hope that the increased opportunity for communication might prevent a nuclear exchange from running its full course. Last but not least, one might also hope to change the international political climate so as to lessen tension, to reduce the incentive for powers that currently do not have nuclear weapons to acquire them and to increase the possibility for agreement by the superpowers on other meaningful arms-control measures.

It is reasonable to expect that successful negotiations might to some degree achieve all these objectives except the first: the reduction of potential damage. To focus on any one objective, or combination of objectives, however, is to obscure the immediate problem. In spite of the restraint of the U.S. in its choices regarding strategic weapons development and deployment during the first two-thirds of this decade, it now appears that in the absence of some understanding between the U.S. and the U.S.S.R. the action-reaction sequence that impels the arms race will not be broken. Therefore the immediate objective of any negotiations must be simply to bring that sequence to a halt, or to moderate its pace so that there will be a better chance of ending the arms race than is offered by continuing the policies of the past two decades.

In retrospect, controlling or reversing the growth of strategic capabilities could have been accomplished more easily a few years ago, when the possibility of ABM-system deployment seemed to be the main factor that would trigger another round in the arms race. Now the prospect of ABM systems is more troublesome because of technological advances. In addition, there are the two other stimuli already discussed: the pos-

sibility of effective counterforce capabilities as a result of the development of MIRV's, and the possibility that the Chinese nuclear capability may serve as a catalyst to the Russian-American action-reaction phenomenon.

Obviously, short of destroying China by nuclear attack, there is little the U.S. can do about Chinese capabilities except to make sure that it does not give them more weight in its thinking than they deserve. This leaves the option of trying to break the ABM-MIRV chain by focusing on the control of MIRV's or ABM defenses.

Whereas one might hope to limit both, if a choice must be made the focus should clearly be on the control of ABM defenses. Verification of compliance would be relatively simple and could probably be accomplished without intrusive inspection. In addition, the incentive to acquire MIRV's for penetrating defenses would be eliminated, although the incentive to acquire them for counterforce purposes would remain.

The problems of verifying compliance with an agreement to control MIRV's would be much more difficult. Moreover, if an ABM system were deployed, there would be great pressure to abrogate or violate any agreement prohibiting MIRV deployment because MIRV's offer high assurance for penetrating defenses. Although reversing the MIRV decision would be difficult, reversing the Sentinel one would present less of a problem.

To be attractive to the U.S.S.R. any proposal to limit defenses would almost certainly have to be coupled with an agreement to limit, if not reduce, inventories of deployed strategic offensive forces. In principle this should not be difficult, since it need not involve serious verification problems.

Complicating any attempt to reach an understanding with the U.S.S.R. on the strategic balance, however, is the fact that the American and Russian positions are not symmetrical. The U.S. has allies and bases around the periphery of the U.S.S.R., whereas the latter has neither near the U.S., unless one counts Cuba. It is clear that a Pandora's box of complications could be opened by any attempt in the context of negotiations on the strategic balance to deal with the threat to America's allies posed by short-range Russian delivery systems, and with the potential threat to the U.S.S.R. of systems in Europe that could reach the U.S.S.R. even though they are primarily tactical in nature. One may hope that initial understandings will not have to include specific agreements on such

thorny issues as foreign bases and dual-purpose systems.

Virtually all the above is based on the premise that for the foreseeable future each side will probably insist on maintaining substantial deterrent capabilities. For some time to come there will unfortunately be little basis for expecting negotiations with the U.S.S.R. to result in a strategic balance with each side relying on a few dozen weapons as a deterrent. The difficulties and importance of verification of compliance at such low levels, the problem of China, the existence of large numbers of tactical nuclear weapons on both sides and the general political climate all militate against this. At the other extreme, negotiations would almost necessarily fail if either party based its negotiating position on the expectation that it might achieve a significant damage-limiting capability with respect to the other.

Thus the range of possible agreement is quite narrow. There is a basis for hope, if both sides can accept the fact that for some time the most they can expect to achieve is a strategic balance at quite high, but less rapidly escalating, force levels, and if both recognize that breaking the action-reaction cycle should be given first priority in any negotiations, and also in unilateral decisions.

There will be risks in negotiating arms limitation. These must be weighed not against the risks that might characterize the peaceful world in which everyone would like to live, or even against the risks of the present. Rather, the risks implicit in any agreement must be weighed against the risks and costs that in the absence of agreement one will probably have to confront in the 1970's.

Whether the superpowers strive to curtail the strategic arms race through mutual agreement or through a combination of unilateral restraint and improved dialogue, they should not do so in the mistaken belief that the bases for the Russian-American confrontation of the past two decades will soon be eliminated. Many of the sources of tension have their origins deep in the social structures and political institutions of the two countries. Resolution of these differences will not be accomplished overnight. Restraining the arms race, however, may shorten the time required for resolution of the more basic conflicts between the two superpowers, it may increase the chances of survival during that period, and it may enable the U.S.S.R. and the U.S. to work more effectively on the other large problems that confront the two societies.

# Biographies and Bibliographies

## I. SCIENTISTS AND SOCIETY

### 1. The History and the Present State of Science Fiction

*The Author*

JAMES R. NEWMAN was a member of the Board of Editors of *Scientific American* and editor of its book department until his death in 1966. He graduated from Columbia Law School in 1929 and later developed a love for mathematics and science. He practiced law, served the Government in many capacities during World War II, and after the war was counsel to the Senate Committee on Atomic Energy and helped draft the Atomic Energy Act. He held a Guggenheim Fellowship in 1946 and 1947. His books include *Science and Sensibility, What is Science?,* and *World of Mathematics.*

### 2. Communicating with Intelligent Life Elsewhere in the Universe

*The Author*

See the preceding biographical note.

### 3. The History of Man's Picture of the Universe

*The Author*

I. BERNARD COHEN is a professor of the history of science and of general education at Harvard University. He held a Guggenheim fellowship in 1956 and a National Science Foundation fellowship in 1960–61. Cohen is a past editor of *Isis,* the history of science journal, and is the author of several books, including *Birth of a New Physics* and *Franklin and Newton.*

### 4. The Scientists Who Made the Atomic Bomb

*The Author*

ROBERT R. WILSON moved so many times in his childhood that his school record lists seven grammar schools and five high schools in various parts of Wyoming, Colorado and California. This resulted in poor grades which made it difficult for him to enter college. Yet before joining the freshman class of the University of California he had already built a rudimentary laboratory in which he made high-vacuum pumps of his own design. Wilson majored in electrical engineering, but a near-failing grade in freshman physics challenged him to study physics instead, against the advice of his teachers and in spite of the handicap of poor health. By his senior year he was doing original research in gaseous discharges under Ernest O. Lawrence, and had discovered an original way to study the time lag of the spark discharge. As a graduate student at Berkeley, Wilson turned to nuclear physics and was the first to work out the theory of the cyclotron. In 1940 Wilson joined the faculty of Princeton University. There he invented a way to separate uranium isotopes, with the result that at the age of 28 he found himself in charge of a $500,000 project employing 50 people. Wilson and his staff were soon moved to Los Alamos, where he led the cyclotron group. Eventually he became head of the Experimental Nuclear Physics Division and mayor of the community. Since World War II Wilson has taught at Harvard University and Cornell University. At Cornell he built the first strong-focusing synchrotron and has experimented with thermonuclear reactors, none of which have worked so far. He is one of the founders of the Federation of American Scientists. In his spare time he likes to carve sculptures in wood.

### 5. The Psychological Wreckage of Hiroshima and Nagasaki

*The Author*

J. BRONOWSKI is a mathematician and author who directed the development of new processes for Britain's National Coal Board. Born in Poland and raised in Germany, Bronowski studied mathematics at the University of Cambridge, receiving his Ph.D. in 1933. From then until 1942, when he entered the service of the British Government, Bronowski taught mathematics at University College, Hull. The Government assigned him to the assessment of bomb damage and later to work in the field of operations research. Bronowski joined the Chiefs of Staff mission to Japan in 1945, served as Head of Projects with UNESCO in 1948 and came to the U.S. in 1953 as Carnegie Visiting Professor at the Massachusetts Institute of Technology. He joined the National Coal Board in 1950 as Director of the Coal Research Establishment. His books include *The Poet's Defence, William Blake: A Man without a Mask* and *The Western Intellectual Tradition* (with Bruce Mazlish). Bronowski is also a senior fellow of the Salk Institute for Biological Studies.

## 6. The Role of Two Scientists in Government

### The Author

P. M. S. BLACKETT, professor of physics at the Imperial College of Science and Technology of the University of London since 1953, is distinguished both as a physicist and as an adviser to the British Government on military and scientific policy. A graduate of the University of Cambridge and a Fellow of the Royal Society, he won the Nobel Prize in physics in 1948. He went into physics from the Royal Navy, a graduate of the Royal Naval College at Dartmouth and a veteran of the Battle of Jutland. From 1934 to the beginning of World War II Blackett served on the Aeronautical Research Committee chaired by Sir Henry Tizard, which developed Britain's radar defense system; his work during the war was instrumental in defeating the German submarine campaign. Since the war he has played a leading role in public discussion of military questions; his writing include *Atomic Weapons and East-West Relations*, published by Cambridge University Press in 1965.

## 7. The Probability of Human Survival

### The Author

See biographical note to Article 1.

# II. THE ROOTS OF SOCIAL BEHAVIOR

## 8. Love in Infant Monkeys

### The Author

HARRY F. HARLOW is George Cary Comstock Professor of Psychology and head of the Primate Laboratory at the University of Wisconsin. He received his A.B. from Stanford University in 1927 and his Ph.D. from the same institution in 1930, the year in which he joined the Wisconsin faculty. Harlow is a past president of the American Psychological Association.

### Bibliography

THE DEVELOPMENT OF AFFECTIONAL RESPONSES IN INFANT MONKEYS. Hary F. Harlow and Robert R. Zimmermann in *Proceedings of the American Philosophical Society*, Vol. 102, pages 501–509; 1958.

THE NATURE OF LOVE. Harry F. Harlow in *American Psychologist*, Vol. 12, No. 13, pages 673–685; 1958.

## 9. Early Experience and Emotional Development

### The Author

VICTOR H. DENENBERG is professor of psychology at Purdue University. Denenberg acquired a B.A. at Buckness University in 1949 and an M.S. and a Ph.D. at Purdue in 1951 and 1953. From 1952 to 1954 he did research at George Washington University on problems and methods of training military personnel. He joined the Purdue faculty in 1954.

### Bibliography

AN ATTEMPT TO ISOLATE THE CRITICAL PERIODS OF DEVELOPMENT IN THE RAT. Victor H. Renenberg in *Journal of Comparative and Physiological Psychology*, Vol. 55, No. 5, pages 813–815; October, 1962.

EFFECTS OF ENVIRONMENTAL COMPLEXITY AND SOCIAL GROUPINGS UPON MODIFICATION OF EMOTIONAL BEHAVIOR. Victor H. Denenberg and John R. C. Morton in *Journal of Comparative and Physiological Psychology*, Vol. 55, No. 2, pages 242–246; April, 1962.

EFFECTS OF MATERNAL FACTORS UPON GROWTH AND BEHAVIOR OF THE RAT. Victor H. Denenberg, Donald R. Ottinger and Mark W. Stephens in *Child Development*, Vol. 33, No. 1, pages 65–71; March 1962.

## 10. Opinions and Social Pressure

### The Author

SOLOMON E. ASCH is professor of psychology at Rutgers University. He was born in Warsaw in 1907, came to the U.S. in his youth and graduated from the College of the City of New York in 1928. After taking his M.A. and Ph.D. from Columbia University, he taught at Brooklyn College, the New School for Social Research, and Swarthmore before joining the Rutgers faculty.

### Bibliography

EFFECTS OF GROUP PRESSURE UPON THE MODIFICATION AND DISTORTION OF JUDGMENTS. S. E. Asch in *Groups, Leadership and Men*, edited by Harold Guetzkow. Carnegie Press, 1951.

SOCIAL LEARNING AND IMITATION. N. E. Miller and J. Dollard. Yale University Press, 1941.

SOCIAL PSYCHOLOGY. Solomon E. Asch. Prentice-Hall, Inc., 1952.

## 11. Cognitive Dissonance

### The Author

LEON FESTINGER is professor of psychology at the New School for Social Research. Festinger took his B.S. in psychology at the College of the City of New York in 1939. He received M.A. and Ph.D. degrees from the State University of Iowa, where he specialized in the field of child behavior, in 1940 and 1942 respectively. He remained at Iowa as a research associate until 1943 and for the next two years served as senior statistician on the Committee on Selection and Training of Aircraft Pilots at the University of Rochester. From 1945 to 1948 he taught at the Massachusetts Institute of Technology, and he was program director of the Research Center for Group Dynamics at the University of Michigan until 1951. Festinger went to Stanford from the University of Minnesota, where he had been professor of psychology since 1951. In 1959 the American Psychological Association awarded Festinger its Distinguished Scientific Contribution Award.

### Bibliography

COGNITIVE CONSEQUENCES OF FORCED COMPLIANCE. Leon Festinger and James M. Carlsmith in The Journal of Abnormal and Social Psychology, Vol. 58, No. 2, pages 203–210; March, 1959.

PREPARATORY ACTION AND BELIEF IN THE PROBABLE OCCURRENCE OF FUTURE EVENTS. Ruby B. Yaryan and Leon Festinger in The Journal of Abnormal and Social Psychology, Vol. 63, No. 3, pages 603–606; November 1961.

A THEORY OF COGNITIVE DISSONANCE. Leon Festinger. Row, Peterson & Company, 1957.

WHEN PROPHECY FAILS. Leon Festinger, Henry W. Riecken and Stanley Schachter. University of Minnesota Press, 1956.

## 12. The Anthropology of Manners

### The Author

EDWARD T. HALL, JR., formerly professor of anthropology at the Foreign Service Institute of the Department of State, has just joined American University. He dates his first cross-cultural experiences from the age of four, when his family lived for a year in New Mexico, a meeting place of Spanish, Indian and North European cultures. Hall has degrees in anthropology from three universities: Arizona, Denver and Columbia. Taking linguistic analysis as a model, he and his associate George L. Trager have been trying to pick out the building blocks in other phases of culture, keeping their eyes open for simple units of behavior equivalent to what phonemes are in language. During the war Hall served as an officer in a Negro engineers' regiment in Europe and the Southwest Pacific. Later he worked as an anthropologist for the U. S. Commercial Company in the Truk Islands in Micronesia and taught anthropology at the University of Denver and at Bennington College, where he collaborated with the psychiatrist Erich Fromm.

### Bibliography

CULTURE AND COMMUNICATION. G. L. Trager and Edward T. Hall in Explorations, No. 3, pages 137–149; August, 1954.

THE CHRYSANTHEMUM AND THE SWORD: PATTERNS OF JAPANESE CULTURE. Ruth Benedict. Houghton-Mifflin Company, 1946.

KINESICS AND COMMUNICATION. Ray Birdwhistell in Explorations, No. 3, pages 31–41; August, 1954.

## 13. Experiments in Group Conflict

### The Author

MUZAFER SHERIF is professor of psychology and director of the Institute of Group Relations at the University of Oklahoma. As a student at Istanbul University he was greatly impressed by the writings of William James. He came to the U.S. to study psychology at Harvard University, where he took an M.A. in 1932, and at Columbia University, where he earned his Ph.D. under the direction of Gardner Murphy in 1935 with a thesis entitled "A Study of Some Social Factors in Perception." If Sherif's interest in perception owes something to the inspiration of James, his concern with group conflict derives in part from his eyewitness experience of war and revolution in his native Turkey, of mass hysteria in Germany in 1932 and of social decay in prewar France. These occurrences made a considerable impression on him, he says, leading him to form the idea of a social psychology "which would embody the main features of actual life events, pointing if possible to realistic solutions of such problems." From Columbia Sherif returned to Turkey and taught psychology at Ankara University. A State Department fellowship brought him to Princeton University in 1945; in 1947 he went to Yale as a Rockefeller fellow and began the experiments which he describes in his article.

### Bibliography

GROUPS IN HARMONY AND TENSION. Muzafer Sherif and Carolyn W. Sherif. Harper & Brothers, 1953.

## 14. The Effects of Observing Violence

### The Author

LEONARD BERKOWITZ is professor of psychology at the University of Wisconsin. A graduate of New York University, Berkowitz received a Ph.D. in psychology from the University of Michigan in 1951. From 1951 to 1955 he did research on bomber-crew effectiveness for the U.S. Air Force at the Human Resources Research Center in San Antonio, Texas. A member of the Wisconsin faculty since 1955, Berkowitz spent the academic year 1964–1965 doing research at the University of Oxford in England. He is the author of Aggression: A Social Psychological Analysis, published by McGraw-Hill in 1962.

*Bibliography*

AGGRESSION: A SOCIAL PSYCHOLOGICAL ANALYSIS. Leonard Berkowitz. McGraw-Hill Book Company, 1962.

EFFECTS OF FILM VIOLENCE ON INHIBITIONS AGAINST SUBSEQUENT AGGRESSION. Leonard Berkowitz and Edna Rawlings in *Journal of Abnormal and Social Psychology*, Vol. 66, No. 5, pages 405–412 ;May, 1963.

ENHANCEMENT OF PUNITIVE BEHAVIOR BY AUDIO-VISUAL DISPLAYS. Richard H. Walters, Edward Llewellyn Thomas and C. William Acker in *Science*, Vol. 136, No. 3519, pages 872–873; June 8, 1962.

IMITATION OF FILM-MEDIATED AGGRESSIVE MODELS. Albert Bandura, Dorothea Ross and Sheila A. Ross in *Journal of Abnormal and Social Psychology*, Vol. 66, No. 1, pages 3–11; January, 1963.

TELEVISION IN THE LIVES OF OUR CHILDREN. Wilbur Schramm, Jack Lyle and Edwin B. Parker. Stanford University Press, 1961.

# III. POPULATION AND HETEROGENEITY PROBLEMS

## 15. Population Control in Animals

### The Author

V. C. WYNNE-EDWARDS is Regius Professor of Natural History at Marischal College of the University of Aberdeen. A graduate of Rugby School and the University of Oxford, he did research at the Marine Biological Laboratory in Plymouth from 1927 to 1929. After obtaining an M.A. from Oxford he taught zoology for a year at the University of Bristol before joining the faculty of McGill University in 1930. While in Canada he participated in the Mac-Millan Baffin Island expedition in 1937 and organized the Canadian Fisheries Research Board expeditions to the Mackenzie River in 1944 and to the Yukon Territory in 1945; he was also a member of the Baird expedition to Central Baffin Island in 1950. He was appointed to his present post in 1946.

### Bibliography

ANIMAL DISPERSION IN RELATION TO SOCIAL BEHAVIOR. V. C. Wynne-Edwards. Hafner Publishing Company, 1962.

THE LIFE OF VERTEBRATES. John Z. Young. Oxford University Press, 1950.

THE NATURAL REGULATION OF ANIMAL NUMBERS. David Lack. Oxford University Press, 1954.

## 16. Population

### The Author

KINGSLEY DAVIS is professor of sociology and director of International Population and Urban Research at the University of California at Berkeley. He was born and reared in West Texas, then a highly rural region, where he acquired a strong preference for open spaces rather than cities. He was graduated from the University of Texas and received a master's degree in philosophy there; in 1936 he obtained a Ph.D. in sociology at Harvard University. Later he held a postdoctoral fellowship from the Social Science Research Council for advanced study in demography; more recently he was a fellow at the Center for Advanced Study in the Behavioral Sciences and a senior postdoctoral fellow of the National Science Foundation. His interest in population has taken him to Europe, Latin America, India, Pakistan and 10 countries in Africa; he has also served as U.S. representative to the Population Commission of the United Nations. Davis has been designated as chairman of the National Research Council's newly created Behavioral Sciences Division. Before going to the Oakland-San Francisco metropolitan area in 1955 he taught for seven years at Columbia University, a juxtaposition that moved him to write: "For a man who dislikes large cities, I have spent much of my adult life in major metropolitan areas."

### Bibliography

FAMILY PLANNING, STERILITY AND POPULATION GROWTH. Ronald Freedman, Pascal K. Whelpton and Arthur A. Campbell. McGraw-Hill Book Co., Inc., 1959.

POPULATION CONTROL. *Law and Contemporary Problems*, Vol. 25, No. 3, pages 377–629; Summer, 1960.

POPULATION GROWTH AND ECONOMIC DEVELOPMENT IN LOW-INCOME COUNTIES: A CASE STUDY OF INDIA'S PROSPECTS. Ansley J. Coale and Edgar M. Hoover. Princeton University Press, 1958.

THE POPULATION OF JAPAN. Irene B. Taeuber. Princeton University Press, 1958.

PROSPERITY AND PARENTHOOD: A STUDY OF FAMILY PLANNING AMONG THE VICTORIAN MIDDLE CLASSES. J. A. Banks. Routledge & Kegan Paul Limited, 1954.

SOCIAL STRUCTURE AND FERTILITY: AN ANALYTIC FRAMEWORK. Kingsley Davis and Judith Blake in *Economic Development and Cultural Change*, Vol. 4, No. 3, pages 211–235; April, 1956.

## 17. Population Density and Social Pathology

### The Author

JOHN B. CALHOUN is research psychologist in the Laboratory of Psychology at the National Institute of Mental Health. Calhoun obtained his B.S. at the University of Virginia in 1939 and his M.S. and Ph.D. at Northwestern University in 1942 and 1943 respectively. After teaching at Emory and Ohio State universities Calhoun in 1946 joined a group at John Hopkins University that was studying the behavior of Norway rats. In 1949 he went to the Roscoe B. Jackson Memorial Laboratory in Bar Harbor, Me., to continue his research. From 1951 to 1955 he did research in the neuropsychiatry department of the Walter

Reed Army Institute of Medical Research and spent 1962 at the Center for Advanced Study in Behavioral Sciences in Palo Alto, Calif.

*Bibliography*

THE HUMAN POPULATION. Edward S. Deevey, Jr., in *Scientific American*, Vol. 203, No. 3, pages 194–204; September, 1960.

A METHOD FOR SELF-CONTROL OF POPULATION GROWTH AMONG MAMMALS LIVING IN THE WILD. John B. Calhoun in *Science*, Vol. 109, No. 2831, pages 333–335; April 1, 1949.

POPULATIONS OF HOUSE MICE. Robert L. Strecker in *Scientific American*, Vol. 193, No. 6, pages 92–100; December, 1955.

THE SOCIAL ASPECTS OF POPULATION DYNAMICS. John B. Calhoun in *Journal of Mammalogy*, Vol. 33, No. 2, pages 139–159; May, 1952.

SOCIAL WELFARE AS A VARIABLE IN POPULATION DYNAMICS. John B. Calhoun in *Cold Spring Harbor Symposia on Quantitative Biology*, Vol. 22, pages 339–356; 1957.

## 18. Cybernetics

### The Author

NORBERT WIENER was emeritus professor of mathematics at the Massachusetts Institute of Technology before his death in 1964. Wiener joined the staff at M.I.T. in 1919, interrupting his teaching there in 1926–27, when he was awarded a Guggenheim fellowship, and again in 1935–36, when he went to Tsing Hua University in China as a visiting professor.

*Bibliography*

A MATHEMATICAL THEORY OF COMMUNICATION. C. E. Shannon in *The Bell System Technical Journal*, Vol. 27, No. 3, pages 379–423; July, 1948.

CYBERNETICS. Norbert Wiener. John Wiley, 1948.

## 19. Feedback

### The Author

ARNOLD TUSTIN was, until his retirement, professor of electrical engineering at Imperial College of Science and Technology, University of London. Previously, he was on the staff at Birmingham University. He is the author of *DC Machines for Control Systems*, published in 1952, and *The Mechanisms of Economic Systems*, published in 1955.

*Bibliography*

CYBERNETICS. Norbert Wiener in *Scientific American*, Vol. 179, No. 5, pages 14–19; November, 1948. (See page 119 of this volume.)

FUNDAMENTALS OF AUTOMATIC CONTROL. G. H. Farrington. John Wiley & Sons, Inc., 1951.

AN INTRODUCTION TO THE THEORY OF CONTROL IN MECHANICAL ENGINEERING. R. H. Macmillan. Cambridge University Press, 1951.

## 20. The Culture of Poverty

### The Author

OSCAR LEWIS is professor of anthropology at the University of Illinois. Although he was born in New York City, he grew up on a small farm in upstate New York. He was graduated from the City College of the City of New York in 1936 and received a Ph.D. in anthropology from Columbia University in 1940. Before going to the University of Illinois in 1948 he taught at Brooklyn College and Washington University. Lewis has spent 25 years in the study of other cultures, including the Blackfoot Indians of Canada, Mexican peasants and city dwellers and low-income Puerto Ricans in San Juan and New York. Among his extensive writings are several books, including *Five Families* and *The Children of Sánchez*.

*Bibliography*

BLUE-COLLAR WORLD: STUDIES OF THE AMERICAN WORKER. Edited by Arthur B. Shostak and William Gomberg. Prentice-Hall, Inc., 1964.

MENTAL HEALTH OF THE POOR: NEW TREATMENT APPROACHES FOR LOW-INCOME PEOPLE. Frank Riessman, Jerome Cohen and Arthur Pearl. The Free Press, 1964.

THE OTHER AMERICA: POVERTY IN THE UNITED STATES. Michael Harrington. The Macmillan Company, 1962.

THE URBAN CONDITION: PEOPLE AND POLICY IN THE METROPOLIS. Edited by Leonard J. Duhl. Basic Books, Inc., Publishers, 1963.

## 21. Abortion as a Disease of Societies

### The Author

See biographical note to Article 1.

## 22. Abortion

### The Authors

CHRISTOPHER TIETZE AND SARAH LEWIT are on the staff of the Bio-Medical Division of the Population Council in New York. Tietze, a native of Austria, holds an M.D. from the University of Vienna. Since he came to the U.S. in 1938 his professional interests have shifted to medical statistics and demography, with a special concern for human fertility and its control. Before joining the Population Council, where he is now associate director of the Bio-Medical Division, he was affiliated with the School of Hygiene at Johns Hopkins University, the Department of State and the National Committee on Maternal Health. In addition to his work with the Population Council he is serving as a member of Governor Rockefeller's Committee to Study Abortion in New York State. Sarah Lewit, who in private life is Mrs. Tietze, was graduated from Brooklyn College of the City of New York and served as a statistician in several Government agencies, including a period as acting chief of the Marriage and Divorce Section of the National Office of Vital Statistics. For the past 10 years she has been working with her husband as a research associate.

## Bibliography

ABORTION IN AMERICA: MEDICAL, PSYCHIATRIC, LEGAL, ANTHROPOLOGICAL AND RELIGIOUS CONSIDERATIONS. Edited by Harold Rosen. Beacon Press, 1967.

THE CASE FOR LEGALIZED ABORTION NOW. Edited by Alan F. Guttmacher. Diablo Press, 1967.

ABORTION AND THE CATHOLIC CHURCH: A SUMMARY HISTORY. John T. Noonan, Jr., in *Natural Law Forum*, Vol. 12, pages 85–131; 1967.

LEGAL ABORTION IN EASTERN EUROPE. Malcolm Potts in *The Eugenics Review*, Vol. 59, No. 4, pages 232–250; December, 1967.

## 23.  Squatter Settlements

### The Author

WILLIAM MANGIN is professor of anthropology and chairman of the department of anthropology at Syracuse University. He was graduated from that university in 1948, obtained a Ph.D. from Yale University in 1954 and has been on the Syracuse faculty most of the time since then. During the same period he has worked extensively in Peru; his activities have included a study of fiestas and the use of alcohol from 1951 to 1953 and an investigation of squatters from 1957 to 1959. From 1962 to 1964 he was deputy director of the Peace Corps in Peru. He spends a few days each month working as a research associate at the Center for Studies in Education and Development in the Graduate School of Education at Harvard University. Mangin writes: "I have been particularly interested in migration to cities and in problems of education in urban areas. I was the coordinator of a school for dropouts and boys having troubles in school in Syracuse. I was a losing candidate for the Syracuse Board of Education in 1965 (I came close)."

### Bibliography

BARRIERS AND CHANNELS FOR HOUSING DEVELOPMENT IN MODERNIZING COUNTRIES. John Turner in *Journal of the American Institute of Planners*, Vol. 33, No. 3, pages 167–181; May, 1967.

CONTEMPORARY CULTURES AND SOCIETIES OF LATIN AMERICA: A READER IN THE SOCIAL ANTHROPOLOGY OF MIDDLE AND SOUTH AMERICA AND THE CARIBBEAN. Edited by Dwight B. Heath and Richard N. Adams. Random House, 1965.

SQUATTER SETTLEMENTS IN LATIN AMERICA. William Mangin in *Latin American Research Review*, Vol. 2, No. 4; Fall, 1967.

URBANIZATION IN LATIN AMERICA. Edited by Philip N. Hauser. International Documents Service, Columbia University Press, 1961.

THE USES OF LAND IN CITIES. Charles Abrams in *Scientific American*, Vol. 213, No. 3, pages 150–160; September, 1965.

## 24.  The Renewal of Cities

### The Author

NATHAN GLAZER is professor of sociology at the University of California at Berkeley. Born in New York City, he was graduated from the City College of the City of New York in 1944; he received a master's degree from the University of Pennsylvania and a doctor's degree from Columbia University. From 1944 to 1953 he was on the staff of *Commentary*, and he then spent several years as an editor or editorial adviser with book-publishing firms in New York. Glazer has also taught sociology at Bennington College and Smith College and spent a year as an urban sociologist with the Housing and Home Finance Agency in Washington. He is the author of *American Judaism* and *The Social Basis of American Communism*, coauthor (with David Riesman and Reuel Denney) of *The Lonely Crowd* and coauthor (with Riesman) of *Faces in the Crowd*.

### Bibliography

THE DEATH AND LIFE OF GREAT AMERICAN CITIES. Jane Jacobs. Random House, 1961.

THE FEDERAL BULLDOZER: A CRITICAL ANALYSIS OF URBAN RENEWAL, 1949–1962. Martin Anderson. The M.I.T. Press, 1964.

THE FUTURE OF OLD NEIGHBORHOODS: REBUILDING FOR A CHANGING POPULATION. Bernard J. Frieden. The M.I.T. Press, 1964.

PLANNING AND POLITICS: CITIZEN PARTICIPATION IN URBAN RENEWAL. James Q. Wilson in *Journal of the American Institute of Planners*, Vol. 29, No. 4, pages 242–249; November, 1963.

THE SOCIAL IMPLICATIONS OF URBAN REDEVELOPMENT. Peter Marris in *Journal of the American Institute of Planners*, Vol. 28, No. 3, pages 180–186; August, 1962.

THE URBAN VILLAGERS: THE COMMUNITY LIFE OF ITALIAN-AMERICANS. Herbert J. Gans. The Free Press, 1962.

## 25.  The Social Power of the Negro

### The Author

JAMES P. COMER is a fellow in psychiatry at the Yale University School of Medicine. He writes: "My father, who died in 1955, was from rural Alabama and worked as a laborer and janitor. My mother was born in rural Mississippi and worked as a domestic before my birth and later as an elevator operator. Both were undereducated. They reared five children, all of whom hold postgraduate degrees." Comer was educated in the racially integrated schools of East Chicago, received a bachelor's degree at Indiana University in 1956 and was graduated from the Howard University College of Medicine in 1960. For two years he was a fellow in public health at Howard. In 1964 he took a masters degree in public health at the University of Michigan, joining the psychiatric residency program at Yale in the same year. "My interest in race relations," he says, "developed at an early age, in part from both troublesome and satisfying experiences as a Negro youngster in a low-income family in a racially integrated community." He adds that work as a volunteer in an agency concerned with social rehabilitation of families with problems influenced his decision "to train in psychiatry and to focus on preventive and social aspects."

## Bibliography

BLACK BOURGEOISIE. Franklin Frazier. The Free Press of Glencoe, 1957.

BLACK CARGOES: A HISTORY OF THE ATLANTIC SLAVE TRADE. Daniel P. Mannix and Malcolm Cowley. The Viking Press, 1962.

LAY MY BURDEN DOWN: A FOLK HISTORY OF SLAVERY. Benjamin A. Botkin. The University of Chicago Press, 1945.

NORTH OF SLAVERY. Leon P. Litwack. The University of Chicago Press, 1961.

SLAVERY. Stanley M. Elkins. The University of Chicago Press, 1959.

THE STRANGE CAREER OF JIM CROW. C. Vann Woodward. Oxford University Press, 1958.

## 26. Sickle Cells and Evolution

### The Author

ANTHONY C. ALLISON is a postgraduate fellow at the University of Oxford, where he is engaged in research on cell metabolism in the Medical Research Council Laboratories, directed by the famous biochemist H. A. Krebs. Allison was brought up on an estate in Kenya and was fluent in two African languages before he could speak English. "It was inevitable in these circumstances that I should become interested in anthropology and natural history. After qualifying in medicine at Oxford I turned to research in human genetics, which helped to satisfy these interests." His field work has taken him on expeditions to Lapland, Syria and most parts of Africa. In 1954 he was a research fellow at the California Institute of Technology.

### Bibliography

SICKLE-CELL ANEMIA. George W. Gray in Scientific American, Vol. 185, No. 2, pages 56–59; August, 1951.

PROTECTION AFFORDED BY SICKLE-CELL TRAITS AGAINST SUBTERTIAN MALARIAL INFECTION. A. C. Allison in British Medical Journal, Vol. 1, pages 290–301; February, 1954.

## 27. Teacher Expectations for the Disadvantaged

### The Authors

ROBERT ROSENTHAL AND LENORE F. JACOBSON are respectively professor of social psychology at Harvard University and principal of an elementary school in the South San Francisco Unified School District. Rosenthal was graduated from the University of California at Los Angeles in 1953 and obtained a Ph.D. there in 1956. He taught at the University of Southern California, U.C.L.A., the University of North Dakota and Ohio State University before going to Harvard in 1962. Miss Jacobson was graduated from San Francisco State College in 1946, received a master's degree at Sacramento State College in 1951 and obtained the degree of Ed.D. from the University of California at Berkeley in 1966. She writes: "The arts were my prime interest aside from teaching school until I began my doctoral work. Since then my extra-job energies have been devoted to educational research, and I suspect that I'm permanently hooked."

### Bibliography

COVERT COMMUNICATION IN THE PSYCHOLOGICAL EXPERIMENT. Robert Rosenthal in Psychological Bulletin, Vol. 67, No. 5, pages 356–367; May, 1967.

THE EFFECT OF EXPERIMENTER BIAS ON THE PERFORMANCE OF THE ALBINO RAT. Robert Rosenthal and Kermit L. Fode in Behavioral Science, Vol. 8, No. 3, pages 183–189; July, 1963.

EXPERIMENTER EFFECTS IN BEHAVIORAL RESEARCH. Robert Rosenthal. Appleton-Century-Crofts, 1966.

PYGMALION IN THE CLASSROOM: TEACHER EXPECTATION AND PUPIL' INTELLECTUAL DEVELOPMENT. Robert Rosenthal and Lenore Jacobson. Holt, Rinehart and Winston, Inc., 1968.

SOCIAL STRATIFICATION AND ACADEMIC ACHIEVEMENT. Alan B. Wilson in Education in Depressed Areas, edited by A. Harry Passow. Teachers College Press, Columbia University, 1966.

# IV.  WHAT PRICE PROGRESS?

## 28. Radiation and Human Mutation

### The Author

H. J. MULLER, the late professor of genetics at Indiana University, found that mutations could be accelerated by X-rays. This discovery, which he made in 1927, won him the Nobel prize in physiology and medicine for 1946. Muller was born in New York City in 1890 and was educated at Columbia University, where he took a Ph.D. in zoology in 1916. From 1915 to 1936 he taught zoology, first at the Rice Institute, then at the University of Texas. In 1933 to 1937 he worked at the Institute of Genetics in Moscow as senior geneticist, but he later became a fierce foe of the Soviet system. After spending some years at the University of Edinburgh and at Amherst College, he joined Indiana University in 1945. More than anyone else he alerted the nation to the danger of indiscriminate use of high energy radiation in medicine. He died in 1967.

### Bibliography

NATIONAL SURVEY OF CONGENITAL MALFORMATIONS RESULTING FROM EXPOSURE TO ROENTGEN RADIATION. Stanley H. Macht and Philip S. Lawrence in American Journal of Roentgenology and Radiation Therapy. Vol. 73, No. 3, page 442–446; March, 1955.

OUR LOAD OF MUTATIONS. H. J. Muller in American Journal of Human Genetics, Vol. 2, No. 2, pages 111–176; June 1950.

X-RAY INDUCED MUTATIONS IN MICE. W. L. Russell in *Cold Spring Harbor Symposia on Quantitative Biology*, Vol. 16, pages 327–336; 1951.

## 29. The Effects of Smoking

### The Author

E. CUYLER HAMMOND has since 1946 been director of the Statistical Research Section of the American Cancer Society. Hammond received a B.S. from Yale University in 1935 and an Sc.D. from Johns Hopkins University in 1938. He was associate statistician in the division of industrial hygiene of the National Institutes of Health until 1942; as a major in the Army Air Force from 1942 to 1946 he served first as chief of the statistics department of the School of Aviation Medicine at Randolph Field in Texas and later as assistant chief of the statistics division of the Office of the Air Surgeon in Washington. Hammond, who has been studying the effects of smoking for more than a decade, was professor of biometry and director of statistical studies in the graduate school of Yale University from 1953 to 1958.

### Bibliography

CHANGES IN BRONCHIAL EPITHELIUM IN RELATION TO SEX, AGE, RESIDENCE, SMOKING AND PNEUMONIA. Oscar Auerbach, A. P. Stout, E. Cuyler Hammond and Lawrence Garfinkel in *The New England Journal of Medicine*, Vol. 267, No. 3; July 19, 1962.

LUNG CANCER AND OTHER CAUSES OF DEATH IN RELATION TO SMOKING. Richard Doll and A. Bradford Hill in *British Medical Journal*, Vol. 2, No. 5001, pages 1071–1081; November 10, 1956.

SMOKING AND DEATH RATES—REPORT ON FORTY-FOUR MONTHS OF FOLLOW-UP OF 187,783 MEN. I: TOTAL MORTALITY. E. Cuyler Hammond and Daniel Horn in *The Journal of the American Medical Association*, Vol. 166, No. 10, pages 1159–1172; March 8, 1958. II: DEATH RATES BY CAUSE. E. Cuyler Hammond and Daniel Horn in *The Journal of the American Medical Association*, Vol. 166, No. 11, pages 1294–1308; March 15, 1958.

SMOKING AND HEALTH: SUMMARY AND REPORT OF THE ROYAL COLLEGE OF PHYSICIANS OF LONDON ON SMOKING IN RELATION TO CANCER OF THE LUNG AND OTHER DISEASES. Pitman Publishing Corporation, 1962.

SMOKING: ITS INFLUENCE ON THE INDIVIDUAL AND ITS ROLE IN SOCIAL MEDICINE. C. van Proosdij. Elsevier Publishing Company, 1960.

TOBACCO CONSUMPTION AND MORTALITY FROM CANCER AND OTHER DISEASES. Harold F. Dorn in *Acta Unio Internationalis contra Cancrum*, Vol. 16, No. 7, pages 1653–1665; 1960.

## 30. The Hallucinogenic Drugs

### The Author

FRANK BARRON, MURRAY E. JARVIK AND STERLING BUNNELL, JR. do research on this subject in New York and California. Barron is a professor of psychology at the University of California, Santa Cruz. A graduate of La Salle College in Philadelphia, he received an M.A. from the University of Minnesota in 1948 and a Ph.D. from the University of California at Berkeley in 1950. He has taught at Bryn Mawr College, Harvard University, Wesleyan University and the University of California. Jarvik is associate professor of pharmacology at the Albert Einstein College of Medicine and attending physician at Bellevue Hospital in New York. He was graduated from the City College of the City of New York in 1944 and subsequently acquired an M.A. in psychology from the University of California at Los Angeles in 1945, an M.D. from the University of California School of Medicine in 1951 and a Ph.D. in psychology from the University of California at Berkeley in 1952. He has taught and done research in the fields of pharmacology, psychology and neurophysiology at various institutions. Bunnell is a resident in psychiatry at the Mount Zion Medical Center in San Francisco. He received an M.D. from the University of California School of Medicine in 1958 and is currently working on a Ph.D. in neurophysiology at the University of California at Berkeley.

### Bibliography

THE CLINICAL PHARMACOLOGY OF THE HALLUCINOGENS. Erik Jacobsen in *Clinical Pharmacology and Therapeutics*, Vol. 4, No. 4, pages 480–504; July-August, 1963.

LYSERGIC ACID DIETHYLAMIDE (LSD-25) AND EGO FUNCTIONS. G. D. Klee in *Archives of General Psychiatry*, Vol. 8, No. 5, pages 461–474; May, 1963.

PROLONGED ADVERSE REACTIONS TO LYSERGIC ACID DIETHYLAMIDE. S. Cohen and K. S. Ditman in *Archives of General Psychiatry*, Vol. 8, No. 5, pages 475–480; May, 1963.

THE PSYCHOTOMIMETIC DRUGS: AN OVERVIEW. Jonathan O. Cole and Martin M. Katz in *The Journal of the American Medical Association*, Vol. 187, No. 10, pages 758–761; March, 1964.

## 31. Pleasure Centers in the Brain

### The Author

JAMES OLDS is engaged in physiological and behavioral studies at the California Institute of Technology. He was born in Chicago in 1922, attended the University of Wisconsin, St. Johns College at Annapolis, and Amherst College, with three years out for the U.S. Army, and went to Harvard for his Ph.D. in social psychology. His wife received a doctorate in philosophy at Radcliffe in the same year. Olds later studied physiological psychology for two years with D. O. Hebb at McGill University. There he began the series of experiments discussed in his article.

### Bibliography

THE BEHAVIOR OF ORGANISMS: AN EXPERIMENTAL ANALYSIS. B. F. Skinner. Appleton-Century-Crofts, Inc., 1938.

DIENCEPHALON. Walter Rudolf Hess. Grune & Stratton, Inc., 1954.

PSYCHOSOMATIC DISEASE AND THE "VISCERAL BRAIN": RECENT DEVELOPMENTS BEARING ON THE PAPEZ THEORY

OF EMOTION. Paul D. MacLean in *Psychosomatic Medicine*, Vol. 11, pages 338–353; 1949.

## 32. An Indictment of the Wide Use of Pesticides

### The Author

LAMONT C. COLE is professor of zoology at Cornell University. Despite a boyhood passion for snakes, he graduated from the University of Chicago as a physicist. His return to the animal kingdom resulted from a trip down the Colorado River with A. M. Woodbury of the University of Utah, who inspired him to study ecology. Cole's chief interest is now in natural populations. He has taught at Cornell since 1948. Before that he occupied the late Alfred Kinsey's post in entomology at Indiana University, which he had taken over when Kinsey "turned to the study of bigger and better things."

## 33. Third Generation Pesticides

### The Author

CARROLL M. WILLIAMS is Bussey Professor of Biology at Harvard University. After his graduation from the University of Richmond in 1937 he went to Harvard and successively obtained master's and doctor's degrees in biology and, in 1946, an M.D. He joined the Harvard faculty in 1946 and became full professor in 1953. From 1959 to 1961 he was chairman of the biology department. He has been a member of the National Academy of Sciences since 1961. Williams' studies of insects have won a number of awards, including the George Ledlie Prize of $1,500, which is given every two years to the member of the Harvard faculty who has made "the most valuable contribution to science, or in any way for the benefit of mankind."

### Bibliography

THE EFFECTS OF JUVENILE HORMONE ANALOGUES ON THE EMBRYONIC DEVELOPMENT OF SILKWORMS. Lynn M. Riddiford and Carroll M. Williams in *Proceedings of the National Academy of Sciences*, Vol. 57, No. 3, pages 595–601; March, 1967.

THE HORMONAL REGULATION OF GROWTH AND REPRODUCTION IN INSECTS. V. B. Wigglesworth in *Advances in Insect Physiology: Vol. II*, edited by J. W. L. Bement, J. E. Treherne and V. B. Wigglesworth. Academic Press Inc., 1964.

SYNTHESIS OF A MATERIAL WITH HIGH JUVENILE HORMONE ACTIVITY. John H. Law, Ching Yuan and Carroll M. Williams in *Proceedings of the National Academy of Sciences*, Vol. 55, No. 3, pages 576–578; March, 1966.

## 34. The Control of Air Pollution

### The Author

A. J. HAAGEN-SMIT is professor of bio-organic chemistry at the California Institute of Technology. He was born in the Netherlands in 1900 and obtained an A.B. and a Ph.D. from the University of Utrecht in 1922 and 1929 respectively. After seven years of teaching organic chemistry at Utrecht he joined the faculty of Harvard University in 1936; the following year he went to Cal Tech. In 1950 he received the Fritzsche Award of the American Chemical Society for his work on essential oils and in 1958 he won the Chambers Award of the Air Pollution Control Association. He is a consultant to the Los Angeles Air Pollution District, California's State Motor Vehicle Pollution Control Board and various other state and county agencies dealing with atmospheric sanitation.

### Bibliography

AIR POLUTION: VOLS. I AND II, edited by Arthur C. Stern. Academic Press, 1962.

AIR POLLUTION CONTROL. William L. Faith. John Wiley & Sons, Inc., 1959.

PHOTOCHEMISTRY OF AIR POLLUTION. Philip A. Leighton. Academic Press, 1961.

WEATHER MODIFICATION AND SMOG. M. Neiburger in *Science*, Vol. 126, No. 3275, pages 637–645; October, 1957.

## 35. The Safety of the American Automobile

### The Author

DAVID HAWKINS, Professor of Philosophy at the University of Colorado, was born in El Paso, Texas, in 1913. He studied first at Stanford University and then at the University of California at Berkeley, where he received his Ph.D. degree in 1940. He then taught for several years at those two institutions. From 1943 to 1946 he was associated with the Los Alamos Scientific Laboratories, first as an administrative assistant and then as Historian. After teaching for a year at George Washington University, he joined the faculty of the University of Colorado in 1947. He has had extensive teaching experience and a long-standing interest in moral philosophy, on the one hand, and in science and applied mathematics, on the other. "In C. P. Snow's language," he says, "I have mixed loyalties between the 'two cultures'."

## 36. Thermal Pollution and Aquatic Life

### The Author

JOHN R. CLARK is assistant director of the Sandy Hook Marine Laboratory of the U.S. Bureau of Sport Fisheries and Wildlife. "My forte," he writes, "has truly been sea work; I've probably spent 1,000 days at sea in the past 20 years, and before that in Seattle I worked summers as a commercial fisherman to get through college." Clark was graduated from the University of Washington in 1949 with a degree in fisheries science. For 10 years after that he worked at the Fisheries Laboratory of the U.S. Bureau of Commercial Fisheries in Woods Hole, Mass. Clark was a founder of the American Littoral Society and is now its president. He wishes to note that his article expresses his personal views and is not to be regarded as an official statement of the U.S. Department of the Interior.

## Bibliography

THE PHYSIOLOGY OF FISHES. Edited by Margaret E. Brown. Academic Press, Inc., 1957.

THE PHYSIOLOGY OF CRUSTACEA, VOL. I: METABOLISM AND GROWTH. Edited by Talbot H. Waterman. Academic Press, Inc., 1960.

FISH AND RIVER POLLUTION. J. R. Erichsen Jones. Butterworths, 1964.

A FIELD AND LABORATORY INVESTIGATION OF THE EFFECT OF HEATED EFFLUENTS ON FISH. J. S. Alabaster in *Fishery Investigations, Ministry of Agriculture, Fisheries, and Food*, Series I, Vol. 6, No. 4; 1966.

THERMAL POLLUTION—1968. Subcommittee on Air and Water Pollution of the Committee on Public Works. U.S. Government Printing Office, 1968.

# V.  WAR: THE ANGUISH OF RENUNCIATION

## 37.  The Fighting Behavior of Animals

### The Author

IRENAUS EIBL-EIBESFELDT is research associate in ethology at the Max Planck Institute for the Physiology of Behavior in Germany. He was born in Vienna in 1928 and studied zoology at the University of Vienna, acquiring his doctorate there in 1949. Later that year he joined the Institute for Comparative Behavior, where he worked with the noted student of animal behavior Konrad Z. Lorenz. When the institute was moved to West Germany in 1951 (becoming the Max Planck Institute in the process), Eibl-Eibesfeldt moved with it. Since then his study of symbiotic relationships among fishes has taken him all over the world. In 1953 he accompanied the Xarifa Expedition of the International Institute for Submarine Research to the Caribbean and the Galápagos Islands. His memorandum on the rapid destruction of flora and fauna in the Galápagos led to his being sent by UNESCO in 1957 to make a survey of the conditions there. The results of his recommendations were the creation of the Darwin Foundation in Brussels and the establishment of a biological station in the Galápagos. The following year Eibl-Eibesfeldt accompanied another Xarifa Expedition, this time to the Indian Ocean. He is the author of the book *Galápagos*.

### Bibliography

AGGRESSION. J. P. Scott. University of Chicago Press, 1958.

KAMPF UND PAARBILDUNG EINIGER CICHLIDEN. Beatrice Oehlert in *Zeitschrift für Tierpsychologie*, Vol. 15, No. 2, pages 141–174; August, 1958.

STUDIES ON THE BASIC FACTORS IN ANIMAL FIGHTING. PARTS I–IV. Zing Yang Kuo in *The Journal of Genetic Psychology*, Vol. 96, Second Half, pages 201–239; June, 1960.

STUDIES ON THE BASIC FACTORS IN ANIMAL FIGHTING. PARTS V–VII. Zing Yang Kuo in *The Journal of Genetic Psychology*, Vol. 97, Second Half, pages 181–225; December, 1960.

ZUM KAMPF- UND PAARUNGSVERHALTEN EINIGER ANTILOPEN. Fritz Walther in *Zeitschrift für Tierpsychologie*, Vol. 15, No. 3, pages 340–380; October, 1958.

## 38.  Preventing a Third World War

### The Author

See biographical note to Article 1.

## 39.  The Use and Misuse of Game Theory

### The Author

ANATOL RAPOPORT is professor and senior research mathematician at the Mental Health Research Institute of the University of Michigan. Rapoport was born in Russia, educated in Chicago's public schools and trained in music at the Vienna State Academy of Music, which gave him degrees in composition, piano and conducting. For the next four years he gave concerts in Europe, the U.S. and Mexico. In 1937 (at the age of 26) he enrolled as a freshman at the University of Chicago, and in 1941 he received his Ph.D. in mathematics. Following service in the Air Force as a liaison officer with the Soviet Air Force in Alaska during World War II, Rapoport taught mathematics for a year at the Illinois Institute of Technology, was research associate and later assistant professor of mathematical biophysics at the University of Chicago from 1947 to 1954, and spent a year at the Center for Advanced Study in the Behavioral Sciences. He went to Michigan in 1955.

### Bibliography

THE COMPLEAT STRATEGYST: BEING A PRIMER ON THE THEORY OF GAMES OF STRATEGY. J. D. Williams. McGraw-Hill Book Co., Inc., 1954.

INTRODUCTION TO THE THEORY OF GAMES. J. C. C. McKinsey. The Rand Corporation. McGraw-Hill Book Co., Inc., 1952.

STRATEGY AND MARKET STRUCTURE: COMPETITION, OLIGOPOLY AND THE THEORY OF GAMES. Martin Shubik. John Wiley & Sons, Inc., 1959.

THE STRATEGY OF CONFLICT. Thomas C. Schelling. Harvard University Press, 1960.

THEORY OF GAMES AND STATISTICAL DECISIONS. David Blackwell and M. A. Girshick. John Wiley & Sons, Inc., 1954.

THEORY OF GAMES AS A TOOL FOR THE MORAL PHILOSOPHER. R. B. Braithwaite. Cambridge University Press, 1955.

## 40.  Thermonuclear War

### The Author

See biographical note to Article 1.

## 41. Steps Toward Disarmament

### The Author

See biographical note to Article 6.

### Bibliography

CRITIQUE OF SOME CONTEMPORARY DEFENCE THINKING. P. M. S. Blackett in *Encounter*, Vol. 16, No. 4, pages 9–17; April, 1961.

THE LIMITS OF DEFENSE. Arthur I. Waskow. Doubleday & Company, Inc., 1962.

## 42. The Detection of Underground Explosions

### The Author

SIR EDWARD BULLARD is professor of geophysics at the University of Cambridge. A Fellow of the Royal Society and a foreign associate of the U.S. National Academy of Sciences, Sir Edward has spent much of his career at Cambridge, of which he is a graduate. During interludes away from the university he has served as chairman of the physics department at the University of Toronto and director of the British National Physical Laboratory. He is interested in most aspects of the study of the solid earth, particularly in the geology of the oceans and in the earth's magnetic field. Sir Edward has attended many conferences on the banning of nuclear tests by treaty; the first was the meeting of scientific experts from several nations at Geneva in 1958.

### Bibliography

THE DETECTION AND RECOGNITION OF UNDERGROUND EXPLOSIONS. United Kingdom Atomic Energy Authority, 1965.

A DISCUSSION ON RECENT ADVANCES IN THE TECHNIQUE OF SEISMIC RECORDING AND ANALYSIS. *Proceedings of the Royal Society of London*, Series A, Vol. 290, No. 1422, pages 287–476; March 1, 1966.

PROLIFERATION. Andrew Martin and Wayland Young in *Disarmament and Arm Control*, Vol. 3, No. 2, pages 107–134; Autumn, 1965.

## 43. National Security and the Nuclear-Test Ban

### The Authors

JEROME B. WIESNER AND HERBERT F. YORK have been engaged for most of their professional lifetimes in consultation on this country's military policy and in active development of the weapons. Wiesner, now dean of science at the Massachusetts Institute of Technology, was chairman of the President's Science Advisory Committee and special assistant to the President for science and technology during the Kennedy Administration. York, now chancellor of the University of California at San Diego, was the first director of the Livermore Laboratory, organized in 1952 when the Truman Administration decided to proceed with the development of thermonuclear weapons. Wiesner received a B.S. in 1937 from the University of Michigan and was proceeding with his graduate education there when the outbreak of war called him into service at the Radiation Laboratory of the Massachusetts Institute of Technology. In 1945 he joined the staff of the Los Alamos Scientific Laboratory. Returning to M.I.T. in 1946 as a member of the department of electrical engineering, he completed his doctoral work, the degree being conferred by the University of Michigan in 1950, and was promoted to professional rank that year. At M.I.T. he also served as director of the Research Laboratory of Electronics and continued his work in military technology as a member of the Army Science Advisory Committee. York's graduate education was similarly interrupted by war service. A graduate of the University of Rochester in 1942, he went to work under the late Ernest O. Lawrence at the Radiation Laboratory of the University of California at Berkeley. He received a Ph.D. there in 1949 and continued as a member of the staff until he assumed responsibility for the new Livermore Laboratory. In 1958, under the Eisenhower Administration, he was called to Washington as chief scientist of the Advanced Research Projects Agency of the Department of Defense and director of defense research and engineering in the office of the Secretary of Defense. Both Wiesner and York have continued their Government service as members of the President's Science Advisory Committee.

### Bibliography

THE EFFECTS OF NUCLEAR WEAPONS. Edited by Samuel Glasstone. United States Atomic Energy Commission, April, 1962.

STEPS TOWARD DISARMAMENT. P. M. S. Blackett in *Scientific American*, Vol. 206, No. 4, pages 45–53; April, 1962.

WORLD PEACE THROUGH WORLD LAW. Louis B. Sohn and Grenville Clark. Harvard University Press, 1960.

A WORLD WITHOUT WAR. Walter Millis, Reinhold Niebuhr, Harrison Brown, James Real and William O. Douglas. Washington Square Press, Inc., 1961.

## 44. The Shelter-Centered Society

### The Author

ARTHUR I. WASKOW is a member of the staff of the Peace Research Institute in Washington. Waskow was graduated from Johns Hopkins University in 1954 and has studied and taught American history at the University of Wisconsin, where he took his M.A. in 1956 and his Ph.D. in 1962. Before joining the Peace Research Institute, Waskow had been legislative assistant to Congressman Robert W. Kastenmeier of Wisconsin since October, 1959. Waskow is the author of *The Limits of Defense*, an analysis of disarmament and U.S. defense policy.

### Bibliography

THE LIMITS OF DEFENSE. Arthur I. Waskow. Doubleday & Company, Inc., 1962.

## 45. A "Light" ABM System

### The Author

ROBERT S. MC NAMARA is the only man to have served as Secretary of Defense under two presidents. He was appointed to the post by President Kennedy in 1961, and, after Kennedy's death, continued under President Johnson until his resignation in 1967. In the early 1940's he taught business administration at Harvard, leaving to join the Army Air Force during the Second World War. After the war he became an executive of the Ford Motor Company; he was appointed president in 1960, but served only a few months before joining Kennedy's cabinet.

## 46. Anti-Ballistic-Missile Systems

### The Authors

RICHARD L. GARWIN AND HANS A. BETHE are respectively director of applied research at the Thomas J. Watson Research Center of the International Business Machines Corporation and professor of physics at Cornell University. Garwin was graduated from the Case Institute of Technology in 1947 and received master's and doctor's degrees in physics from the University of Chicago. He has served as a consultant to the Los Alamos Scientific Laboratory and the President's Science Advisory Committee, of which he was a member from 1962 to 1965. Bethe, who won the Nobel prize in physics "for his contributions to the theory of nuclear reactions," has been at Cornell since 1935. He was chief of the theoretical physics division at Los Alamos during World War II and has continued to be a consultant to the Los Alamos laboratory as well as to several firms involved in atomic energy. From 1956 to 1959 he was a member of the President's Science Advisory Committee. In 1958 and 1959 he was a member of the U.S. delegation at the discussions in Geneva on the discontinuation of nuclear-weapons tests.

### Bibliography

ADDRESS BY HONORABLE ROBERT S. MC NAMARA, SECRETARY OF DEFENSE, BEFORE UNITED PRESS INTERNATIONAL EDITORS AND PUBLISHERS, SAN FRANCISCO, SEPTEMBER 18, 1967. Office of Assistant Secretary of Defense (Public Affairs), 1967. (See page 337 of this volume.)

THE EFFECTS OF NUCLEAR WEAPONS. Edited by Samuel Glasstone. United States Department of Defense and the United States Atomic Energy Commission, April, 1962.

STATEMENT OF SECRETARY OF DEFENSE ROBERT S. MC-NAMARA BEFORE THE SENATE ARMED SERVICES COMMITTEE ON THE FISCAL YEAR 1969-73 DEFENSE PROGRAM AND 1969 DEFENSE BUDGET. U.S. Government Printing Office; January 22, 1968.

## 47. The Dynamics of the Arms Race

### The Author

GEORGE W. RATHJENS is vice-president for research at Cornell University. After his graduation from Yale University in 1946 he took his Ph.D. at the University of California at Berkeley in 1951. He taught chemistry at Columbia University from 1950 to 1953. From 1953 to 1958 he was on the staff of the Weapons Systems Evaluation Group in the Department of Defense. After a year as a research fellow at Harvard University he returned to Washington as a member of the staff of the special assistant to the President for science and technology. In 1961 he was chief scientist in the Advanced Research Projects Agency of the Department of Defense, becoming deputy director of the agency later that year. From 1962 to 1965 he held various administrative posts with the U.S. Arms Control and Disarmament Agency, and from 1965 until he went to M.I.T. he was director of the Weapons Systems Evaluation Division of the Institute for Defense Analyses. Much of the material in his article also appeared in *The Future of the Strategic Arms Race: Options for the 1970's*, a pamphlet by Rathjens that was published by the Carnegie Endowment for International Peace.

### Bibliography

THE EFFECTS OF NUCLEAR WEAPONS. Edited by Samuel Glasstone. United States Department of Defense and the United States Atomic Energy Commission, April, 1962.

DEBATE: THE ANTIBALLISTIC MISSILE. Edited by Engene Rabinowitch and Ruth Adams. Bulletin of the Atomic Scientists, 1968.

THE ESSENCE OF SECURITY. Robert S. McNamara. Harper & Row, Publishers, 1968.

ANTI-BALLISTIC-MISSILE SYSTEMS. Richard L. Garwin and Hans A. Bethe in *Scientific American*, Vol. 218, No. 3, pages 21-31; March, 1968.

THE FUTURE OF THE STRATEGIC ARMS RACE: OPTIONS FOR THE 1970's. George W. Rathjens. Carnegie Endowment for International Peace, 1969.

# Index